Reagents for Organic Synthesis

ADVISORS

Fieser and Fieser's
Reagents for Organic Synthesis

VOLUME SIXTEEN

Mary Fieser
Harvard University

A WILEY-INTERSCIENCE PUBLICATION
John Wiley & Sons, Inc.
NEW YORK / CHICHESTER / BRISBANE / TORONTO / SINGAPORE

ISBN 0-471-52721-1
ISSN 0271-616X

Printed in the United States of America

10 9 8 7 6 5 4 3 2 1

Printed and bound by Courier Companies, Inc.

PREFACE

This volume of Reagents includes reports on synthetic use of reagents published for the most part in 1989 and 1990. As in previous volumes, the advisors have provided invaluable help. Scott Virgil, Greg Reichard, and Mark Bilodeau have read and improved large portions of the manuscript.

MARY FIESER

Cambridge, MA
December 10, 1991

CONTENTS

Reagents for Organic Synthesis

A

Acyl(carbonyl)cyclopentadienyl(triphenylphosphine)iron, 12, 1–2; **14**, 1–2

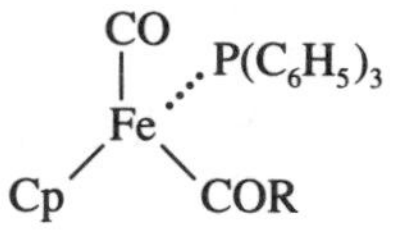

$$\underset{Cp}{\overset{\overset{\displaystyle CO}{|}}{\underset{}{Fe}}}\overset{P(C_6H_5)_3}{\underset{COR}{}}$$

Review. Davies[1] has reviewed use of the chiral iron auxiliary [CpFe-(CO)P(C$_6$H$_5$)$_3$] for effecting asymmetric reactions of an acyl group, including alkylation, aldol reactions, Michael additions.

[1] S. G. Davies, *Aldrichim. Acta*, **23**, 31 (1990).

Alkylaluminum halides.

Reaction of ester enolates with epoxides.[1] Lithium enolates of esters do not open epoxides, but the aluminum enolates do. Li to Al exchange can be effected with diethylaluminum chloride. The less substituted O–C bond is cleaved and the *syn*-diastereomer predominates. Reactions of optically active epoxides proceed with high

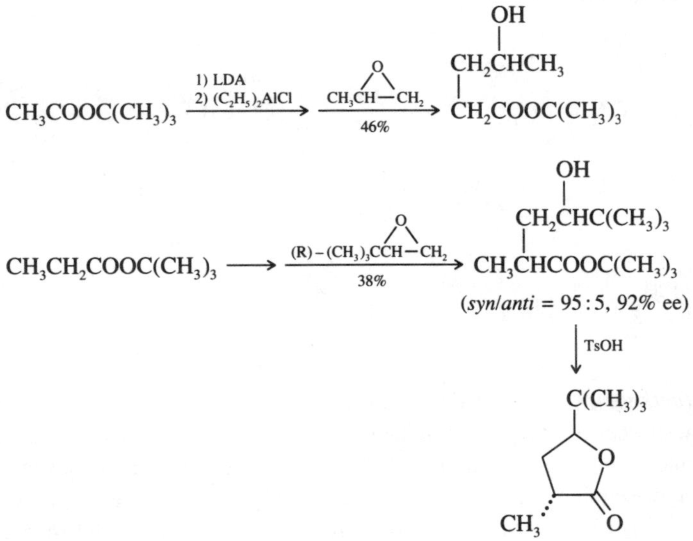

$$CH_3COOC(CH_3)_3 \xrightarrow[\text{2) (C_2H_5)_2AlCl}]{\text{1) LDA}} \xrightarrow[46\%]{CH_3CH\overset{O}{-}CH_2}$$

$$\overset{OH}{\underset{CH_2CHCH_3}{|}}$$
$$CH_2COOC(CH_3)_3$$

$$CH_3CH_2COOC(CH_3)_3 \longrightarrow \xrightarrow[38\%]{(R)-(CH_3)_3C\overset{O}{CH}-CH_2}$$

$$\overset{OH}{\underset{CH_2CHC(CH_3)_3}{|}}$$
$$CH_3CHCOOC(CH_3)_3$$

(*syn/anti* = 95 : 5, 92% ee)

↓ TsOH

$$\underset{CH_3}{\overset{C(CH_3)_3}{\bigcirc}}O$$

enantioselectivity. Hydrolysis and cyclization of the products provide a route to *trans*-2,4-disubstituted γ-lactones.

Intramolecular cyclization of a double bond with dienones.[2] Treatment of the trienone **1** with a typical Lewis acid, $C_2H_5AlCl_2$, results in a non-photochemical [2+2] cycloaddition to form a cyclobutane ring (**2**). In contrast, use of the strong acidic Amberlyst 15 resin results in the bicyclic product **3**.

3

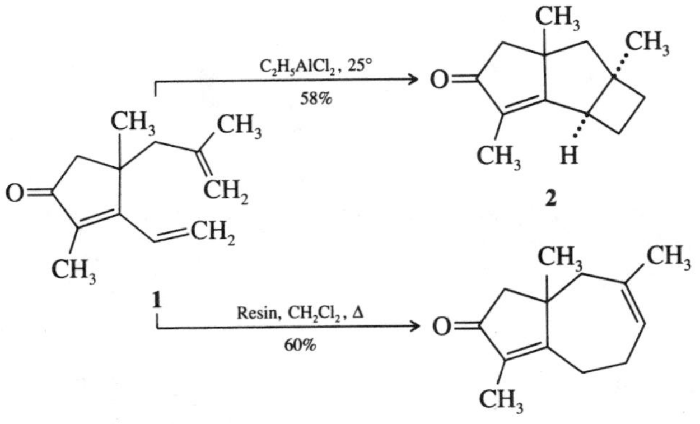

These cyclizations are also useful for formation of six-membered rings, as shown for cyclization of **4** to **5** and **6**.

4		**5**		**6**
$C_2H_5AlCl_2$		45%		0%
Amberlyst 15	85%		1:1	

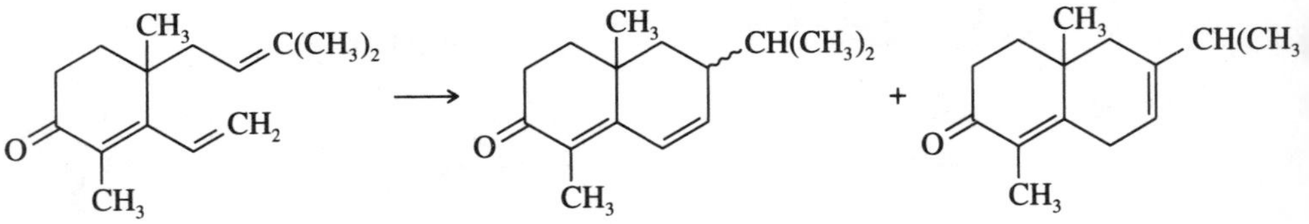

Cycloalkenones.[3] Cycloalkenones can be prepared by a retro Diels–Alder reaction of norbornenes of type **1**, conducted at 25–70° in the presence of CH_3AlCl_2 (1 equiv.) and a reactive dienophile, usually maleic anhydride or fumaronitrile. The [4+2]cycloreversion was used to prepare 12-oxophytodienoic acid (**4**), which epimerizes at C_{13} to the *trans*-isomer on brief exposure to acid. The precursor norbornene **3** was prepared from the known dienone **2** as shown. Treatment of **3** at room

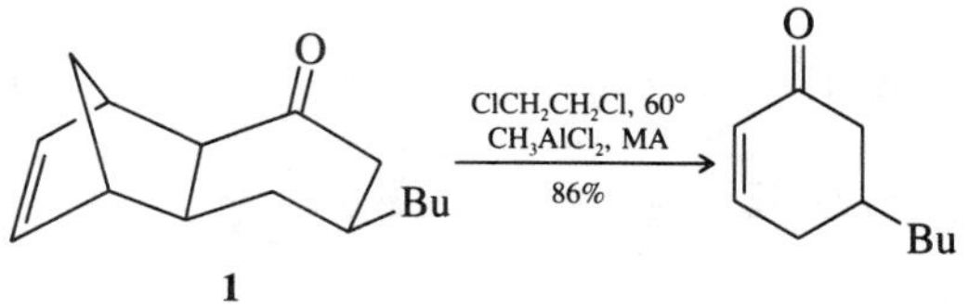

1

temperature with $C_2H_5AlCl_2$ and fumaronitrile in $ClCH_2CH_2Cl$ provides **4** in 60% yield.

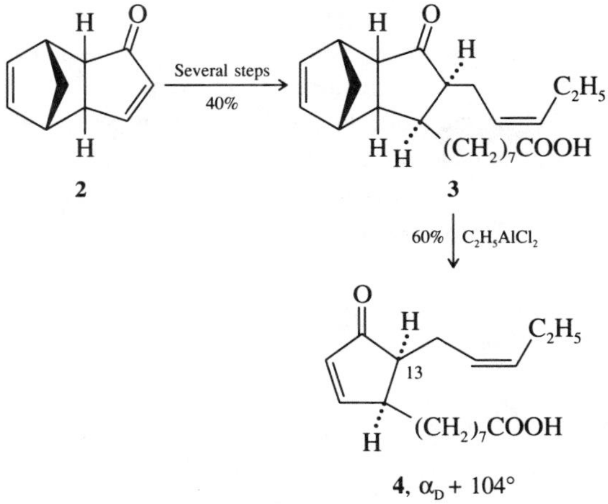

2   3

4, $\alpha_D + 104°$

[1] T.-J. Sturm, A. E. Marolewski, D. S. Rezenka, S. K. Taylor, *J. Org.*, **54**, 2039 (1989).
[2] G. Majetich and V. Khetani, *Tetrahedron Letters*, **31**, 2243 (1990).
[3] P. A. Grieco and N. Abood, *J. Org.*, **54**, 6008 (1989).

Alkyllithiums.

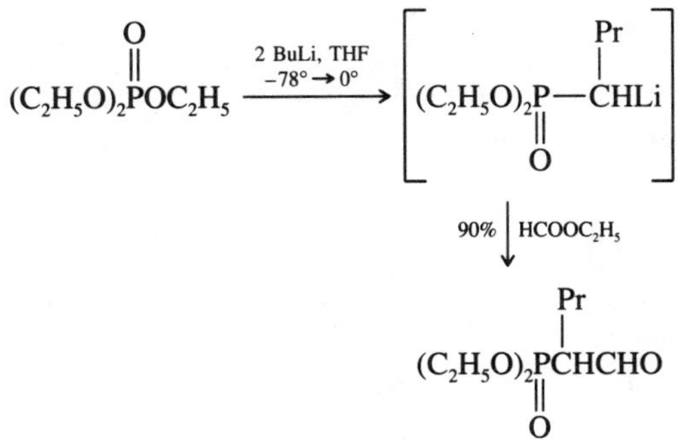

α-Lithiophosphonates.[1] Reaction of triethyl phosphate with a primary alkyllithium (RCH_2Li, 2 equiv.) in THF at $-78 \to 0°$ results in an α-lithioalkylphosphonate, the intermediate in Wittig–Horner type reactions. Both linear and branched (such as i-BuLi) primary alkyllithiums can be employed, but the reaction with methyllithium is too slow and inefficient to be useful. This one-step alkylation/metallation provides a general route to α-formylalkylphosphonates.

[1] M.-P. Teulade and P. Savignac, *Tetrahedron Letters*, **28**, 405 (1987); P. Savignac and C. Patois, *Org. Syn.*, submitted (1990).

Alkyllithium/Cerium(III) chloride.

vic.-Diamines. Reetz et al.[1] reported a stereoselective route to *vic.*-diamines from N,N-dibenzylamino aldehydes, readily available from amino acids. The corresponding N-benzyl aldimines (**2**) undergo addition of alkyllithium compounds in ether to provide the chelation-controlled adducts. Yields are generally improved by use of RLi/CeCl$_3$ (equation I). The diastereoselectivity can be reversed by change of the aldimine group to an N-tosyl group by reaction of the aldehyde with N-sulfinyl-p-toluenesulfonamide. These N-tosyl derivatives react with Grignard reagents to give

mainly the nonchelation-controlled adducts (equation II). Since many amino acids are accessible in both the (S)- and (R)-form, it is possible to prepare the enantiomeric *vic.*-diamines.

[1] M. T. Reetz, R. Jaeger, R. Drewlies, and M. Hübel, *Angew. Chem. Int. Ed.*, **30**, 103 (1991).

Alkylmercury chlorides.

Reductive alkylation of α,β-enones. This reaction can be effected by a photostimulated free-radical reaction with RHgCl and a base (KI/K$_2$S$_2$O$_8$, NaI, DABCO).[1]

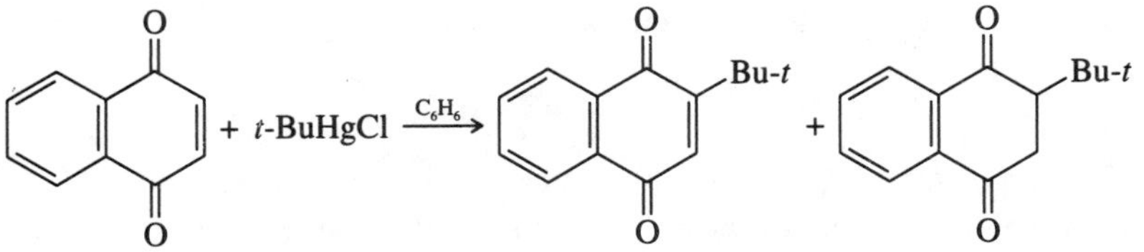

	34%	14%
hv	9%	52%
hv + NaI	—	95%
hv + $(NH_4)_2S_2O_8$		

[1] G. A. Russell, B. H. Kim, and S. V. Kulkarni, *J. Org.*, **54**, 3768 (1989).

Allenyl chloromethyl sulfone (1).

Preparation:

$$HC \equiv CCH_2OLi + ClCH_2SCl \xrightarrow[-LiCl]{} CH_2 = C = CHSOCH_2Cl \xrightarrow[\text{overall}]{\text{[O]}}$$

$$CH_2 = C = CHSO_2CH_2Cl$$

1, m.p. 39°

Homologation of **1,3-dienes.**[1]   The reagent is particularly useful for cyclohomologation of 1,3-dienes. Thus the adduct **a** of a Diels–Alder reaction of **1** with 1,2-bis(methylene)cyclohexane undergoes Ramberg–Bäcklund rearrangement to the triene **2**. Repetition of the sequence provides the tetraene **3**. The sequence is applicable to cyclopentadiene, furan, and 2,3-dimethyl-1,3-butadiene.

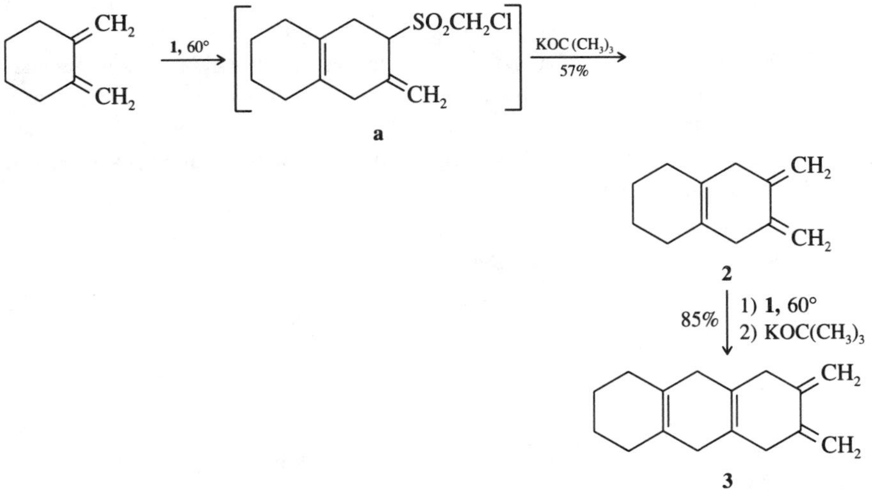

[1] E. Block and D. Putman, *Am. Soc.*, **112**, 4072 (1990).

Allenylsilanes.

Furan annelation.[1] Allenylsilanes substituted with an alkyl group at C_1 react with acyl chlorides and $AlCl_3$ (1 equiv. each) in CH_2Cl_2 at $-20°$ to form furans in 50–80% yield.

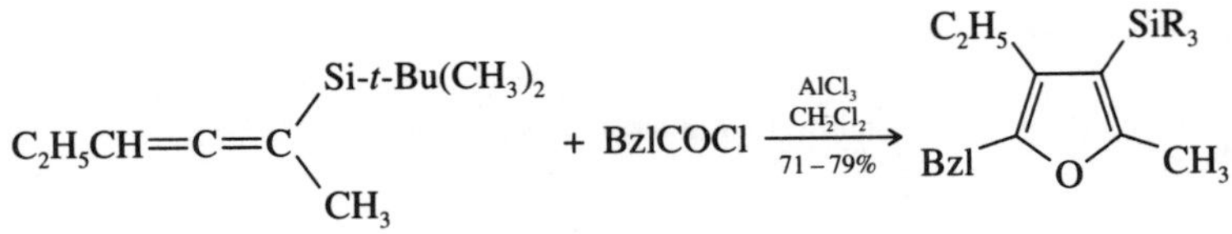

[1] R. L. Danheiser, E. J. Stoner, H. Koyama, D. S. Yamashita, and C. A. Klade, *Am. Soc.*, **111**, 4407 (1989).

B-Allylbis(2-isocaranyl)borane,

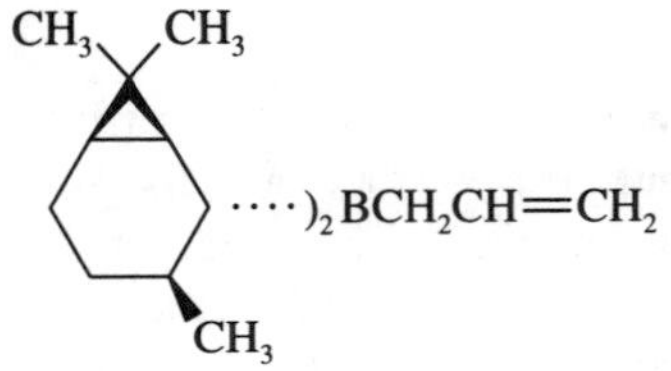

The borane is obtained by hydroboration $[BH_3 \cdot S(CH_3)_2]$ of $(+)$-2-carene followed by methanolysis and reaction with allylmagnesium bromide.

Asymmetric allylation of RCHO.[1] The reagent converts aldehydes to homoallylic alcohols in 94–99% ee, which is significantly higher than the enantioselectivities obtained with B-allyl(diisopinocampheyl)borane. The enantiomer of $(+)$-2-carene is not known, but B-allylbis(4-isocaranyl)borane (**2**) behaves as an enantiomer of **1** in allylation of aldehydes.

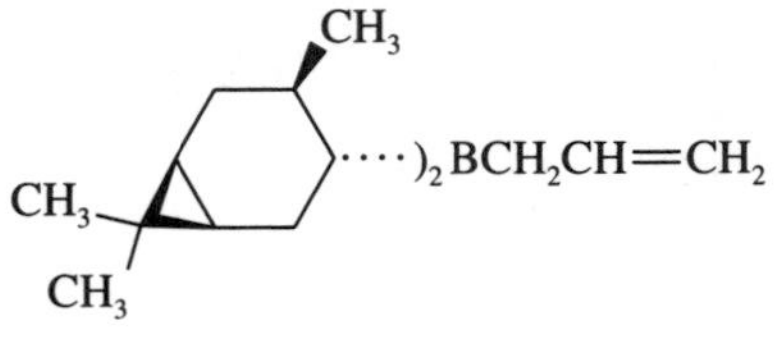

[1] H. C. Brown, R. S. Randad, K. S. Bhat, M. Zaidlewicz, and U. S. Racherla, *Am. Soc.*, **112**, 2389 (1990).

Allyltitanium triphenoxide, $CH_2=CHCH_2Ti(OC_6H_5)_3$ **(1).**

Selective cleavage of oxiranes.[1] This reagent is more effective than allyltitanium triisopropoxide for selective cleavages of oxiranes at the more-substituted carbon atom.

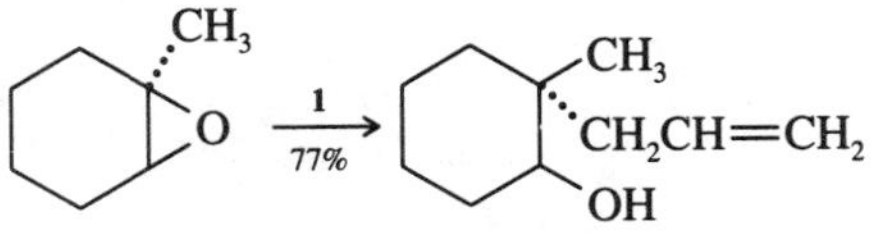

(*trans/cis* = 10 : 1)

[1] T. Tanaka, T. Inoue, K. Kamei, K. Murakami, and C. Iwata, *J.C.S. Chem. Comm.*, 906 (1990).

Allyltributyltin, $Bu_3SnCH_2CH=CH_2$ **(1).**

Allylation of quinones.[1] This reaction has been effected with allylsilanes mediated by $TiCl_4$, but allyltin reagents activated by BF_3 etherate are now preferred. The allylated hydroquinones can be oxidized to quinones directly by $FeCl_3$.

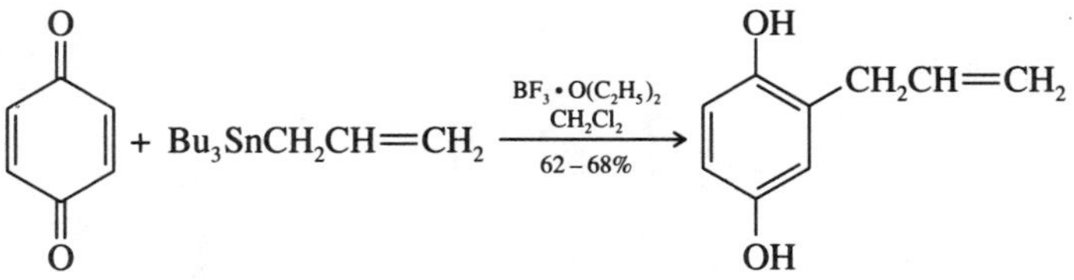

A further advantage is that the tin reagent is prepared in quantitative yield by coupling of tributyltin chloride and allyl chloride in the presence of Mg turnings (1 equiv.) in THF under ultrasound irradiation.

1,2,3-syn-*Triols.* Allyltributyltin, activated by $ZnCl_2$, adds to acylsilanes bearing an α-alkoxy-β-silyloxy group with high *syn*-selectivity. Protiodesilylation of the

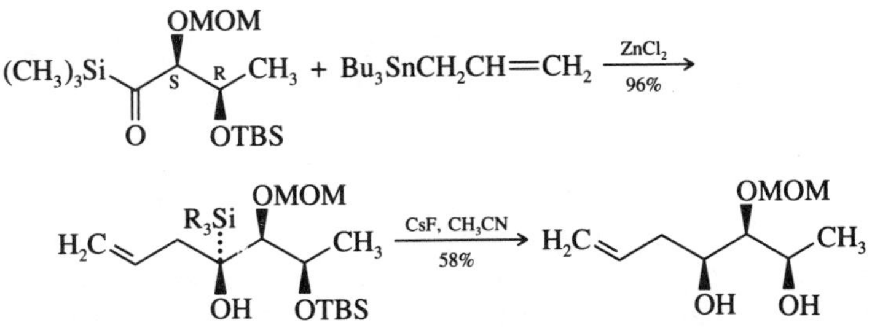

(*syn/anti* = 91 : 9)

product affords the monoprotected derivative of a *syn*-1,2,3-triol.[2] The required acylsilanes can be prepared by Sharpless OsO_4-catalyzed dihydroxylation of C_1-alkoxy crotylsilanes.[3]

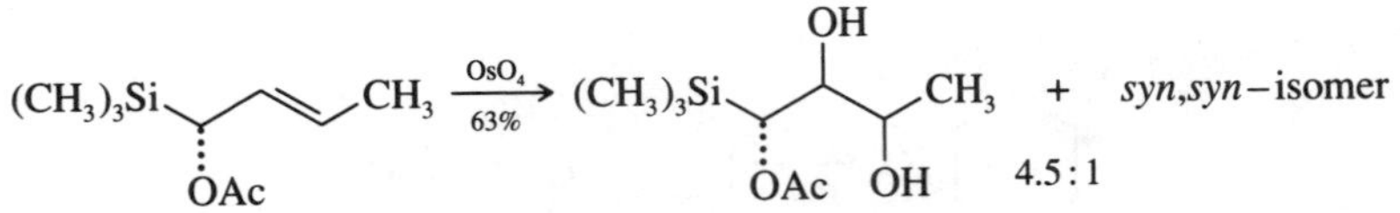

γ-Alkoxyallyltins.[4] The reaction of γ-alkoxyallyltins with α-alkoxy aldehydes (**14, 16**) has been extended to reactions with chiral components. Highly stereoselective additions to provide *syn*-1,2-diols can be obtained, depending on the chirality of the tin reagent and the choice of the Lewis acid catalyst. Thus addition of (R)-**1** to the aldehyde **2** catalyzed by BF_3 etherate provides the homoallylic alcohol **3** as a 92:8 mixture. The reaction of (S)-**1** with **2** catalyzed by BF_3 etherate results in a 67:33

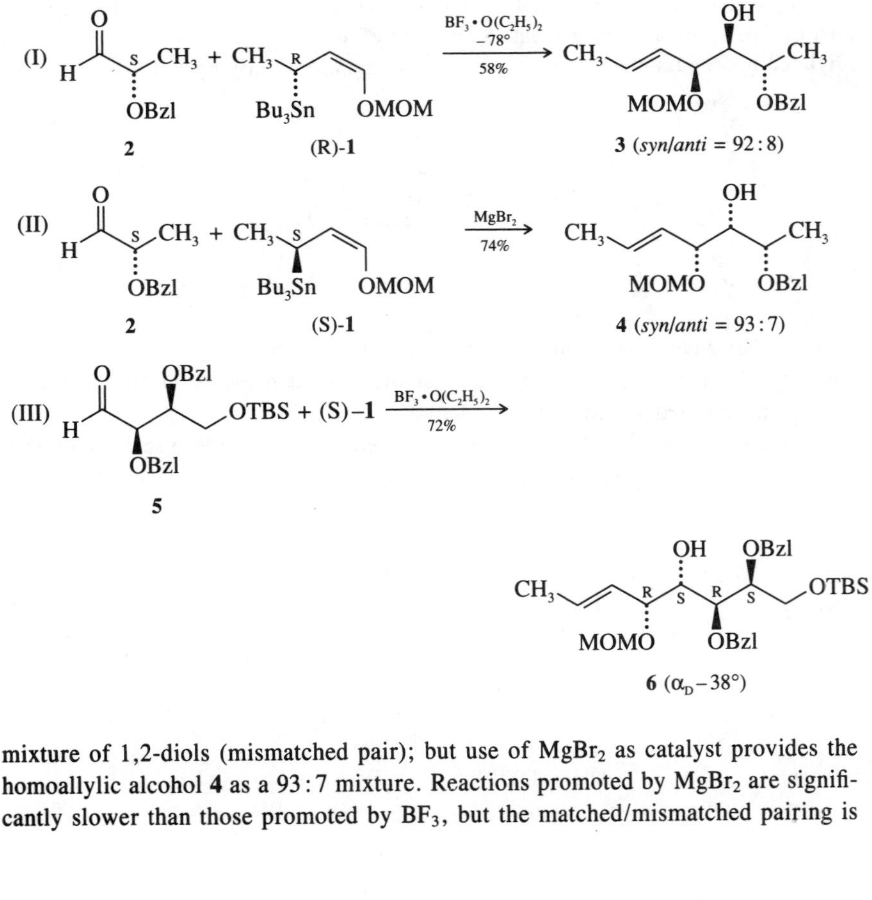

mixture of 1,2-diols (mismatched pair); but use of $MgBr_2$ as catalyst provides the homoallylic alcohol **4** as a 93:7 mixture. Reactions promoted by $MgBr_2$ are significantly slower than those promoted by BF_3, but the matched/mismatched pairing is reversed. It is also noteworthy that the products from either (R)- or (S)-**1** have an (E)-double bond.

This diastereoselective reaction can be extended to an aldehyde such as **5**, prepared from L-diethyl tartrate. Reaction of **5** with (S)-**1** catalyzed by BF_3 etherate affords a single product **6** in 72% yield (equation III).

[1] Y. Naruta and K. Maruyama, *Org. Syn.*, submitted (1989); Y. Naruta, Y. Nishigaichi, and K. Maruyama, *ibid.*, (1989).

[2] P. F. Cirillo and J. S. Panek, *J. Org.*, **55**, 6071 (1990).

[3] *Idem, Am. Soc.*, **112**, 4873 (1990).

[4] J. A. Marshall and G. P. Luke, *J. Org.*, **56**, 483 (1991).

Allyltrifluorosilane, $CH_2{=}CHCH_2SiF_3$.

Allylation of α-hydroxy ketones.[1]   An allyltrifluorosilane and triethylamine (1:1) react with α-hydroxy ketones in refluxing THF to form tertiary homoallyl

(I) $CH_3\overset{O}{\overset{\|}{C}}CH(OH)CH_3 + CH_3$⟍⟋$SiF_3$ $\xrightarrow[74\%]{N(C_2H_5)_3}$

(E/Z = 97:3)

(II) $C_6H_5\overset{O}{\overset{\|}{C}}CH(OH)C_6H_5 + (CH_3)_2C{=}CHCH_2SiF_3$ $\xrightarrow[75\%]{N(C_2H_5)_3}$

(100% de)

alcohols with predominant *syn*-selectivity of the hydroxyl group, probably via a pentacoordinate allylsilicate. In fact, only one diastereomer is formed on reaction of benzoin with (3-methyl-2-butenyl)trifluorosilane (equation II).

[1] K. Sato, M. Kira, and H. Sakurai, *Am. Soc.*, **111**, 6429 (1989).

Alumina.

Hydrohalogenation of alkenes and alkynes.   In the presence of Al_2O_3 or SiO_2, hydrogen halides or their precursors add to alkenes or alkynes at useful rates and often in quantitative yield.[1]

$$C_5H_{11}CH_2CH{=}CH_2 \xrightarrow[86\%]{(COCl)_2,\ Al_2O_3} C_5H_{11}CH_2\underset{\underset{Cl}{|}}{C}HCH_3$$

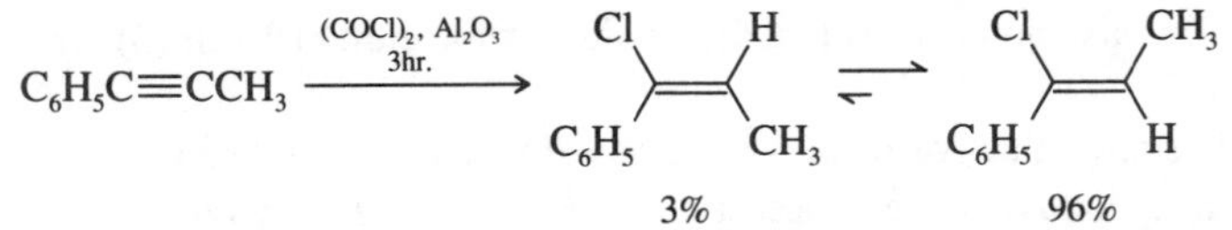

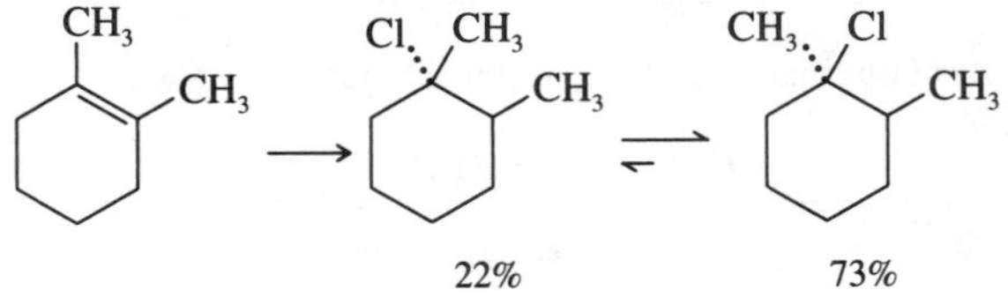

$C_6H_5C\equiv CCH_3$ $\xrightarrow[\text{3hr.}]{(COCl)_2,\ Al_2O_3}$

3%   96%

22%   73%

Under these conditions alkynes and alkenes undergo *syn*-addition initially, but rearrange to the more stable adduct.

[1] P. J. Kropp, K. A. Daus, S. D. Crawford, M. W. Tubergen, K. D. Kepler, S. L. Craig, and V. P. Wilson, *Am. Soc.*, **112**, 7433 (1990).

Aluminum chloride.

2-Methyl-1,3-cyclopentanedione.   This dione is commercially available, but it can be prepared in one step by reaction of succinic acid with propionyl chloride and aluminum chloride (3 equiv. each) in refluxing nitromethane.[1]

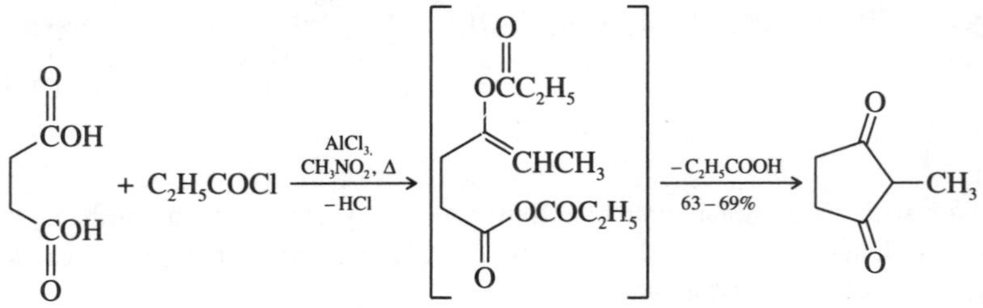

Enol ethers via a Diels–Alder reaction.   Reaction of the diene **1** with the benzylideneaniline **2** catalyzed by $AlCl_3$ results in the adducts **3** and **4** as the primary products, evidently formed by a Diels–Alder reaction.[2]

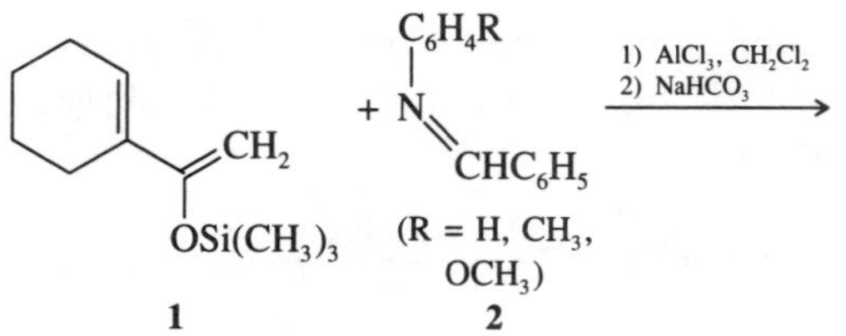

1) $AlCl_3$, CH_2Cl_2
2) $NaHCO_3$

(R = H, CH_3, OCH_3)

1   2

	3 (exo)	4 (endo)
−78°	70:30	
20°	2:98	

Friedel–Crafts alkylation.   Reaction of arenes with acid chlorides in CH_2Cl_2 with $AlCl_3$ (1 equiv.) and $(C_2H_5)_3SiH$ (2.5–3 equiv.) results in the alkylated arene by deoxygenation of the intermediate acylated arene. Yields of 95% are obtainable, and this procedure avoids the problem of polyalkylation observed in regular Friedel–Crafts reactions.[3]

[1] P. G. Meister, M. R. Sivik, and L. A. Paquette, *Org. Syn.*, submitted (1989).
[2] L. Le Coz, L. Wartski, J. Seyden-Penne, P. Charpin, and M. Nierlich, *Tetrahedron Letters*, **30**, 2795 (1989).
[3] A. Jaxa-Chamiec, V. P. Shah, and L. I. Kruse, *J.C.S. Perkin I*, 1705 (1989).

Aluminum(III) iodide.
Reductive dehalogenation of β-halo ketones.[1]   α-Bromo or α-chloro ketones undergo reductive dehalogenation on reaction with freshly prepared AlI_3 in refluxing CH_3CN (80–95% yield). The reaction probably involves an aluminum enolate since addition of benzaldehyde results in an aldol condensation.

[1] D. Konwar, R. C. Boruah, and J. S. Sandhu, *Synthesis*, 337 (1990); H. N. Borah, R. C. Boruah, and J. S. Sandhu, *J.C.S. Chem. Comm.*, 154 (1991).

3-Amino-2-hydroxybornanes (1).
The bornanes **1a** and **1b** are prepared from 3-*endo*-2-*endo*-**1a** and 3-*exo*-2-*exo*-**1b** and (S)-N-(benzyloxycarbonyl)proline, respectively.

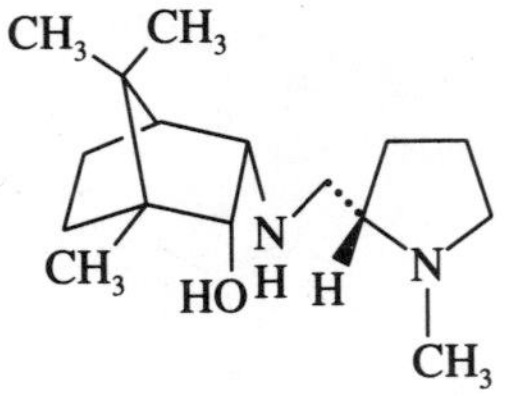

endo,endo-**1a**

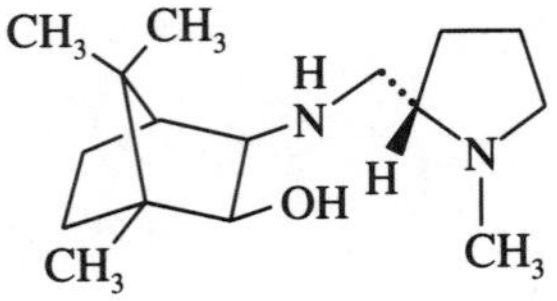

exo,exo-**1b**

Enantioselective addition of $(C_2H_5)_2Zn$ to RCHO. These are the first secondary amino alcohols known to effect efficient enantioselective addition of dialkylzincs to aromatic and aliphatic aldehydes.

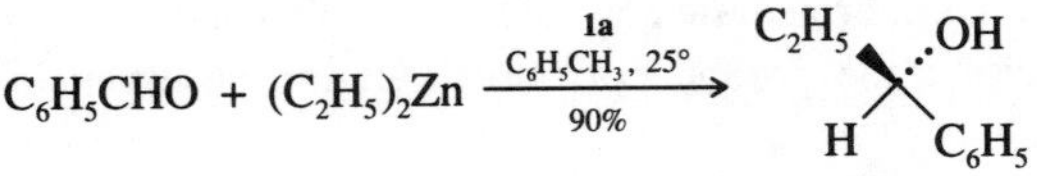
C6H5CHO + (C2H5)2Zn
1a
C6H5CH3, 25°
90%
C2H5
OH
H
C6H5

R, 92% ee

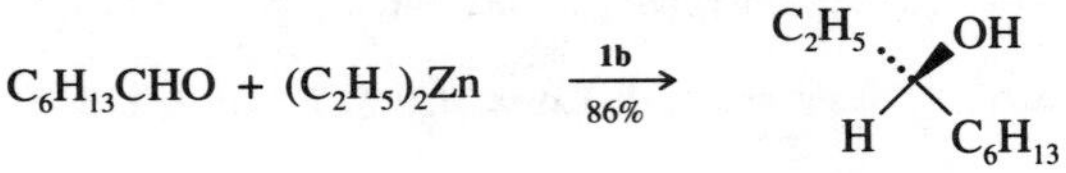
C6H13CHO + (C2H5)2Zn
1b
86%
C2H5
OH
H
C6H13

S, 74% ee

[1] K. Tanaka, H. Ushio, and H. Suzuki, *J.C.S. Chem. Comm.*, 1700 (1989).

(S)-1-Amino-2-methoxymethyl-1-pyrrolidine (SAMP).

***Diastereoselective synthesis of* 1,2-*diols*.**[1] The key step in an asymmetric synthesis of 1,2-diols is an enantioselective silylation of SAMP hydrazones **2** with

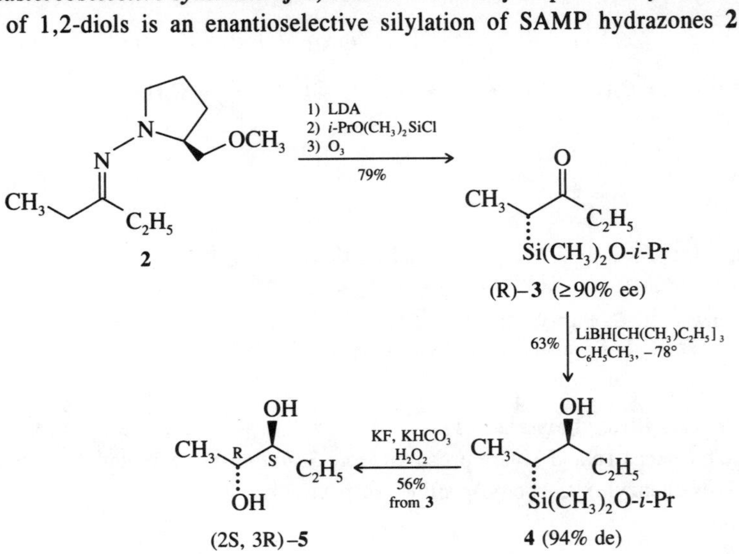
1) LDA
2) i-PrO(CH3)2SiCl
3) O3
79%
2
(R)–3 (≥90% ee)
63%
LiBH[CH(CH3)C2H5]3
C6H5CH3, –78°
KF, KHCO3
H2O2
56%
from 3
(2S, 3R)–5
4 (94% de)

isopropoxydimethylsilyl chloride to furnish, after ozonolysis, α-silyl ketones (**3**) in 90% ee. Stereoselective reduction of the carbonyl group followed by oxidative cleavage of the C–Si bond (**12**, 243–245) furnishes *vic*-diols (**5**) in ~90% de.

[1] D. Enders and S. Nakai, *Ber.*, **124**, 219 (1991).

Anthracene.

Diethyl methylidenemalonate.   This reagent (**3**) polymerizes easily and can be prepared by depolymerization of oligomers, but is more readily available in sufficiently pure form from diethyl malonate (**1**) by conversion to the Diels–Alder adduct **2** from anthracene, paraformaldehyde, and **1**. When heated at 190–200° with maleic anhydride, the adduct **2** decomposes to **3** in an overall yield of ~50%.

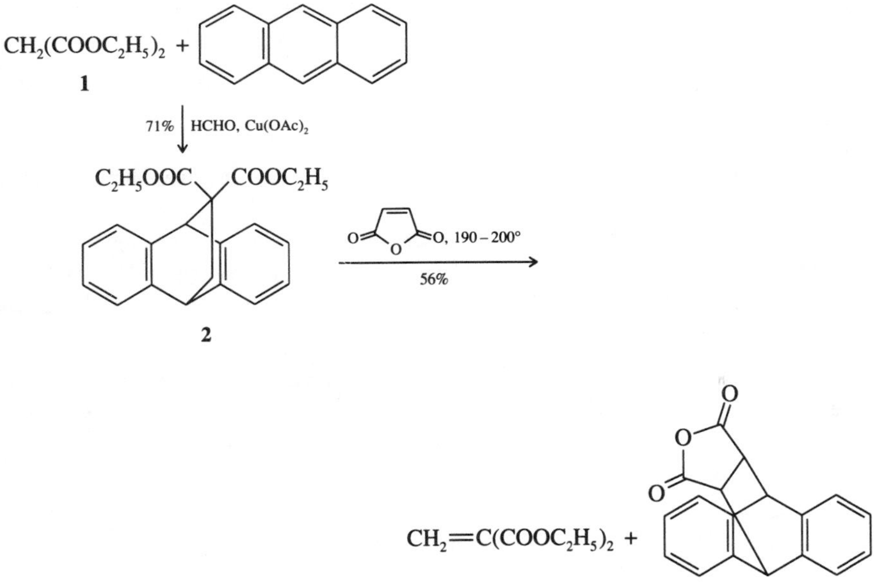

[1] C. J. C. DeCock, J.-L. De Keyser, J. H. Poupaert, and P. Dumont, *Org. Syn.*, submitted (1989).

Arene(tricarbonyl)chromium complexes.

Stereoselective 1,3-dipolar cycloaddition of nitrones.[1]   The cycloaddition of the nitrone **1** with an electron-rich alkene such as ethyl vinyl ether or vinyltrimethylsilane provides a regioselective route to 3,5-disubstituted isoxazolidines (**2**) (**12**, 566), but with low stereoselectivity.

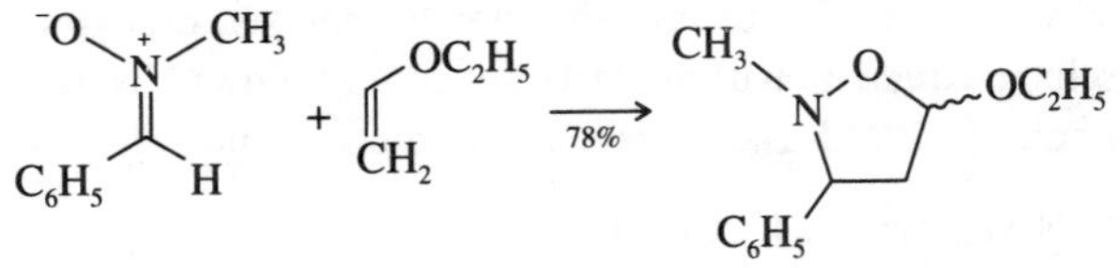

1   **2** (*cis/trans* = 1 : 1)

In contrast, the nitrone **3**, derived from tricarbonylchromium-complexed benzaldehyde, undergoes these 1,3-dipolar cycloadditions to give essentially only the *cis*-
disubstituted isoxazolidine (**2**) after decomplexation with CAN in methanol. How-

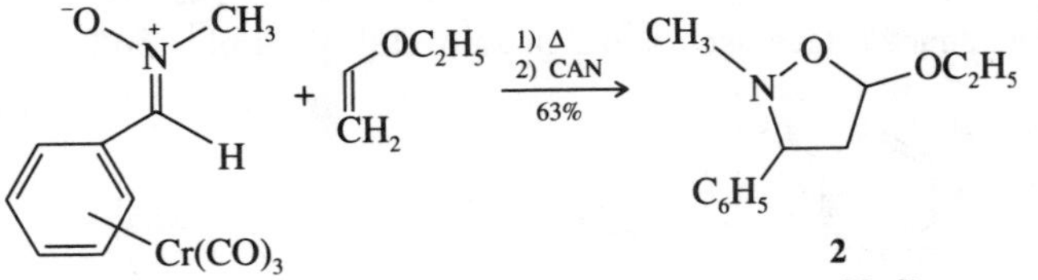

3   **2**
                                         (*cis*, >98 : 2)

ever, cycloaddition of **3** with acrylonitrile, an electron-poor alkene, shows no im-
provement in diastereoselectivity.

Silyl dienol ethers.[2] The isomerization of 1,3-dienes by (naphthalene)chro-
mium(CO)$_3$ (**14, 25**) has been extended to (silyloxy)methylbutadienes. Thus (1Z)-1-
(silyloxymethyl)-butadienes such as **1** are isomerized exclusively to silyl dienol

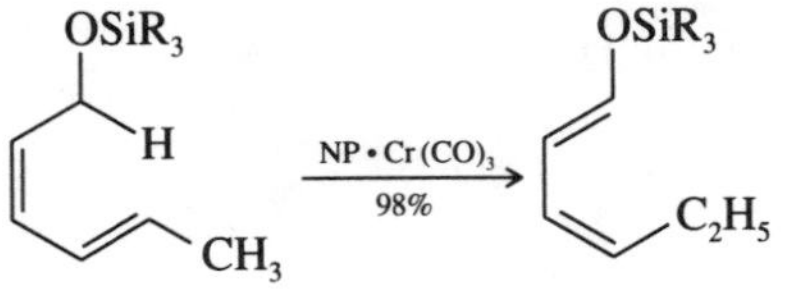

(Z)-**1**   **2**

ethers (**2**), whereas (E)-**1** is not isomerized. Silyl dienol ethers such as **2** are useful
for inter- and intramolecular Diels–Alder reactions because of their high *endo*-selec-
tivity (equation I).

(I)

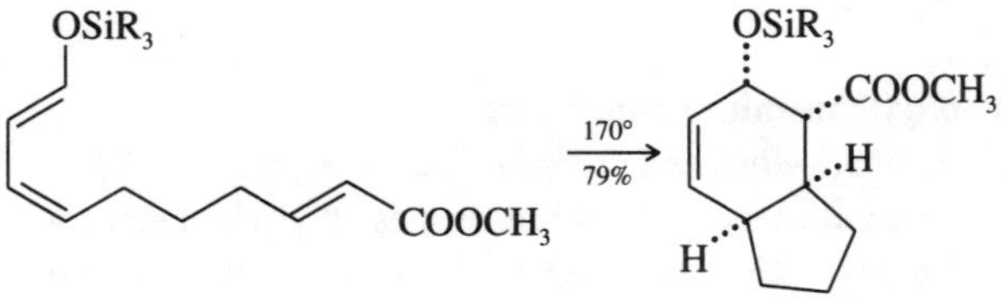

The precursors (**1**) to the silyl dienol ethers are prepared as a mixture of (E)- and (Z)-isomers by Wittig reactions with R_3SiOCH_2CHO.

α-Substituted benzyl alcohols.[3] Reaction of the complex **1** of *o*-triisopropylsilylbenzaldehyde with L-valinol followed by chromatography and hydrolysis provides two optically active complexes [(−) and (+)-**2**)] in about equal amounts. The complex (+)-**2** can be converted into either (+)- or (−)-**3** (α-methylbenzyl alcohol) by

1 (−)−**2** (29%) (+)−**2** (31%)

(+) **3** (100% ee)

(−)−**3** (100% ee)

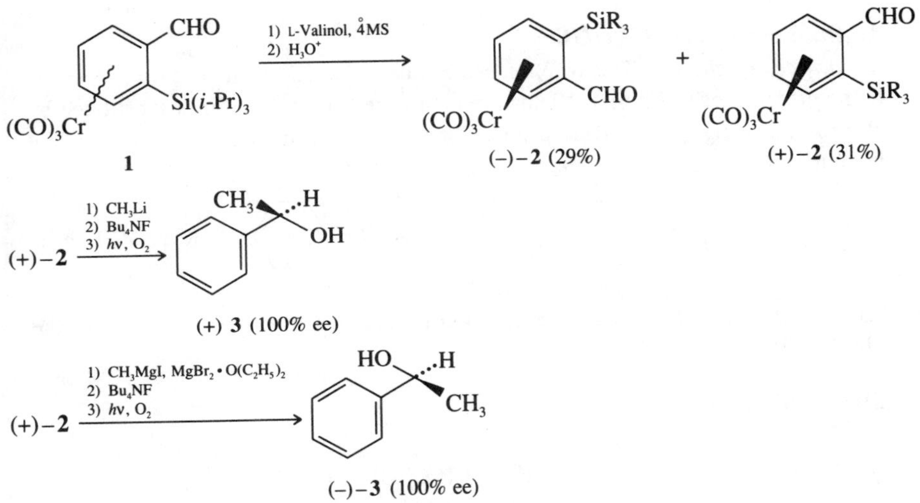

reaction with CH_3Li or CH_3MgI. The difference in selectivity between CH_3Li and CH_3MgI is attributed to the greater Lewis acidity of the latter reagent.

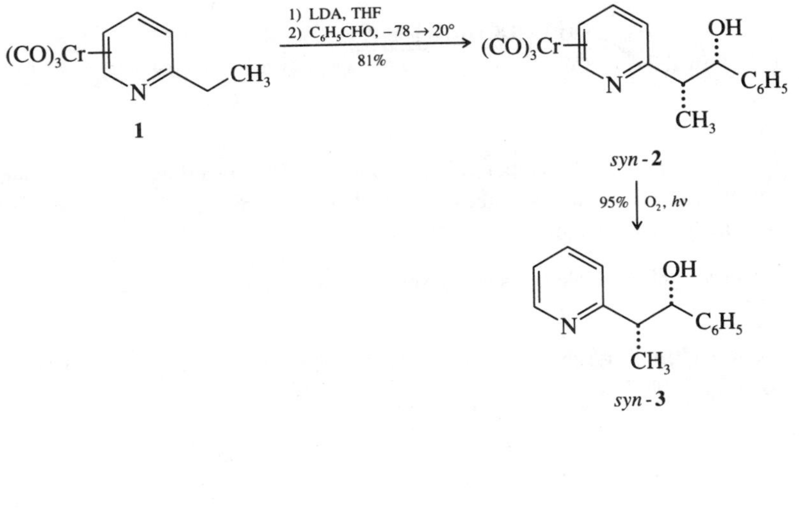

Aldol-type reactions.[4] The lithium anion generated with LDA from 2-ethylpyridine chromium tricarbonyl (**1**) reacts with nonenolizable aldehydes to give a single aldol-type product (**2**) shown to be the *syn*-diastereomer. The same reaction, when

1 *syn*-**2**

syn-**3**

carried out on uncomplexed 2-ethylpyridine, results in *syn*- and *anti*-**3**, with some preference for *anti*-**3** (48% de).

[1] C. Mukai, W. J. Cho, I. J. Kim, and M. Hanaoka, *Tetrahedron Letters*, **31**, 6893 (1990).
[2] M. Sodeoka, H. Yamada, and M. Shibasaki, *Am. Soc.*, **112**, 4906 (1990).
[3] S. G. Davies and C. L. Goodfellow, *Synlett*, 59 (1989).
[4] S. G. Davies and M. R. Shipton, *Synlett*, 25, (1991).

Aryllead triacetates, $ArPb(OAc)_3$.

Amine arylation.[1]   This reaction can be effected at 25° with aryllead triacetates in the presence of $Cu(OAc)_2$. Yields are high in reactions of anilines substituted by electron-releasing groups. Only primary linear aliphatic amines are arylated, albeit in poor yield.

$$CH_3O{-}\langle\rangle{-}NH_2 \xrightarrow[91\%]{\substack{p\text{-}CH_3OC_6H_4Pb(OAc)_3, \\ Cu(OAc)_2, CH_2Cl_2}} CH_3O{-}\langle\rangle{-}NHC_6H_4OCH_3\text{-}p$$

[1] D. H. R. Barton, D. M. X. Donnelly, J.-P. Finet, and P. J. Guiry, *Tetrahedron Letters*, **30**, 1377 (1989).

(S)-Aspartic acid,

$$\underset{H_2N}{\overset{H\cdots}{\diagdown}}\!\!\!\!\!\!\!\underset{COOH}{\overset{CH_2COOH}{\diagup}}$$

Asymmetric amino-acid synthesis.[1]   The (S)-1-*t*-butyl 4-methyl N-benzyloxy-carbonylaspartate (**1**), prepared in 80% yield from the 4-methyl ester of aspartic acid, undergoes diastereoselective alkylation at the β-carbon (LDA or lithium hexa-

$$\underset{BocHN}{\overset{H\cdots}{\diagdown}}\!\!\!\!\!\!\underset{COOC(CH_3)_3}{\overset{CH_2COOCH_3}{\diagup}} \xrightarrow[50\%]{\substack{1) \text{ LHMDS} \\ 2) \text{ BzlBr}}} \underset{BocHN}{\overset{H\cdots}{\diagdown}}\!\!\!\!\!\!\underset{COOC(CH_3)_3}{\overset{\overset{\displaystyle Bzl}{\diagdown}\!\!\!\!-COOCH_3}{\diagup}}$$

$$\textbf{1} \qquad\qquad\qquad\qquad\qquad \textbf{2}\ (5:1)$$

methyldisilazide) with a number of electrophiles. This reaction offers a potential route to optically active acids, since the methyl ester can be selectively hydrolyzed by LiOH in aqueous methanol.

[1] J. E. Baldwin, M. G. Moloney, and M. North, *J.C.S. Perkin I*, 833 (1989).

Azidotributyltin, Bu_3SnN_3 (1).   The azide is prepared from Bu_3SnCl and NaN_3 in refluxing THF (86% yield).

1,2-*Azido alcohols*.[1]   This azide is considerably more reactive for cleavage of epoxides than azidotrimethylsilane, which requires a Lewis acid promotor. Although DMF enhances the reactivity of the silyl azide, it lowers the reactivity of the stannyl azide. The reactivity of the latter azide is also decreased by neighboring ester, acetonide, or ether groups.

$$C_6H_5 \overset{O}{\triangle} + 1 \xrightarrow[86\%]{60°} C_6H_5\overset{OH}{\underset{|}{C}}HCH_2N_3 + C_6H_5\overset{N_3}{\underset{|}{C}}HCH_2OH$$

1 : 14

$$C_6H_5OCH_2 \overset{O}{\triangle} + 1 \xrightarrow{95\%} C_6H_5OCH_2\overset{OH}{\underset{|}{C}}HCH_2N_3 + C_6H_5OCH_2\overset{N_3}{\underset{|}{C}}HCH_2OH$$

17 : 1

[1] S. Saito, S. Yamashita, T. Nishikawa, Y. Yokoyama, M. Inaba, and T. Moriwake, *Tetrahedron Letters*, **30**, 4153 (1989).

Azidotrimethylsilane.

***Reaction with (R)-epoxystyrene* (1).**[1]   Reaction of $N_3Si(CH_3)_3$ with (R)-1 in CH_2Cl_2 catalyzed by $Al(O$-i-$Pr)_3$ provides (S)-2 as the product in 90% ee. Use of other solvents also provides (S)-2, but in lower enantioselectivity. Use of $Ti(O$-i-$Pr)_4$ as the catalyst also provides 2 as the major product, but the enantioselectivity de-

	(R)-1		(S)-2		3
+ Al(O-*i*-Pr)₃	50%		90% ee	93:7	
+ Ti(O-*i*-Pr)₄	56%		43% ee		

pends on the solvent. Use of THF provides (R)-2 in 63–99% ee; use of CH_2Cl_2, hexane, or ether provides (S)-2 in 31–72% ee.

[1] K. Sutowardoyo, M. Emziane, and D. Sinou, *Tetrahedron Letters*, **30**, 4673 (1989).

B

Barium manganate, BaMnO$_4$.

Selective oxidation of diols.[1] BaMnO$_4$ is handicapped by insolubility in organic solvents and by instability to acids. It is a more useful oxidant when supported on Al$_2$O$_3$ and CuSO$_4 \cdot$5H$_2$O. This supported reagent efficiently oxidizes benzylic, allylic, and *sec*-alcohols, but oxidation of saturated primary alcohols is negligible. Its main value as opposed to that of supported KMnO$_4$ is for the selective oxidation of primary–secondary diols such as **1** and **2**.

$$\underset{\mathbf{1}}{C_6H_5CH{=}CHCHCH_2CH_2OH} \; \overset{\underset{\text{85\%}}{\overset{C_6H_6,\,25°}{}}}{\xrightarrow{\text{BaMnO}_4/\text{Al}_2\text{O}_3/\text{CuSO}_4.}} \; C_6H_5CH{=}CHCCH_2CH_2OH$$

with OH on the carbon of **1** and O (=O) on the product carbonyl.

$$\underset{\mathbf{2}}{CH_3CHCH_2CH_2OH} \; \xrightarrow{65\%} \; CH_3COCH_2CH_2OH$$

with OH on the CH of **2**.

[1] K. S. Kim, S. Chung, I. H. Cho, and C. S. Hahn, *Tetrahedron Letters*, **30**, 2559 (1989).

1,2-Benzenedimethanol, (1).

with structure: benzene ring bearing two adjacent CH$_2$OH groups.

Protection of $\diagdown$C$=$O. This diol (**1**) reacts sluggishly with carbonyl compounds under acid catalysis, but the orthoformate 2,3-methoxy-1,5-dihydro-3*H*-2,4-benzodioxepine, derived from **1**, reacts to form 2,4-benzodioxepines (**3**). The car-

$$\mathbf{1} + HC(OCH_3)_3 \xrightarrow[\text{86\%}]{\text{TsOH, DME}} \quad \mathbf{2},\ \text{m.p. } 44° \xrightarrow[\text{TsOH}]{\text{R}^1\text{COR}^2 \atop \text{90–95\%}} \quad \mathbf{3}$$

2, m.p. 44°

3

$$\mathbf{3} \xrightarrow[\text{quant.}]{\text{H}_2,\ \text{Pd(0)}, \atop 25°} R^1COR^2$$

18

bonyl compound is regenerated by Pd-catalyzed hydrogenation. These protective groups should be useful in the case of acid-sensitive carbonyl compounds.

[1] N. Machinaga and C. Kibayashi, *Tetrahedron Letters*, **30**, 4165 (1989).

[1,2-Benzenediolato-*o*,*o*']oxotitanium,

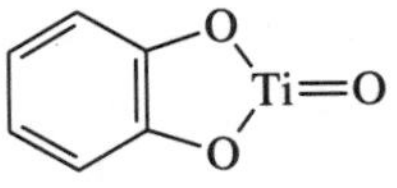

(**1**).

The reagent is prepared by reaction of catechol with $(i\text{-PrO})_2\text{Ti}{=}\text{O}$ followed by removal of the 2-propanol by-product.

Michael reactions.[1] This reagent serves as a nonacidic catalyst for Michael reactions of ketene silyl acetals with α,β-enones, and is more effective than dialkoxy-titanium oxides, $(\text{RO})_2\text{Ti}{=}\text{O}$. The reaction proceeds at $-78°$ in various solvents, CH_2Cl_2, ether, and toluene.

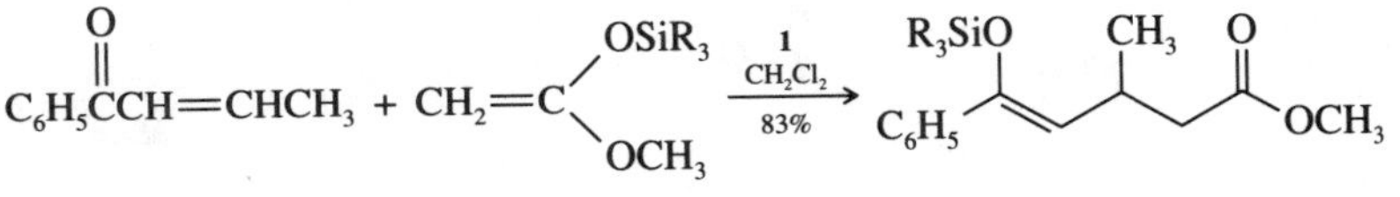

β-Hydroxy esters.[2] This Ti complex (**1**) is an efficient catalyst for an aldol-type reaction of activated aldehydes with ketene silyl acetals to form β-hydroxy esters as the O-silyl ethers.

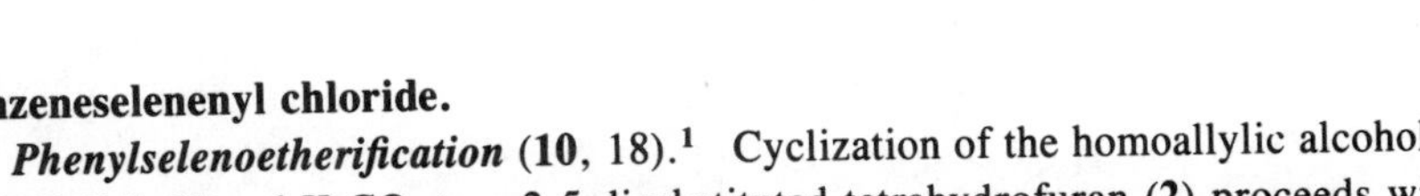

[1] T. Mukaiyama and R. Hara, *Chem. Letters*, 1171 (1989).
[2] R. Hara and T. Mukaiyama, *ibid.*, 1909 (1989).

Benzeneselenenyl chloride.
  Phenylselenoetherification (**10**, 18).[1] Cyclization of the homoallylic alcohol **1** with C_6H_5SeCl and K_2CO_3 to a 2,5-disubstituted tetrahydrofuran (**2**) proceeds with

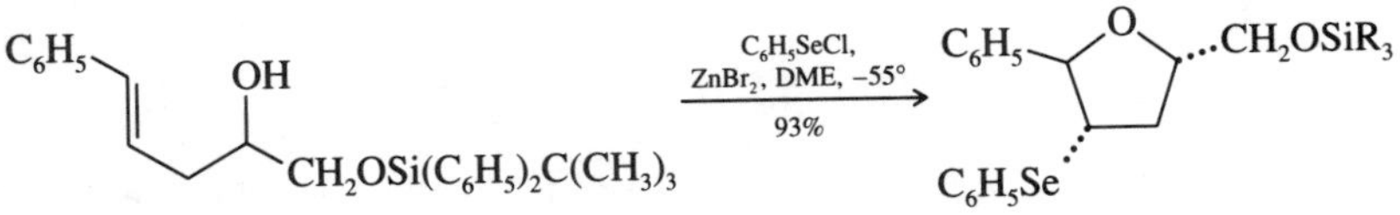

much higher *trans*-selectivity in THF or DME than in CH_2Cl_2, but yields are poor because of low conversion even after addition of more reagent. After some experimentation, the combination of C_6H_5SeCl and $ZnBr_2$ was found to give the highest yields with high *trans*-selectivity.

Cyclization of allylic ureas. The reaction of C_6H_5SeCl (1) and silica gel with allylic ureas in $CHCl_3$ at 25° provides 2-oxazolines in 30–90% yield. The C_6H_5Se

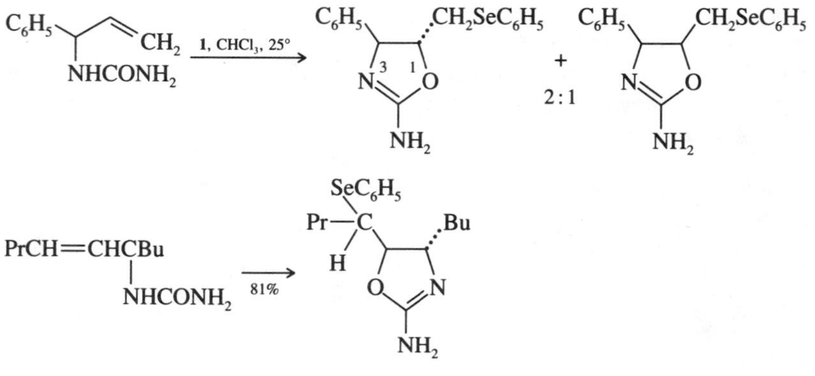

group of these products is removed in high yield by $(C_6H_5)_3SnH$. Hydrolysis of the products is known to provide 1,2-amino alcohols.[2]

The same cyclization of allylic O-methylisoureas provides imidazolines or 5,6-dihydro-1,3-oxazines (when catalyzed by benzeneselenenyl triflate).[3]

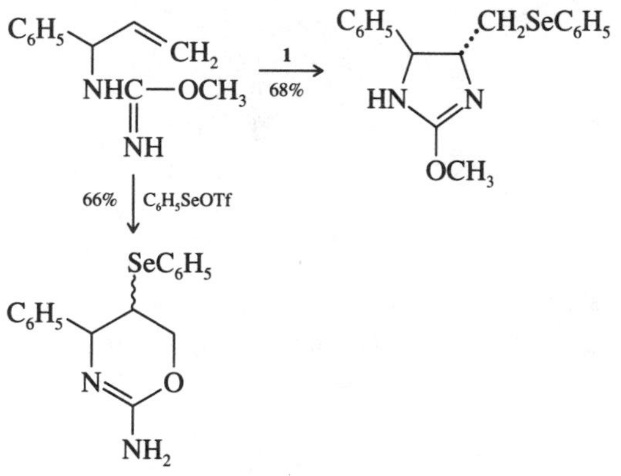

Acyl selenides.[4] The triethylammonium salt of carboxylic acids reacts with C_6H_5SeCl and Bu_3P in THF to form selenoesters in 62–85% yield. These products on reaction with Bu_3SnH (AIBN) furnish acyl radicals.

$$RCOOH \xrightarrow[\substack{62-85\%}]{\substack{1)\ N(C_2H_5)_3,\ CH_2Cl_2, \\ 2)\ C_6H_5SeCl,\ Bu_3P,\ THF}} RCOSeC_6H_5$$

[1] S. H. Kang, T. S. Hwang, W. J. Kim, and J. K. Lim, *Tetrahedron Letters*, **31**, 5917 (1990).
[2] C. Betancor, E. I. León, T. Prange, J. A. Salazar, and E. Suárez, *J.C.S. Chem. Comm.*, 450 (1989).
[3] R. Freire, E. I. León, J. A. Salazar, and E. Suárez, *ibid.*, 452 (1989).
[4] D. Batty and D. Crich, *Synthesis*, 273 (1990).

Benzeneselenenyl trifluoromethanesulfonate, $C_6H_5SeOSO_2CF_3$. The triflate is prepared from C_6H_5SeCl and AgOTf in CH_2Cl_2.

Selenolactonization.[1] Chlorinated by-products can be formed when selenolactonization is effected with the usual reagent, C_6H_5SeCl (**8**, 26–28; **13**, 26). Reaction of γ,δ- and δ,ε-unsaturated acids with C_6H_5SeOTf proceeds even at $-78°$ to give γ- and δ-lactones, respectively, in high yield.

$(trans/cis = 10:1)$

Selenoetherification.[2] C_6H_5SeOTf also converts alkenes substituted by an hydroxy group in the 5- or 6-position into tetrahydrofurans or -pyrans in generally high yield.

$(cis/trans = 1:1)$

[1] S. Murata and T. Suzuki, *Chem. Letters*, 849 (1987).
[2] *Idem, Tetrahedron Letters*, **25**, 4297, 4415 (1987).

Benzotriazole.

Enamines.[1] Enamines are generally prepared by condensation of the carbonyl compound with a secondary amine with removal of the water formed. In a new method, an aldehyde and a secondary amine are added to benzotriazole in ether at room temperature (3-Å molecular sieves can be used). On treatment of the product with sodium hydride, sodium benzotriazolate is formed with release of an enamine.

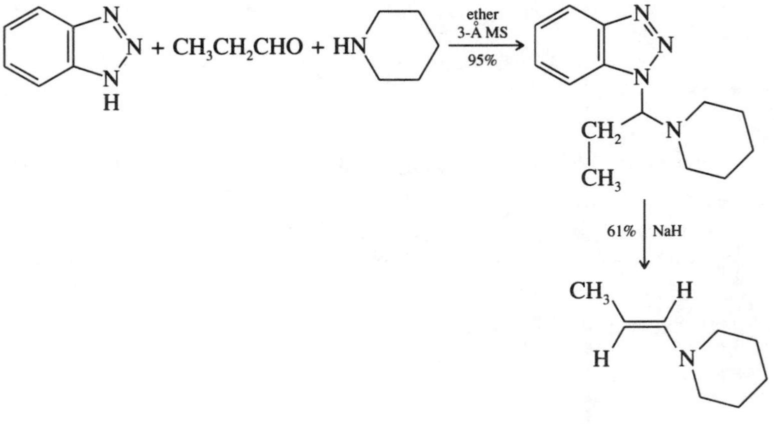

[1] A. R. Katritzky, Q.-H. Long, P. Lue, and A. Jozwiak, *Tetrahedron*, **46**, 8153 (1990).

Benzylamine/Potassium cyanide.

Strecker α-amino nitrile synthesis.[1] A new variation of NH_3/HCN for the Strecker synthesis uses $BzlNH_2$, KCN, and HOAc in CH_3OH.

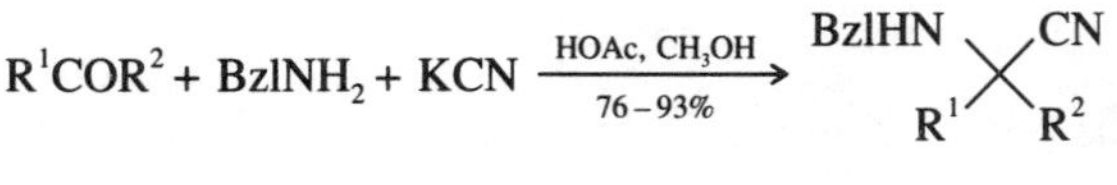

[1] M. P. Georgiadis and S. A. Haroutounian, *Synthesis*, 616 (1989).

O-Benzylhydroxylamine, $C_6H_5CH_2ONH_2$ (1).

1,2-cis-Amino alcohols.[1] α-Hydroxy ketones can be converted to 1,2-*cis*-amino alcohols in two steps. Reaction with O-benzylhydroxylamine provides the

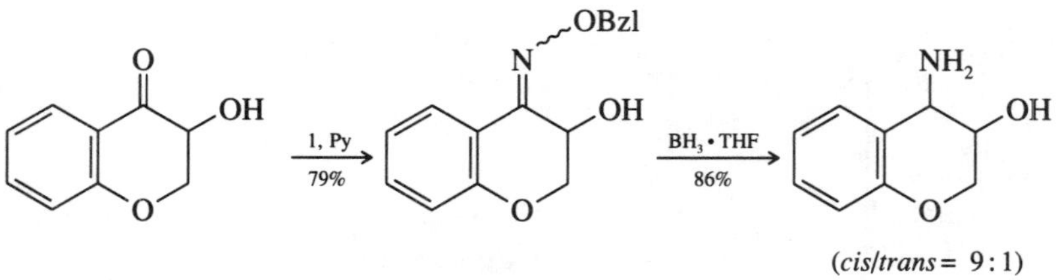

($cis/trans = 9:1$)

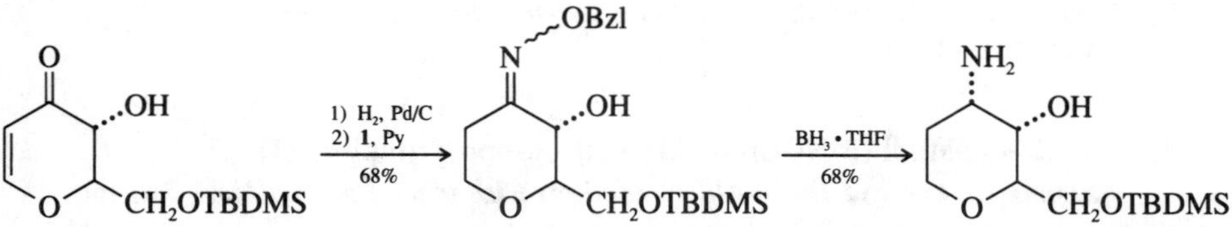

(cis/trans = 96:4)

corresponding α-hydroxy benzyloximes (*cis*- and *trans*-mixtures). These oximes are reduced by $BH_3 \cdot THF$ stereoselectivity to *cis*-1,2-amino alcohols.

[1] A. K. Ghosh, S. P. McKee; and W. M. Sanders, *Tetrahedron Letters*, **32**, 711 (1991).

Benzyltriethylammonium borohydride–Chlorotrimethylsilane.

Reduction of alkenes to alcohols. This combination (**1**) of $C_6H_5CH_2(C_2H_5)_3$-$\overset{+}{N}BH_4{}^-$ and $ClSi(CH_3)_3$ (1:1) converts alkenes to the corresponding *anti*-Markovnikov alcohols in good yield. $NaBH_4$–$ClSi(CH_3)_3$ in DME is less effective.

$$CH_3(CH_2)_6CH\text{==}CH_2 \xrightarrow{\text{1, } CH_2Cl_2,\ 0°} CH_3(CH_2)_6CH_2CH_2OH + CH_3(CH_2)_7CH_3$$

72%   18%

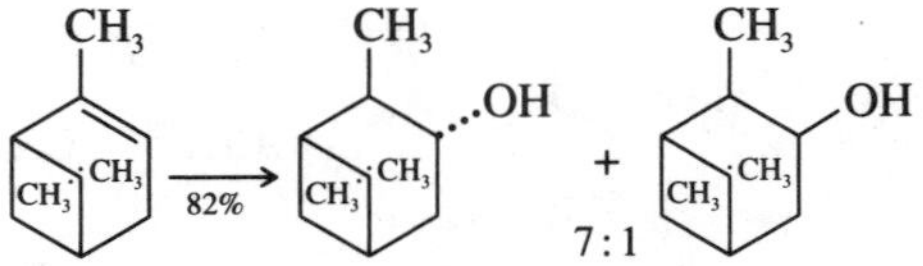

[1] S. Baskaran, V. Gupta, N. Chidambaram, and S. Chandrasekaran, *J.C.S. Chem. Comm.*, 903 (1989).

Benzyne.

Phenanthridinones.[1] Benzyne, generated *in situ* by $Pb(OAc)_4$ oxidation of 1-aminobenzotriazole (**1**),[2] undergoes [4+2]cycloaddition to cyclic vinyl isocyanates to form phenanthridinones.

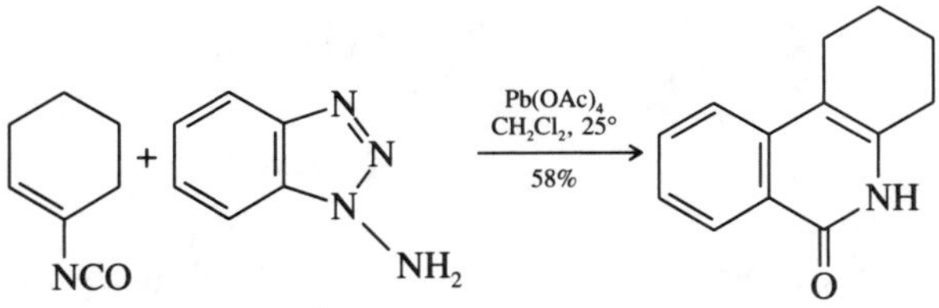

1

[1] J. H. Rigby, D. D. Holsworth, and K. James, *J. Org.*, **54**, 4019 (1989).
[2] H. Hart and D. Ok, *ibid.*, **51**, 979 (1986).

1,1′-Bi-2,2′-naphthol (BINOL)-**Dichlorodiisopropoxytitanium (1), 15**, 26–27.

Asymmetric ene and Diels–Alder reactions with methyl glyoxalate. The reaction of methyl glyoxalate with isoprene catalyzed by the BINOL–titanium complex (R)-**1** provides not only the expected ene product (**2**), but also the Diels–Alder product (**3**), both in 97% ee (equation I). This chiral titanium complex is also an

2 (97% ee) **3** (97% ee)

2,6-*cis*-**4**
(96% ee)

5 (90% ee)

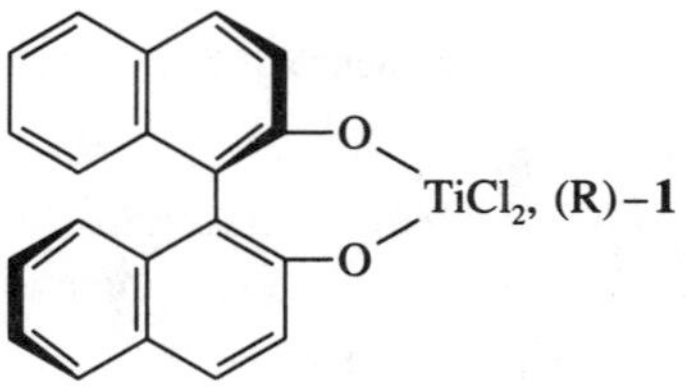

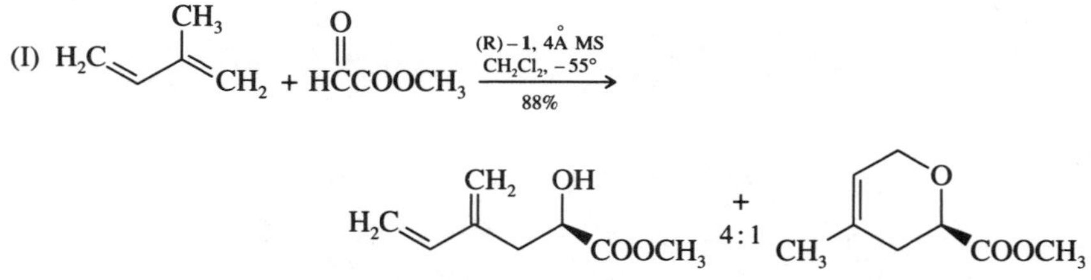

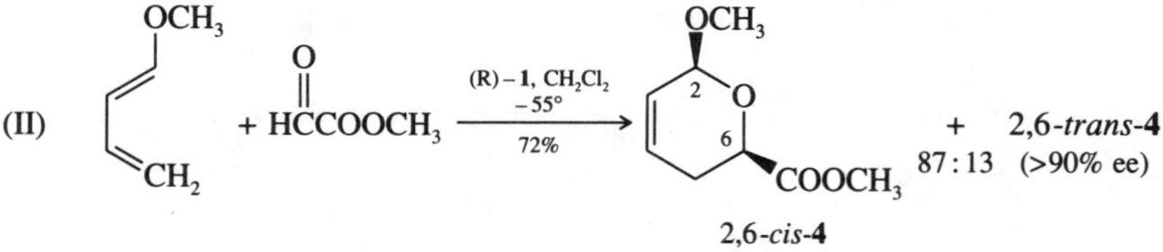

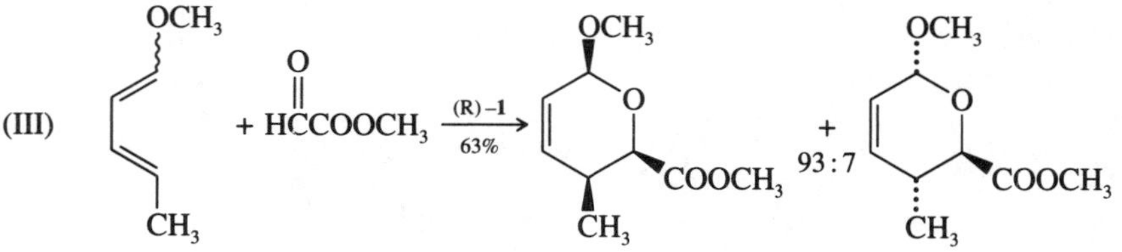

effective catalyst for a Diels–Alder reaction of methyl glyoxalate with 1-methoxy-1,3-butadiene to provide the *cis*-adduct **4** in 96% ee (equation II). Dihydropyrans of this type are useful precursors to δ-lactones (10, 258). The cycloaddition of methyl glyoxalates with 1-methoxy-1,3-pentadiene catalyzed by R-1 proceeds with high stereoselectivity at three stereogenic centers (equation III).

[1] M. Terada, K. Mikami, and T. Nakai, *Tetrahedron Letters*, 32, 935 (1991).

(R) or (S)-1,1'-Binaphthyl-2,2'-diyl hydrogen phosphate (1), 12, 49.

Asymmetric hydrocarboxylation of styrenes.[1]   Use of (S)- or (R)-1 as a chiral ligand in the palladium-catalyzed hydrocarboxylation of *p*-isobutylstyrene (2) results in (S)- or (R)-2 (ibuprofen) in 83–84% ee. Similar enantioselectivity obtains in hydrocarboxylation of a 2-vinylnaphthalene to form naproxen.

[1] H. Alper and N. Hamel, *Am. Soc.*, 112, 2803 (1990).

Bipyridine–Palladium(II) acetate.

Hydrogenation catalyst.[1]   A supported form (1) of this complex is prepared by treatment of H-montmorillonite (Fluka) with $SOCl_2$, followed by BuLi, and then bipyridine. The resulting montmorillite bipyridine is then treated with $Pd(OAc)_2$.

Catalyzed hydrogenation of alkynes, alkenynes, and alkadienes.[1]   This catalyst effects highly *cis*-selective hydrogenation of triple bonds of alkynes and alkenynes, with easy recovery of the complex by filtration. It also effects only 1,2-addition in hydrogenation of even hindered 1,4-substituted 1,3-butadienes.

[1] B. M. Choudary, G. V. M. Sharma, and P. Bharathi, *Angew. Chem. Int. Ed.*, 28, 465 (1989).

Bis(acetonitrile)dichloropalladium.

(E)-Vinylsilanes.[1]   These silanes (2) are obtained in good to high yield by allylic transposition of C_1-acyloxy (E)-crotylsilanes (1) with this Pd(II) catalyst.[2] Moreover, the rearrangement of optically pure substrates can proceed with complete preservation of optical purity.

$$(S)-1 \longrightarrow (R)-2$$

This rearrangement is also effected by a number of Lewis acids such as BF$_3$ etherate, but with less stereoselectivity.

Carbonylative coupling of vinyl triflates with organotins (**12**, 470–471).[3] The final step in a synthesis of the macrocyclic diterpene jatrophone (**2**) was effected by carbonylative coupling of a vinyl triflate with vinyltin catalyzed by bis(acetonitrile)-dichloropalladium.

Alkoxylation and acetylation of activated alkenes.[4] Alcohols can add to activated double bonds in the presence of this Pd(II) catalyst. Addition of LiCl is required to effect addition of acetic acid.

$$\underset{\text{O}}{\overset{\text{O}}{C_6H_5\overset{\|}{C}CH=CH_2}} + CH_3OH \xrightarrow[\text{97%}]{\underset{\text{CH}_2\text{Cl}_2,\ 25°}{\text{PdCl}_2(\text{CH}_3\text{CN})_2}} C_6H_5\overset{\text{O}}{\overset{\|}{C}}CH_2CH_2OCH_3$$

$$C_2H_5\overset{\text{O}}{\overset{\|}{C}}CH=CHCH_3 + BzlOH \xrightarrow[\text{89%}]{} C_2H_5\overset{\text{O}}{\overset{\|}{C}}CH_2\overset{\text{CH}_3}{\overset{|}{C}}HOBzl$$

[1] J. S. Panek and M. A. Sparks, *J. Org.*, **55**, 5564 (1990).
[2] L. E. Overman, *Angew. Chem. Int. Ed.*, **23**, 579 (1984).
[3] A. C. Gyorkos, J. K. Stille, and L. S. Hegedus, *Am. Soc.*, **112**, 8465 (1990).
[4] T. Hosokawa, T. Shinohara, Y. Ooka, and S.-I. Murahashi, *Chem. Letters,* 2001 (1989).

Bis(acetylacetonate)cobalt(II), Co(acac)$_2$.

Hydration of alkenes. This reaction can be effected with oxygen and Co(acac)$_2$ and a secondary alcohol (2-propanol, cyclopentanol). The reactive species is assumed to be $HOO(acac)_2Co(III)$. A mixture of a secondary alcohol, a ketone, and an alkane is obtained, with the alcohol as the major product.[1] On further investigation of this hydration, use of bis(trifluoroacetylacetonate)cobalt(II) was shown to markedly improve the yield of the alcoholic product. The solvent *sec*-alcohol is oxidized during this reaction to a ketone. In fact, a variety of secondary alcohols are oxidized to ketones by oxygen and $Co(tfa)_2$ (20 mole %) at 75° in toluene containing 4-Å MS in 80–100% yield.[2]

$$CH_3(CH_2)_7CH{=}CH_2 \xrightarrow[\text{Co(II)}]{\text{O}_2,\ \text{R}_2\text{CHOH}}$$

	$CH_3(CH_2)_7CHOHCH_3$ +	$CH_3(CH_2)_8CH_3$+	$CH_3(CH_2)_7COCH_3$
$Co(acac)_2$	45%	22%	7%
$Co(CH_3COCH_2COCF_3)_3$	81%	13%	2%

Coupling of α,β-*unsaturated nitriles, amides, or esters with RCHO.*[3]   This coupling can be effected with $Co(acac)_2$ as catalyst and $C_6H_5SiH_3$ as the hydrogen source. Thus the reaction of benzaldehyde with acrylonitrile in 1,2-dichloroethane with $C_6H_5SiH_3$ (2 equiv.) and a catalytic amount of $Co(acac)_2$ at 70° provides the β-silyloxy nitrile as the major product (equation I).

$$\text{(I)}\ \ C_6H_5CHO + CH_2{=}CHCN \xrightarrow[\text{ClCH}_2\text{CH}_2\text{Cl, 70°}]{\text{C}_6\text{H}_5\text{SiH}_3,\ \text{Co(acac)}_2}$$

$$\underset{\underset{\displaystyle CN}{|}}{\overset{\overset{\displaystyle C_6H_5SiH_2O}{|}}{C_6H_5CHCHCH_3}} \xrightarrow[70\%]{H_3O^+} \underset{\underset{\displaystyle CN}{|}}{\overset{\overset{\displaystyle OH}{|}}{C_6H_5CHCHCH_3}}$$

The same coupling of aldehydes and α,β-unsaturated amides and esters can also be effected. In these couplings the best catalyst is bis(dipivaloylmethanato)cobalt (II), $Co(dpm)_2$.[4]

[1] T. Mukaiyama, S. Isayama, S. Inoki, K. Kato, T. Yamada, and T. Takai, *Chem. Letters*, 515 (1989).
[2] T. Yamada and T. Mukaiyama, *ibid.*, 519 (1989).
[3] S. Isayama and T. Mukaiyama, *ibid.*, 2005 (1989).
[4] F. A. Cotton and R. H. Soderberg, *Inorg. Chem.*, 3, 1 (1964).

Bis(acetylacetonato)nickel(II)–Diiosbutylaluminum hydride.

Hydrosilylative cyclization of 1,7-diynes.[1]   Hydrosilylation of 1,7-diynes with $HSiX_3$ promoted by a Ni(0) catalyst, prepared from $Ni(acac)_2$ and DIBAH, results in a 1,2-dialkylidenecyclohexane with a (Z)-vinylsilane group. An internal 1,7-diyne undergoes a similar cyclization, but in lower yield. An unsymmetrical diyne forms a single product in which the silyl group is introduced exclusively into the terminal acetylene group.

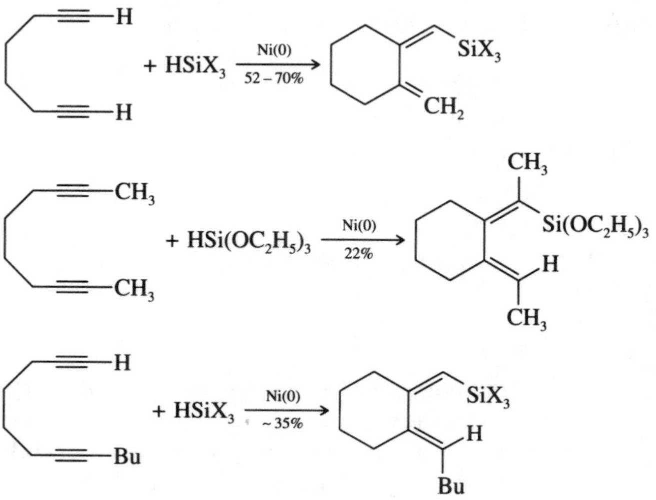

The exocyclic silyl dienes are useful for Diels–Alder reactions. The silyl group can also undergo Pd-catalyzed coupling with aryl iodides as well as oxidation to a hydroxyl group.

[1] K. Tamao, K. Kobayashi, and Y. Ito, *Am. Soc.*, **111**, 6478 (1989).

1,4-Bis(bromomagnesio)pentane (1), 13, 138–140.

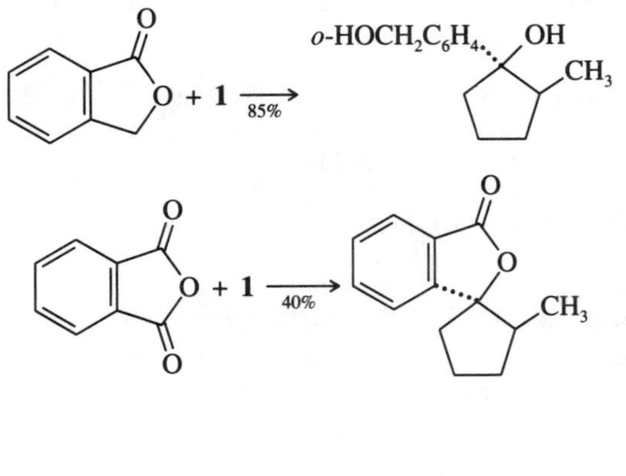

Reaction with lactones, anhydrides. Canonne *et al.*[1] have observed high stereoselectivity in reaction of **1** with aromatic lactones and anhydrides resulting in the *trans*-isomer in the case of the former substrates but in the *cis*-isomer in reaction with the latter substrates.

[1] P. Canonne, R. Boulanger, and M. Bernatchez, *Tetrahedron*, **45**, 2525 (1989).

Bis(cyclooctadiene)nickel, $Ni(COD)_2$.

Cyclization of dienynes.[1]   This Ni(0) catalyst in combination with a triarylphosphite, particularly tri-*o*-biphenyl phosphite, permits intramolecular cycloaddition at 25° of [4 + 2] dienynes, in which the diene and the alkyne are separated by 3- and 4-atom units. This reaction is a useful route to products containing a cyclohexadiene group, which are oxidized to an arene by DDQ.

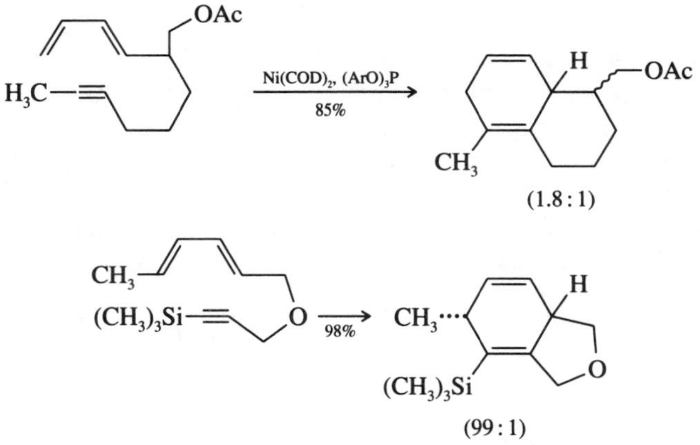

(1.8 : 1)

(99 : 1)

Cyclization of diynes to iminocyclopentadienes.   1,6-1,8-Diynes cyclize with 2,6-dimethylphenyl isocyanide when treated with $Ni(COD)_2$, 1 equiv., to form bicy-

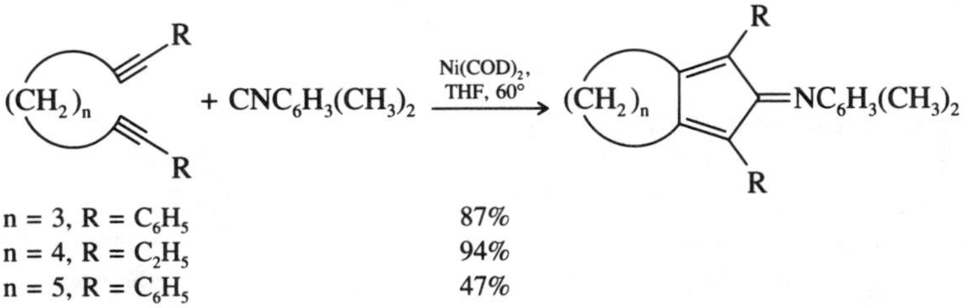

n = 3, R = C_6H_5 87%
n = 4, R = C_2H_5 94%
n = 5, R = C_6H_5 47%

clic iminocyclopentadienes. R can be alkyl, aryl, or trimethylsilyl, but not H.[2]

[1] P. A. Wender and T. E. Jenkins, *Am. Soc.*, **111**, 6432 (1989).
[2] K. Tamao, K. Kobayashi, and Y. Ito, *J. Org.*, **54**, 3517 (1989).

Bis(dialkylamino)magnesium, $[R_2N]_2Mg$.

Amides.[1] Carboxylic acids can be converted directly into amides by reaction with a bis(dialkylamino)magnesium, prepared *in situ* by reaction of $2R_2NH$ with *n*-butyl-*sec*-butylmagnesium (Alfa).

[1] R. Sanchez, G. Vest, and L. Despres, *Syn. Comm.*, **19**, 2909 (1989).

(−)-1,1′-Bis(2,4-dicyanonaphthalene), **(1)**

The reagent was obtained in about 70% ee from (−)-1,1′-bis(2-naphthoic acid).

Enantioselective triplex Diels–Alder reaction.[1] 1,3-Cyclohexadiene is known to undergo Diels–Alder reactions with electron-rich dienes when irradiated with arene sensitizers. The reaction is presumed to involve a triplex involving the sensitizer (**15**, 129). The first enantioselective example of this Diels–Alder reaction has been achieved using (−)-**1** as sensitizer. Thus irradiation of 1,3-cyclohexadiene with *trans*-β-methylstyrene in the presence of (−)-**1** at −65° provides the *endo*-adduct **2** in 15% ee (equation I). The enantioselectivity decreases with an increase of the temperature, being 1% at 25°.

2, 15% ee

[1] J.-I. Kim and G. B. Schuster, *Am. Soc.*, **112**, 9635 (1990).

Bis(1,3-diketonato)nickel(II) complexes.

Oxygenation of RCHO to RCOOH.[1] This oxidation can be effected by oxygen when catalyzed by nickel complexes such as bis(acetylacetonato)nickel, $Ni(acac)_2$. However, the highest conversion and yields are obtained with the complex prepared from 1,3-di(*p*-methoxyphenyl)-1,3-propanedionate, $Ni(dmp)_2$. Ether or an alcohol

$(p\text{-}CH_3OC_6H_4C)_2CH_2$ (Hdmp)

is a poor solvent for this oxidation, but ketones or esters are appropriate. Highest yields are obtained with straight-chain aldehydes (83–87%), but secondary and even tertiary aldehydes are oxidized in reasonable yield. Benzaldehyde is oxidized to benzoic acid in 79% yield.

Epoxidation of alkenes by O_2. This epoxidation can be effected with O_2 and a reductant when catalyzed by a bis(1,3-diketonato)nickel(II) complex. Reductants can be a primary or secondary alcohol[2] or, preferably, an aldehyde (which is converted to an acid.[3] The most efficient catalyst is Ni(dmp)$_2$, although Ni(acac)$_2$ is almost as

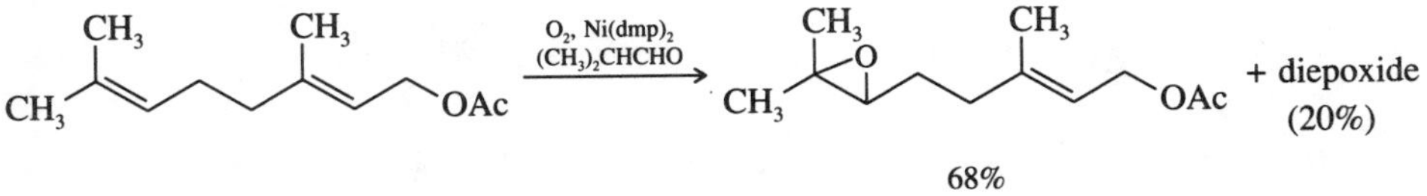

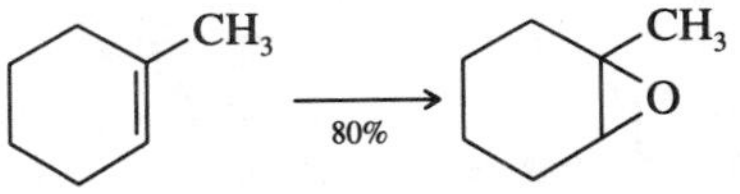

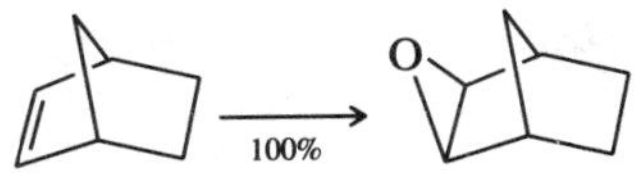

68%

80%

100%

satisfactory. This epoxidation proceeds in high yield with styrene derivatives, 1,1-disubstituted terminal alkenes, norbornenes, and various trisubstituted alkenes, but is generally less effective for 1,2-disubstituted internal alkenes. Of various aldehydes, isobutyraldehyde or pivaldehyde is more satisfactory than butyraldehyde.

Oxygenation of silyl enol ethers.[4] Oxygenation of a silyl enol ether under the conditions cited above results in a silyloxy epoxide, which rearranges spontaneously to an α-silyloxy ketone. The preferred Ni catalyst for this epoxidation is bis(3-methyl-2,4-pentanedionato)nickel(II), Ni(mac)$_2$. The α-silyloxy ketone is converted

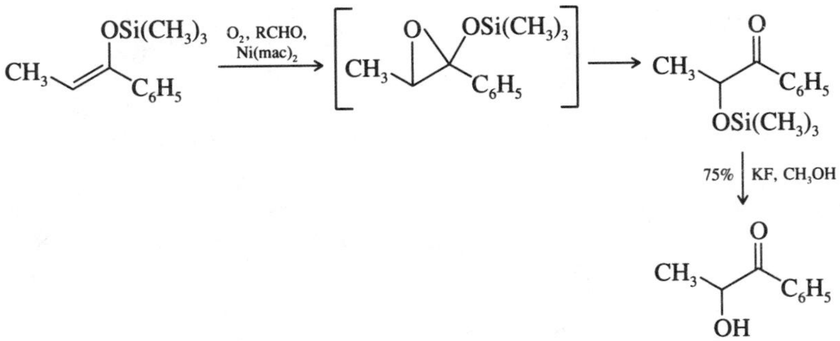

$$ \underset{\text{Bzl}}{\overset{\text{OSi(CH}_3)_3}{\diagdown}}\diagup\text{OCH}_3 \quad \xrightarrow[77\%]{} \quad \underset{\underset{\text{OH}}{|}}{\text{BzlCHCOOCH}_3} $$

to an α-hydroxy ketone by KF in CH_3OH. The same sequence when applied to a silyl ketene acetal gives an α-hydroxy ester.

[1] T. Yamada, O. Rhode, T. Takai, and T. Mukaiyama, *Chem. Letters*, 5 (1991).
[2] T. Mukiyama, T. Takai, T. Yamada, and O. Rhode, *ibid.*, 1657, 1661 (1990).
[3] T. Yamada, T. Takai, O. Rhode, and T. Mukaiyama, *ibid.*, 1 (1991).
[4] T. Takai, T. Yamada, O. Rhode, and T. Mukaiyama, *ibid.*, 281 (1991).

Bis(dimethylaluminum)selenide, $[(CH_3)_2Al]_2Se$ (**1**). The selenide is obtained *in situ* by reaction of $[(CH_3)_3Si]_2Se$ with $(CH_3)_2AlCl$ (2 equiv.) in toluene.

Selenoketones.[1] Selenoketones can be prepared by reaction of dialkyl ketones with **1** and trapped by a Diels–Alder reaction with a diene. A sterically hindered selenoketene, selenofenchone, has been isolated from this reaction.

[1] M. Segi, T. Koyama, T. Nakajima, S. Suga, S. Murai, and N. Sonoda, *Tetrahedron Letters*, **30**, 2095 (1989).

2,2'-Bis(diphenylphosphino)-1,1'-binaphthyl (1, BINAP).
Isomerization of allylic amines (**11**, 53–54; **12**, 56–57).[1] The asymmetric isomerization of allylamines to enamines effected with ruthenium complexes of (R)- and (S)-**1** is applicable to C_5-isoprenoids, even to ones with an allylic dialkylamino group at one end and an allylic O-function at the other end. Thus it can be used to

isomerize **3** to the optically pure 4-benzyloxy-3-methylbutanal **4**, a useful building block for isoprenoid homologs. In some cases 2,2'-bis(diphenylphosphino)-6,6'-di-methylbiphenyl (BIPHEMP, **2**) is somewhat superior to BINAP as the chiral component.

The (R)- and (S)-aldehydes **4** were used to prepare the four (E)-stereoisomers of vitamin K, all of which have essentially identical biological activity.

$$CH_3\quad P(C_6H_5)_2$$
$$CH_3\quad P(C_6H_5)_2 \quad \textbf{2}$$

Asymmetric hydroboration of $C_6H_5CH=CH_2$.[2] The reaction of cate-cholborane with styrene provides, after oxidation, 2-phenylethanol. Hydroboration catalyzed by a cationic rhodium catalyst, $[Rh(COD)_2]^+BF_4^-$ and dppb, provides 1-phenylethanol (99:1) in 86% yield. A catalytic asymmetric hydroboration is possible with BINAP. Use of (−)-BINAP as ligand and a temperature of −78° provides 1-phenylethanol in 96% ee (equation I).

(I) $C_6H_5CH=CH_2 + HB$ $\xrightarrow[\text{91%}]{\substack{1)\ [Rh(COD)_2]^+BF_4^-,\ (-)-\textbf{1} \\ 2)\ H_2O_2,\ NaOH}}$

(96% R)

Stereoselective hydrogenation.[3] Hydrogenation of the cyclic β-keto ester **2** catalyzed by an (R)-BINAP-Ru complex (**1**) proceeds with high *anti*-diastereoselectivity to give a 99:1 mixture of **3** and **4**. On the other hand, catalytic hydrogenation of the acyclic keto ester **5a** proceeds with no appreciable resolution to give a 1:1

2 $\xrightarrow[\substack{(R)-BINAP-Ru\ \textbf{(1)}\\100\%}]{H_2,}$ **3**, (1R, 2R), 92% ee   +   **4**, (1R, 2S), 93% ee

99:1

mixture of *syn*- and *anti*-**6**. However, the presence of an amide or carbamate group at C_2 results in extensive kinetic resolution. Thus **5b** is hydrogenated to N-acetyl L-threonine (**6b**) in 98% ee. In these asymmetric hydrogenations, the configuration at C_3 is determined by the chirality of the BINAP ligand and the C_2 configuration depends on the substrate.

Ordinarily, the yield in kinetic resolutions does not exceed 50%. The high yield in the present case depends on racemization of the substrate being more rapid than hydrogenation.

Asymmetric hydrogenation of allylic alcohols (14, 38–40).[4] Full details are available for the preparation of $Ru(OAc)_2[(R)$- or (S)-BINAP], the catalyst for asym-

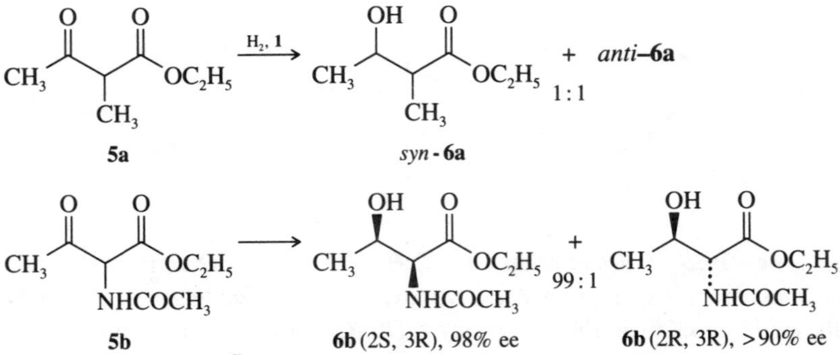

5a   syn-6a   + anti-6a   1:1

5b   6b (2S, 3R), 98% ee   6b (2R, 3R), >90% ee   99:1

metric hydrogenation of allylic alcohols without effect on other double bonds in the substrate. Ru(OCOCF$_3$)$_2$(BINAP) also shows high catalytic activity.

Asymmetric hydrogenation of β-oxo esters.[5] The preferred catalyst for this reaction is RuCl$_2$[BINAP], prepared by reaction of Ru(OAc)$_2$[BINAP], with HCl (2 equiv.). Details for the preparation of the catalyst and use in asymmetric hydrogena-

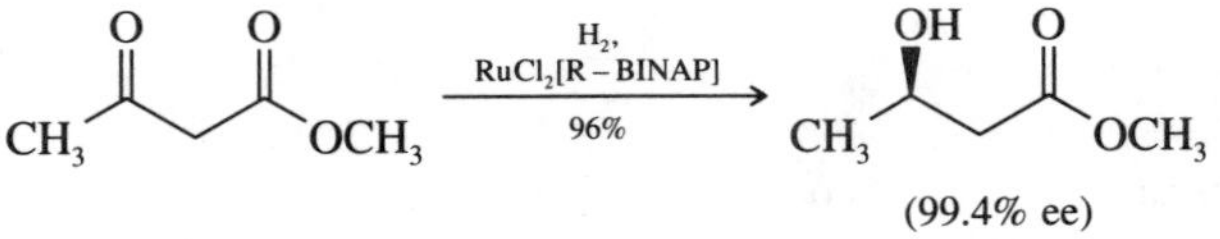

(99.4% ee)

tion of methyl 3-oxobutanoate to methyl (R)-3-hydroxy butanoate in 99.4% ee and 96% yield are available. This asymmetric hydrogenation is applicable to 3-oxo esters with various substituents at C$_2$ and C$_3$ (23 examples are listed in the report with optical yields of 88–100%).

Asymmetric hydrogenation of 4-oxo esters.[6] A Ru catalyst prepared *in situ* by reaction of Ru(OAc)$_2$(BINAP) with HCl (2 equiv.) effects hydrogenation of 4-oxo esters to 4-hydroxy esters, which are converted to γ-lactones when heated with acetic acid in toluene. If the Ru is complexed with (R)- or (S)-BINAP, the lactones are obtained in >98% ee.

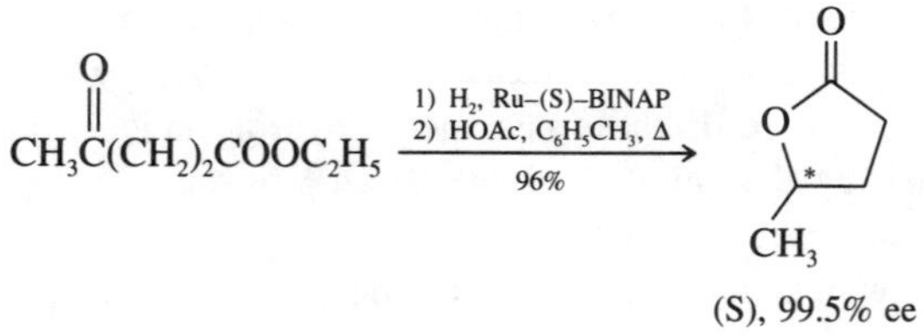

CH$_3$C(CH$_2$)$_2$COOC$_2$H$_5$ 1) H$_2$, Ru–(S)–BINAP 2) HOAc, C$_6$H$_5$CH$_3$, Δ   96%

(S), 99.5% ee

The sequence can be extended to preparation of optically active phthalides (**1 → 2**).

Asymmetric Heck-type arylation.[7]   A catalyst prepared *in situ* from Pd(OAc)$_2$ and (R)-BINAP effects a highly enantioselective arylation of 2,3-dihydrofuran with aryl triflates with diisopropylethylamine as base and benzene as solvent. Two products (**2** and **3**) are formed with opposite configurations. The highly enantioselective formation of the major product (**2**) is ascribed to a kinetic resolution step which converts the enantiomer of **2** into **3**.

Use of phenyl iodide for this reaction results in racemic **2** and **3** in 23 and 2% yield, respectively, probably because the BINAP ligand in the catalyst is displaced by the iodide.

Optically active *cis*-decalins can be obtained from substrates such as **4** by a Heck-type reaction with PdCl$_2$/(R)-**1** (1:1) as the catalyst.[8] Addition of various silver salts improves the yield and enantioselectivity. For cyclization of **4** to **5**, the highest enantioselectivity was observed by use of Ag$_3$PO$_4$ and CaCO$_3$ (2 equiv. of each) with 1-methyl-2-pyrrolidinone (NMP) as solvent (60°).

Asymmetric intramolecular hydrosilylation.[9]   The intramolecular hydrosilylation of allylic alcohols (**14**, 137) can be enantioselective when catalyzed by Rh(I) complexed with either (R)-BINAP or (R,R)-DIOP. The enantioselectivity is dependent on the groups attached to silicon, being higher with a phenyl than with a methyl group. Highest enantioselectivity (93% ee) was obtained with the di(3,5-xylyl)silyl ether, ROSiH[C$_6$H$_3$(CH$_3$)$_2$-3,5]$_2$.

syn/anti = 95:5
71% ee

Review.[10] Noyori has reviewed homogeneous asymmetric catalysis, particularly by complexes of (R)- and (S)-BINAP with ruthenium. He suggests possible reasons for the high efficiency of Ru–BINAP complexes in catalytic hydrogenation (40 references).

[1] R. Schmid and H.-J. Hansen, *Helv.*, **73**, 1258 (1990).
[2] T. Hayashi, Y. Matsumoto, and Y. Ito, *Am. Soc.*, **111**, 3426 (1989).
[3] R. Noyori, T. Ikeda, T. Ohkuma, M. Widhalm, M. Kitamura, H. Takaya, S. Akutagawa, N. Sayo, T. Saito, T. Taketomi, and H. Kumobayashi, *ibid.*, **111**, 9134 (1989).
[4] H. Takaya, T. Ohta, S. Inoue, and R. Noyori, *Org. Syn.*, submitted (1990).
[5] M. Kitamura, N. Sayo, and R. Noyori, *Org. Syn.*, submitted (1990).
[6] T. Ohkuma, M. Kitamura, and R. Noyori, *Tetrahedron Letters*, **31**, 5509 (1990).
[7] F. Ozawa, A. Kubo, and T. Hayashi, *Am. Soc.*, **113**, 1417 (1991).
[8] Y. Sato, M. Sodeoka, and M. Shibasaki, *J. Org.*, **54**, 4738 (1989); *idem, Chem. Letters*, 1953 (1990).
[9] K. Tamao, T. Tohma, N. Inui, O. Nakayama, and Y. Ito, *Tetrahedron Letters*, **31**, 7333 (1990).
[10] R. Noyori, *Chem. Soc. Rev.*, **18**, 187 (1989).

Bis(1,3-bistrimethylsilylcyclopentadienyl)chloroyttterbium(III), Cp₂YbCl (1). Preparation.[1]

Mukaiyama aldol reaction (**6**, 590–591). This reaction is generally effected with TiCl₄ in stoichiometric amounts as the promotor. This lanthanide complex is also effective and can be used as a catalyst if trimethylsilyl chloride is also present.[2] Yields are >80% in the case of aromatic aldehydes, and are >50% in the case of

(I) C₆H₅CHO +

(Z)

(syn/anti = 78:22)

(II) C₆H₅CHO +

(E)

(anti/syn = 81:19)

aliphatic aldehydes, but highly hindered aldehydes do not react at all. α,β-Enals react to give only the 1,2-adduct. Use of the complex also shows diastereoselectivity in the reactions of the (E)- and (Z)-isomers of silyl enol ethers of propionate esters with benzaldehyde (equations I and II).

[1] M. F. Lappert, A. Singh, J. L. Atwood, and W. Hunter, *J.C.S. Chem. Comm.*, 1190 (1981).
[2] L. Gong and A. Streitwieser, *J. Org.*, **55**, 6235 (1990).

Bis(N-isopropylsalicylaldiminato)copper(II), 1.

(1)

Conjugate addition of Grignard reagents.[1] CuI (or CuBr), particularly when activated by ClSi(CH$_3$)$_3$ and HMPA, is known to catalyze conjugate addition of Grignard reagents to α,β-enones (**14**, 88). Surprisingly, CuBr$_2$ is as effective as CuBr when used in combination with these two activators (3 equiv. each).

However, neither CuI nor CuBr$_2$ activated by ClSi(CH$_3$)$_3$ and HMPA is useful as a catalyst for conjugate addition of CH$_3$MgBr to α,β-unsaturated esters. Even the CH$_3$Cu·BF$_3$ complex is not useful in this reaction. Surprisingly, the Cu(II) complex **1** is remarkably efficient (equation I).

$$\text{(I) } PrCH{=}CHCOOC_2H_5 + CH_3MgBr \xrightarrow[\text{99\%}]{\substack{\text{1, ClSi(CH}_3)_3 \\ \text{Ether/THF, } -46°}} PrCHCH_2COOC_2H_5 \atop \overset{\displaystyle CH_3}{\vert}$$

Bis(salicylidene)ethylenediaminatocopper(II) in combination with ClSi(CH$_3$)$_3$ can also promote conjugate addition to α,β-unsaturated esters, but is less effective than **1**.

[1] H. Sakata, Y. Aoki, and I. Kuwajima, *Tetrahedron Letters*, **31**, 1161 (1990).

2,4-Bis(4-methoxyphenyl)-1,3-dithia-2,4-diphosphetane-2,4-disulfide
(Lawesson's reagent, **1**).

Thiomidization.[1] Reaction of the phthalimide **2** with Lawesson's reagent (**1**) provides the thioimide **3** as the only product, probably because of steric effects. Unlike **2**, which does not undergo cyclization under a variety of basic conditions, the thioimide **3** undergoes cyclization to magallanesine (**4**) in 76% yield when refluxed in DMF dimethyl acetal, (CH$_3$O)$_2$CHN(CH$_3$)$_2$. Magallanesine is the first known isoindolobenzazocine alkaloid. For a related cyclization to a benzazapin see rhodium(II) carboxylates, this volume.

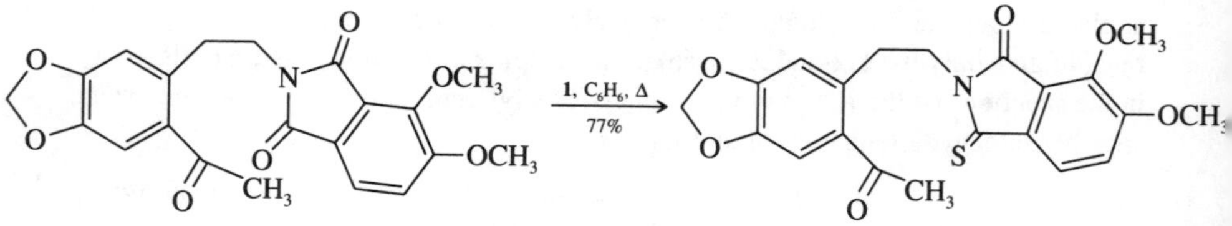

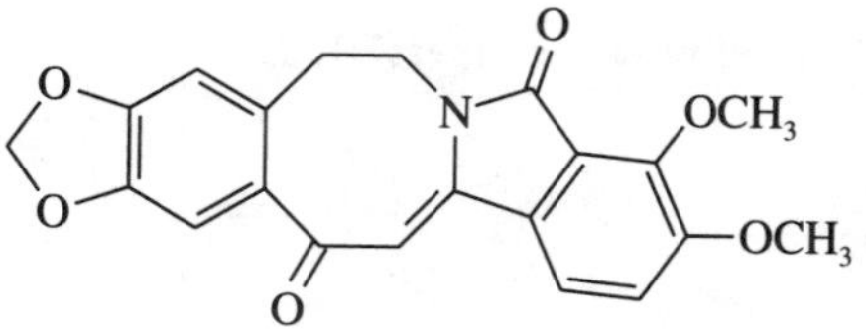

[1] F. G. Fang, G. B. Feigelson, and S. J. Danishefsky, *Tetrahedron Letters*, **30**, 2743 (1989).

Bis(oxazolines).

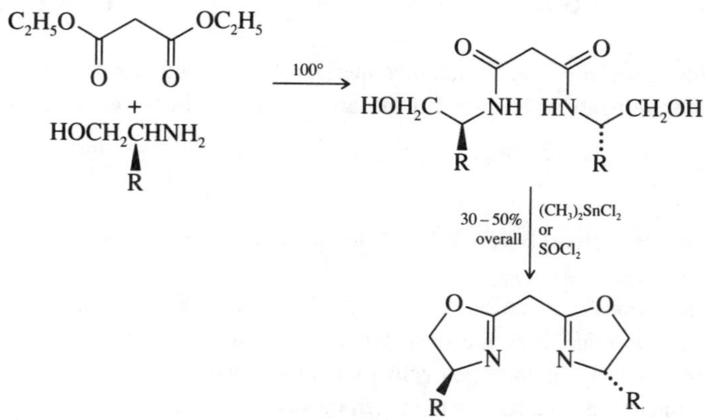

Asymmetric cyclopropanation. Three laboratories have reported that copper complexes of chiral bis(oxazolines) are effective catalysts for asymmetric cyclopropanation of alkenes with diazoacetates. Bis(oxazolines) such as **1** are readily available by condensation of α-amino alcohols with diethyl malonate followed by cyclization, effected with dichlorodimethyltin or thionyl chloride. Cyclopropanation of styrene with ethyl diazoacetate catalyzed by copper complexes of type **1** indicates that both the diastereoselectivity and enantioselectivity increases with the size of the R group, being highest when R=C(CH$_3$)$_3$ as in the product obtained from *t*-leucinol. A (CH$_3$)$_2$COH group is also highly effective, but this bis(oxazoline) is less available. The ester group of the diazoacetate also effects the stereoselectivity of cyclopropanation. The *trans* : *cis* selectivity as well as the enantioselectivity are particularly high

(I) $C_6H_5CH=CH_2$ + N_2CHCO_2R' $\xrightarrow{\text{Cu} \cdot \text{1a}}$

		trans		*cis*
R' = C$_2$H$_5$	84%	90% ee	75:25	77% ee
R' = *l*-Men.*	72%	98% ee	86:14	96% ee

in catalyzed reactions of *l*-menthyl diazoacetate (equation I).[1,2] This asymmetric cyclopropanation is generally applicable to monosubstituted and to 1,1-disubstituted alkenes, but not to 1,2-disubstituted alkenes.

Highly efficient catalytic asymmetric cyclopropanation can be effected with copper catalysts complexed with ligands of type **2**.[3] These bis(oxazolines) are prepared by reaction of dimethylmalonyl dichloride with an α-amino alcohol. As in the case of ligands of type **1**, particularly high stereoselectivity obtains when R is *t*-butyl. Cyclopropanation of styrene with ethyl diazoacetate catalyzed by copper complexed with

2a, R=C(CH$_3$)$_3$

2a results in a mixture of *trans*- and *cis*-cyclopropyl esters, both in >97% ee (equation II). Increase in the bulk of the ester group increases the *trans*-selectivity. The ester derived from 2,6-di-*t*-butyl-4-methylphenol (BHT) forms essentially only the *trans*-cyclopropyl ester in 99% ee. Esters of the type cannot be hydrolyzed, but can

(II) $C_6H_5CH=CH_2$ + N_2CHCO_2R' $\xrightarrow{\text{Cu} \cdot \text{2a}}$

R' = C$_2$H$_5$	77%	99% ee	73:27	97% ee
R' = BHT	85%	99% ee	94:6	

be reduced by LiAlH$_4$ to the corresponding primary alcohol (76%, >99% ee). The *gem*-dimethyl groups at the methylene bridge, which prevent deprotonation at that position, are apparently the optimal ligands for asymmetric cyclopropanation.

Enantioselective Diels–Alder reaction.[4] Highly stereoselective Diels–Alder reactions can be achieved by use of the 4,4'-diphenylbis(oxazoline) **2b**, prepared from (+)-phenylglycinol, as a chiral, bidentate ligand for iron salts. Thus reaction of FeI_3 with **2b** and I_2 in CH_3CN forms a complex presumed to be I-FeI_3, which can catalyze reaction of 3-acryloyl-1,3-oxazolidin-2-one with cyclopentadiene at $-50°$ to give the *endo*-adduct in 95% yield. The product is the 2R-enantiomer (82% ee).

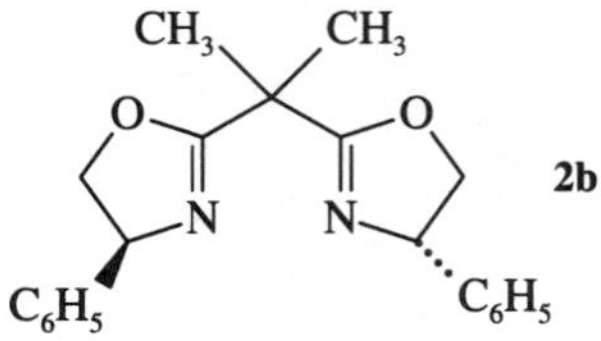

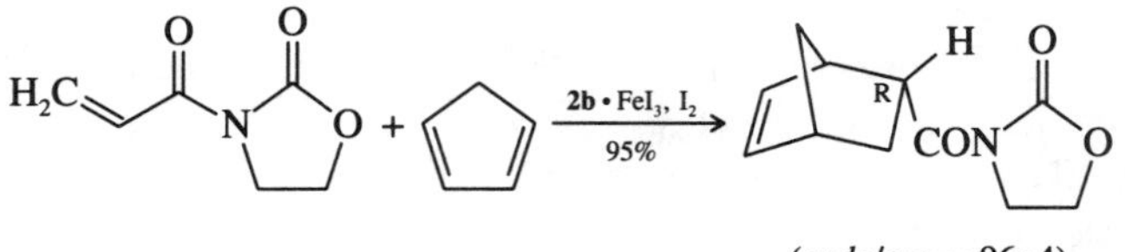

$$(endo/exo = 96:4)$$
(R), 82.2% ee

Asymmetric transfer hydrogenation.[2] Bis(oxazolines) lacking a methylene bridge such as **3** are prepared by reaction of dimethyl oxalate with α-amino alcohols,

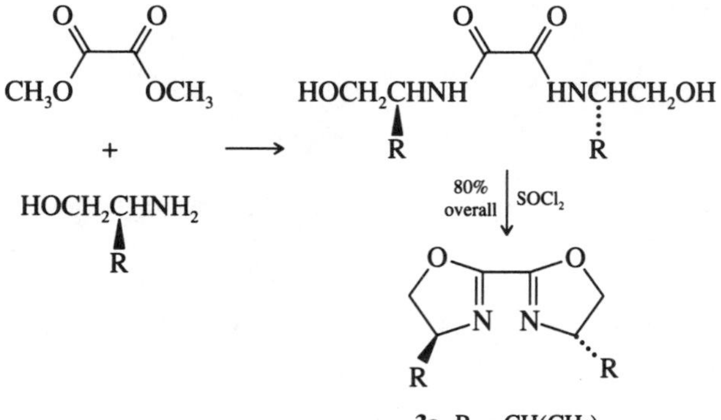

3a, R = $CH(CH_3)_2$

followed by cyclization with $SOCl_2$. Bis(oxazolines) of this type when complexed with $[Ir(COD)Cl]_2$ can effect catalytic enantioselective transfer reduction of alkyl aryl ketones with *i*-PrOH. Dialkyl ketones are reduced without any stereoselectivity.

Preliminary results suggest that ligands of type **3** are useful for Pd-catalyzed nucleophilic enantioselective substitution of allylic acetates by the sodium salt of dimethyl malonate.

$$\text{C}_6\text{H}_5\text{COR}' + i\text{-PrOH} \xrightarrow{\text{Ir} \cdot \textbf{3a}} \underset{\text{Ar}}{\overset{\text{OH}}{\bigwedge}} \text{R}'$$

R' = CH$_3$	89%	58% ee
R' = CH(CH$_3$)$_2$	70%	91% ee

[1] R. E. Lowenthal, A. Abiko, and S. Masamune, *Tetrahedron Letters*, **31**, 6005 (1990).
[2] D. Müller, G. Umbricht, B. Weber, and A. Pfaltz, *Helv.*, **74**, 232 (1991).
[3] D. A. Evans, K. A. Woerpel, M. M. Hinman, and M. M. Faul, *Am. Soc.*, **113**, 726 (1991).
[4] E. J. Corey, N. Imai, and H.-G. Zhang, *ibid.*, **113**, 728 (1991).

Bis(oxazolinyl)pyridines.

A typical member (**1a**) of this group of chiral C$_2$-symmetrical ligands is prepared by reaction of pyridine-2,6-dicarboxylic acid chloride with L-valinol (60% yield).

Enantioselective hydrosilylation of ketones.[1]   The complex formed from RhCl$_3$ and **1a** when activated by a silver salt, AgBF$_4$ or AgOTf, is an effective catalyst for enantioselective hydrosilylation of ketones with diphenylsilane to provide, after acidic hydrolysis, (S)-secondary alcohols. In all cases, addition of free ligand (4–6 mole %) improves the enantioselectivity markedly. The highest enantioselectivity

obtains in reduction of 1-tetralone (99% ee) and of methyl aryl ketones, but even dialkyl ketones are reduced with significant enantioselectivity (>60% ee).

Replacement of the isopropyl groups of **1a** by *sec*-butyl groups has little effect on the enantioselectivity, but replacement by *t*-butyl groups or phenyl groups lowers the enantioselectivity and also the catalytic activity.

[1] H. Nishiyama, M. Kondo, T. Nakamura, and K. Itoh, *Organometallics*, **10**, 500 (1991).

Bis(pentamethylcyclopentadienyl)lanthanum hydride dimer, $(Cp_2'LaH)_2$ (1).

Intramolecular hydroamination of amino alkenes.[1]   This lanthanide effects cyclization of amino alkenes in hydrocarbon solvents to five- and six-membered nitrogen heterocycles.

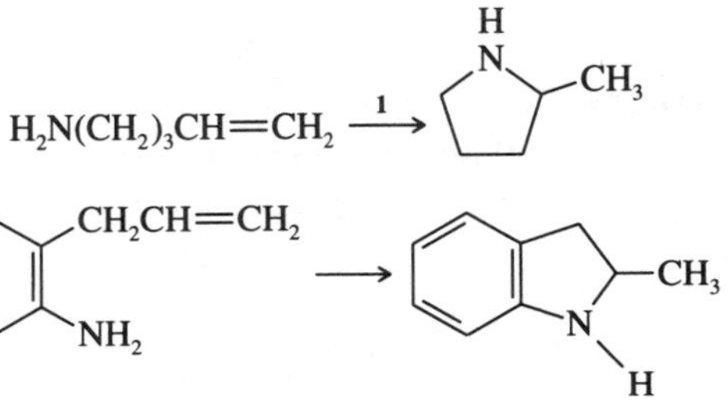

[1] M. R. Gagné and T. J. Marks, *Am. Soc.*, **111**, 4108 (1989).

Bis(2,2,6,6-tetramethylpiperidino)magnesium, $(TMP)_2Mg$.   This stable magnesium diamide is obtained by reaction of 2,2,6,6-tetramethylpiperidine, TMPH, in refluxing THF with dibutylmagnesium.

Directed magnesiation.   *Ortho*-metallation is not limited to lithium but can be effected with magnesium diamides, which are more stable than lithium amides. Even esters can be metallated with these bases (equations I, II).

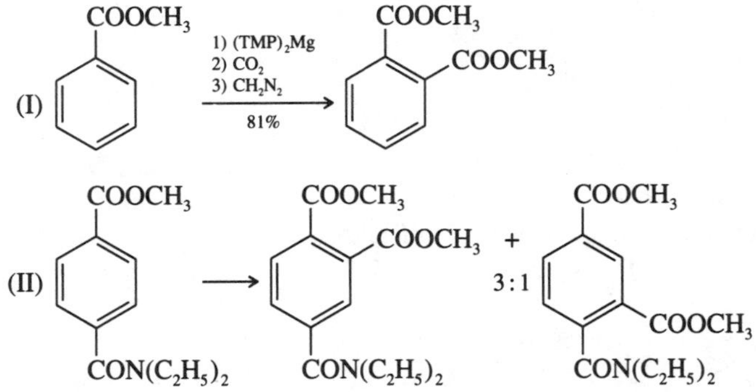

[1] P. E. Eaton, C.-H. Lee, and Y. Xiong, *Am. Soc.*, **111**, 8016 (1989).

1,8-Bis(trimethylsilyl)-2,6-octadiene, $[(CH_3)_3SiCH_2CH=CHCH_2]_2$ (1).

This disilyloctadiene is obtained as a mixture of (Z,Z)- and (Z,E)-isomers by reductive dimerization of a mixture of butadiene and $ClSi(CH_3)_3$ with lithium metal (55% yield). The main by-product is 1,4-bis(trimethylsilyl)-2-butene.

Substituted cyclopentanes.[1] The central methylene groups of this diene can undergo substitution by an electrophile followed by intramolecular cyclization to give substituted pentanes. In particular, the reaction with acyl chlorides provide a useful route to 1-alkyl-2,5-divinylcyclopentanols. The reaction usually results in *dl*- and *meso*-isomers, but it can be stereoselective, as shown in the second example.

(53%)   (32%)

(*dl*/*meso* = 11.5:1)

[1] A. Tubul and M. Santelli, *Tetrahedron*, **44**, 3982 (1988); *idem, Org. Syn.*, submitted (1990).

Bis(trimethylsilyl)peroxide, $(CH_3)_3SiOOSi(CH_3)_3$ (1).

Preparation.[1] Details are available for the preparation of DABCO complexed with 2 equiv. of H_2O_2 and for reaction of this complex with $ClSi(CH_3)_3$ to give the peroxide **1** on a large scale. This preparation is relatively safe because 35% H_2O_2 is adequate for preparation of the complex, but the product should be handled behind safety shields.

[1] P. Dembech, A. Ricci, G. Seconi, and M. Taddei, *Org. Syn.*, submitted (1990).

Bis(triphenylarsine)palladium(II) acetate, $[(C_6H_5)_3As]_2Pd(OAc)_2$ (1). Preparation.[1]

Cyclization of allyl propargyl ethers.[2] This Pd complex is far superior in re-

1a R = H   73%   **2**
1b = C_6H_5 96%

spect to yield and ease of workup for cyclization of the ethers **1a** or **1b** to 1,3-dienes **2** and for the cyclization of the more elaborate substrate **3**.

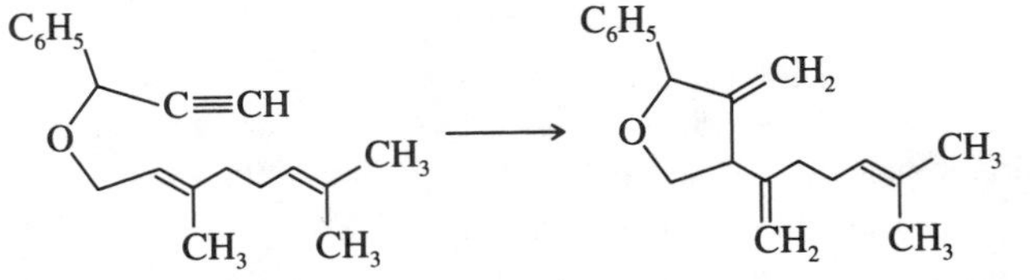

3   4

[1] T. A. Stephenson, S. M. Morehouse, A. R. Powell, J. P. Heffer, and G. Wilkinson, *J.C.S.*, 3632 (1965).
[2] B. M. Trost, E. D. Edstrom, and M. B. Carter-Petillo, *J. Org.*, **54**, 4489 (1989).

Boron trifluoride etherate.

1,2-*Amino alcohols*.   Imines, particularly those derived from methyl 4-amino-benzoate and an aryl or aliphatic aldehyde such as **1**, couple with (1-alkoxypropen-3-

$$\text{(I)} \quad p\text{-}CH_3OOCC_6H_4N{=}CHC_6H_5 + Bu_3SnCH_2CH \overset{(Z)}{=} CHOTHP \xrightarrow[74\%]{BF_3 \cdot O(C_2H_5)_2}$$

1   **2**

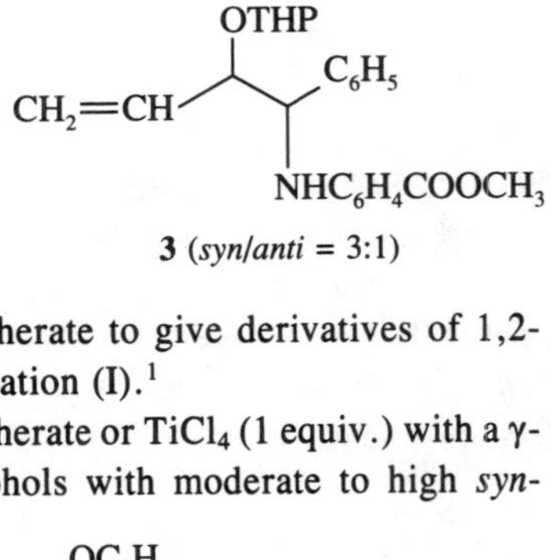

3 (*syn/anti* = 3:1)

yl)tributylstannanes (**2**) in the presence of BF_3 etherate to give derivatives of 1,2-amino alcohols with *syn*-selectivity (3–10:1), equation (I).[1]

The reaction of α-ethoxy carbamates and BF_3 etherate or $TiCl_4$ (1 equiv.) with a γ-alkoxy allyltin reagent results in 1,2-amino alcohols with moderate to high *syn*-

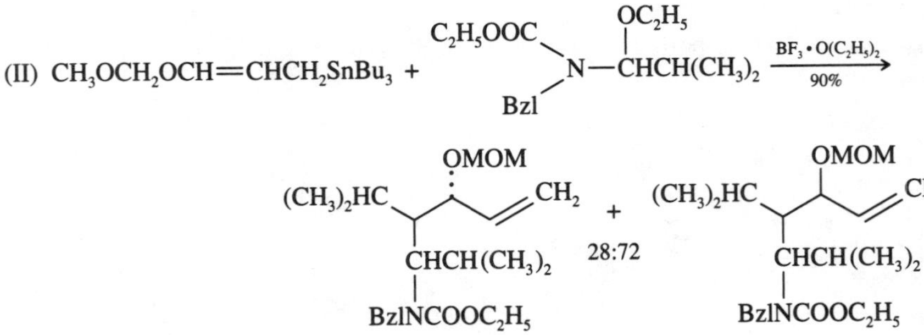

$$\text{(II)} \quad CH_3OCH_2OCH{=}CHCH_2SnBu_3 + \overset{C_2H_5OOC}{\underset{Bzl}{\diagdown}}N{-}\overset{OC_2H_5}{\overset{|}{C}H}CH(CH_3)_2 \xrightarrow[90\%]{BF_3 \cdot O(C_2H_5)_2}$$

28:72

diastereoselectivity (equation II). This procedure provides a stereoselective synthesis of **4**, a precursor to statine.[4]

$$\underset{\underset{\text{Bzl}}{|}}{\overset{\overset{\text{OC}_2\text{H}_5}{|}}{\text{C}_2\text{H}_5\text{OOCN}-\text{CHCH}_2\text{CH(CH}_3)_2}} + \text{CH}_3\text{OCH}_2\text{OCH}=\text{CHCH}_2\text{SnBu}_3 \xrightarrow[84\%]{\text{BF}_3 \cdot \text{O(C}_2\text{H}_5)_2}$$

$$\underset{\underset{\text{OMOM}}{|}}{\overset{\overset{\text{BzlNCOOC}_2\text{H}_5}{|}}{(\text{CH}_3)_2\text{CHCH}_2\text{CH}-\text{CHCH}=\text{CH}_2}} \xrightarrow[\substack{\text{2) HCl, THF (43\%)} \\ \text{3) Na/NH}_3 \text{ (27\%)}}]{\text{1) 9-BBN; H}_2\text{O}_2\text{ (46\%)}}$$

(*syn*/*anti* = 100:0)

$$\underset{(\text{CH}_3)_2\text{CHCH}_2 \qquad (\text{CH}_2)_2\text{OH}}{\text{structure}}$$

4

C-Glycosidation.[3] Protected pyranosides undergo stereoselective C-glycosidation when treated with various C_1-oxygenated allylsilanes promoted by BF_3 etherate. Thus 1-acetyl-2,3,4,6-tetrabenzylglucopyranose (**1**) reacts with 1-acetoxy-2-propenylsilane in $ClCH_2CH_2Cl$ at 10° promoted by BF_3 etherate to provide the C_1-glycoside **3** in 74% yield as a mixture of α/β-isomers in the ratio 10:1. Similar results obtain with other oxygenated allylic silanes and activated glycals.

1   **2**

3 (α/β = 10:1)

Polyols. Rychnovsky[4] has used the BF_3-catalyzed coupling of oxiranes with an alkyllithium of Ganem (**12**, 68) to obtain polyol chains. Thus the tetraalkyltin **1** reagent prepared from a β-hydroxy aldehyde reacts with the oxirane **2** in the presence of BF_3 etherate (2.5 equiv.) to give the protected decanehexol **3**.

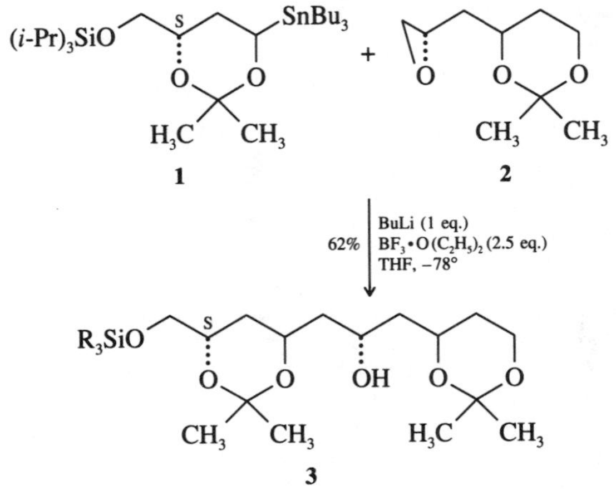

$$
\begin{array}{c}
\textbf{1} \quad + \quad \textbf{2}
\end{array}
$$

BuLi (1 eq.)
62% | BF$_3$·O(C$_2$H$_5$)$_2$ (2.5 eq.)
THF, $-78°$

3

Aldol reactions with enol silyl ethers[5] (Mukaiyama reaction).   Although TiCl$_4$ is commonly used to effect aldol reactions with enol silyl ethers, BF$_3$ etherate is more effective for aldol reactions of tetrasubstituted enol ethers such as (Z)- and (E)-3-methyl-2 (trimethylsilyloxy)-2-pentene (**1**). These enolates react with aldehydes in the same sense as the corresponding metal enolates, but with generally higher levels

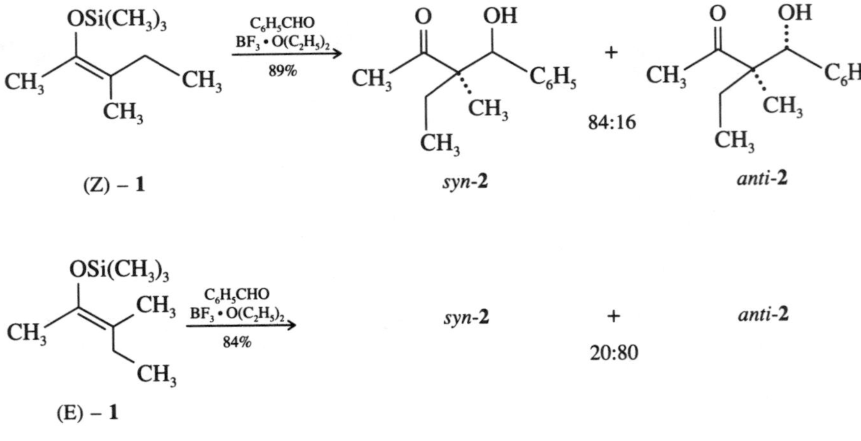

of diastereoselectivity, and higher yields. Since (Z)- and (E)-**1** differ only by a methyl substituent, their behavior should reflect the intrinsic stereoselectivity of (Z)- and (E)-enols.

Diastereoselective aldol reactions.   Heathcock and Flippin[6] have observed that BF$_3$ etherate can improve *syn*-selectivity in the reaction of an enolate (**1**) of pinacolone with 2-phenylpropanal (equation I). Thus, the reaction of the *t*-butyldimethylsilyl enolate in the presence of BF$_3$ etherate is more *syn*-selective than that of the

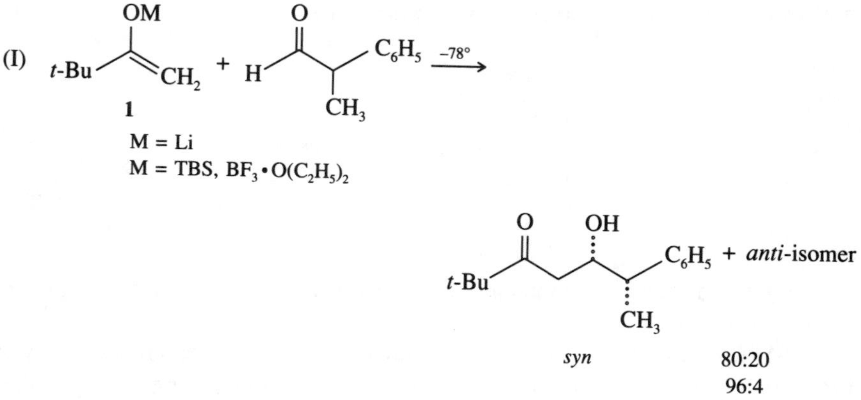

(I)

$$M = Li$$
$$M = TBS, BF_3 \cdot O(C_2H_5)_2$$

syn   80:20
96:4

lithium enolate. An even more striking example of this effect has been reported by Evans and Gage[7] in reactions of the enolates of **1** with the aldehyde **2** (equation II). Surprisingly, the reaction of the lithium enolate in this case results mainly in the *anti*-aldol adduct **3**. The desired *syn, syn*-adduct **3** is obtained in 80% yield by reaction of

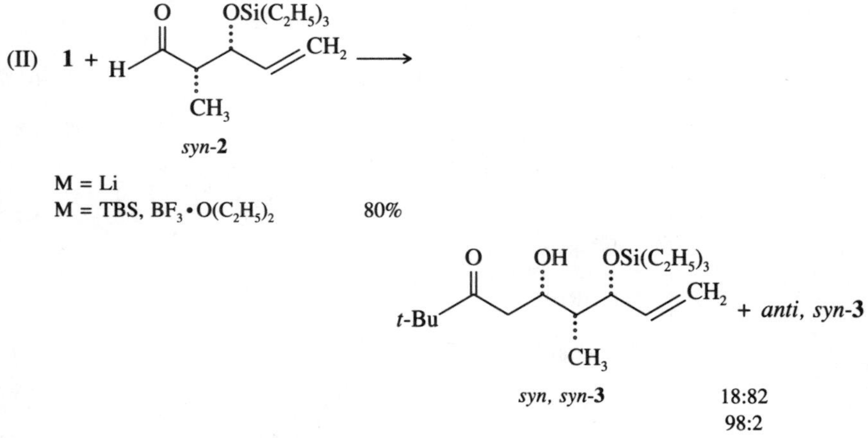

(II) **1** +

syn-**2**

$$M = Li$$
$$M = TBS, BF_3 \cdot O(C_2H_5)_2 \qquad 80\%$$

syn, syn-**3**   18:82
98:2

the *t*-butyldimethylsilyl ether and BF_3 etherate. In this example the Lewis acid can even reverse the aldehyde diastereofacial selectivity.

[1] M. A. Ciufolini and G. O. Spencer, *J. Org.*, **54**, 4739 (1989).
[2] Y. Yamamoto and M. Schmid, *J.C.S. Chem. Comm.*, 1310 (1989).
[3] J. S. Panek and M. A. Sparks, *J. Org.*, **54**, 2034 (1989).
[4] S. D. Rychnovsky, *ibid.*, **54**, 4982 (1989).
[5] S. Yamago, D. Machii, and E. Nakamura, *ibid.*, **56**, 2098 (1991).
[6] C. H. Heathcock and L. A. Flippin, *Am. Soc.*, **105**, 1667 (1983).
[7] D. A. Evans and J. R. Gage, *Tetrahedron Letters*, **31**, 6129 (1990).

N-(Bromomagnesio)-2,2,6,6-tetramethylpiperidide,

(BrMgTMP)

This reagent is prepared by reaction of the tetramethylpiperidine with C_2H_5MgBr in refluxing THF.

(E)-Silyl enol ethers.[1]   In theory it should be possible to obtain four stereoisomeric aldols from a single chiral ketone such as 1 by controlling the geometry of the enolate (E or Z) and the diastereofacial selectivity of the carbonyl group by chelation. The two possible *syn*-diols have been obtained as expected from the (Z)-lithium and boron enolates of 1. Because of the presence of four methyl groups α to the metal, this Grignard reagent deprotonates 1 to the (E)-magnesium enolate, which

reacts with various aldehydes to form *anti*-aldols (92–95 : 8 : 5) in 75–85% yields (equation I). To obtain the other possible *anti*-aldol it is necessary to convert the (E)-magnesium enolate into a nonchelating metal enolate. Transmetallation can be effected with triisopropoxytitanium chloride, HMPA, and sonication. The resulting (E)-Ti enolate provides as the major product the *anti*-aldol which is the minor product from the Mg-enolate. The (E)-selectivity observed with BrMgTMP in reaction with 1 is not general and depends markedly on the presence of an α-alkoxy group.

[1] N. A. Van Draanen, S. Arseniyadis, M. T. Crimmins, and C. H. Heathcock, *J. Org.*, **56**, 2499 (1991); C. H. Heathcock, *Aldrichim. Acta*, **23**, 99 (1990).

9-Bromo-9-phenylfluorene, 13, 48–49.

Full details are available for preparation of this reagent, which is useful for protection of amino groups and which is not commercially available.[1]

[1] T. F. Jamison, W. D. Lubell, J. M. Dener, M. J. Krisché, and H. Rapoport, *Org. Syn.*, submitted (1990).

N-Bromosuccinimide.

Radical oxidation of silyl ethers.[1] Primary benzylic silyl ethers are oxidized by NBS catalyzed by AIBN to the corresponding aldehydes in 70–87% yield. Yields are lower in a similar oxidation of secondary benzylic silyl ethers because the corresponding α-bromo ketone is a by-product.

$$RC_6H_4CH_2OSi(CH_3)_3 \xrightarrow[70-87\%]{\text{NBS, AIBN}} RC_6H_4CHO$$

Aliphatic primary silyl ethers are oxidized, probably via an aldehyde or acetal, to the simple esters formed from two equivalents of the silyl ethers (equation I). When a mixture of an aliphatic aldehyde and a primary silyl ether are oxidized under the

$$(I)\ 2C_5H_{11}CH_2OSi(CH_3)_3 \xrightarrow[80\%]{\text{NBS, AIBN}} C_5H_{11}COOC_6H_{13}$$

same conditions, a mixed ester is obtained in quantitative yield (equation II). Similar oxidation of a mixture of an aromatic aldehyde and a silyl ether results in a mixed ester only when catalyzed by trimethylsilyl triflate (equation III).

$$(II)\ C_6H_{13}CHO + C_{12}H_{25}OSi(CH_3)_3 \xrightarrow[92\%]{\text{NBS, AIBN}} C_6H_{13}COOC_{12}H_{25}$$

$$(III)\ C_6H_5CHO + C_{12}H_{25}OSi(CH_3)_3 \xrightarrow[95\%]{\substack{\text{NBS, AIBN,} \\ (CH_3)_3SiOTf}} C_6H_5COOC_{12}H_{25}$$

RCHO → RCOBr.[2] Reaction of aldehydes with NBS in the presence of AIBN provides acyl bromides in good yield. The products, without isolation, can be converted into amides.

[1] I. E. Markó, A. Mekhalfia, and W. D. Ollis, *Synlett*, 345, 347 (1990).
[2] I. E. Markó and A. Mekhalfia, *Tetrahedron Letters*, **31**, 7237 (1990).

N-Bromosuccinimide–Pyridinium poly(hydrofluoride).

$R_2C{=}O \rightarrow R_2CF_2$. This reaction has been carried out directly with diethylaminosulfur trifluoride (DAST, **6**, 183–184). It can also be effected by reaction of hydrazones of aldehydes or ketones with bromine fluoride, BrF, generated *in situ* from NBS and Py(HF)$_n$. Yields from reactions of hydrazones of ketone are in the range 30–95%; the yield from reaction of the hydrazone of an aldehyde was 28%.

[1] G. K. Surya Prakash, V. Prakash Reddy, X.-Y. Li, and G. A. Olah, *Synlett*, 594 (1990).

Bromotrimethylsilane.

trans-2,5-Diaryltetrahydrofurans[1] (**14**, 240–241).   In a synthesis of the platelet activating factor known as L-659,989 (**1**), Merck chemists have used Corey's route to *trans*-2,5-diaryltetrahydrofurans, but have introduced some improvements. The most significant is the conversion of the intermediate lactol (**2**) to the α-bromo ether with BrSi(CH$_3$)$_3$. This step is markedly improved if conducted on the *t*-butyldimethylsilyl ether, in which case the by-product is a disiloxane rather than (CH$_3$)$_2$SiOH, which on dimerization forms water, which can interfere with the subsequent

1, >98% ee
(*trans/cis* = 40:1)

Grignard reactions. This anomeric activation should be useful for synthesis of C-glycosides.

The *trans*-selectivity of the Grignard coupling can be increased by addition of copper salts such as CuCN or dilithium tetrachlorocuprate.

[1] A. S. Thompson, D. M. Tschaen, P. Simpson, D. McSwine, W. Russ, E. D. Little, T. R. Verhoeven, and I. Shinkai, *Tetrahedron Letters*, **31**, 6953 (1990).

1-Bromovinyldimethylsilyl chloride,
(1, b. p. 80°/120 mm).

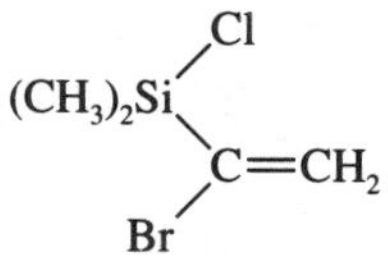

$$(CH_3)_2Si \overset{Cl}{\underset{Br}{\diagdown}} C = CH_2$$

The reagent is prepared from $[(CH_2=CH)Si(CH_3)_2]_2O$ by bromination and dehydrobromination to give $[(CH_2=CBr)Si(CH_3)_2]_2O$, which is then silylated with CH_3SiCl_3.

Hydroacylation and hydrovinylation of allyl alcohols.[1]   A typical allyl alcohol (2) is silylated by 1 to provide 3. This product undergoes radical cyclization to give 4

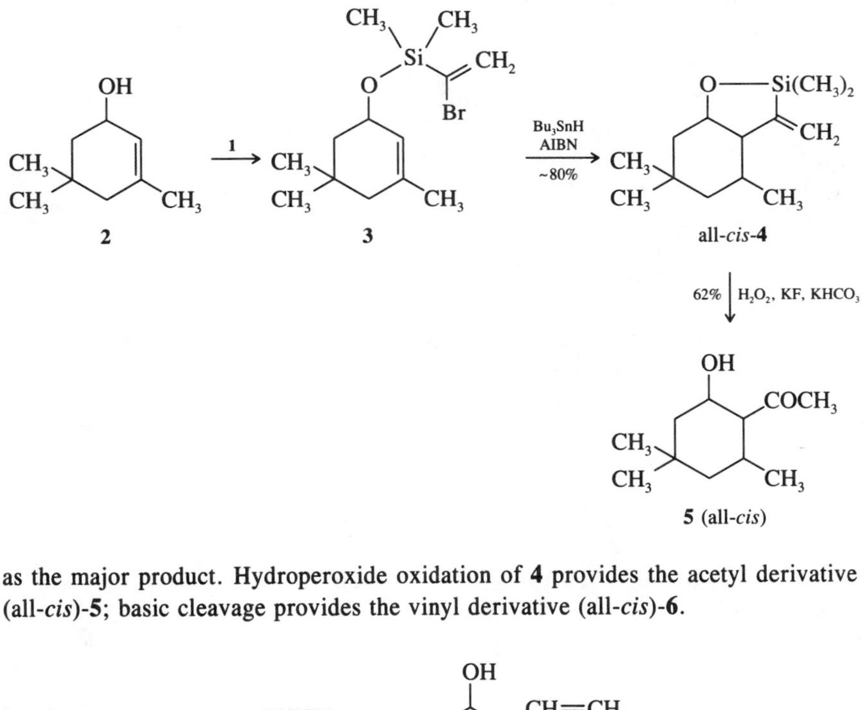

as the major product. Hydroperoxide oxidation of 4 provides the acetyl derivative (all-*cis*)-5; basic cleavage provides the vinyl derivative (all-*cis*)-6.

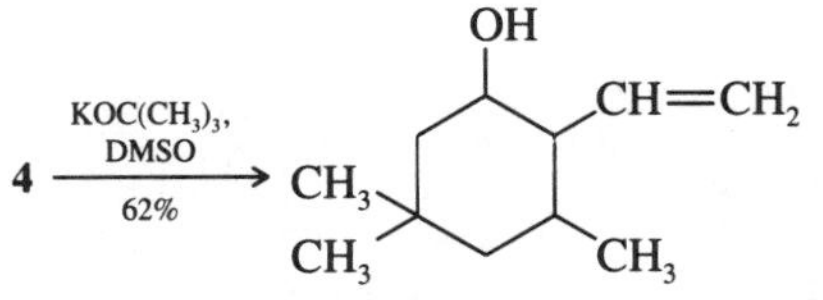

[1] K. Tamao, K. Maeda, T. Yamaguchi, and Y. Ito, *Am. Soc.*, **111**, 4984 (1989).

Butadiene–Iron carbonyl complexes.

Review.   Grée[1] has reviewed the synthetic uses of these complexes since their first preparation (1930) until the present (109 references). The complexes undergo electrophilic substitution at the terminal nonsubstituted carbon. They can function as a protective group during hydrogenation of a remote double or triple bond. Decomplexation is usually effected with oxidizing agents. Several complexes are now available in chiral forms, and the bulky $Fe(CO)_3$ group should be effective for diastereoselective reactions.

[1] R. Grée, *Synthesis*, 341 (1989).

(R,R)-(−)-2,3-Butanediol, 11, 84–85; 12, 80–81.

Chiral acetals of ArCHO.[1] The chiral acetal (1) of $Cr(CO)_3$-complexed benzaldehyde undergoes a highly stereoselective reaction with $(CH_3)_3Al$ and $TiCl_4$ (1 equiv.) to provide (R,R,R)-2 in 72% yield and 97.5% de. A subsequent Ritter

reaction[2] and decomplexation provides (R)-N-acetyl-1-phenylethylamine (3) and the (R,R)-butanediol. The displacement of O by a methyl group thus proceeds with retention, whereas the same reaction with the uncomplexed acetal proceeds with inversion.

Addition of metal acetylides to carbonyl compounds.[3] The overall strategy for an enantioselective synthesis of the highly strained epoxy cyclononadiyne core (8) of the antitumor agent neocarzinostatin involves regio- and enantioselective consecutive reaction of a chiral epoxy diyne (4) with the carbonyl groups of α-formylcyclopentenone. Enantioselective addition to the ketone group was achieved by protection of formyl group as the chiral acetal, formed with (2R,3R)-2,3-butanediol, and of the double bond by a 2-naphthylthio group, formed by 1,4-addition of 2-naphthalenethiol to 2. This reaction provides a 1:1 mixture of the two *trans*-diastereomers, which can be separated by crystallization to give (2R,3R)-3 in about 50% yield. The sodium acetylide of 4, prepared by reaction of the monoprotected epoxy diyne with

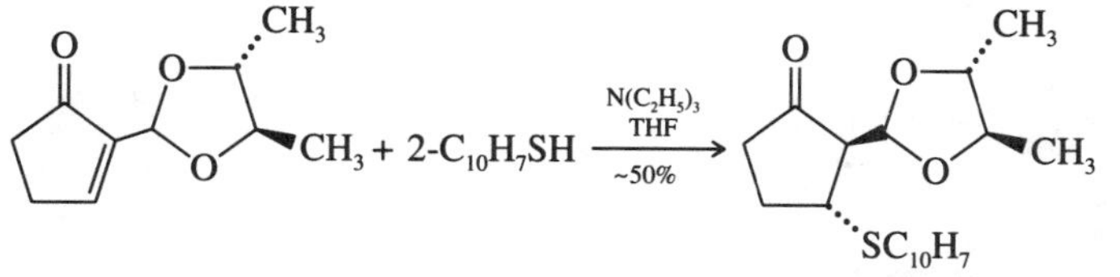

2   (2R, 3R)-**3**

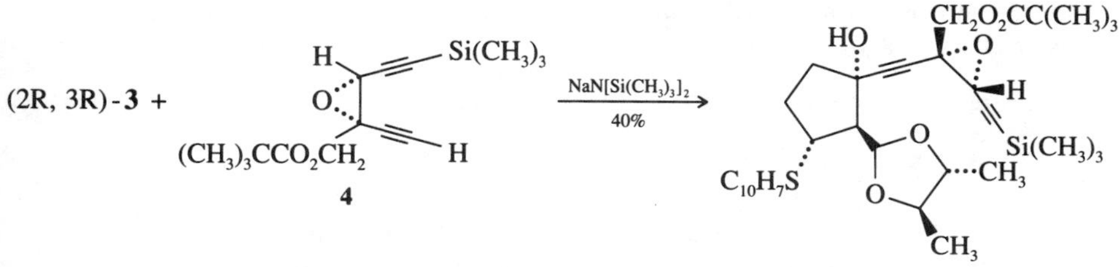

(2R, 3R)-**3** +

5 (18:1)

NaN[Si(CH₃)₃]₂, reacts with (2R,3R)-**3** to form an 18:1 mixture of **5** and the epimeric alcohol, separated by flash chromatography to provide **5** in about 40% yield.

Sulfoxide elimination of the arylthio group of **5**, deprotection of the silylacetylene (KF), acetal hydrolysis (CSA in CH₃CN–H₂O), and silylation of the free hydroxyl group provides the aldehyde **6** in 80% overall yield from **5**. On treatment with LiN[Si(CH₃)₃]₂ and CeCl₃ (3 equiv.) in THF at 0°, **6** cyclizes to **7**, obtained as a

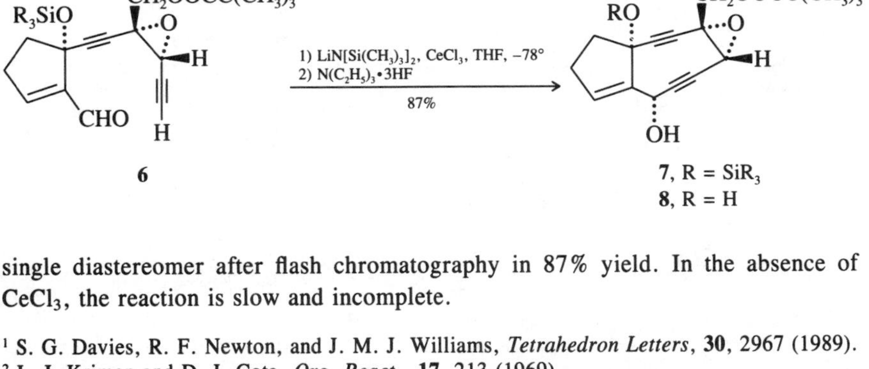

6

1) LiN[Si(CH₃)₃]₂, CeCl₃, THF, –78°
2) N(C₂H₅)₃·3HF
87%

7, R = SiR₃
8, R = H

single diastereomer after flash chromatography in 87% yield. In the absence of CeCl₃, the reaction is slow and incomplete.

[1] S. G. Davies, R. F. Newton, and J. M. J. Williams, *Tetrahedron Letters*, **30**, 2967 (1989).
[2] L. I. Krimen and D. J. Cota, *Org. React.*, **17**, 213 (1969).
[3] A. G. Meyers, P. M. Harrington, and E. Y. Kuo, *Am. Soc.*, **113**, 694 (1991).

t-Butyl hydroperoxide.

Stereospecific epoxidation of hydroxy enones.[1] Epoxidation of the racemic hydroxy enones **1** with H₂O₂ in a basic medium shows only a slight preference for the

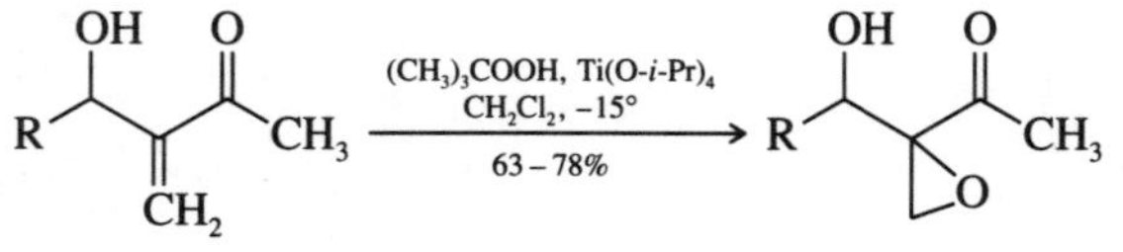

1   **2** (100% *syn*)

anti-epoxides. In contrast, epoxidation with *t*-butyl hydroperoxide and Ti(O-*i*-Pr)$_4$ results only in the *syn*-epoxides (**2**). The presence of a free allylic hydroxyl group is essential for this oxidation.

[1] M. Bailey, I. E. Markó, W. D. Ollis, and P. R. Rasmussen, *Tetrahedron Letters*, **31**, 4509 (1990).

t-Butyl hydroperoxide—Dialkyl tartrate—Titanium(IV) isopropoxide.

L-Hexoses.[1]   The groups of Masamune and of Sharpless[1] have synthesized all possible eight L-hexoses from the four-carbon allylic alcohol, 4-benzhydryloxy-(E)-2-butene-1-ol, (C$_6$H$_5$)$_2$CHOCH$_2$CH=CHCH$_2$OH (**1**). The first step is asymmetric epoxidation with an L-(+)-tartrate in >95% ee. The epoxide is opened regioselec-tively with benzenethiol in a basic medium to give a thioether diol, which is con-verted into the acetonide **2** in 65% yield. A Pummerer reaction on **2** followed by reduction provides the *erythro*-aldehyde **3**, whereas basic hydrolysis with epimeriza-tion at C$_2$ results in the *threo*-aldehyde **4**.

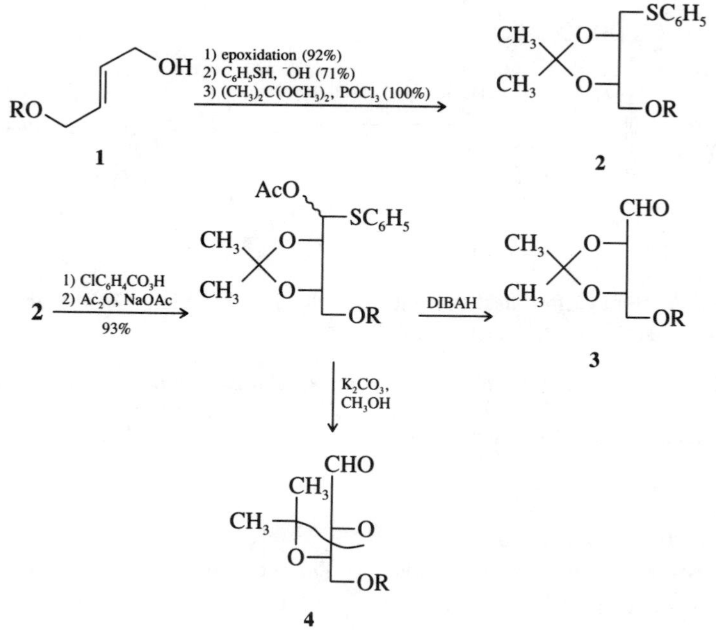

A second cycle involves conversion of **3** and **4** into the two-carbon homologated (E)-allylic alcohols by a Wittig reaction with $(C_6H_5)_3P{=}CHCHO$ followed by $NaBH_4$ reduction. Each alcohol is then subjected to asymmetric epoxidation with an L-(+)- and a D-(−)-tartrate. Four epoxides are obtained in this way, each with diastereoselectivities of >20:1. Cleavage of the epoxides as in the first cycle provide the eight L-hexoses, all in high optical purity. The methodology demonstrates the value of a "reagent-controlled' strategy.

[1] S. Y. Ko, A. W. M. Lee, S. Masamune, L. A. Reed, III, K. B. Sharpless, and F. J. Walker, *Tetrahedron*, **46**, 245 (1990).

Butyllithium–Sodium *t*-butoxide.

Metalation of 2-substituted 1,3-dithianes.[1] This combination (1:1) is more efficient than BuLi/TMEDA for this metalation at −78°. The active species may be BuNa. Moreover, the resulting anion undergoes facile reaction with various electrophiles, even epoxides.

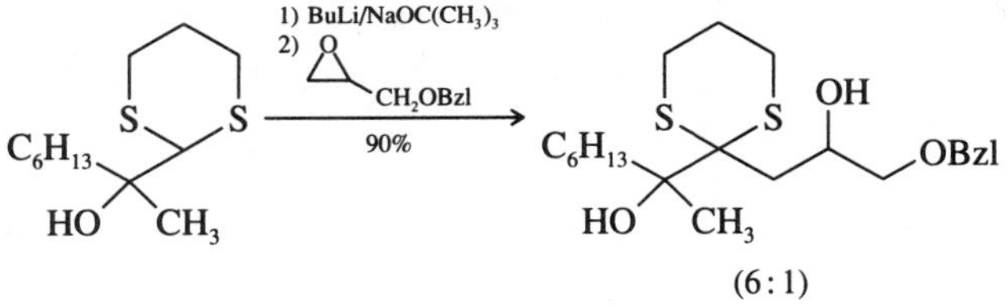

(6:1)

[1] B. H. Lipshutz and E. Garcia, *Tetrahedron Letters*, **31**, 7261 (1990).

Butyllithium–Tetramethylethylenediamine.

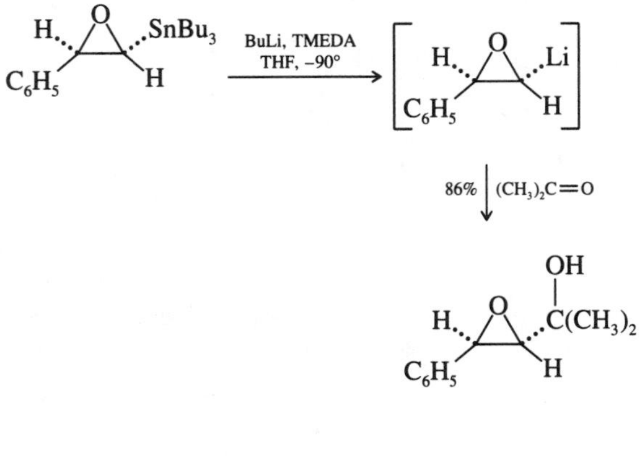

Oxiranyllithium reagents.[1] Epoxides stabilized by a silyl group can be lithiated at the activated position (**7**, 45–46), but lithiation of nonstabilized epoxides results mainly in products derived from alkoxycarbenes. However, transmetalation of oxiranyltin compounds with butyllithium and TMEDA (2 equiv.) at $-90°$ provides the corresponding oxiranyllithium reagents in high yield. In the absence of TMEDA, the major products are α,α'-dialkoxy alkenes.

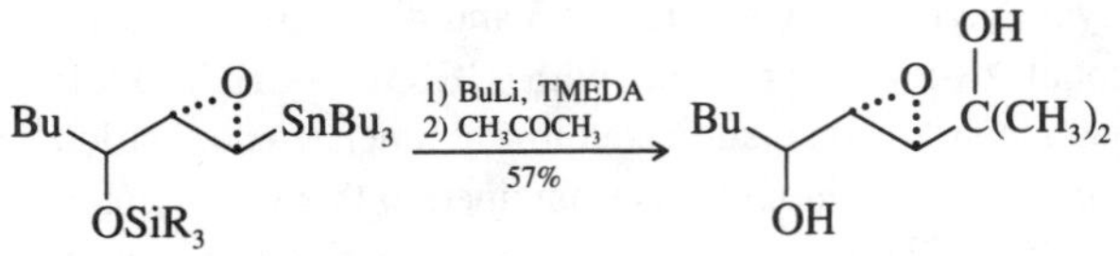

[1] P. Lohse, H. Loner, P. Acklin, F. Sternfeld, and A. Pfaltz, *Tetrahedron Letters*, **32**, 615 (1991).

sec-Butyllithium.

Anthraquinone synthesis.[1]   The original anthraquinone synthesis (**10**, 75) from benzamides and benzaldehydes involving a tandem *ortho*-lithiation can be improved by use of an *ortho*-bromobenzaldehyde as the second component. In this version, the second lithiation involves halogen–metal exchange, which results in higher yields. In the example cited here, the yield was only 15% in the absence of the bromine substituent on the aldehyde.

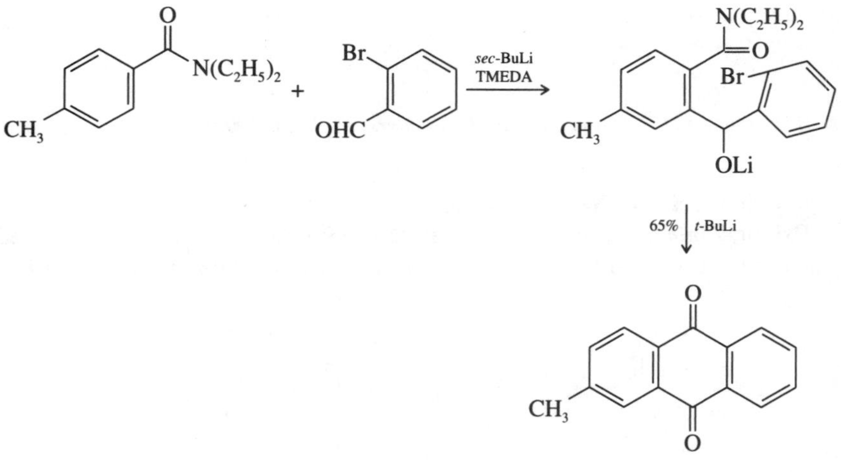

[1] X. Wang and V. Snieckus, *Synlett.*, 313 (1990).

t-Butyllithium.

Primary alkyllithiums.[1]   These alkyllithiums are usually prepared by reaction of alkyl chlorides or bromides with lithium metal, but can be obtained in 85–95%

$$C_6H_5(CH_2)_2I + 2 \text{ } t\text{-BuLi} \xrightarrow{\text{ether, pentane}} C_6H_5(CH_2)_2Li + t\text{-BuI}$$

(90–95%)

yield by reaction of the alkyl iodides with *t*-butyllithium (2 equiv.) in ether at −78°.[1,2]

[1] W. F. Bailey and E. R. Punzalan, *J. Org.*, **55**, 5404 (1990); E. Negishi, D. R. Swanson, and C. J. Rousset, *ibid.*, **55**, 5406 (1990).
[2] E. Negishi, D. R. Swanson, C. J. Rousset, W. F. Bailey, E. R. Punzalan, and J. J. Patricia, *Org. Syn.*, submitted (1990).

t-Butyl methyl sulfone.

Cyclopropanation of 1-alkenes.[1] *t*-Butylsulfonylmethyllithium, generated *in situ* from this sulfone with methyllithium, serves as a methylene transfer reagent under catalysis with Ni(acac)$_2$. Terminal alkenes and cyclic 1,2-disubstituted alkenes are cyclopropanated in 70–95% yield, but open-chain 1,2-disubstituted alkenes do not react with this reagent. Other alkyl methyl sulfones are much less effective for this purpose.

[1] Y. Gai, M. Julia, and J.-N. Verpeaux, *Synlett*, 56 (1991).

t-Butyl thionitrate, *t*-BuSNO$_2$.

α-*Oximino ketones.*[1] This reagent is generally superior to sodium nitrite or other nitrites for α-oximation of ketones. It is also prone to afford dioximines in reactions with ketones having two α-methylene groups.

$$C_6H_5\overset{\text{O}}{\overset{\|}{C}}CH_2CH_3 \xrightarrow[97\%]{\substack{t\text{-BuSNO}_2 \\ 0°}} C_6H_5\overset{\text{O}}{\overset{\|}{C}}-\underset{\underset{\text{NOH}}{\|}}{C}CH_3$$

[1] Y. H. Kim, Y. J. Park, and K. Kim, *Tetrahedron Letters*, **30**, 2833 (1989).

C

Camphorsulfonic acid (CSA).

Cyclization of hydroxy epoxides to tetrahydrofurans or tetrahydropyrans.[1] CSA is the most efficient and convenient catalyst for this intramolecular epoxide opening. The cyclization of *trans*-epoxides is particularly useful because it can lead to either 5- or 6-membered cyclic ethers as shown in equation (I) and (II).

Formation of 5-membered cyclic ethers is favored in cyclization of *cis*-hydroxy epoxides (equation III). The cyclization can be extended to bicyclic or even polycyclic ethers.

[1] K. C. Nicolaou, C. V. C. Prasad, P. K. Somers, and C.-K. Hwang, *Am. Soc.*, **111**, 5330 (1989).

10-Camphorsultam (Oppolzer's auxiliary), 1, 13, 62; 14, 71–72.

This reagent is now commercially available (Aldrich), but can be prepared with much less expense by reduction of (−)-(10-camphorsulfonyl)imine (2).[1]

syn-Aldols.[2] N-Acylsultams such as the N-propionylsultam 2 on reaction of the lithium or boron enolate with aldehydes furnish *syn*-aldols as the major product. The absolute configuration depends on the enolate counterion. Reaction of boron enolates results in (2R,3S)-aldols (3), whereas reaction of lithium or tin(IV) enolates results in (2S,3R)-aldols (4). Hydroperoxide-assisted hydrolysis of 3 or 4 followed by esterification results in recovery of the auxiliary sultam and methoxycarbonyl aldols 5 and

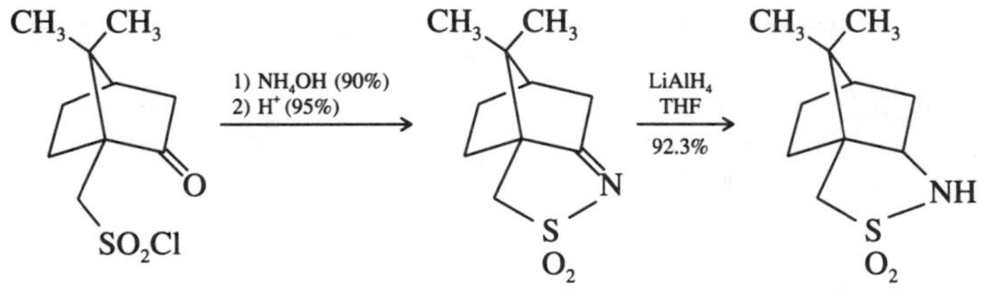

2   **1**, m.p. 183–184°,
                         $[\alpha]_D$ –30.7°

6. The practical advantage of this chiral auxiliary is that the aldols **3** and **4** can be obtained in optically pure form by flash chromatography and crystallization.

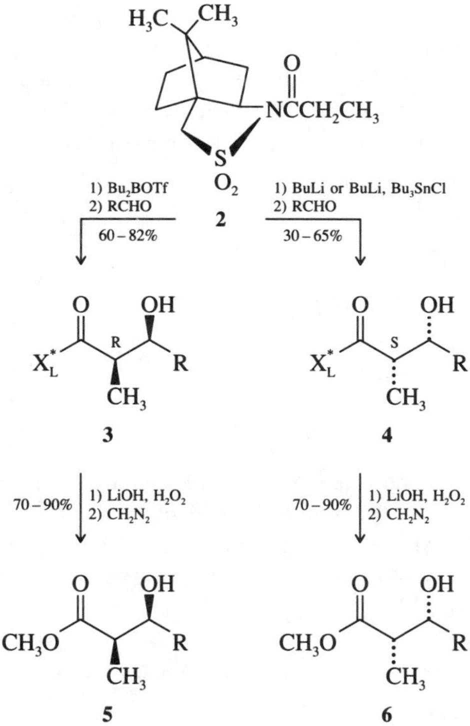

3   **4**

5   **6**

This sultam can also be used to obtain enantiomerically pure (2R,2S)-*anti*-aldols (**9**). Thus reaction of **2** with *t*-butyldimethylsilyl triflate (TBDMSOTf) and $N(C_2H_5)_3$ provides the (Z)-O-silyl-N,O-ketene acetal **7**, which reacts with aldehydes in the

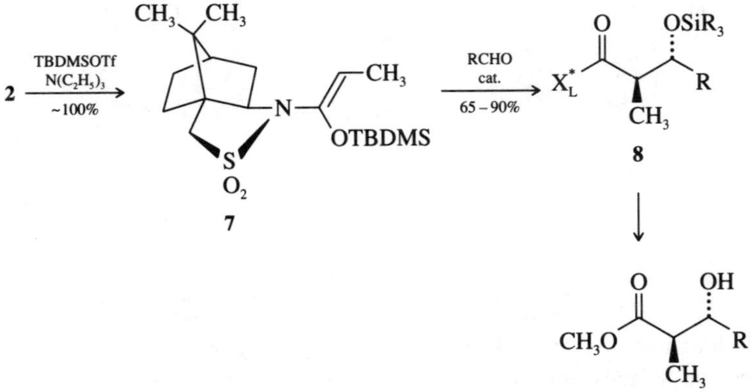

8

9 (>99% ee)

presence of a Lewis acid catalyst (TBDMSOTf, $ZnCl_2$, or $TiCl_4$) to give almost exclusively *anti*-aldols (8) with the (2R,3S)-configuration.

Asymmetric radical reactions. Curran *et al.*[3] report several asymmetric reactions of radicals derived from Oppolzer's camphorsultam. Thus the reaction of the iodosultam 1 with allyltributyltin initiated by triethylborane provides an epimeric

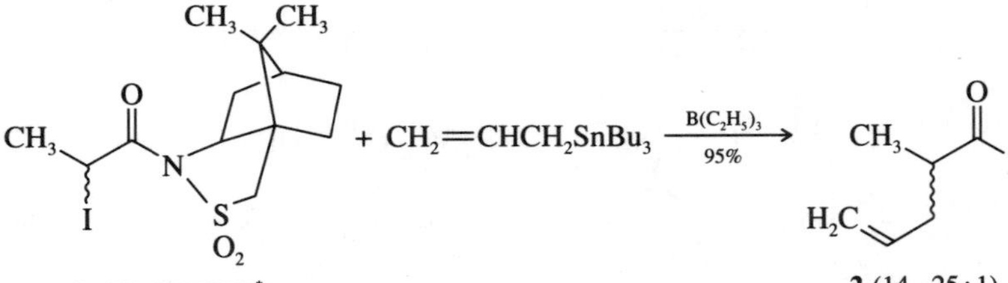

1 $(CH_3CHICOX_L^*)$

2 (14–25:1)

mixture of products in the ratio 14–25:1 at 0–25°. At lower temperatures, only one diastereomer is formed, but the reaction is too slow to be useful.

A combination of allylation and addition to an acryloylsultam (3) furnishes an adduct 4 as an 11:1 mixture of diasteromers (equation I).

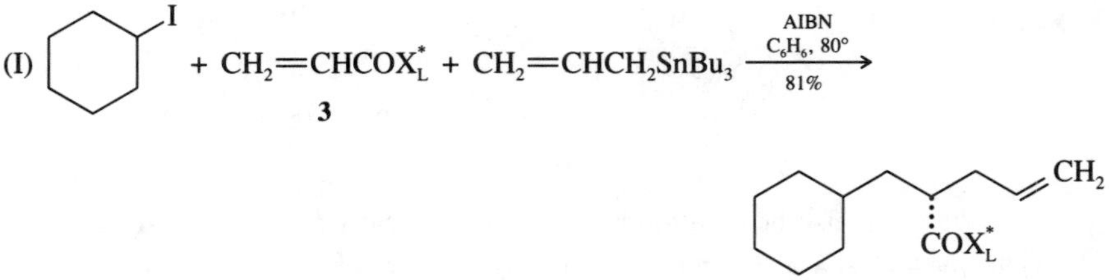

(I)

3

4 (11:1)

Asymmetric atom-transfer annelations can also be effected. Thus irradiation of butynyl iodide with the acryloyl sultam **3** in the presence of $Bu_3SnSnBu_3$ at 80° followed by deiodination provides **5** as the major product.

3   **5**

[1] M. C. Weismiller, J. C. Towson, and F. A. Davis, *Org. Syn.*, submitted (1989).
[2] W. Oppolzer, J. Blagg, I. Rodriguez, and W. Walther, *Am. Soc.*, **112**, 2767 (1990).
[3] D. P. Curran, W. Shen, J. Zhang, and T. A. Heffner, *ibid.*, **112**, 6738 (1990).

(10-Camphorylsulfonyl)oxaziridines (1), 13, 64–65; 14, 72.

Complete details for synthesis of (+)- or (−)-**1** from (1S)- or (1R)-10-camphor-sulfonic acid in 77% yield are now available. In general, this oxaziridine is less active than other N-sulfonyloxaziridines, but it is the preferred reagent for hydroxylation of lithium enolates of esters, amides, and ketones in 50–95% ee.[1]

(+)–**1**

Enantioselective α-hydroxylation of carbonyl compounds.[2]   Useful enantioselectivity (60–95% ee) obtains in the oxidation of enolates of a number of carbonyl compounds (ketones, esters, amides) with the simplest member of this series, (+)- or (−)-**1**. This reagent, however, is not useful for enantioselective oxidations resulting in tertiary α-hydroxyl ketones. For this purpose, the 8,8-dichloro derivative (**2**) of (+)-**1** is markedly superior, as shown in equation (I). This derivative can also be

(I)

$NaN[Si(CH_3)_3]_2, -78°$

(+)–**1**   (16–30% ee)
(+)–**2**   (>95% ee)

more effective than (+)-**1** for enantioselective hydroxylation of acyclic ketones. Apparently electronic as well as steric factors play a role in the enantioselectivity of hydroxylation with these oxaziridines.

The antibiotic (+)-kjellmanianone (**2**) has been prepared by asymmetric hydroxylation of the sodium enolate of the β-keto ester **1** with several (camphoryl)oxaziridines. The highest enantioselectivity (68.5% ee) was obtained by use of the *p*-(trifluoromethyl)benzyl derivative **3**.[3]

| **1** | | **2** (68.5% ee) |

3

[1] J. C. Towson, M. C. Weismiller, G. S. Lal, A. C. Sheppard, and F. A. Davis, *Org. Syn.*, submitted (1989).
[2] F. A. Davis and M. C. Weismiller, *J. Org.*, **55**, 3715 (1990).
[3] B.-C. Chen, M. C. Weismiller, F. A. Davis, D. Boschelli, J. R. Empfield, and A. B. Smith, III, *Tetrahedron*, **47**, 173 (1991).

Carbomethoxymethanesulfonyl chloride, $CH_3O_2CCH_2SO_2Cl$ (**1**), b. p. 78–80°/ 0.5 mm.

Preparation from methyl thioglycolate:

$$HSCH_2COOCH_3 \xrightarrow[\substack{67\%}]{\substack{Cl_2,\ CH_2Cl_2,\ O_2 \\ 30°}} \textbf{1}$$

β-*Sultams*.[1]   Reaction of **1** with imines and a base (usually pyridine) in THF results in 4-carbomethoxy-1,2-thiazetidine 1,1-dioxides (**2**) in 20–93% yield. All products have the *trans*-orientation. The N-unsubstituted β-sultams can be obtained by use of phenylselenenylethyl as the R group. This R group can be replaced by hydrogen in 70–90% yield by selenoxide elimination to an enamide, which is then treated with I_2 and Na_2SO_3.[2]

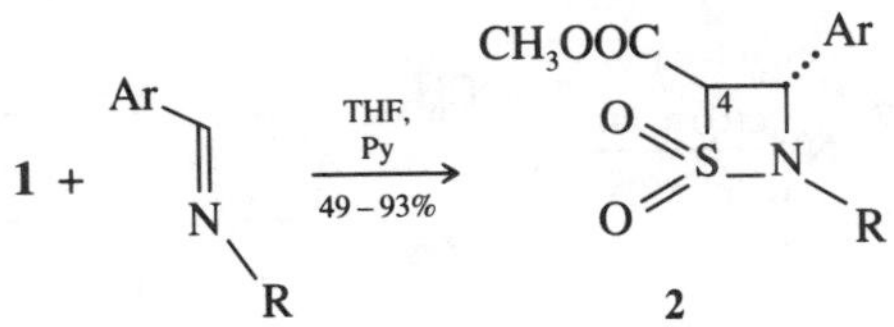

Unlike β-lactams, unsubstituted β-sultams can undergo selective monoalkylation at C_4 or dialkylation at N_2 and C_4 when the dianion is treated with a large excess of the electrophile.

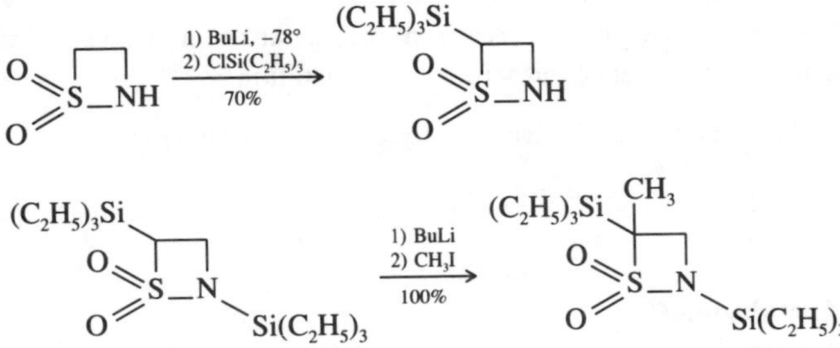

[1] M. J. Szymonifka and J. V. Heck, *Tetrahedron Letters*, **30**, 2869, 2873 (1989).
[2] J. V. Heck and B. G. Christensen, *ibid.*, **22**, 5027 (1981).

Carbon suboxide, $O{=}C{=}C{=}C{=}O$ (1).
  4-Hydroxy-2H-pyranones-2.   These heterocycles are formed on reaction of silyl enol ethers with carbon suboxide in ether at $-20 \rightarrow 25°$.[1]

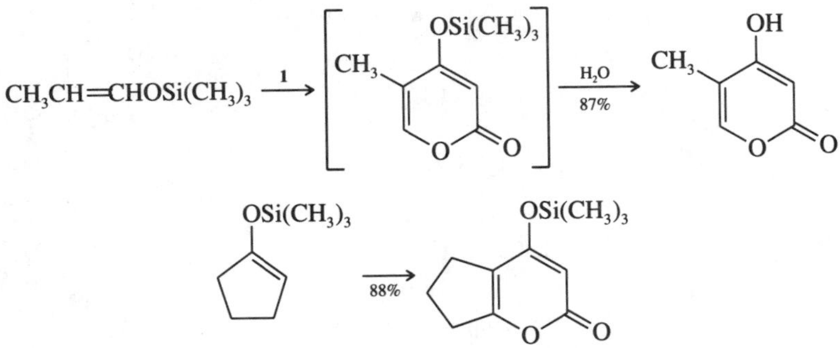

[1] L. Bonsignore, S. Cabiddu, G. Loy, and D. Secci, *Heterocycles*, **29**, 913 (1989).

1,1'-Carbonylbis(3-methylimidazolium) triflate (1, m.p. 78–80°). The reagent is prepared by reaction of 1,1'-carbonyldiimidazole with methyl triflate in CH_3NO_2 at

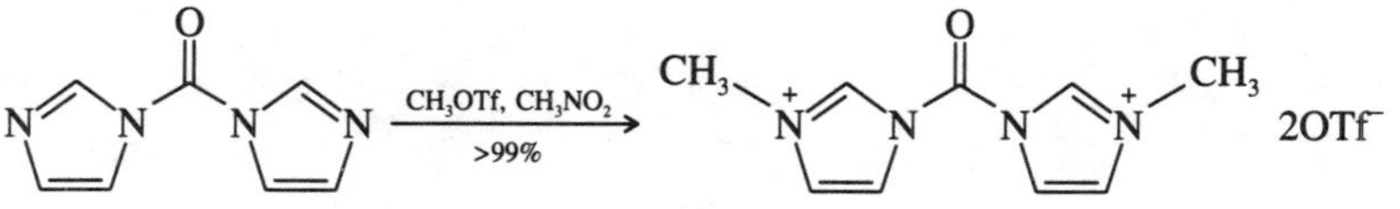

1

10°. It can be isolated as a white, moisture-sensitive solid, but is usually generated in solution just before use.

Amino acylations.[1]   This salt is far more reactive, particularly for O-acylation, than N,N'-carbonyldiimidazole. Thus it effects esterification of N–Cbz protected amino acids with even hindered alcohols such as *l*-menthol in 98% yield without need of a base and, consequently, free from racemization. It also can effect coupling of amino acids in high yields and without racemization.

[1] A. K. Saha, P. Schultz, and H. Rapoport, *Am. Soc.*, **111**, 4856 (1989).

Carbonyldiimidazole (1).

Glycosidation.[1]   The anomeric C_1-hydroxyl group of glycoses reacts with carbonyldiimidazole to form 1-imidazolylcarbonyl glycosides, which form glycosides

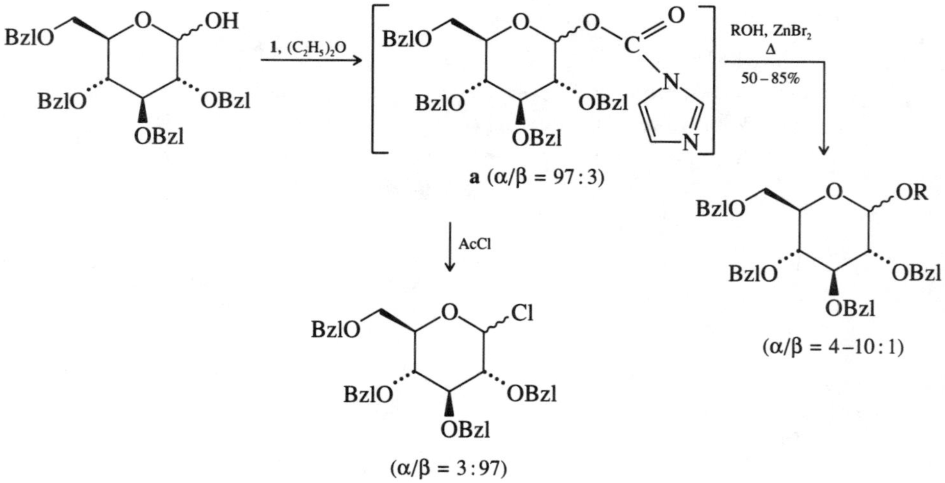

on reaction with an alcohol and zinc bromide. The intermediate also can react with acetyl chloride to give an anomeric chloride, with almost complete inversion of configuration.

[1] M. J. Ford and S. V. Ley, *Synlett*, 255 (1990).

Catecholborane.

syn-1,3-*Diols*.[1] β-Hydroxy ketones are reduced by excess catecholborane (CB) to 1,3-diols with moderate to high *syn*-selectivity. The need for at least 2 equiv. of the borane suggests that formation of a boronic ester preceeds reduction of

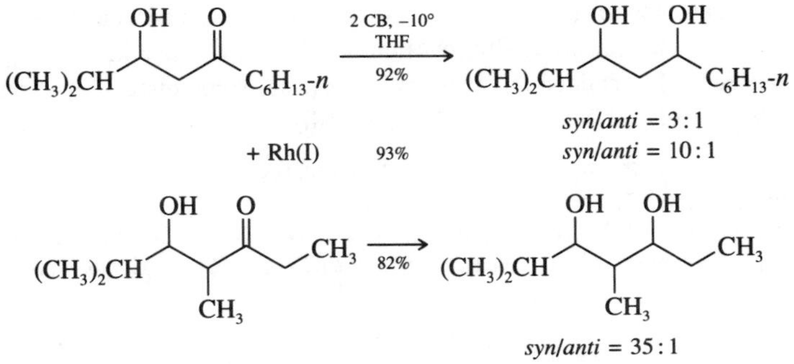

$syn/anti = 3:1$
$syn/anti = 10:1$

$syn/anti = 35:1$

the keto group. In some cases the diastereoselectivity is enhanced by catalysis with $ClRh[P(C_6H_5)_3]_3$.

1,4-*Reduction*.[2] This borane can effect efficient conjugate reduction of α,β-enones that can adopt an S-*cis*-conformation at 25° in THF. In the case of β-ionone

$(syn/anti = 10:1)$

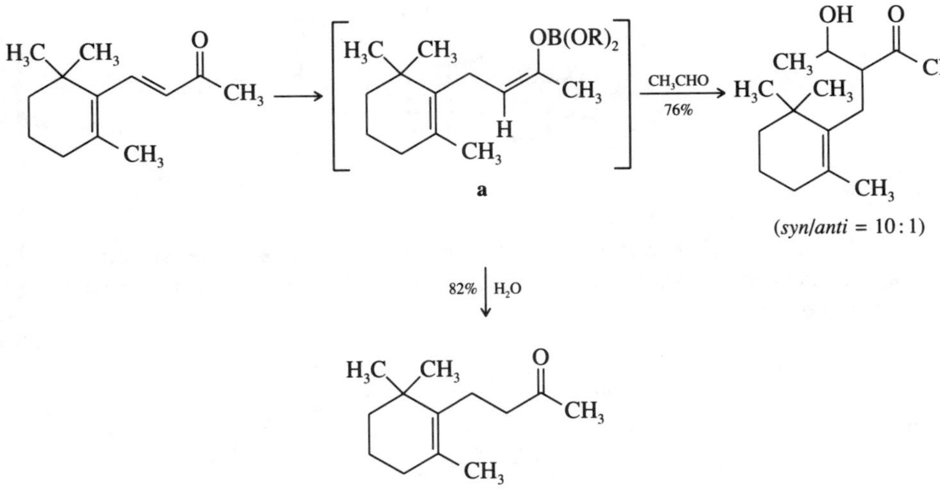

a single (Z) enolate (**a**) is formed as an intermediate, which undergoes *syn*-selective aldolization (equation I).

α,β-Unsaturated amides, imides, and esters can also be reduced by CB, but only if catalyzed by chlorotris(triphenyl)rhodium (**15**, 90–91). This catalyzed reduction can be effected even at −20° and in yields of 55–82%.

1,4-*Hydroboration* of 1,3-*dienes*.[3] This reaction, catalyzed by Pd(0) or $Rh_4(CO)_{12}$, provides (Z)-allylic boronates in about 85% yield. Hydroboration of a 1,3-enyne gives an allenic boronate (~55% yield).

Catalysis with LiBH₄.[4] Hydroboration of alkenes with catecholborane is generally slow, but can be affected at room temperature in about an hour when catalyzed by a small amount of $LiBH_4$. The catecholborane can be generated *in situ* by reaction of BH_3 with catechol; after evolution of H_2 stops, the alkene (1 equiv.) and $LiBH_4$ (0.1 equiv.) are added. Hydroboration is usually complete after stirring for 1 hour at 25°. Yields (after oxidation) are generally almost quantitative.

[1] D. A. Evans and A. H. Hoveyda, *J. Org.*, **55**, 5190 (1990).
[2] D. A. Evans and G. C. Fu, *ibid.*, **55**, 5678 (1990).
[3] M. Satoh, Y. Nomoto, N. Miyaura, and A. Suzuki, *Tetrahedron Letters*, **30**, 3789 (1989).
[4] A. Arase, Y. Nunokawa, Y. Masuda, and M. Hoshi, *J.C.S. Chem. Comm.*, 205 (1991).

Cerium ammonium nitrate (CAN).

1,4-*Diones*.[1] CAN effects cross-coupling between 1,2-disubstituted silyl enol ethers and a 1-substituted silyl enol ether to give a 1,4-dione. The reaction involves oxidation of **1** to a β-oxo radical, $R^1\dot{C}HCOR^2$, which adds to the 1-substituted silyl enol ether (**2**) to form an adduct that is oxidized to the dione.

Cyclization of unsaturated enol silyl ethers.[2] Oxidation of certain δ,ε- and ε,ρ-unsaturated enol silyl ethers with either CAN or $Cu(OTf)_2$ can result in radical cyclization.

(*cis/trans* = 20:1)

[1] E. Baciocchi, A. Casu, and R. Ruzziconi, *Tetrahedron Letters*, **30**, 3707 (1989).
[2] B. B. Snider and T. Kwon, *J. Org.*, **55**, 4786 (1990).

Cerium(III) chloride.

Addition of RMgBr to carbonyls.[1] Cerium(III) chloride (also $LaCl_3$, $PrCl_3$, $NdCl_3$) promotes addition of Grignard reagents to ketones by suppression of enolization, aldol condensation, and reduction. Enhanced reactivity is observed regardless of whether the $CeCl_3$ (1 equiv.) is first added to the Grignard reagent or to the mixture with the ketone. THF or THF/ether is the solvent of choice rather than $C_6H_5CH_3$, CH_2Cl_2, or DME. Adducts can be prepared by this simple expedient in yields as high as 95% even though the yield is zero in the absence of a lanthanoid salt. $CeCl_3$ can also retard 1,4-addition to α,β-enones and thereby improve the yield of 1,2-adducts. $CeCl_3$ is also useful for enhancing the reactivity of Grignard reagents with esters, amides, and nitriles.

Allylsilanes from esters (**14**, 76–77). Lee et al.[2] find that this conversion is highly dependent on strictly anhydrous conditions. In particular, completely anhy-

$$(CH_3O)_2CHCOOCH_3 + 2(CH_3)_3SiCH_2MgCl \xrightarrow[40\%]{\substack{1) \text{ CeCl}_3 \\ 2) \text{ SiO}_2}} (CH_3O)_2CHCCH_2Si(CH_3)_3$$

with the $=CH_2$ group.

$$(CH_3O)_2CH(CH_2)_3COOCH_3 + 2(CH_3)_3SiCH_2MgCl \xrightarrow{42\%} (CH_3O)_2CH(CH_2)_3CCH_2Si(CH_3)_3$$

with the $=CH_2$ group.

$$\text{(lactone)} + 2(CH_3)_2SiCH_2MgCl \xrightarrow{74\%} HO(CH_2)_3CCH_2Si(CH_3)_3$$

with the $=CH_2$ group.

drous $CeCl_3$ should be prepared by rigorous drying of the heptahydrate salt at 150° for 7 hours. They have extended the original method to functionalized esters.[3]

These products are useful for [3+2]annelations with O-silyl enolates to provide six- and seven-membered rings.[3]

$$\text{(silyl enol ether)} + (CH_3O)_2CHCH_2CCH_2Si(CH_3)_3 \xrightarrow[53\%]{\text{AlCl}_3}$$

[1] T. Imamoto, N. Takiyama, K. Nakamura, T. Hatajima, and Y. Kamiya, *Am. Soc.*, **111**, 4392 (1989).

[2] T. V. Lee, J. A. Channon, C. Cregg, J. R. Porter, F. S. Roden, and H. Y.-L. Yeoh, *Tetrahedron*, **45**, 5877 (1989).

[3] T. V. Lee, R. J. Boucher, J. R. Porter, and C. J. M. Rockell, *ibid.*, **45**, 5887 (1989).

Cerium(III) chloride–Chlorotrimethylsilane.

Addition of NaX or Bu₄NX to 1-alkyn-3-ones.[1]   In the presence of these two reagents, sodium or ammonium salts undergo conjugate addition to acetylenic carbonyl compounds. This reaction provides *trans*-β-halovinyl ketones or N,N-diethyl *cis*-β-haloacrylamides.

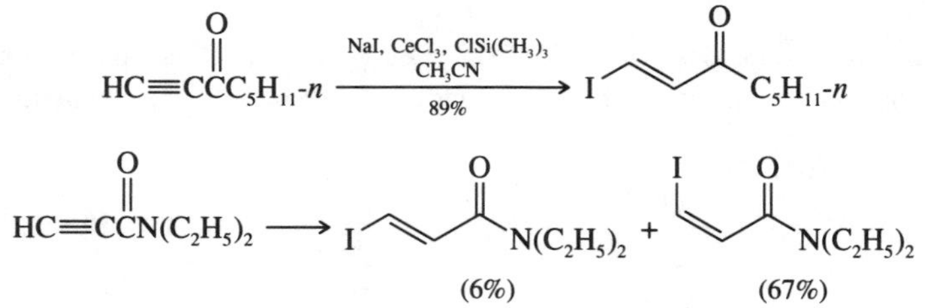

[1] T. Fujisawa, A. Tanaka, and Y. Ukaji, *Chem. Letters*, 1255 (1989).

Cerium(III) chloride–Tin(II) chloride.

[3+4] and [3+2] Cycloadditions of α,α'-dibromo ketones.[1]   This combination is a catalyst for cycloaddition of α,α'-dibromo ketones to furans and to 1,3-

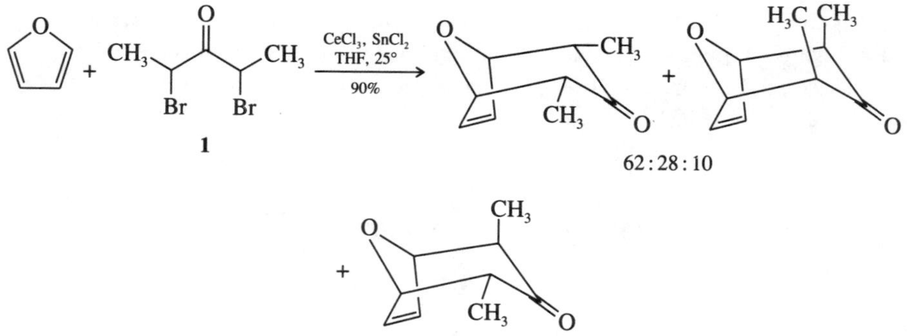

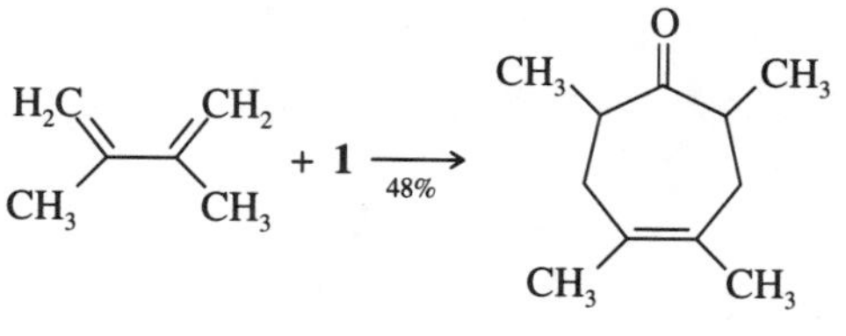

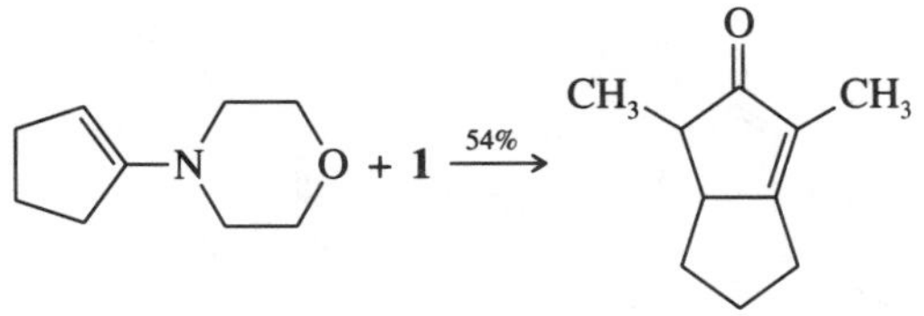

dienes or enamines at 25°. These reactions may involve a cerium α-bromo enolate,

$$\underset{OCeCl_2}{\underset{|}{CH_3CH\cdots\overset{+}{C}\cdots CHCH_3}}$$

which is then converted to an oxyallyl cation, $CH_3CH\cdots\overset{+}{C}\cdots CHCH_3$.

[1] S. Fukuzawa, M. Fukushima, T. Fujinami, and S. Sakai, *Bull. Chem. Soc. Japan*, **62**, 2348 (1989).

Cerium(IV) trifluoromethanesulfonate, $Ce(OSO_2CF_3)_4$.

Oxidation.[1] This reagent is prepared by reaction of CAN with K_2CO_3 (2 equiv.) to form $Ce(CO_3)_2$, which is then treated with trifluoromethanesulfonic acid (4 equiv.). This oxidant is effective for oxidation of benzylic alcohols to aldehydes (72–92% yield), and of alkylarenes to aldehydes or ketones (65–70% yield).

[1] T. Imamoto, Y. Koide, and S. Hiyama, *Chem. Letters*, 1445 (1990).

Cesium fluoride.

Rearrangement of aryl propargyl ethers.[1] Claisen rearrangement of the naphthyl propargyl ether **1** at 215° results in the benzopyran **2** as the only isolable product (40% yield). Addition of 1 equiv. of CsF results in the benzofuran **3** as the major product, presumably formed via an α-allenyl ketone (**a**). Related salts such as KF, RbF, or BaF$_2$ are completely ineffective. This modified Claisen rearrangement provides a route to *o*-hydroxy aldehydes such as **4** from a phenol.

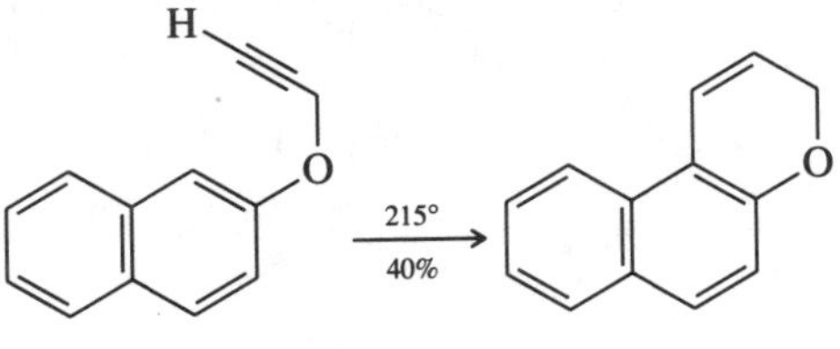

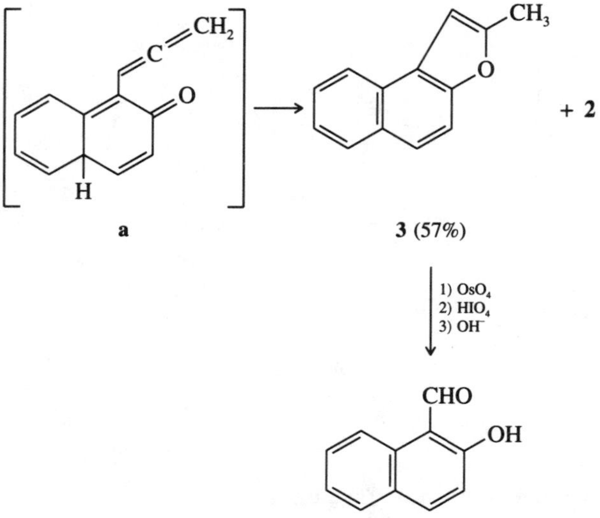

a   **3** (57%)   **+ 2**

1) OsO₄
2) HIO₄
3) OH⁻

4

[1] H. Ishii, T. Ishikawa, S. Takeda, S. Ueki, M. Suzuki, and T. Harayama, *Chem. Pharm. Bull.*, **38**, 1775 (1990).

μ-Chlorobis(cyclopentadienyl)(dimethylaluminum)-μ-methylenetitanium
(Tebbe reagent, **1**).

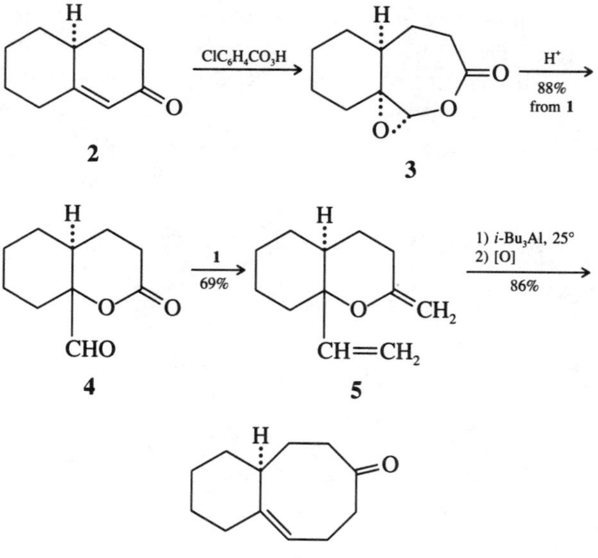

Two-carbon ring expansion. Paquette[1] has reported a method for expansion of a cyclohexenone ring by insertion of a $-CH_2CH_2-$ group between the carbonyl group and the double bond. The initial step involves oxidation of the enone with 3 equiv. of m-chloroperbenzoic acid, which effects epoxidation and a Baeyer–Villiger reaction to give an epoxy lactone (**3**), which rearranges in an acid medium to the aldehydo lactone **4**. Both carbonyl groups of **4** undergo methylenation on reaction with excess Tebbe reagent. The resulting allyl vinyl ether (**5**) undergoes a Claisen rearrangement to a cyclooctenol, which is oxidized to **6**.

This ring enlargement was applied successfully to testosterone. In this case the Claisen rearrangement was markedly retarded by the angular methyl group.

[1] C. M. G. Philippo, N. H. Vo, and L. A. Paquette, *Am. Soc.*, **113**, 2762 (1991).

B-Chloro-9-borabicyclo[3.3.1]nonane (**1**, B-Cl-9-BBN); **Dicyclohexylchlorobor-**
ane (**2**, Chx$_2$BCl). Preparation.[1]

(E)- or (Z)-Enol borinates. These enols are useful for stereoselective preparation of *anti*- or *syn*-aldols, respectively, and have usually been obtained stereoselectively by variation of the alkyl group attached to boron triflates. They can also be prepared in essentially quantitative yield by reaction with dialkylboron chlorides at

$0°$ in the presence of t-amines, $N(C_2H_5)_3$ or i-Pr$_2$NC$_2$H$_5$. Of these reagents, the most useful for stereoselective formation of enol borinates from alkyl ethyl ketones are **1** and **2**, which result in the (Z)- and (E)-isomers, respectively (equation I).

[1] H. C. Brown, N. Ravindran, and S. U. Kulkarni, *J. Org.*, **44**, 2417 (1979).
[2] H. C. Brown, R. K. Dhar, and R. K. Bakshi, P. K. Pandiarajan, and B. Singaram, *Am. Soc.*, **111**, 3441 (1989).

Chloro(chloromethyl)dimethylsilane, $ClCH_2Si(CH_3)_2Cl$ (**1**).

Corticosteroids. Upjohn chemists[1] have devised a new synthesis of corticoids from 17β-cyanohydrins (**2**), readily available from 17-keto steroids. Silylation of the cyanohydrin **2** with **1** provides the silyl ether **3**, which on treatment with LDA cyclizes to **a**, which is hydrolyzed in an acidic medium to a 21-chloro corticoid **4**. This product can be converted to a corticoid acetate (**5**) or reduced to a 17α-hydroxy-pregnane (**6**). One advantage of this route is that it is compatible with a Δ^4-3-keto group in the starting material.

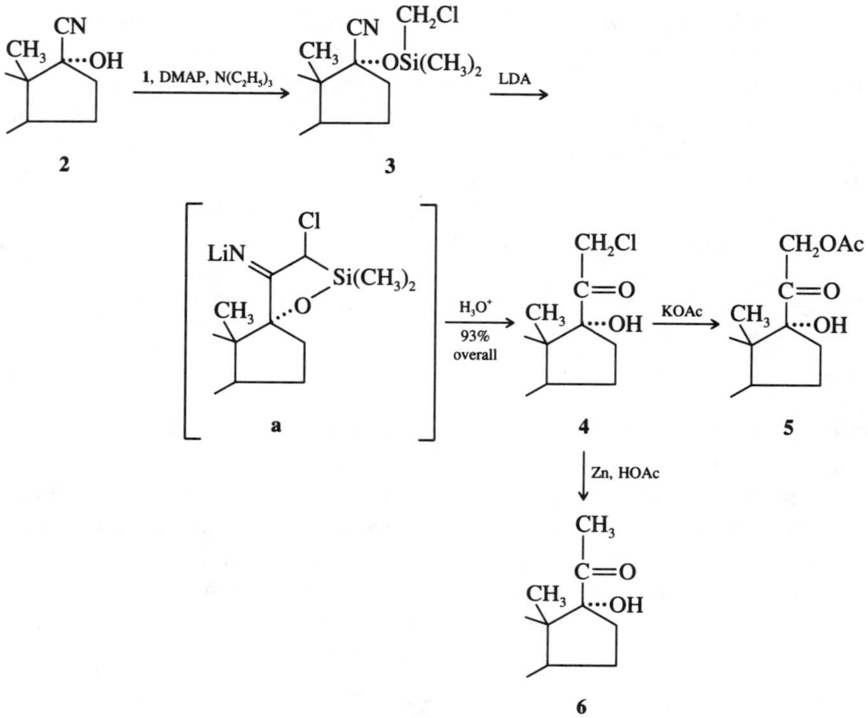

[1] D. A. Livingston, J. E. Petre, and C. L. Bergh, *Am. Soc.*, **112**, 6449 (1990).

Chlorodi(cyclopentadienyl)hydridozirconium (Schwartz reagent), $Cp_2Zr(H)Cl$ (**1**).

Preparation (**14**, 81).[1]   Full details are available for the preparation by reduction of zirconocene dichloride with $LiAlH_4$ followed by a methylene chloride wash to convert the unwanted Cp_2ZrH_2 to **1**. Overall yield is 77–92%.

Generation **in situ.**   Lipshutz[2] recommends lithium triethylborohydride for reduction of Cp_2ZrCl_2 to **1**, because the co-product is $(C_2H_5)_3B$, a relatively weak

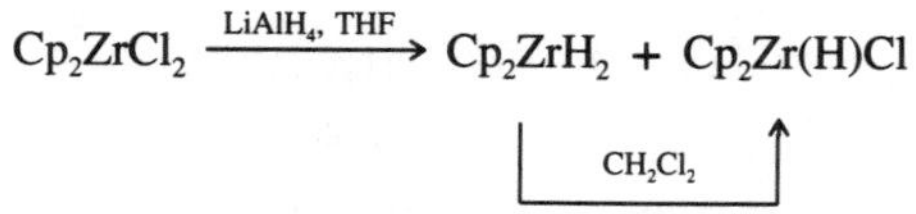

$$Cp_2ZrCl_2 \xrightarrow{\text{LiAlH}_4, \text{ THF}} Cp_2ZrH_2 + Cp_2Zr(H)Cl$$

$$CH_2Cl_2$$

Lewis acid. Consequently **1** can be prepared and used *in situ* for hydrozirconation of even sensitive alkynes in yields comparable to or even higher than those obtained with isolated reagent. A further advantage of *in situ* preparation is that the reagent is more active when freshly prepared.

Vinyl cuprates.[3,4] An attractive route to vinyl cuprates involves transmetallation of vinylzirconates, available by hydrozirconation of 1-alkynes with the Schwartz reagent in THF at 25°. Transmetallation can be effected with CH_3Li (3 equiv.)

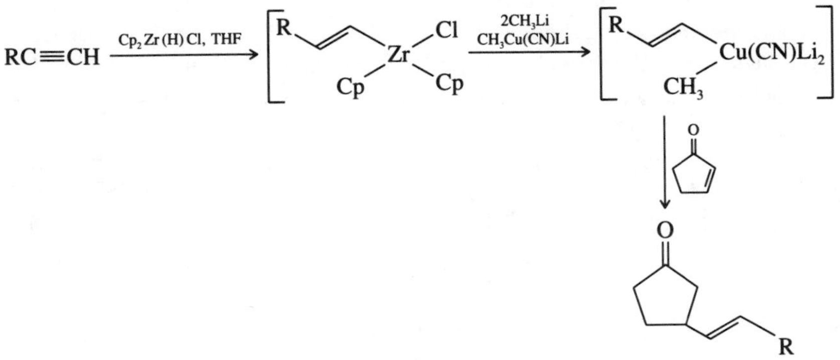

followed by $CuCN \cdot LiCl$ or with CH_3Li (2 equiv.) and $CH_3Cu(CN)Li$. These vinyl cuprates undergo 1,4-addition to cyclic and acyclic enones in high yield. This route is particularly useful for preparation of prostaglandins.

Conjugate addition of vinylzirconiums.[5] (E)-Vinylzirconium reagents are readily prepared by reaction of **1** with 1-alkynes. The product undergoes conjugate addition to α,β-enones in the presence of a nickel catalyst.

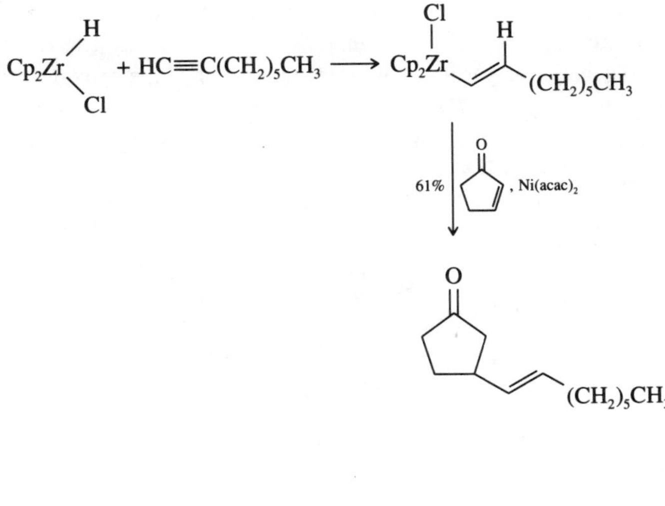

[1] S. L. Buchwald, S. J. LaMaire, R. B. Nielsen, B. T. Watson, and S. M. King, *Org. Syn.*, submitted (1990).
[2] B. H. Lipshutz, R. Keil, and E. L. Ellsworth, *Tetrahedron Letters*, **31**, 7257 (1990).
[3] B. Lipshutz and E. L. Ellsworth, *Am. Soc.*, **112**, 7440 (1990).
[4] K. A. Bakiak, J. R. Behling, J. H. Dygos, K. T. McLaughlin, J. S. Ng, V. J. Kalish, S. W. Kramer, and R. L. Shone, *ibid.*, **112**, 7441 (1990).
[5] R. C. Sun, M. Okabe, D. L. Coffen, and J. Schwartz, *Org. Syn.*, submitted (1990).

Chlorodi(cyclopentadienyl)titanium(III), 15, 81–82.

Reduction and deoxygenation of epoxides. The radical formed on reaction of epoxides with Cp$_2$TiCl (**1**) can be trapped by a H-atom donor such as cyclohexa-1,4-diene (**2**) to provide an alcohol. This reduction is not useful for reduction of mono-

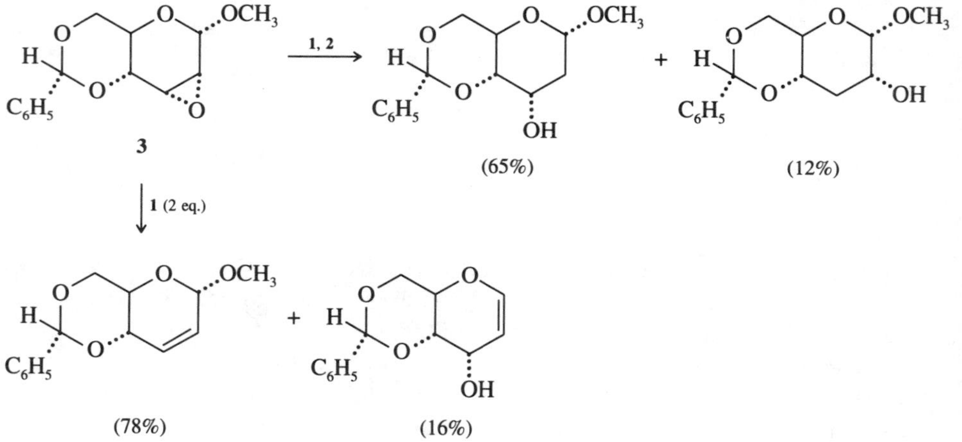

substituted terminal epoxides. In this case, deoxygenation to a 1-alkene is also observed, and is the major reaction if 2 equiv. of Cp$_2$TiCl is used. This radical reduction and deoxygenation is particularly useful for carbohydrate epoxides such as **3**.

Cyclization of ω-epoxy alkenes[2] (**15**, 82). A modified version of this radical cyclization using ω-epoxy vinylstannanes results in a cyclopentane with an exocyclic

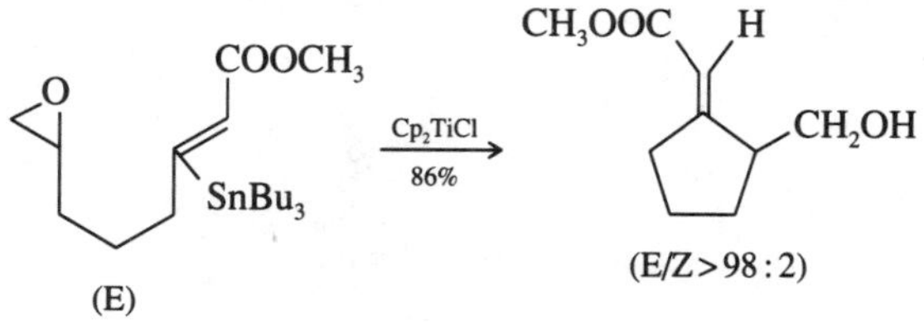

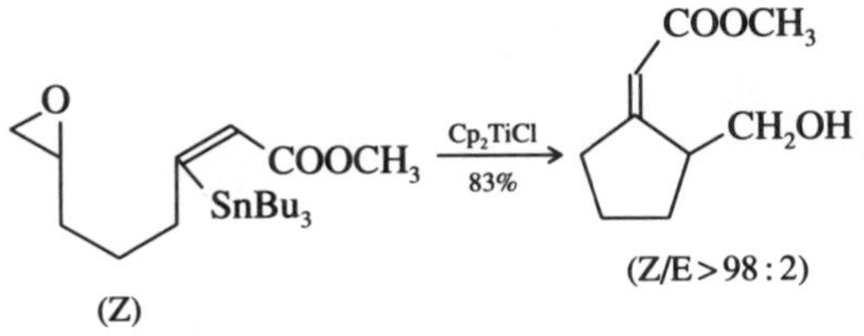

(Z/E > 98 : 2)

(Z)

double bond as well as a hydroxymethyl group. The cyclization involves practically complete retention of the stereochemistry of the double bond.

[1] T. V. RajanBabu, W. A. Nugent, and M. S. Beattie, *Am. Soc.*, **112**, 6408 (1990).
[2] T. B. Lowinger and L. Weiler, *Can. J. Chem.*, **68**, 1636 (1990).

Chlorodi(cyclopentadienyl)methylzirconium, $Cp_2Zr(CH_3)Cl$ (**1**).
  Preparation.[1]
  Allylic amines.[2] Reaction of **1** with lithium trimethylsilylbenzylamide generates the imine complex (**a**), a zirconaaziridine, which couples with terminal or symmetrical alkynes to form metallapyrrolines (**b**). Methanol cleaves Zr–C and Zr–N bonds to provide (Z)-allylic amines in 48–75% yield.
  An enantioselective version of this allylamine synthesis employs the chiral (S,S)-dimethylzirconium derivative **3**, prepared from (S,S)-[1,2-ethylenebis(tetrahydro-1-indenyl)]zirconium dichloride.[3] Displacement of one methyl group by triflic acid followed by reaction with a lithium anilide results in a zirconaaziridine (**a**) with loss of methane. Reaction of **a** with a symmetrical alkyne provides a metallapyrroline (**b**), which is hydrolyzed to an (S)-allylamine (**4**) in 90–99% ee. The method tolerates variation in the lithium anilide. Terminal alkynes do not react with **a**, but 1-trimeth-

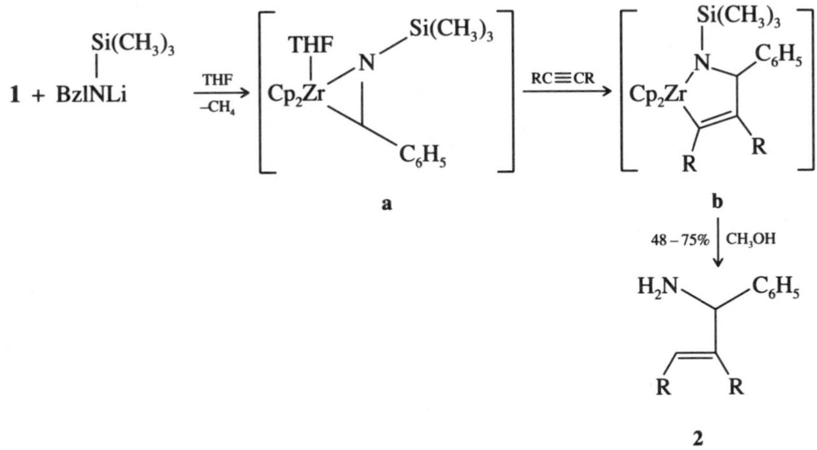

2

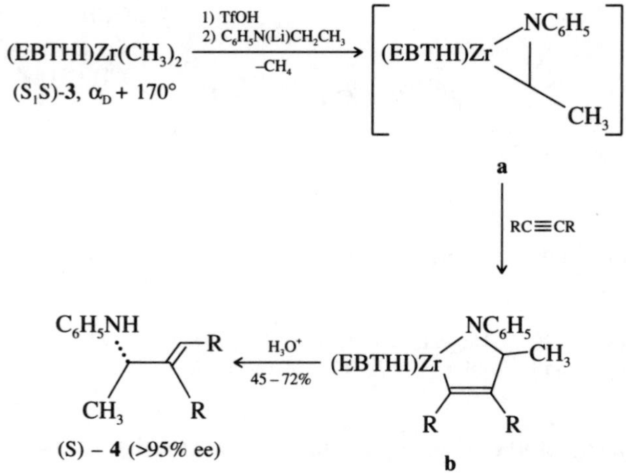

ylsilyl- and 1-phenylacetylene react regioselectively to give products in which these terminal groups are adjacent to the metal. Hydrolysis of the metallapyrrolines (**b**) affords optically active (Z)-allylamines with the (S)-configuration.

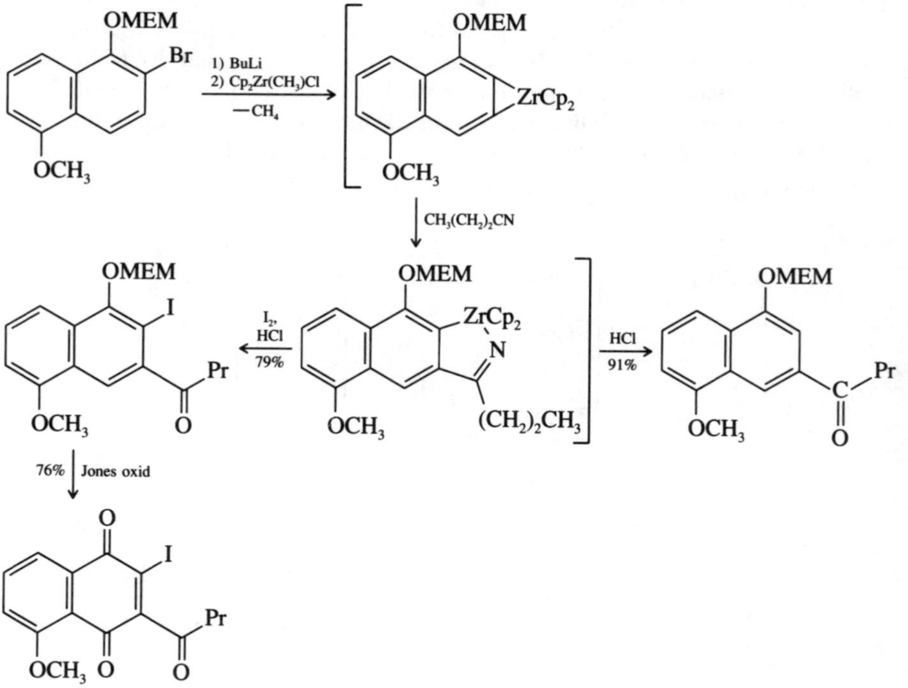

Naphthalyne zirconocene complex.[4] Buchwald[1] has extended the preparation and use of benzyne zirconocene complexes (**14**, 133–134) to a similar naphthalyne complex as a route to substituted naphthalenes and naphthoquinones.

This naphthalene synthesis has several attractive features. The intermediate organometallics need not be isolated; it is regiospecific; and aliphatic, aromatic, or heterocyclic acyl groups can be introduced efficiently.

[1] P. C. Wailes, H. Weigold, and A. P. Bell, *J. Organomet. Chem.*, **33**, 181 (1971).
[2] S. L. Buchwald, B. T. Watson, M. W. Wannamaker, and J. C. Dewan, *Am. Soc.*, **111**, 4486 (1989).
[3] R. B. Grossman, W. M. Davis, and S. L. Buchwald, *ibid.*, **113**, 2321 (1991).
[4] S. M. King and S. L. Buchwald, *ibid.*, **113**, 258 (1991).

B-Chlorodiisopinocampheylborane, Ipc_2BCl (**1**), 13, 72; **14**, 82.

Reduction of $RCOSiR_3$. Acylsilanes are reduced by (−)-**1** to (R)-α-silyl alcohols in 96–98% ee in isolated yields of 60–65%.

$$i\text{-PrCOSi(CH}_3)_3 \xrightarrow[64\%]{(-)\text{-}1} $$

(98% ee)

[1] J. A. Soderquist, C. L. Anderson, E. I. Miranda, I. Rivera, and G. W. Kabalka, *Tetrahedron Letters*, **31**, 4677 (1990).

Chlorodiphenylphosphine, $(C_6H_5)_2PCl$ (**1**).

Alkenes from diols. Reaction of *vic*-diols with two secondary hydroxyls or one primary and one secondary alcohol with chlorodiphenylphosphine (2 equiv.), imidazole (4 equiv.), and iodine (2 equiv.) results in alkenes. The reaction presumably involves a *vic*-iododiphenylphosphinate, which can be isolated in some cases and converted to an alkene with zinc in acetic acid.

$$C_9H_{19}CH_2\underset{\overset{|}{OH}}{CHCH_2OH} \xrightarrow[85\%]{1} C_9H_{19}CH_2CH{=}CH_2$$

[1] Z. Liu, B. Classon, and B. Samuelsson, *J. Org.*, **55**, 4273 (1990).

2-Chloro-1-methylpyridinium iodide (1), 8, 95–96.

$RCH_2COOH \rightarrow RCH=C=O.$[1]   Mukaiyama's reagent has been used for macro-lactonization of ω-hydroxy carboxylic acids (**14**, 117–118). However, reaction of the substrate **2** with **1** and $N(C_2H_5)_3$ in refluxing CH_3CN does not lead to the expected

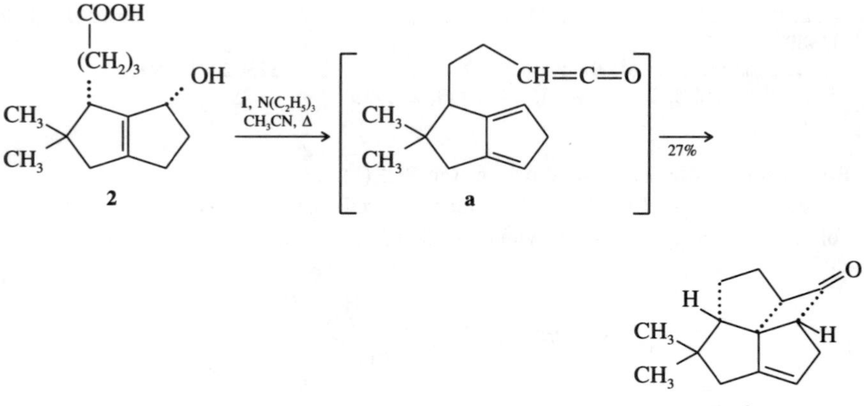

lactone but to **3**, presumably via a [2+2]ketene–alkene cyclization. Similarly the alkenoic acids **4** are converted to cycloadducts **5** in reasonable yield. However, there is no evidence that macrolactonization with **1** proceeds through a ketene intermediate.

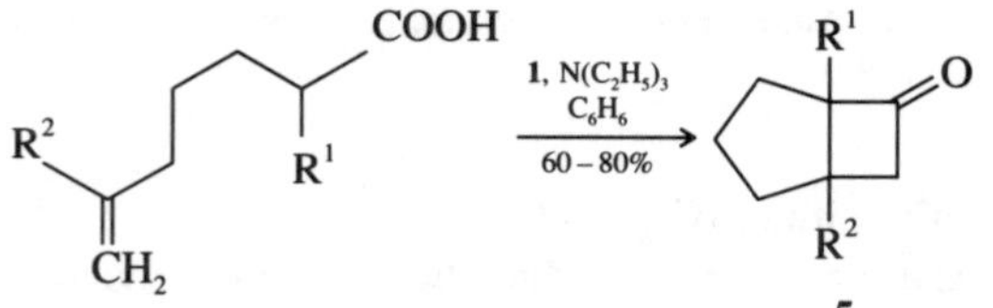

β-Lactams.[2]   Carboxylic acids and imines condense to form β-lactams when treated with 2-chloro-N-methylpyridinium iodide (1 equiv.) and tripropylamine (3 equiv.) in CH_2Cl_2. The stereoselectivity depends upon the temperature. Reactions conducted at 25° favor formation of *cis*-β-lactones. Yields are increased when the reactions are conducted for 12 hours at reflux, but the *cis*-selectivity decreases. Compared with triethylamine, tributylamine improves the yield and the *cis*-selectivity.

[1] R. L. Funk, M. M. Abelman, and K. M. Jellison, *Synlett*, 36 (1989).
[2] G. I. Georg, P. M. Mashava, and X. Guan, *Tetrahedron Letters*, **32**, 581 (1991).

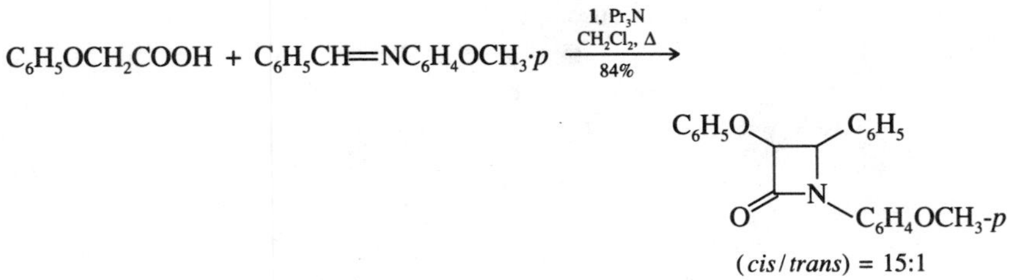

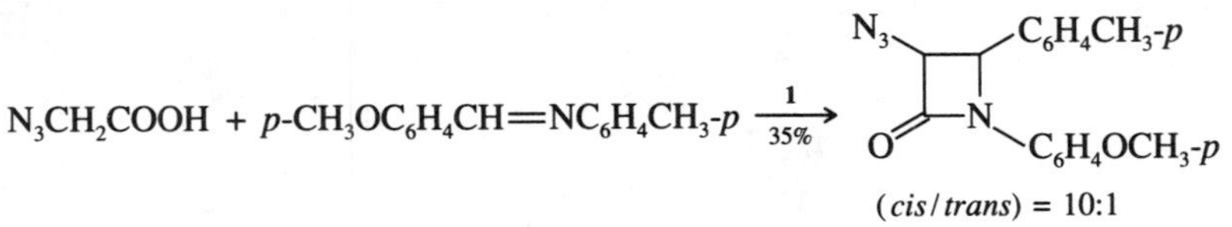

$C_6H_5OCH_2COOH + C_6H_5CH=NC_6H_4OCH_3\text{-}p \xrightarrow[\substack{84\%}]{\substack{1, Pr_3N \\ CH_2Cl_2, \Delta}}$

(cis/trans) = 15:1

$N_3CH_2COOH + p\text{-}CH_3OC_6H_4CH=NC_6H_4CH_3\text{-}p \xrightarrow[\substack{35\%}]{\substack{1}}$

(cis/trans) = 10:1

Chloromethyl trimethylsilyl ether.

Chloromethylation.[1] Chloromethyl methyl ether has been generally used for electrophilic aromatic chloromethylation, but it is highly toxic and now considered a carcinogen. Chloromethylation can be effected by use of a trimethylsilyl ether (1) of a chlorohydrin prepared as shown from trioxane and chlorotrimethylsilane in the presence of stannic chloride in chloroform. This reagent, generated *in situ*, is effective for chloromethylation of styrene in the presence of $SnCl_4$; any excess is easily decomposed by hydrolysis. Bromomethylation is possible by replacement of $ClSi(CH_3)_3$ by $BrSi(CH_3)_3$.

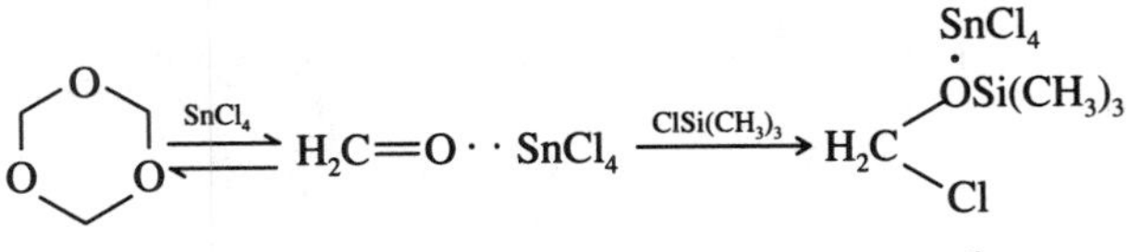

[1] S. Itsuno, K. Uchikoshi, and K. Ito, *Am. Soc.*, **112**, 8187 (1990).

(R)-(−)-Chloromethyl p-tolyl sulfoxide, 1.

This reagent (pure) can be obtained by reaction of (R)-(+)-methyl p-tolyl sulfoxide with NCS in the presence of K_2CO_3 followed by several recrystallizations.

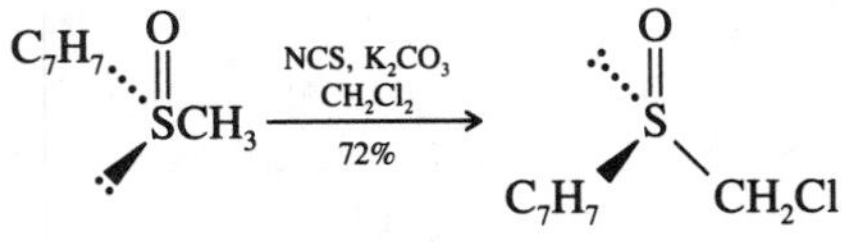

1, α_D −239°

Chiral epoxides.[1]   Alkylation of the anion (LDA) of **1** (97% ee) with 1-iodo-decane provides **2**, which reacts with symmetrical ketones at −40° to give a single (S)-chlorohydrin (**5**) with complete 1,2-asymmetric induction. However, use of an aldehyde or unsymmetric ketones results in formation of two chlorohydrins, which can usually be separated by chromatography. Thus the reaction of **2** with 6-methyl-1-

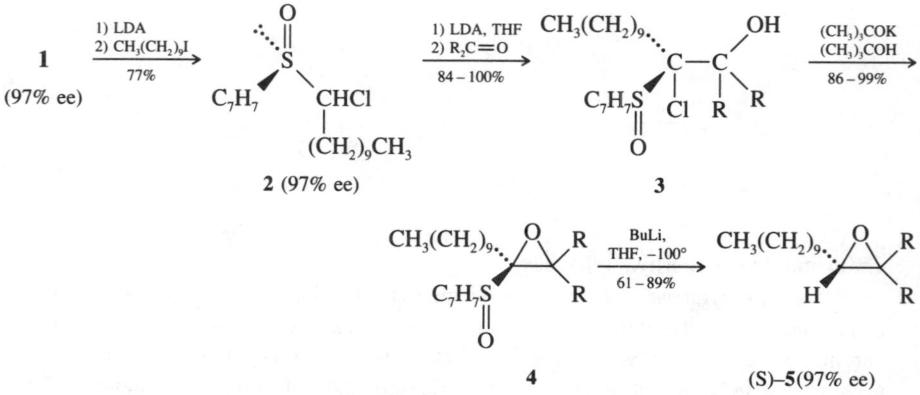

2 (97% ee)

3

4

(S)–**5**(97% ee)

heptanal provides (+)-disparlure (**6**) and (7S,8S)-(−)-*trans*-disparlure (**7**) in a ratio of ~40:50.

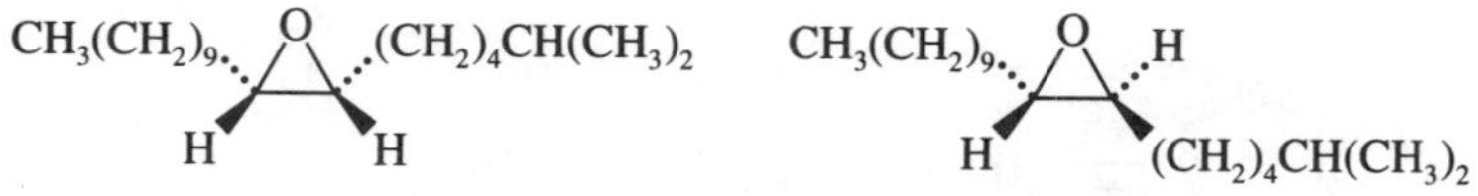

(7R, 8S)−(+)−**6**

(7S, 8S)−(−)−**7**

[1] T. Satoh, T. Oohara, Y. Ueda, and K. Yamakawa, *J. Org.*, **54**, 3130 (1989).

m-Chloroperbenzoic acid.

RSnBu₃ → ROH.[1]   This reaction can be effected by treatment of RSnBu₃ with bromine, which effects cleavage of one butyl group to form an alkylbromodibutyl-stannane. This product is oxidized by basic *m*-chloroperbenzoic acid to an alcohol with retention of configuration.

Dihydroxyacetone side chain of corticoids (**14**, 86–87).   Details for the double hydroxylation of the enol silyl ether **1** are now available.[2] For highest yields of **2**, an excess of oxidant (3 equiv.) is required as well as excess powdered KHCO₃. In the absence of base, the main product is the 17-hydroxy-20-ketopregnane, formed by

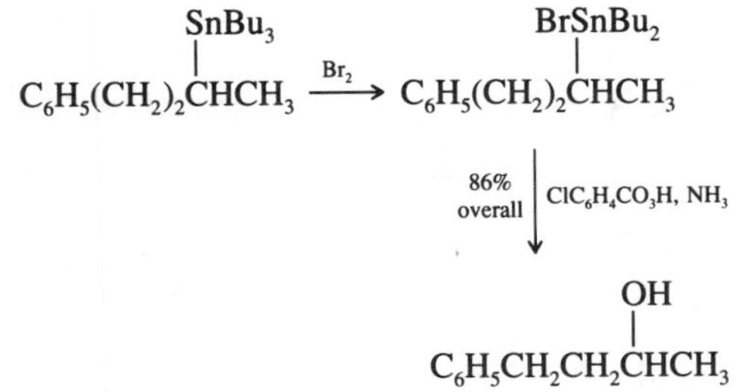

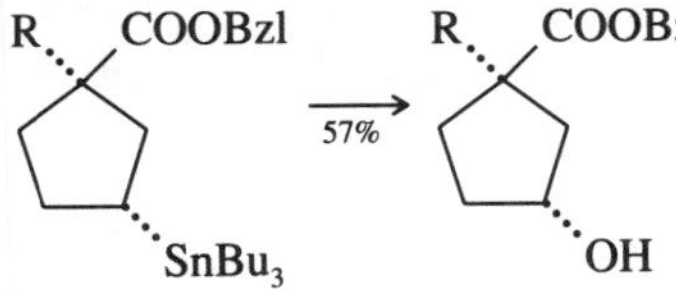

$$\underset{\text{C}_6\text{H}_5(\text{CH}_2)_2\overset{\displaystyle\overset{\text{SnBu}_3}{|}}{\text{C}}\text{HCH}_3}{} \xrightarrow{\text{Br}_2} \underset{\text{C}_6\text{H}_5(\text{CH}_2)_2\overset{\displaystyle\overset{\text{BrSnBu}_2}{|}}{\text{C}}\text{HCH}_3}{}$$

$$\Bigg\downarrow \begin{array}{l}86\% \\ \text{overall}\end{array} \;\; \text{ClC}_6\text{H}_4\text{CO}_3\text{H, NH}_3$$

$$\underset{\text{C}_6\text{H}_5\text{CH}_2\text{CH}_2\overset{\displaystyle\overset{\text{OH}}{|}}{\text{C}}\text{HCH}_3}{}$$

protonolysis of an intermediate epoxide. If **1** is added to a mixture of ClC$_6$H$_4$CO$_3$H (2.5 equiv.) and KHCO$_3$ (excess), the major product is the 17-ketone, probably formed by a Baeyer–Villiger oxidation of **2**. If only 1 equiv. of oxidant is used, the major product is **3**. The oxidation was also used for a synthesis of 16α-methyl-cortexolone (equation I).

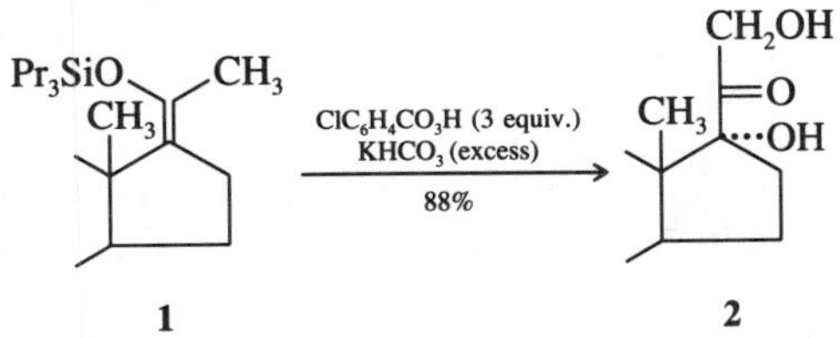

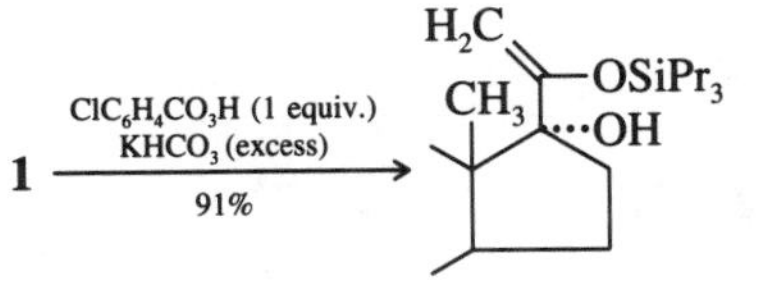

α,α'-Dihydroxy ketones.[3]   Reaction of this peracid in a basic medium with the enol silyl ether of a methyl *sec*-alkyl ketone can result in double hydroxylation via an intermediate allylic alcohol. This abnormal oxidation is favored by a bulky tripropylsilyl group.

Oxidation of ArCH=NN(CH$_3$)$_2$ to ArC≡N.[4]   Aromatic aldehydes are converted to aryl nitriles by oxidation of their dimethylhydrazones with *m*-chloroperbenzoic acid in 46–97% overall yield. The oxidation can also be effected with H_2O_2 catalyzed by SeO_2 or 2-nitrobenzeneseleninic acid. The former method is applicable to aliphatic aldehydes.

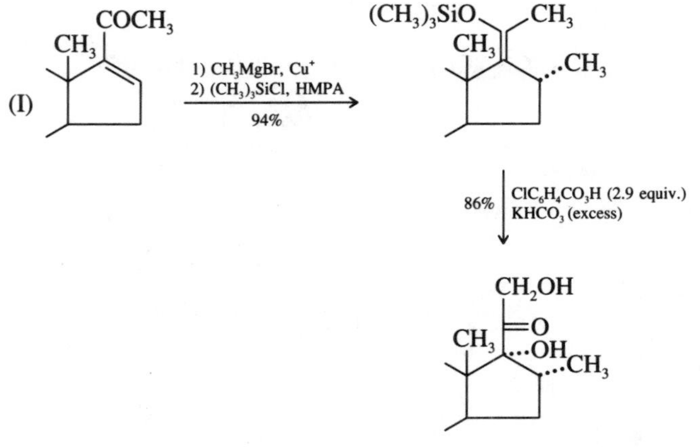

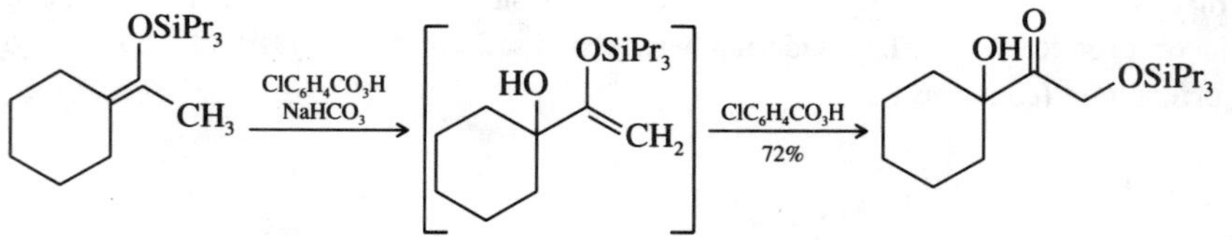

Directed epoxidation.[5] Epoxidation (MCPBA) of γ,δ-unsaturated amides or esters results mainly in a *syn*-epoxide with respect to the amide or ester group.

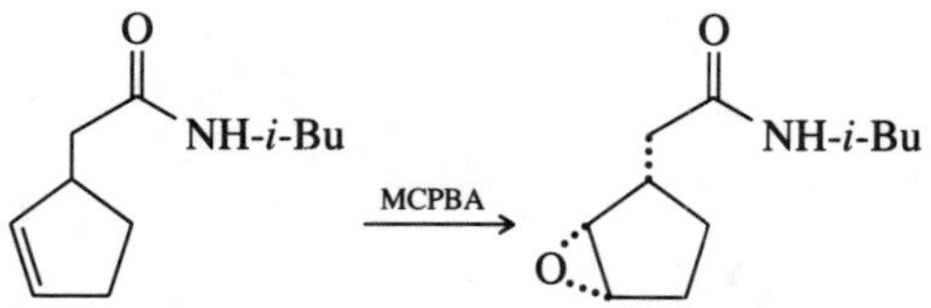

syn (>20:1)

Diastereoselective epoxidations.[6] Epoxidation of the ergoline **1** with *m*-chloroperbenzoic acid provides the α-epoxide (**2**) with high diastereoselectivity and yield. In contrast, epoxidation via the bromohydrin provides the β-oxide (**2**) in equally high diastereoselectivity. Osmylation of **1** also provides the 5α,6α-diol selectively (7:1) with similar preference for the α-face.

Epoxidation in water.[7] Epoxidation of alkenes can be effected with MCPBA in an aqueous solution of $NaHCO_3$ (pH = 8.3) at 20° even when both reactants are

1

$ClC_6H_4CO_3H$ 95–97%

NBS; NaOH   62–71%

α-2 (*exo*)   β-2 (*endo*)

98–99 : 2–1

1–2 : 98–99

insoluble. The reaction can be exothermic, particularly with reactive liquid alkenes (1-octene). Yields are generally about 90–95%.

[1] J. W. Herndon and C. Wu, *Tetrahedron Letters*, **30**, 6461 (1989).
[2] Y. Horiguchi, E. Nakamura, and I. Kuwajima, *Am. Soc.*, **111**, 6257 (1989).
[3] Y. Horiguchi, E. Nakamura, and I. Kuwajima, *Tetrahedron Letters*, **30**, 3323 (1989).
[4] S. B. Said, J. Skarzewski, and J. Mlochowski, *Synthesis*, 223 (1989).
[5] F. Mohamadi and M. M. Spees, *Tetrahedron Letters*, **30**, 1309 (1989).
[6] M. R. Leanna, M. J. Martinelli, D. L. Varie, and T. J. Kress, *ibid.*, **30**, 3935 (1989).
[7] F. Fringuelli, R. Germani, F. Pizzo, and G. Savelli, *ibid.*, **30**, 1427 (1989).

m-Chloroperbenzoic acid–Trifluoroacetic acid.

Baeyer–Villiger oxidation.[1] Since 90% H_2O_2 is no longer readily available owing to hazards in its preparation, alternative reagents for Baeyer–Villiger reactions are particularly desirable. *m*-Chloroperbenzoic acid (80–85%) as such is not always effective, but addition of TFA (1:1) enhances the reactivity and the yield.

[1] S. S. Canan Koch and A. R. Chamberlin, *Syn. Comm.*, **19**, 829 (1989).

$$SC_6H_5$$

Chloro(phenylthio)acetonitrile, $ClCHCN$ (1). This reagent can be prepared by reaction of $C_6H_5SCH_2CN$ with sulfuryl chloride (75% yield).

Esters; macrolides. Trost and Granja[1] have developed a new synthesis of esters based on conversion of **1** into an alkoxy(phenylthio)acetonitrile (**2**), which can un-

$$(I) \quad \underset{1}{\overset{\displaystyle \overset{SC_6H_5}{|}}{ClCHCN}} + CH_3OH \xrightarrow[74\%]{AgNO_3} \underset{2}{\overset{\displaystyle \overset{SC_6H_5}{|}}{CH_3OCHCN}}$$

$$2 + \underset{\displaystyle \overset{\diagup\!\diagdown}{O}}{CH_2{=}CHCHCH_2} \xrightarrow[72\%]{Pd\,(0)} \underset{3}{\overset{\displaystyle \overset{SC_6H_5}{|}}{\underset{\displaystyle \underset{CN}{|}}{CH_3OCCH_2CH{=}CHCH_2OH}}}$$

$$3 \xrightarrow[86\%]{\substack{AgNO_3,\ CH_3OH \\ H_2O}} \underset{4}{CH_3O\overset{\displaystyle \overset{O}{\|}}{C}CH_2CH{=}CH_2OH}$$

dergo Pd(0)-catalyzed alkylation. The potential carbonyl group in the product is then released by treatment with moist silver nitrate to form the ester (equation I).

$$(II) \quad 1 + CH_2{=}CH(CH_2)_9OH \xrightarrow[70\%]{\substack{AgNO_3, \\ CH_3OH}} \underset{}{\overset{\displaystyle \overset{SC_6H_5}{|}}{CH_2{=}CH(CH_2)_9OCHCN}} \xrightarrow[88\%]{O_3}$$

$$\underset{}{\overset{\displaystyle \overset{SC_6H_5}{|}}{O{=}CH(CH_2)_9OCHCN}} \xrightarrow[72\%]{\substack{1)\ CH_2{=}CHMgBr \\ 2)\ CH_3OCOCl}} \underset{5}{\overset{\displaystyle \overset{SC_6H_5}{|}}{\underset{\displaystyle \underset{OCO_2CH_3}{|}}{CH_2{=}CHCH(CH_2)_9OCHCN}}}$$

$$5 \xrightarrow[95\%]{Pd(0)} \mathbf{6} \xrightarrow[82\%]{AgNO_2/SiO_2/H_2O}$$

This use of **1** as a carbonyl synthon has been extended to a synthesis of macrolides by use of a long-chain alcohol for the acylation step, followed by Pd(0)-catalyzed cyclization and deblocking (equation II).

[1] B. M. Trost and J. R. Granja, *Am. Soc.*, **113**, 1044 (1991).

N-Chlorosuccinimide–Dimethyl sulfide (Corey–Kim reagent, 1).

Sulfur ylides.[1] This reagent (**1**) reacts with active methylene compounds to form sulfur ylides in generally good yield.

$$(CH_3)_2\overset{+}{S}-N \overset{O}{\underset{O}{\Big]}} \quad + \quad H_2C(COC_6H_5)_2 \xrightarrow[99\%]{\overset{CH_2Cl_2,}{\underset{}{N(C_2H_5)_3}}} (CH_3)_2S\!=\!C(COC_6H_5)_2$$
$$Cl^-$$

[1] S. Katayama, T. Watanabe, and M. Yamauchi, *Chem. Letters*, 973 (1989).

Chlorotrimethylsilane.

1,2-*Addition of R_2CuLi to* $>C\!=\!O$.[1]   Ordinarily cuprates do not undergo 1,2-addition to ketones, but this reaction can be effected in the presence of added $ClSi(CH_3)_3$ (1–2 equiv.) in THF (but not in ether) to give the silyl ether formed by axial attack in the case of cyclohexanones (equation I).

$$(I) \quad \overset{O}{\underset{CH_3}{\bigcirc}} \quad + \ Bu_2CuLi \xrightarrow[56-75\%]{\overset{ClSi(CH_3)_3}{\underset{}{THF}}} \quad \underset{CH_3}{\overset{(CH_3)_3SiO \quad Bu}{\bigcirc}}$$

$$(78-96:22-4)$$

Organocuprates undergo 1,2-addition to aldehydes in the absence of the silane. But addition of the silane to reactions in THF results in formation of the silyl ether as the major product. Moreover, use of the silane in the case of a chiral aldehyde results in a marked preference for the *syn*-adduct, which can be further enhanced by addition of a crown ether (equation II).

$$(II) \ C_6H_5\underset{CH_3}{\overset{|}{C}}HCHO \ + \ Bu_2CuLi \xrightarrow[71\%]{\overset{ClSi(CH_3)_3}{\underset{}{THF, -70°}}} C_6H_5\underset{CH_3}{\overset{OSi(CH_3)_3}{\diagup}}Bu \quad + \ C_6H_5\underset{CH_3}{\overset{OSi(CH_3)_3}{\diagup}}Bu$$

$$5.3:1$$

Protection of $-COOH$.[2]   Trimethylsilyl esters are useful for temporary protection of carboxylic acid groups during hydroboration of an unsaturated acid. The silyl esters need not be isolated and deprotection occurs spontaneously during the oxidation or iodination step.

[1] S. Matsuzawa, M. Isaka, E. Nakamura, and I. Kuwajima, *Tetrahedron Letters*, **30**, 1975 (1989).
[2] G. W. Kabalka and D. E. Bierer, *Syn. Comm.*, **19**, 2783 (1989).

Chlorotrimethylsilane–Sodium iodide.

Generation of HI in situ.[1] These two reagents in the presence of water or an alcohol generate HI + NaCl + 1/2[(CH$_3$)$_3$Si]$_2$O. Thus, reaction of allylic alcohols

$$(I) \quad CH_3 \diagdown \diagup \diagdown OH + HI \xrightarrow[71\%]{CH_3CN} CH_3 \diagdown \diagup \diagdown I + H_2O$$

with this *in situ* generated HI produces allylic iodides in 55–90% yield. In contrast, the same reaction but with an allylic alcohol having a terminal double bond is accompanied by rearrangement to provide an allylic iodide (equation II).

$$(II) \quad CH_2{=}CHCHOH \xrightarrow[68\%]{} CH_3 \diagdown \diagup \diagdown I$$
$$\qquad\qquad\quad | $$
$$\qquad\qquad CH_3$$

This reaction can also be used to prepare homoallylic alcohols. Reaction of 3-butene-2-ol (1) with ClSi(CH$_3$)$_3$ and NaI in acetonitrile followed by a reaction with a ketone and zinc provides a homoallylic tertiary alcohol (2) in 52% yield.

$$CH_2{=}CHCHCH_3 \xrightarrow[52\%]{\substack{1) \ ClSi(CH_3)_3, \ NaI, \ CH_3CN \\ 2) \ CH_3COPr, \ Zn, \ 4\text{-Å MS}}} CH_2{=}CHCH{-}\overset{\displaystyle OH}{\underset{\displaystyle CH_3}{\overset{|}{\underset{|}{C}}}}{-}Pr$$
$$\qquad | \qquad\qquad\qquad\qquad\qquad\qquad\qquad\qquad\qquad |$$
$$\qquad OH \qquad\qquad\qquad\qquad\qquad\qquad\qquad\qquad\quad CH_3$$
$$\qquad \mathbf{1} \qquad\qquad\qquad\qquad\qquad\qquad\qquad\qquad\qquad\qquad \mathbf{2}$$

[1] T. Kanai, S. Irifune, Y. Ishii, and M. Ogawa, *Synthesis*, 283 (1989).

Chlorotrimethylsilane–Sodium nitrite or nitrate.

Deoximation.[1] This reaction can be effected with this combination of reagents and a phase-transfer catalyst at room temperature in yields generally >90%. The actual reagent is believed to be nitrosyl chloride, NOCl.

[1] J. G. Lee, K. H. Kwak, and J. P. Hwang, *Tetrahedron Letters*, **31**, 6677 (1990).

Chlorotris(triphenylphosphine)rhodium(I).

Hydroboration of alkenes[1] (**15**, 91). The Rh(I)-catalyzed hydroboration provides a highly diastereoselective reaction in a synthesis of a polyether antibiotic. Thus the derivative (1) of an acyclic allylic alcohol is converted to the primary alcohol 2 by hydroboration with catecholborane (CB) catalyzed by ClRh[P(C$_6$H$_5$)$_3$]$_3$ with 94:6 selectivity. Note that hydroboration of 1 with disiamylborane (**12**, 484) proceeds with the opposite selectivity at C$_{10}$ (8:92).

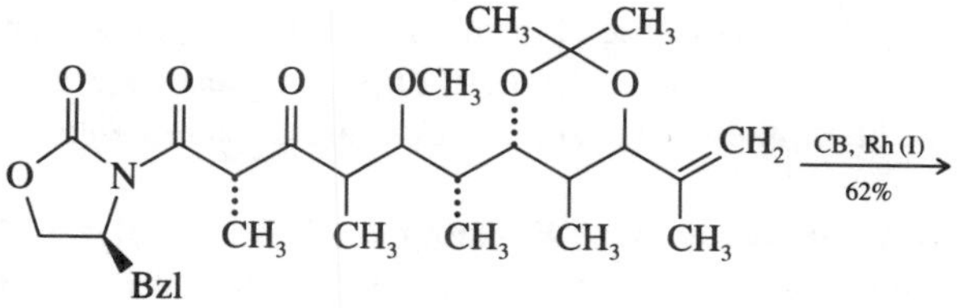

1

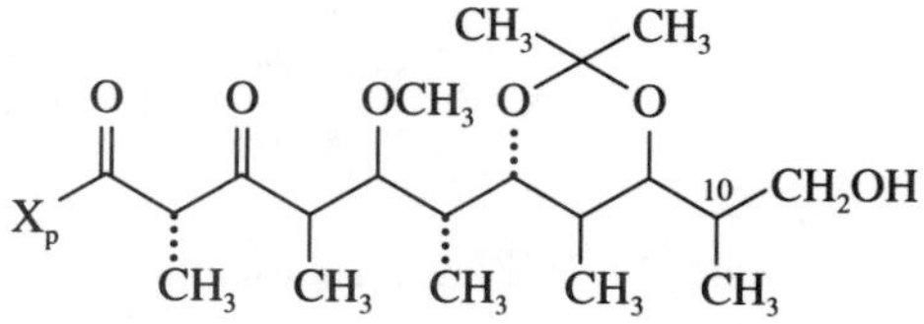

2 (94:6)

Decarbonylation of aldoses.[2]   Although this rhodium complex has been known since 1968 to effect decarbonylation of aldehydes, it has been used for decarbonylation of sugars only recently, probably for lack of a compatible solvent. Actually, this reaction when carried out in N-methyl-2-pyrrolidinone (NMP) at 110–130° is extremely useful in the case of simple aldoses, which are converted to the lower alditol with formation of carbonylchlorobis(triphenylphosphine)rhodium(I). The yields are 75–95%. This method of degradation has the further advantage that protecting groups are not necessary. Deoxyaldoses, particularly 2-deoxyaldoses, are decarbonylated in 75–99% yield. A disadvantage of this reaction is that a full equivalent of the complex is required.

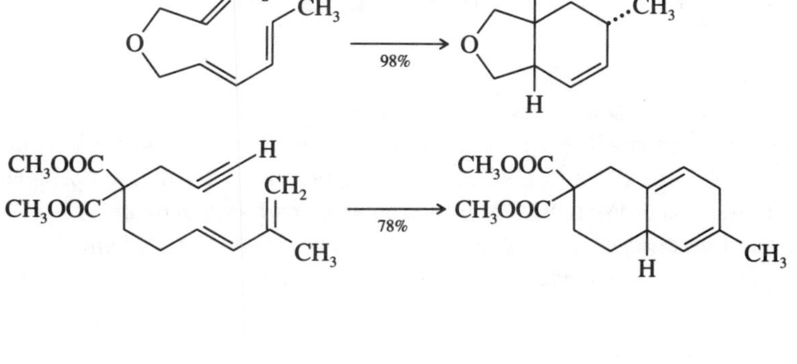

Diels–Alder catalysis.[3]   Several Rh(I) complexes are remarkably efficient catalysts for intramolecular cyclization of electronically neutral dienynes or trienes. The simplest one, $ClRh[P(C_6H_5)_3]_3$, requires use of trifluoroethanol (TFE) as solvent for reasonable rates and high yield, but more elaborate Rh(I) complexes are efficient even in THF at 25°. In addition, these catalyzed reactions are diastereoselective, as shown in the examples.

Cyclization of triynes to benzenes.[4] Wilkinson's catalyst catalyzes [2 + 2 + 2]cycloaddition of 1,6-heptadiynes with monoynes to form substituted benzenes. Intramolecular [2 + 2 + 2]cycloaddition of triynes is also possible with this catalyst.

This [2 + 2 + 2]cycloaddition is useful for synthesis of highly substituted aromatic compounds since substitution reactions with arenes are seldom regiospecific. An example is the synthesis of calomelanolactone (2) from triyne 1.[5]

[1] D. A. Evans and G. S. Sheppard, *J. Org.*, **55**, 5192 (1990).
[2] M. A. Andrews, G. L. Gould, and S. A. Klaeren, *ibid.*, **54**, 5257 (1989).
[3] R. S. Jolly, G. Luedtke, D. Sheehan, and T. Livinghouse, *Am. Soc.*, **112**, 4965 (1990).
[4] R. Grigg, R. Scott, and P. Stevenson, *J.C.S. Perkin I*, 1357, 1365 (1988).
[5] S. J. Neeson and P. J. Stevenson, *Tetrahedron*, **45**, 6239 (1989).

Chromium carbene complexes.

Vinylcyclopentenediones.[1] The Fischer carbene 1 reacts with terminal alkynes (excess) in benzene (70°) to give as the major product a 2-vinylcyclopentene-1,3-dione (2) with incorporation of two molecules of carbon monoxide as well as the alkyne and formation of six C–C bonds. Reaction of an internal alkyne is possible, but the yield of cyclopentenediones is lower (27%, one example).

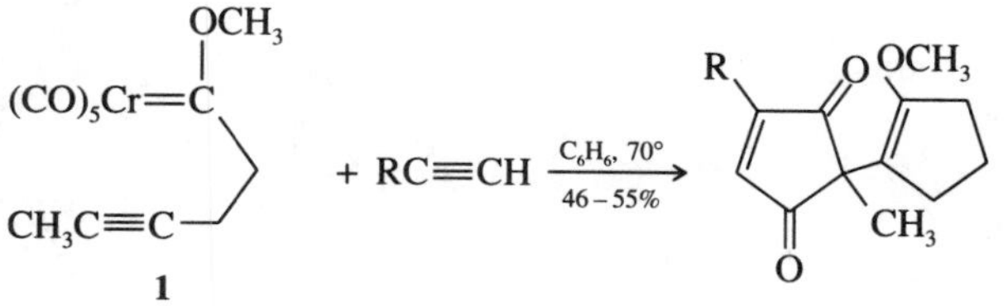

Dihydrobenzenes.[2] Pyranylidene pentacarbonyl chromium complexes such as **1**, prepared as shown, react with electron-rich alkenes, such as enol ethers, to form an adduct **a**, which extrudes $Cr(CO)_6$ to provide dihydrobenzenes (**2**) in high yield. Attempted chromatography of **2** results in aromatization.

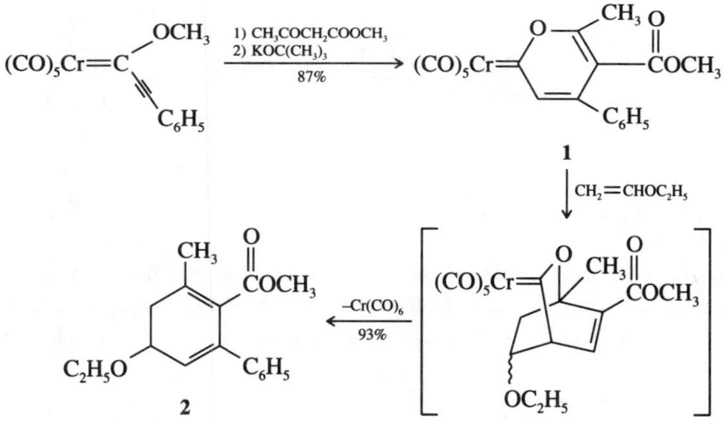

Pyrrole synthesis.[3] Imino carbene complexes such as **1** react with alkynes in hexane at 70°, possibly by a [2 + 2]cycloaddition, to form pyrroles in 80–98% yield. Usually, only a single pyrrole is obtained. This heteroannelation is unusual because carbon monoxide is not incorporated to give a six-membered product. However, O-alkyl imidate carbene complexes such as **3** react with 1-alkynes to form 3-hydroxypyridines as the major product.

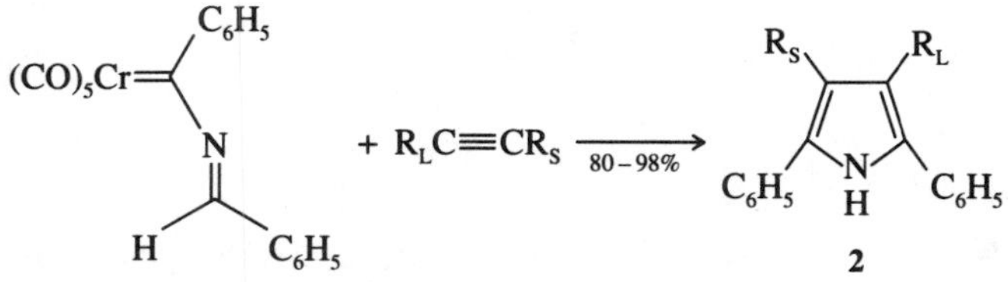

$(CO)_5Cr$=C with C_6H_5 and OCH_3 groups, N=C with C_6H_5 + PrC≡CH ⟶

3

[pyridine/pyridinol product] C_6H_5, Pr, OH, N, C_6H_5

(51%)

+

[pyrrole product] Pr, C_6H_5, N-H, C_6H_5

(4%)

Alkylation.[4] Alkylation of anions of the usual alkoxy carbene complexes is not generally attractive because of low reactivity. However, the anion of dialkylamino chromium carbenes such as **1** (R = H) can be alkylated readily and in a useful yield.

$(CO)_5Cr$=C (N-pyrrolidine), CH_2R

1 (R=H)

1) BuLi, THF, –78°
2) CH_3CH_2I, 0°C
⟶
95%

$(CO)_5Cr$=C (N-pyrrolidine), $CHRC_2H_5$

2

However, forcing conditions are necessary for alkylation of **1**, R = CH_3, and yields are only modest.

2-Amino-γ-butyrolactones.[5] The (dimethylamino)carbene **1** undergoes highly diastereoselective aldol reactions with aldehydes even in the absence of a Lewis acid to form *syn*-adducts **2**, from which the metal unit can be removed by photolysis.

$(CO)_5Cr$=C with $N(CH_3)_2$ and CH_3 + HCOCH(C_6H_5)CH_3

1

BuLi, THF
–78°
⟶
78%

$(CO)_5Cr$= [chain] $(CH_3)_2N$, OH, C_6H_5, CH_3

2 (40:1)

56% | hν, CH_3CN
↓

[butyrolactone product] $(CH_3)_2N$, O, O, CH_3, C_6H_5, H

(*cis/trans*=8:1)

Allylaminocarbenes; indanones.[6] Thermal decarbonylation of the pentacarbonylchromium carbene **1** provides **2**, which reacts with a 1-alkyne to form a complex (**3**) of an indanone (**4**).

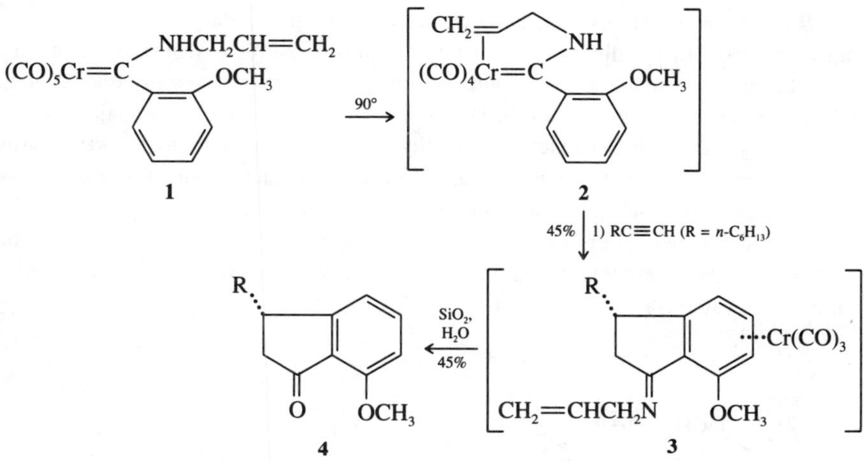

$(CO)_5Cr=C(SCH_3)C_6H_5$ **(1)**.   This complex, a purple oil, is prepared in the usual way from $Cr(CO)_6$, C_6H_5Li, and CH_3SH. It reacts with alkynes only in the presence of BF_3 etherate, Ac_2O, and $N(C_2H_5)_3$ (5 equiv. each) to form acetates of 1,4-

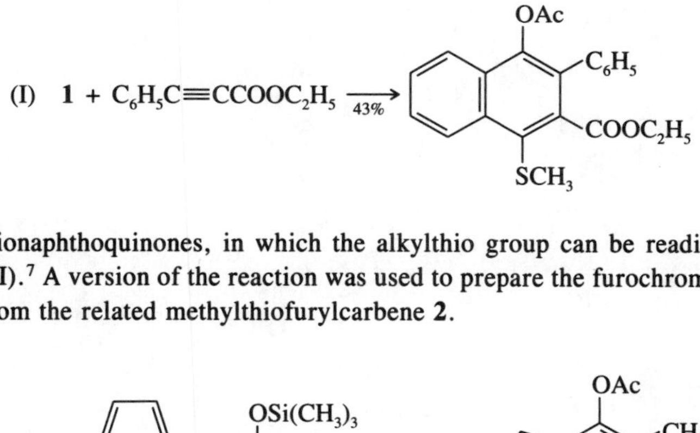

$$(I) \quad 1 + C_6H_5C{\equiv}CCOOC_2H_5 \xrightarrow{43\%}$$

dihydrothionaphthoquinones, in which the alkylthio group can be readily reduced (equation I).[7] A version of the reaction was used to prepare the furochromene visnagan **(3)** from the related methylthiofurylcarbene **2**.

Diels–Alder reactions.[8]   Alkenylchromium carbenes undergo facile Diels–Alder reactions with 1,3-dienes as shown by a typical example (equation I). Thus the vinylchromium complex **1** reacts with isoprene at 25° to give mainly one adduct, which is oxidized by dissolution in DMSO to the methyl ester **3**, the major adduct obtained by a Diels–Alder reaction of methyl acrylate with isoprene. However, use of the carbene rather than the ester analog has some distinct advantages. The rate is significantly faster and the regioselectivity is significantly higher, being comparable to that observed with Lewis acid catalysis. Alkenylchromium carbenes also react with cyclopentadiene with much higher *endo*-selectivity than the corresponding α,β-unsaturated esters.

[1] Y.-C. Xu, C. A. Challener, V. Dragisich, T. A. Brandvold, G. A. Peterson, W. D. Wulff, and P. G. Williard, *Am. Soc.*, **111**, 7269 (1989).
[2] S. L. B. Wang and W. D. Wulff, *ibid.*, **112**, 4550 (1990).
[3] V. Dragisich, C. K. Murray, B. P. Warner, W. D. Wulff, and D. C. Yang, *ibid.*, **112**, 1251 (1990).
[4] W. D. Wulff, B. A. Anderson, and L. D. Isaacs, *Tetrahedron Letters*, **30**, 4061 (1989).
[5] W. D. Wulff, B. A. Anderson, and A. J. Toole, *Am. Soc.*, **111**, 5485 (1989).
[6] K. H. Dötz, H.-G. Erben, and K. Harms, *J.C.S. Chem. Comm.*, 692 (1989).
[7] A. Yamashita, A. Toy, N. B. Ghazal, and C. R. Muchmore, *J. Org.*, **54**, 4481 (1989).
[8] W. D. Wulff, W. E. Bauta, R. W. Kaesler, P. J. Lankford, R. A. Miller, C. K. Murray, and D. C. Yang, *Am. Soc.*, **112**, 3642 (1990).

Chromium(II) chloride.

Nozaki reaction (**12**, 137; **14**, 96; **15**, 95).[1]   An intramolecular version of this reaction can effect cyclization even to a strained nine-membered ring. Thus **1** on treatment with $CrCl_2$ at a low temperature cyclizes mainly to **2**, a mixture of epimers in which a cyclononadiyne ring is fused to a cyclopentene ring. After acetylation of the secondary hydroxyl group, the mixture is dehydrated to **3**, with a nine-membered enediyne ring.

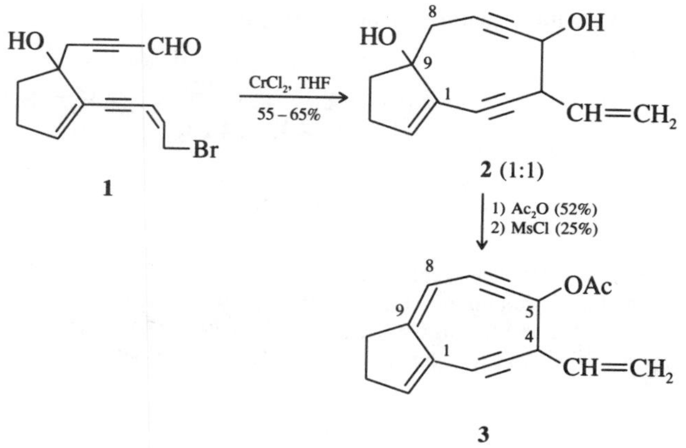

2 (1:1)

1) Ac_2O (52%)
2) MsCl (25%)

3

Alkylchromium(III) reagents, RCr(III).   In the presence of catalytic amounts of vitamin B_{12} or cobalt phthalocyanine (CoPc), $CrCl_2$ reacts with alkyl halides, particularly 1-iodoalkanes, to form an alkylchromium reagent that adds to aldehydes without effect on ketone or ester groups.[2]

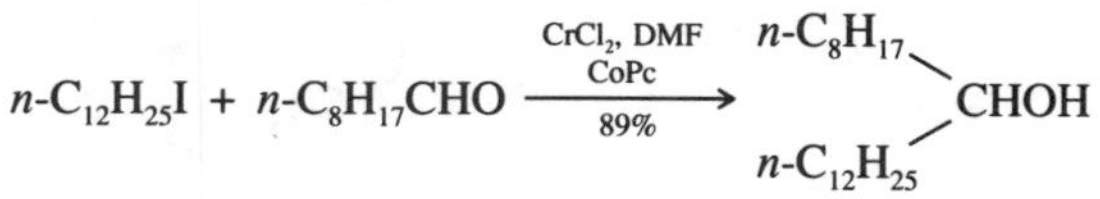

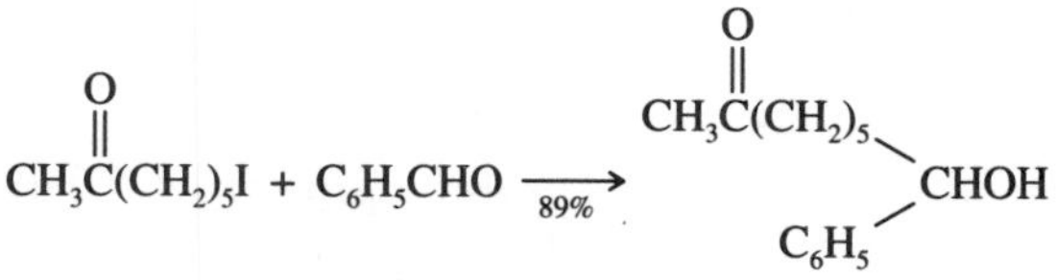

The reaction of a 3-alkyl-1,1-dichloro-2-propene, $R^2CH{=}CHCHCl_2$, with $CrCl_2$ results in an α-chloroallylchromium(III) reagent.[3] This species could react with an aldehyde in theory to form four different 4-chlorohomoallylic alcohols differing in the stereochemistry of the double bond and in *syn/anti*-configurations. In practice this reaction provides essentially only one product, a 2-substituted *anti*-(Z)-4-chloro-3-butene-1-ol.

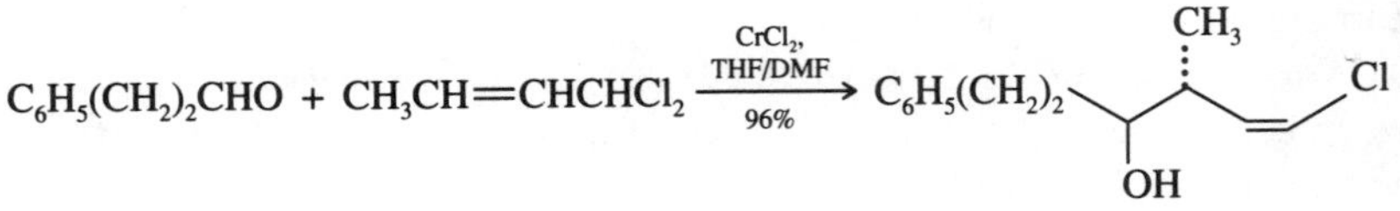

$$C_6H_5(CH_2)_2CHO + CH_3CH{=}CHCHCl_2 \xrightarrow[96\%]{\underset{THF/DMF}{CrCl_2,}} C_6H_5(CH_2)_2\overset{\displaystyle CH_3}{\underset{\displaystyle OH}{\diagdown}}Cl$$

(Z/E = 97:3; *anti/syn* = 97:3)

Reaction of RCHO with 1,3-diene monoepoxides.[4]   The monoepoxides of butadiene or of isoprene after treatment with CrCl$_2$ and LiI (2:1) react with aldehydes to form *cis*-1,3-diols with a quaternary center at C$_2$.

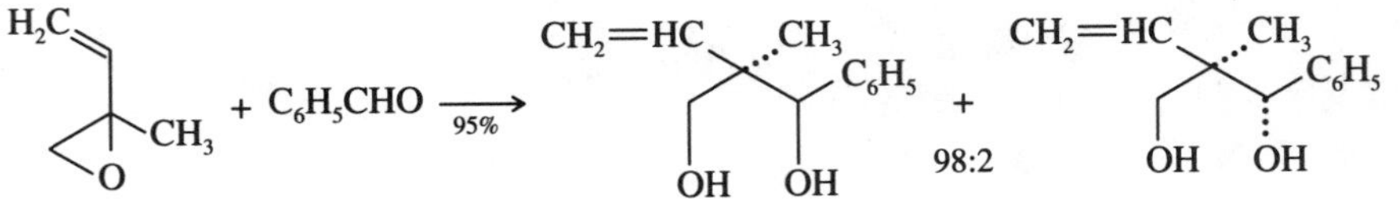

[1] P. A. Wender, J. A. McKinney, and C. Mukai, *Am. Soc.*, **112**, 5369 (1990).
[2] K. Takai, K. Nitta, O. Fujimura, and K. Utimoto, *J. Org.*, **54**, 4732 (1989).
[3] K. Takai, Y. Kataoka, and K. Utimoto, *Tetrahedron Letters*, **30**, 4389 (1989).
[4] O. Fujimura, K. Takai, and K. Utimoto, *J. Org.*, **55**, 1705 (1990).

Cobaloxime(I), 11, 135–136.
   Cyclization of unsaturated epoxides.[1]   Reaction of the epoxy alcohol (**2**) derived from linalool with cobaloxime(I), (**1**), forms the β-hydroxycobaloxime **3**.

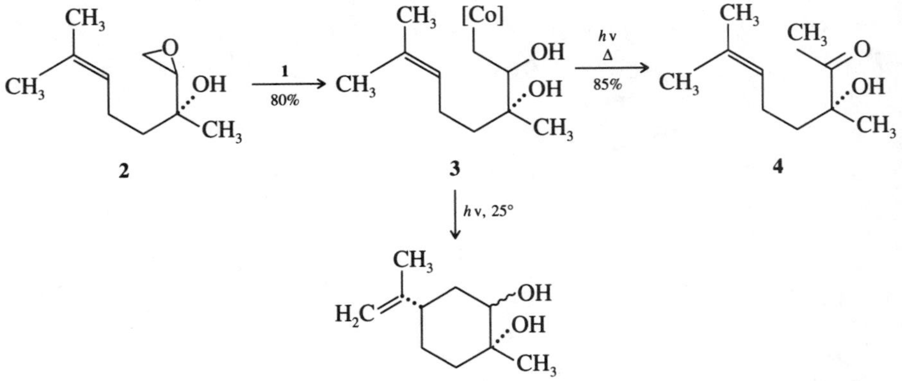

5 (3:2)

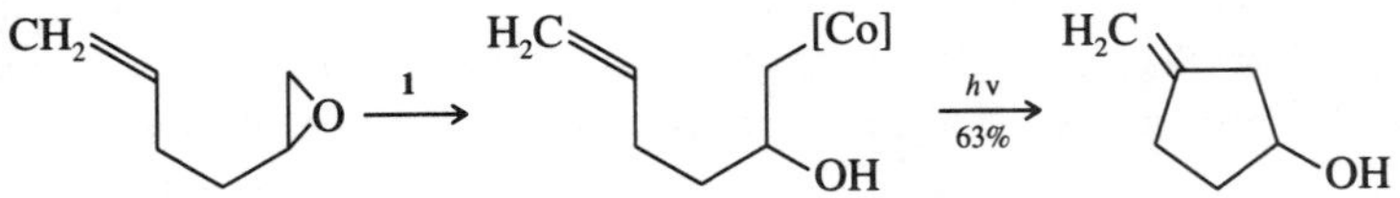

When refluxed in benzene, **3** undergoes dehydrocobaltation to provide the methyl ketone **4**. When irradiated with a sunlamp at 25°, **3** cyclizes to a cyclohexane-1,2-diol (**5**) as a 3:2 mixture of isomers. A similar cyclization to a cyclopentanol is also possible.

[1] D. C. Harrowven and G. Pattenden, *Tetrahedron Letters*, **32**, 243 (1991).

Cobalt(II) acetate, $(CH_3CO_2)_2Co$.

Tetrahydrofurans.[1] $Co(OAc)_2$ promotes a reaction between methyl acetoacetate and terminal alkenes to form tetrahydrofurans. This reaction is similar to a synthesis of dihydrofurans effected with Mn(III) acetate (**14**, 198), but differs in that oxygen is required with the cobalt catalyst. Of further significance, only a single tetrahydrofuran is formed and in relatively good yield (68–71%). A dihydrofuran is also formed, but only in minor amounts.

[1] J. Iqbal, T. K. P. Kumar, and S. Manogaran, *Tetrahedron Letters*, **30**, 4701 (1989); P. Tarakeshwar, J. Iqbal, and S. Manogaran, *Tetrahedron*, **47**, 297 (1991).

Copper(0).

ArCHO → ArC≡N.[1] Aromatic aldehydes are converted to the corresponding nitriles by reaction with copper powder (1.5 equiv.) and NH_4Cl (2 equiv.) at 60° in pyridine under oxygen. Yields are generally 80–99%. The reaction involves *in situ* formation of $CuCl_2$ and NH_3 (equation I). The conversion involves condensation of NH_3 with the aldehyde to form an aldimine, which is then oxidized to a nitrile catalyzed by $CuCl_2$. This method is also applicable to tertiary aldehydes: $(CH_3)_3C$-$CHO → (CH_3)_3CC≡N$ (90%).

$$(I)\ Cu(0) + 2NH_4Cl + \tfrac{1}{2}\ O_2 \longrightarrow CuCl_2 + 2NH_3 + H_2O$$

[1] P. Capdevielle, A. Lavigne, and M. Maumy, *Synthesis*, 451 (1989).

Copper azide.

S_N2-Reactions.[1] Reaction of NaN_3 with a γ-mesyloxy α,β-unsaturated ester such as **1** proceeds as expected with inversion. The reaction with NaN_3–CuI proceeds mainly with retention (*syn-S_N2*).

NaN$_3$, DMSO	86%	100:0
NaN$_3$·CuI, HMPA	90%	9:91

[1] Y. Yamamoto and N. Asao, *J. Org.*, **55**, 5303 (1990).

Copper(I) bromide.

ArBr → ArOCH$_3$.[1] Bromobenzene can be converted to anisole in 95% yield by reaction of NaOCH$_3$ in DMF/CH$_3$OH at 110°, catalyzed by CuBr. The active species is probably Nå[Cu(OCH$_3$)$_2$]$^-$.

[1] H. L. Aalten, G. van Koten, D. M. Grove, T. Kuilman, O. G. Piekstra, L. A. Hulshof, and R. A. Sheldon, *Tetrahedron*, **45**, 5565 (1989).

Copper(I) bromide–Dimethyl sulfide.

Intramolecular Ullmann reaction.[1] A number of antitumor antibiotics consist of polypeptides containing a diphenyl ether group, arising by oxidative coupling of two tyrosine groups. However, several laboratories have reported that an intramolecular Ullmann reaction fails to effect the desired cyclization to 14-membered rings. A detailed study of the intramolecular Ullmann reaction with a model substrate such as **1** reveals that under moderately dilute conditions at least 2 equiv. of CuBr·S(CH$_3$)$_2$ are required. Reactions proceed well in pyridine (130°), but DMSO, DMF, and C$_6$H$_5$Cl are poor solvents. Racemization can be a problem in the case of pyridine, but can be minimized by use of dioxane (110°). The presence of an alkoxy or hydroxy substituent *ortho* to the participating phenol in the coupling has an adverse effect, but does not prevent cyclization. Several attempts to use a 3-iodotyrosine as a partner in

1, R^1, R^2 = H	Py(130°)	58%	**2**
R^1 = OCH$_3$, R^2 = H	Py	46%	
R^1 = OCH$_3$, R^2 = COOCH$_3$	Py	51%	(S/R = 55:45)
R^1 = OCH$_3$, R^2 = COOCH$_3$	Dioxane	31%	(S/R = 96:4)

an intramolecular Ullmann reaction have failed. Application of these model conditions for Ullmann cyclizations has led to a successful synthesis of deoxybouvardin **4** and the *o*-methyl deoxybouvardin **3**.[2]

	R[1]	R[2]	R[3]	R[4]	R[5]	
3	H	CH₃	CH₃	H	H	*O*-methyl deoxybouvardin
4	H	H	CH₃	H	H	deoxybouvardin

[1] D. L. Boger and D. Yohannes, *J. Org.*, **56**, 1763 (1991).
[2] idem, *Am. Soc.*, **113**, 1427 (1991).

Copper(I) chloride–Chlorotrimethylsilane.

Conjugate addition of RMgCl and RMnCl to α,β-enoates.[1] This reaction can generally be effected in good yield in THF at room temperature in the presence CuCl (3%) and (CH₃)₃SiCl (1.2 equiv.). Even β,β-disubstituted enoates undergo this conjugate addition, but yields are poor in the case of CH₃MgBr. Another exception is noted with allylmagnesium halides, which give only the 1,2-adduct. CuCl can be replaced by CuCN with only a minor decrease in yields. (CH₃)₃SiCl (1.2 equiv.) can be replaced by CH₃SiCl₃ (0.45 equiv.).

[1] G. Cahiez and M. Alami, *Tetrahedron Letters*, **31**, 7423, 7425 (1990).

Copper(I) cyanide–Chlorotrimethylsilane.

Michael reactions.[1] Various organozinc reagents bearing remote ester, nitrile, or α-amino acid groups undergo Michael addition to α,β-unsaturated carbonyl compounds in the presence of CuCN, (CH₃)₃SiCl, and HMPA.[1]

$$\text{IZn(CH}_2)_3\text{CN} + \text{CH}_2=\text{CHCOOCH}_3 \xrightarrow[\substack{59\%}]{\substack{\text{CuCN, ClSi(CH}_3)_3 \\ \text{HMPA, 0}°}} \text{NC(CH}_2)_5\text{COOCH}_3$$

$$\text{IZn(CH}_2)_3\text{COOC}_2\text{H}_5 + \text{C}_6\text{H}_5\text{CH}=\text{CHCHO} \xrightarrow[94\%]{} \underset{}{\text{HCCH}_2\text{CH(CH}_2)_3\text{COOC}_2\text{H}_5}$$

[1] Y. Tamaru, H. Tanigawa, T. Yamamoto, and Z. Yoshida, *Angew. Chem. Int. Ed.*, **28**, 351 (1989).

Copper(I) iodide.

Allylation of 1-*alkynes*.[1] Cu(I)-promoted allylation of 1-alkynes was first reported in patents (1957–1959) and has since been markedly improved by use of a 1-alkynylmagnesium halide or by use of phase-transfer conditions (Bu$_4$NCl/NaCO$_3$), both of which allow use of substituted allylic halides. Under the latter conditions, substituted 1,4-enynes can be obtained in 76–95% yield.

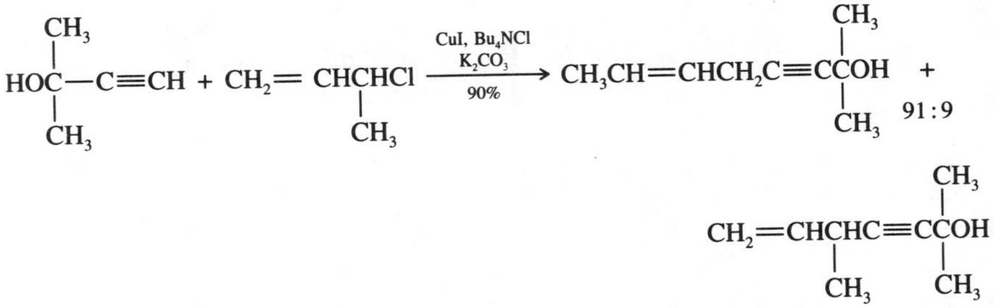

[1] T. Jeffrey, *J.C.S. Chem. Comm.*, 909 (1988); *idem*, *Tetrahedron Letters*, **30**, 2225 (1989).

Copper(I) iodide–Tetrakis(triphenylphosphine)palladium(0).

Coupling of vinyl bromides with alkynes.[1] The β-bromide of ethyl (Z)-2,3-dibromopropenoate (**1**) couples selectively with trimethylsilylacetylene in the presence of CuI, Pd(0), and a base to provide the bromoenyne **2** in 86–90% yield.

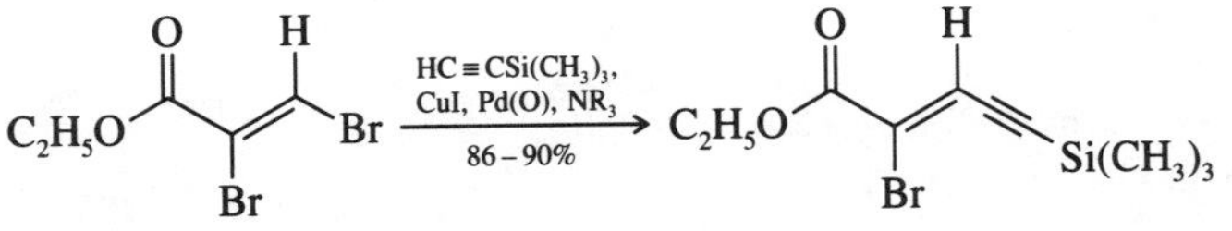

Bromide **2** undergoes similar coupling with a variety of 1-alkynes to provide (Z)-enediynes in high yield. This coupling can provide a variety of enynes and enediynes.

[1] A. G. Myers, M. M. Alauddin, M. A. M. Fuhry, P. S. Dragovich, N. S. Finney, and P. M. Harrington, *Tetrahedron Letters*, **30**, 6997 (1989).

Copper(I) oxide, Cu_2O.

Benzofurans.[1] 2-Substituted benzofurans can be obtained in 65–83% yield by reaction of *o*-iodophenols and 1-alkynes in the presence of Cu_2O suspended in pyridine. This synthesis has been conducted in the past by reaction of copper(I) acetylides with *o*-halophenols, but yields are somewhat lower. In addition, copper(I) acetylides can be explosive.

$$(CH_3)_3C\text{–}\overset{I}{\underset{OH}{\bigcirc}} + CH{\equiv}C(CH_2)_3CH_3 \xrightarrow[82\%]{Cu_2O, Py} (CH_3)_3C\text{–}\overset{O}{\bigcirc}\text{–}(CH_2)_3CH_3$$

[1] G. J. S. Doad, J. A. Barltrop, C. M. Petty, and T. C. Owen, *Tetrahedron Letters*, **30**, 1597 (1989).

Copper(I) trifluoromethanesulfonate, CuOTf.

β-Lactams.[1] CuOTf (1 equiv.), particularly in combination with $CaCO_3$, is effective for cyclization of β-amino thiolesters to *cis*-β-lactams (67–92% yields). Mercury(II) trifluoroacetate has also been used, but it is highly toxic.

$$\xrightarrow[67\%]{CuOTf, CaCO_3, C_6H_5CH_3, reflux}$$

[1] N. Miyachi, F. Kanda, and M. Shibasaki, *J. Org.*, **54**, 3511 (1989).

Crotyltributyltin, **11**, 143; **12**, 146.

$CoCl_2$-Controlled addition to aldehydes.[1] This reaction, when catalyzed by a Lewis acid, $TiCl_4$ or BF_3 etherate, involves an allylic rearrangement, regardless of the order of addition of reactants (**12**, 146). In contrast, if $CoCl_2$ (1 equiv.) is present, the addition produces mainly or even exclusively the linear homoallylic alcohol (α-adduct).

$$n\text{-}C_6H_{13}CHO + CH_3CH{=}CHCH_2SnBu_3 \xrightarrow[70\%]{CoCl_2} CH_3CH{=}CHCH_2\underset{\underset{OH}{|}}{C}HC_6H_{13}\text{-}n$$

[1] J. Iqbal and S. P. Joseph, *Tetrahedron Letters*, **30**, 2421 (1989).

Cyanoarenes.

Diels–Alder reactions.[1]   Both 1,4-dicyanonaphthalene (DCN) and 2,6,9,10-te-tracyanoanthracene (TCA) have been used as sensitizers to effect photochemical [4 + 2]cycloadditions of electron-rich dienes and electron-rich dienophiles, which do not normally undergo thermal cycloadditions. These cycloadditions are known as triplex Diels–Alder reactions because they are postulated to involve as an intermediate a three-membered complex of sensitizer, dienophile, and diene. This reaction is useful for synthesis of bicyclo[2.2.2]octenes from some silyl enol ethers, alkenes, or arylalkynes.

endo - trans

(exo / endo = 1 : 1)

[1] N. Akbulut, D. Hartsough, J.-I. Kim, and G. B. Schuster, *J. Org.*, **54**, 2549 (1989).

Cyanotrimethylsilane–Di-μ-chlorobis(cyclooctadiene)dirhodium.

Cyanation of acetals.   This reaction has been effected with $CNSi(CH_3)_3$ and a Lewis acid catalyst (**11**, 150). It can also be effected under neutral conditions with several transition metal catalysts, in particular, with [Rh(COD)Cl]₂, $CoCl_2$, and $NiCl_2$, listed in the order of decreasing activity. Based upon this reaction, Mukai-yama *et al.*[1] have examined the use of the combination of cyanotrimethylsilane and the rhodium catalyst for general activation of silyl enol ethers or ketene silyl acetals.

Aldol reactions.[2]   This combination of catalysts is effective for promoting reaction of acetals with silyl enol ethers and ketene silyl acetals. It can also promote reaction of aldehydes or imines with ketene silyl acetals. The reactions occur in high yield at 25°; either CH_3CN or THF can be used as the solvent.

[1] T. Mukaiyama, T. Soga, and H. Takenoshita, *Chem. Letters,* 997 (1989).
[2] T. Soga, H. Takenoshita, M. Yamada, and T. Mukaiyama, *Bull. Chem. Soc. Japan,* **63**, 3122 (1990).

Cyanuric fluoride.

Amino acid fluorides.[1]   Amino acids in which the nitrogen group is protected by the acid-stable 9-fluorenylmethoxycarbonyl (FMOC) group react with cyanuric fluoride and pyridine in CH_2Cl_2 to form FMOC amino acid fluorides in 65–75% yield. *t*-Butyl ester and ether groups are also stable to cyanuric fluoride. Hence this reaction is compatible with most amino acids containing an additional OH, COOH, or NH_2 group such as serine, threonine, and aspartic acid. The resulting crystalline acid fluorides are more stable to water and methanol than the acid chlorides, but are more reactive than the chlorides in reactions with another amine. In fact they have been used to prepare a heptapeptide by solid-phase coupling reactions.

[1] L. A. Carpino, D. Sadat-Aalaee, H. G. Chao, and R. H. DeSelms, *Am. Soc.,* **112**, 9651 (1990).

Cyclic sulfates of 1,2-diols (15, 105–107).

1,2-Diamines.[1]   The cyclic sulfate (2) of (R,R)-stilbenediol (1) reacts with benzamidine (3) in refluxing DME to form the imidazoline 4 as a single enantiomer. This product is converted by known steps into (S,S)-stilbenediamine (5) in 44% yield. The overall process involves inversion of both stereogenic centers of 1. The transformation was applied to several aliphatic cyclic sulfates and shown to afford 1,2-diamines in high percent ee.

Aziridines; amino alcohols.[2]   The 1,2-cyclic sulfates (1) of chiral diols react with RNH_2 to form β-amino sulfates (2), which are convertible into aziridines or amino alcohols.

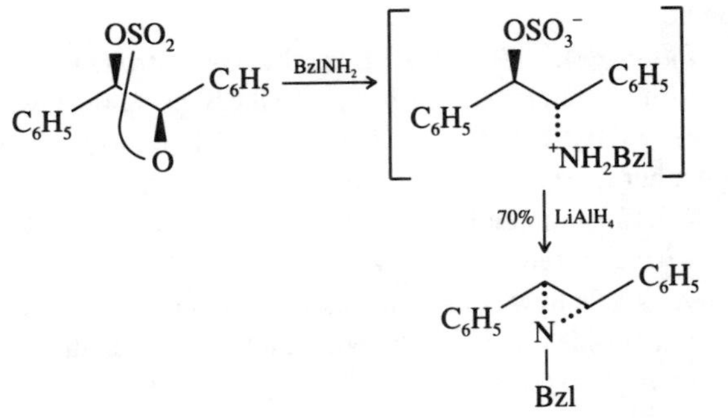

(S, S, > 96% ee)

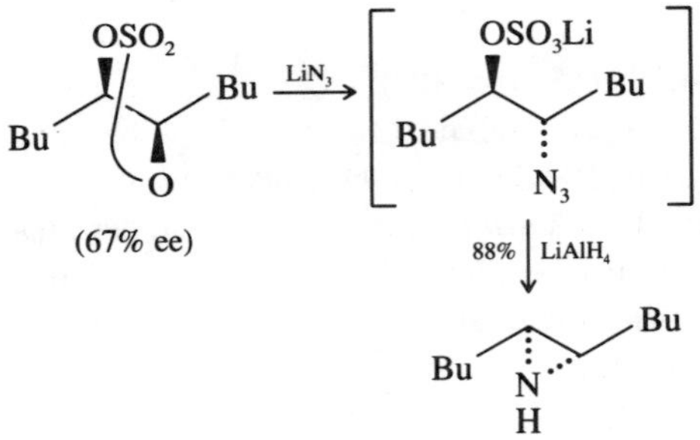

(S, S, 67% ee)

[1] R. Oi and K. B. Sharpless, *Tetrahedron Letters*, **32**, 999 (1991).
[2] B. B. Lohray, Y. Gao, and K. B. Sharpless, *ibid.*, **30**, 2623 (1989).

Cyclobutenediones.

Quinone synthesis (**13**, 97–98; 209–210).   Liebeskind[1] has reviewed the use of organolithiums for preparation of benzoquinones from cyclobutenedione and of the reaction of benzocyclobutenediones with $ClCo[P(C_6H_5)_3]_3$ to form either benzoquinones or naphthoquinones. These routes are particularly useful for regioselective synthesis of highly substituted quinones.

[1] L. S. Liebeskind, *Tetrahedron*, **45**, 3053 (1989).

(−)-*trans*-Cyclohexane-(1R,2R)-disulfonamides (1).

Enantioselective addition of R_2Zn to RCHO.[1]   A Ti-complex, formulated as **2**, prepared from a (−)-*trans*-cyclohexanedisulfonamide, particularly **1**, when com-

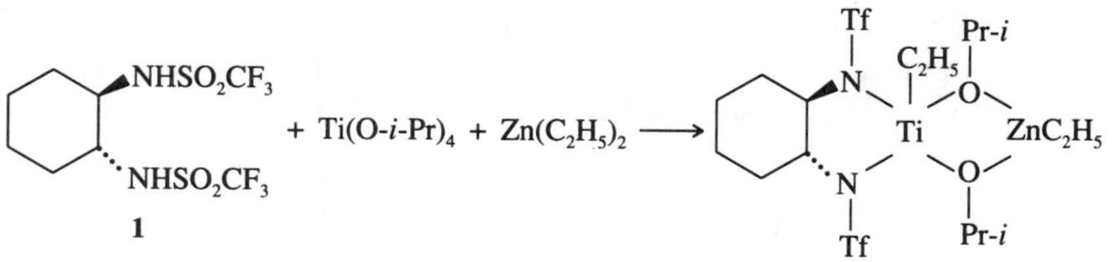

$$C_6H_5CHO + 2 \xrightarrow[97-99\%]{} C_6H_5\overset{OH}{\underset{}{\diagup}}C_2H_5$$

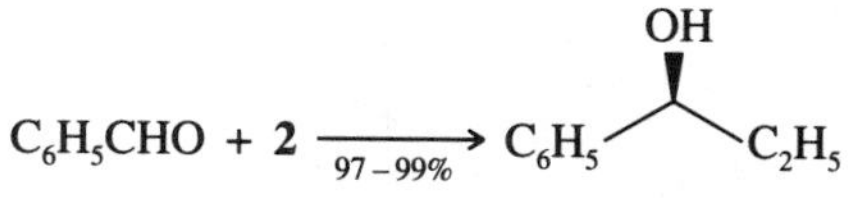

(98–99% ee)

bined with $Ti(O\text{-}i\text{-}Pr)_4$ and diethylzinc effects additions of diethylzinc to aldehydes in enantiomeric excesses approaching 100% even when the complex is used in catalytic amounts.

The process is an interesting example of ligand-accelerated catalysis, but the only substrate reported is benzaldehyde. It is noteworthy that the less-reactive dibutylzinc is almost as effective as diethylzinc.

[1] M. Yoshioka, T. Kawakita, and M. Ohno, *Tetrahedron Letters*, **30**, 1657 (1989).

[(Cyclopentadienyl)dicarbonyl(phenylthio)carbenium]iron   hexafluorophosphate, $Cp(CO)_2Fe^+{=}CHSC_6H_5$ PF_6^- (**1**, m.p., 126°). This stable, crystalline phenylthio iron carbene is obtained in 80% yield by reaction of $NaCpFe(CO)_2$ with $ClCH_2SC_6H_5$ and then with $(C_6H_5)_3C^+PF_6^-$.[1]

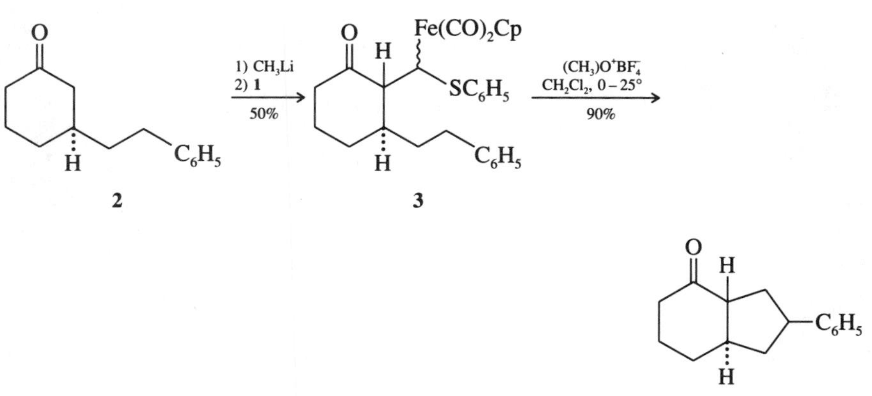

Cyclopentane annulation.[2] This complex is useful for cyclopentane annulation by insertion into C–H bonds. Thus reaction of **1** with the enolate of ketone **2** provides the adduct **3**. Reaction of **3** with trimethyloxonium tetrafluoroborate provides a sulfonium salt which liberates a carbene that inserts into the benzylic C–H bond of **3** to form the cyclopentane-fused product **4**. Cyclopentane annulation has also been ob-

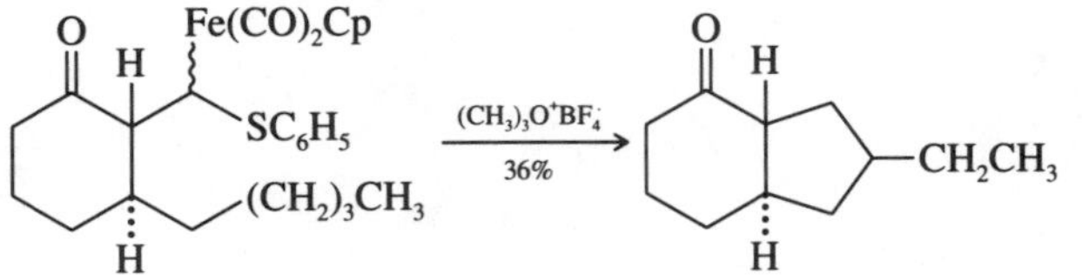

$$\beta:\alpha = 2:1$$

tained by a formal allylic C–H insertion. Insertion into a simple alkyl side chain is possible, but in lower yield (36%, equation I). Insertion into a tertiary C–H bond is particularly facile, as employed for a synthesis of sterpurene (**8**) from the enol silyl ether **5**.

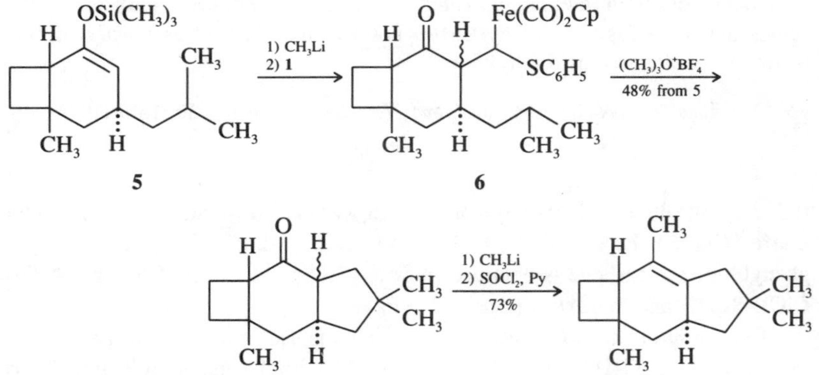

[1] C. Knors, G.-H. Kuo, J. W. Lauher, C. Eigenbrot, and P. Helquist, *Organometallics*, **6**, 988 (1987).

[2] S.-K. Zhao, C. Knors, and P. Helquist, *Am. Soc.*, **111**, 8527 (1989); S.-K. Zhao and P. Helquist, *J. Org.*, **55**, 5820 (1990).

D

3,4-Dialkoxyfurans.

Diels–Alker reactions.[1] Both 3,4-dimethoxy- and 3,4-dibenzyloxyfuran undergo Diels–Alder reactions with reactive dienophiles in refluxing benzene. This cycloaddition can be markedly catalyzed by ZnI_2, which also enhances *endo*-selectiv-

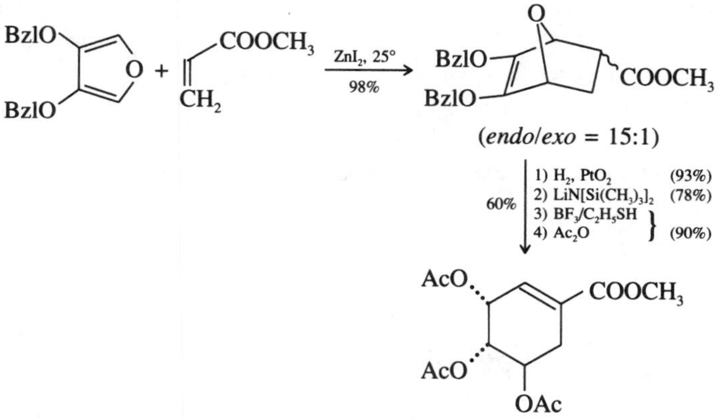

$(endo/exo = 15:1)$

ity. This reaction has been used for an efficient synthesis of methyl triacetylshikimate (**1**).

[1] M. Koreeda, K.-Y. Jung, and J. Ichita, *J.C.S. Perkin I*, 2129 (1989).

1,8-Diazabicyclo[5.4.0]undecene-7 (DBU).

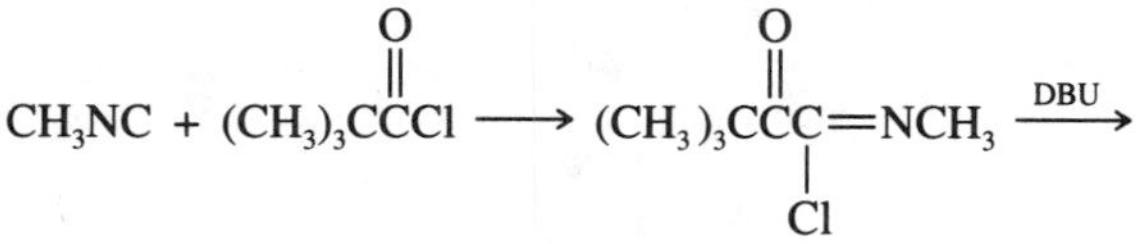

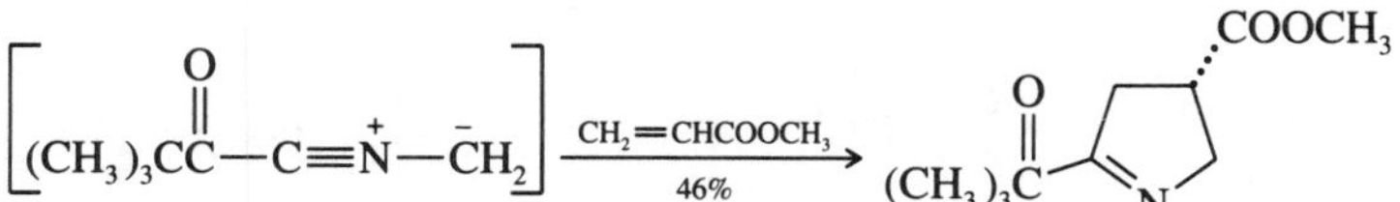

1,3-Dipolar cycloadditions.[1] On treatment with DBU, imidoyl halides **1**, prepared by reaction of acid chlorides with isocyanides, undergo 1,3-dehydrochlorination to form α-ketonitrile ylides (**a**), which can be trapped by electron-deficient alkenes to form Δ^1-pyrrolines in 45–65% yields.

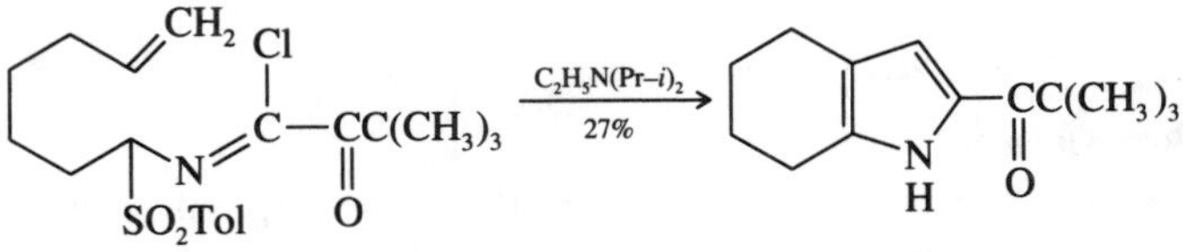

The report includes one example of an intramolecular cycloaddition (equation I).

Deacetylation.[2]   Acetyl groups are cleaved by DBU at 20–80° in methanol or in a mixed solvent, CH_3OH/CH_2Cl_2, if required for solubility of the substrate. In general, primary acetates are hydrolyzed more readily than secondary or tertiary ones. A bonus for this methodology is the selectivity: Esters lacking an α-hydrogen (benzoyl, pivaloyl) are stable to this cleavage. The reaction is probably not hydrolytic since it can proceed in benzene without a protic solvent.

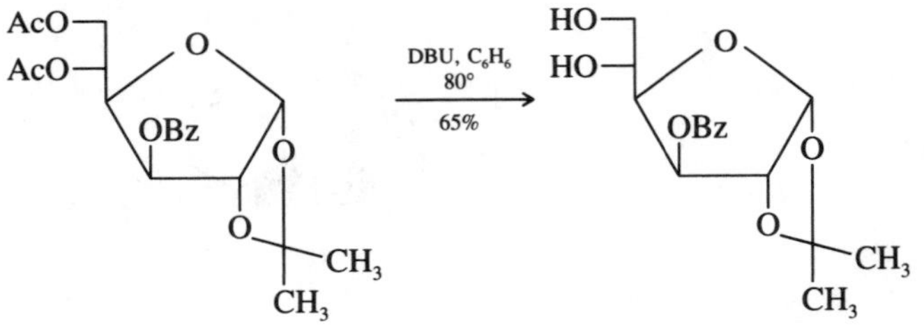

[1] W.-S. Tian and T. Livinghouse, *J.C.S. Chem. Comm.*, 819 (1989).
[2] L. H. B. Baptistella, J. F. dos Santos, K. C. Ballabio, and A. J. Marsaioli, *Synthesis*, 436 (1989).

Dibromodifluoromethane/Zinc, CF_2Br_2/Zn.

Difluorocarbene.[1]   This carbene can be generated from CF_2Br_2 by reaction with zinc in THF catalyzed by I_2. This reagent should prove useful for the generation of difluorocyclopropanes.

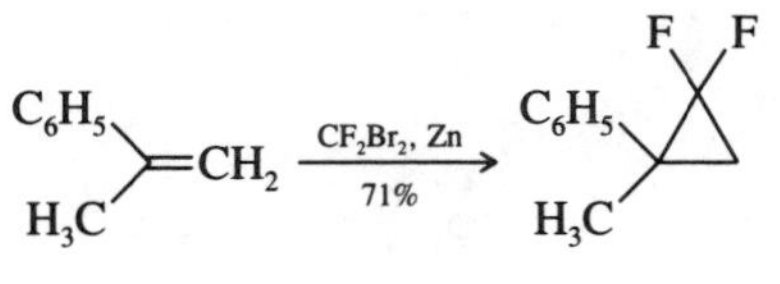

[1] W. R. Dolbier, Jr., H. Wojtowicz, and C. R. Burkholder, *J. Org.*, **55**, 5420 (1990).

Dibromomethane–Zinc/Copper(I) chloride, 13, 93.

Cyclopropanation.[1]   The cyclopropanation of alkenes by this combination can be facilitated by sonication or by Ti(IV) chloride, but acetyl chloride is more effective. The effectiveness of this promotor may result from removal of traces of water and hydroxylic impurities. Yields using this expedient are generally higher (45–85%) than those obtained by the original Simmons–Smith conditions using diiodomethane.

[1] E. C. Friedrich and E. J. Lewis, *J. Org.,* **55**, 2491 (1990).

Dibromomethyl(trimethyl)silane, $(CH_3)_3SiCHBr_2$ **(1).** The reagent is prepared from CH_2Br_2 and $(CH_3)_3SiCl$.[1]

(Z/E = >95:<5)

(Z/E = 91:9)

Alkenylsilanes.[2]   A reagent **(2),** prepared from **1,** $TiCl_4$, zinc, and TMEDA, is efficient for conversion of esters or thiolesters into alkenylsilanes with (Z)-selectivity.

[1] J. Villieras, C. Bacquet, and J. F. Normant, *Bull. Soc. Chem., Fr.,* 1797 (1975).
[2] K. Takai, M. Tezuka, Y. Kataoka, and K. Utimoto, *Synlett,* **1,** 27 (1989).

2,3-Dibromo-1-phenylsulfonyl-1-propene,

(1, m.p. 65°)

This reagent is obtained by addition of Br_2 to phenylsulfonylallene.

Furan annelation.[1]   The reagent reacts with the enolate of dimethyl malonate **(2)** to form furan **3** in good yield. A similar reaction of **1** with a β-diketone **(4)** and $NaOCH_3$ in CH_3OH is accompanied by deacylation to provide furans such as **5.**

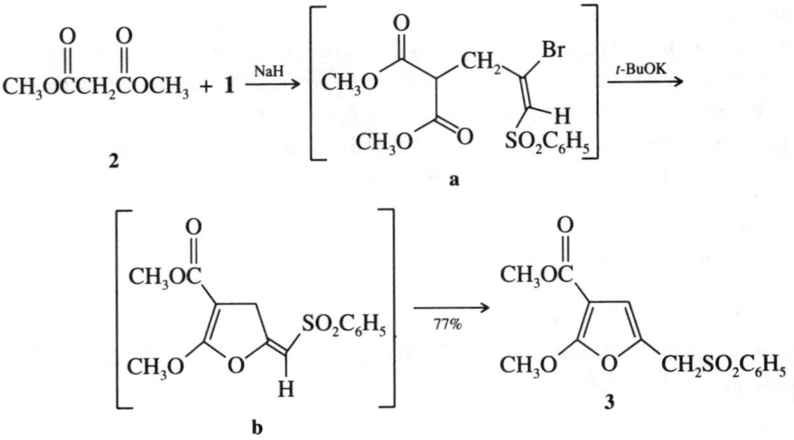

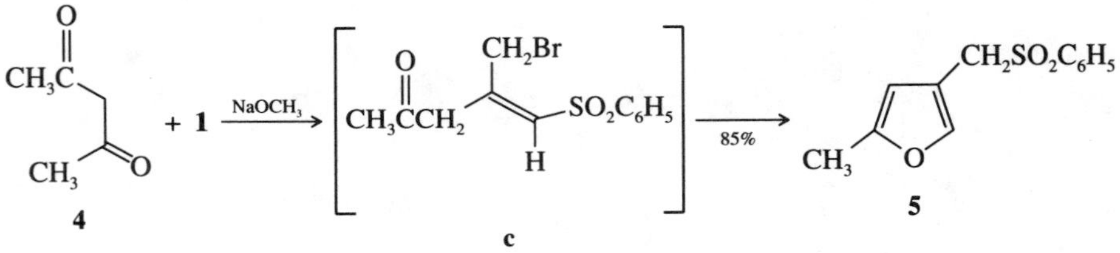

$CH_3OCCH_2COCH_3$ + **1** $\xrightarrow{\text{NaH}}$ [CH_3O ... CH_2 ... Br ... H ... $SO_2C_6H_5$ ... CH_3O] $\xrightarrow{t\text{-BuOK}}$

2

a

[CH_3OC ... $SO_2C_6H_5$ ... CH_3O ... H] $\xrightarrow{77\%}$ CH_3OC ... CH_3O ... $CH_2SO_2C_6H_5$

b

3

CH_3C ... CH_3 + **1** $\xrightarrow{\text{NaOCH}_3}$ [CH_3CCH_2 ... CH_2Br ... $SO_2C_6H_5$ ... H] $\xrightarrow{85\%}$ CH_3 ... $CH_2SO_2C_6H_5$

4

c

5

Reactions of **1** with a cyclic 1,3-diketone results in a 2,3-fused bicyclic furan (equation I).

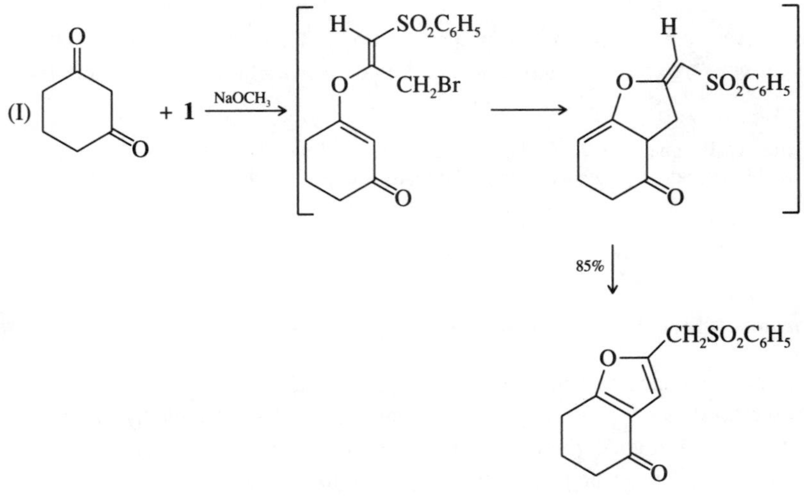

(I) ... + **1** $\xrightarrow{\text{NaOCH}_3}$ [H ... $SO_2C_6H_5$ ... CH_2Br ... O ... O → H ... O ... $SO_2C_6H_5$ ... O] $\xrightarrow{85\%}$ O ... $CH_2SO_2C_6H_5$ ... O

[1] A. Padwa, S. S. Murphree, and P. E. Yeske, *J. Org.*, **55**, 4241 (1990).

Dibutylboryl trifluoromethanesulfonate, Bu_2BOTf **(1).**

Boron enolates of α-benzyloxy esters.[1] The triflate **1** converts alkyl benzyloxy-acetates (**2**) into the boron enolate, which readily undergoes aldol reactions with high *syn*-diastereoselectivity. Somewhat higher *syn*-selectivity obtains with dicyclopentylboryl triflate, whereas use of LDA results in slight *anti*-selectivity (*syn/anti*=34–37:66–63). Diisopropylethylamine is essential for the aldol reaction. *Syn*-**3** is re-

$$C_6H_5CH_2OCH_2COOR^1 + R^2CHO \xrightarrow[\text{EtN}(i\text{-Pr})_2]{1,} \underset{\underset{syn-3}{OH}}{R^2} \overset{OBzl}{\diagdown} COOR^1 + anti-3$$

2

$R^1 = CH_3$	$R^2 = C_6H_5CH_2CH_2$	82%	96:4
$R^1 = C(CH_3)_3$	$R^2 = C_6H_5CH_2CH_2$	76%	99:1
$R^1 = CH_3$	$R^2 = Pr$	84%	97:3

duced by lithium aluminum hydride followed by catalytic hydrogenation (Pd/C) to glycerol derivatives, $R^2CHOHCHOHCH_2OH$.

[1] Y. Sugano and S. Naruto, *Chem. Pharm. Bull.*, **37**, 840 (1989).

Dibutyldicyclopentadienylzirconium, Cp_2ZrBu_2 **(1).** The reagent (a zirconocene equivalent) is prepared by reaction of BuLi (2 equiv.) with Cp_2ZrCl_2 in THF (25°).[1]

Bicyclization of unsaturated imines.[2] Unsaturated hydrazones, particularly N,N-dimethylhydrazones, undergo bicyclization when treated with **1** (1.35 equiv.) to *cis*-cycloalkylhydrazides.

[1] E. Negishi, F. E. Cederbaum, and T. Takahashi, *Tetrahedron Letters*, **27**, 2829 (1986).
[2] M. Jensen and T. Livinghouse, *Am. Soc.*, **111**, 4495 (1989).

2,6-Di-*t*-butyl-4-methoxyphenol (BHA, butylated hydroxyanisole), **13**, 94–95.

1,1,2- *and* 1,2,2-*Trisubstituted dihydronaphthalenes*.[1]   BHA esters (1**) of 1-naphthalenecarboxylic acid react with alkyllithiums to form an adduct (**a**), which is not isolated but reduced with lithium triethylborohydride to the enolate (**b**) of an aldehyde. This intermediate can be alkylated (**c**) and reduced to **2** in 75% overall yield. The same sequence when applied to **3** provides **4** in 93% yield. 1,2-Disubstitu-

ted dihydronaphthalenes are obtained when the alkylation is omitted. The report suggests that the reduction of **a** to **b** involves a ketene intermediate.

[1] K. Tomioka, M. Shindo, and K. Koga, *J. Org.*, **55**, 2276 (1990).

2,6-Di-*t*-butyl-4-methylpyridine (**1**).

Vinyl triflates (**10**, 123).[1]   The most satisfactory conditions for preparation of vinyl triflates from carbonyl compounds using triflic anhydride and catalyzed by the pyridine **1** or the polymer-bound reagent **2** are discussed by Wright and Pulley. In the case of aldehydes, the reaction is best carried out in $CHCl_3$ or $(CH_2Cl)_2$ at 50–70°, and results in a *gem*-bistriflate, which decomposes thermally to the vinyl triflate. These conditions, when applied to a ketone, particularly aldol-prone ones, are usually not useful. The preferred solvent is nonpolar (CCl_4) with a reaction temperature of about 50°. In addition, the Tf_2O should be freshly distilled from P_2O_5.

[1] M. E. Wright and S. R. Pulley, *J. Org.*, **54**, 2886 (1989).

Dibutyl telluride, Bu_2Te. Preparation from $Te(0)$ and $BuCl$.

Tellurium–Wittig reactions.[1]   The reaction of aldehydes with α-bromo esters or ketones, triphenyl phosphite, and a weak base in the presence of a catalytic amount of Bu_2Te results in α,β-unsaturated esters and ketones. The function of the triphenyl phosphite is the regeneration of Bu_2Te from the dibutyl telluroxide formed in the

$$BrCH_2COOCH_3 + Bu_2Te \xrightarrow{K_2CO_3} [Bu_2Te=CHCOOCH_3]$$

$$\xrightarrow[89\%]{C_6H_5CHO} C_6H_5CH\overset{(E)}{=}CHCOOCH_3 + Bu_2TeO$$

$$\downarrow P(OC_6H_5)_3$$

$$Bu_2Te + (C_6H_5O)_3P=O$$

Wittig-type reaction. Yields are $>70\%$, and the products all have the (E)-configuration.

[1] Y.-Z. Huang, L.-L. Shi, S.-W. Li, and X.-Q. Wen, J. C. S. Perkin I, 2397 (1989).

Dibutyltin bistriflate, $Bu_2Sn(OTf)_2$.

Carbonyl activation and deactivation.[1]   Aldehydes, but not ketones, undergo aldol condensation with silyl enol ethers at $-78°$ in the presence of dibutyltin bistriflate. In contrast, the dimethyl acetals of ketones, but not of aldehydes, can undergo this condensation (Mukaiyama reaction) with silyl enol ethers at $-78°$ with almost complete discrimination, which is not observed with the usual Lewis-acid catalysts. Thus dibutyltin bistriflate activates aldehydes, but deactivates acetals of

aldehydes. It is possible to effect selective reactions of ketones admixed with aldehydes by acetalization and addition of other silyl nucleophiles, such as R_3SiH or
R_3SiCN, in the presence of this catalyst. Note that trimethylsilyl triflate can effect
reactions of acetals with silyl enol ethers in competing reactions with aldehydes or
ketones (15, 349), but this Lewis acid does not discriminate between acetals of
aldehydes and of ketones.

Michael addition.[2] This triflate is an effective catalyst for Michael addition of
enol silyl ethers to α,β-enones such as methyl vinyl ketone to provide adducts in 60–
75% yield, equation (I). This variation is useful in Robinson annelations.

[1] T. Sato, J. Otera, and H. Nozaki, *Am. Soc.*, **112**, 901 (1990).
[2] T. Sato, Y. Wakahara, J. Otera, and H. Nozaki, *Tetrahedron Letters*, **31**, 1581 (1990).

Dibutyltin oxide, Bu_2SnO.

Monoacylation of diols.[1] Monoacylation of unsymmetrical 1,2-, 1,3-, and 1,4-
diols can be effected by acylation of the dibutylstannylene derivatives, followed by
quenching with oxalic acid or $ClSi(CH_3)_2C_6H_5$. This process effects monoacylation
of the more-substituted hydroxyl group, even a tertiary one. This method also is

useful in distinguishing between primary and secondary diols, but fails with 1,5-diols
probably because of inability to form the dioxastannane intermediate.

[1] G. Reginato, A. Ricci, S. Roelens, and S. Scapecchi, *J. Org.*, **55**, 5132 (1990).

Dicarbonyl(cyclopentadienyl)cobalt.

[2+2+2]Cycloaddition to a hydrocyclobutaindane.[1] The framework of the
sesquiterpenoid illudol (3) can be obtained in one step and high yield by a [2+2+2]
cycloaddition of the acyclic enediyne 1 mediated by $CpCo(CO)_2$. Transformation of
2 to 3 involves a number of standard reactions. The cycloaddition of 1 to 2 is the first
instance of formation of three rings in a single step and with stereospecific generation
of three contiguous chiral centers.

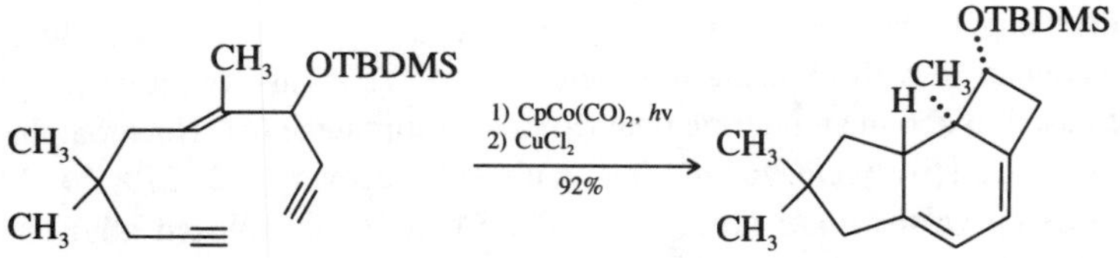

1

2

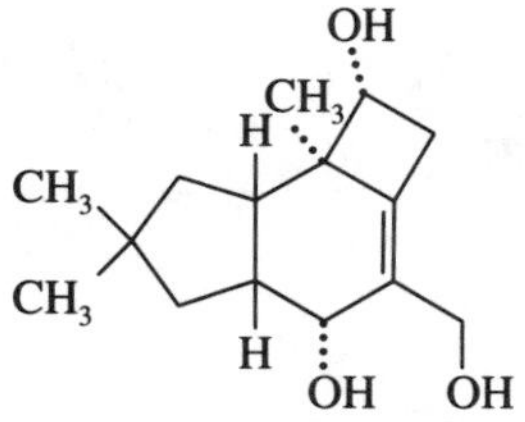

3

[1] E. P. Johnson and K. P. C. Vollhardt, *Am. Soc.*, **113**, 381 (1991).

Di-μ-carbonylhexacarbonyldicobalt, $Co_2(CO)_8$.

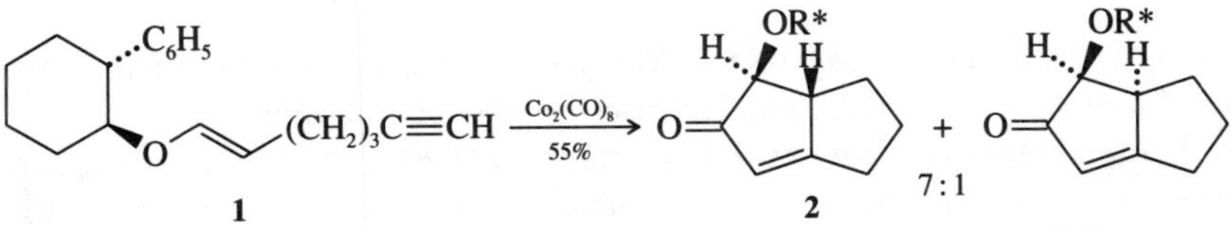

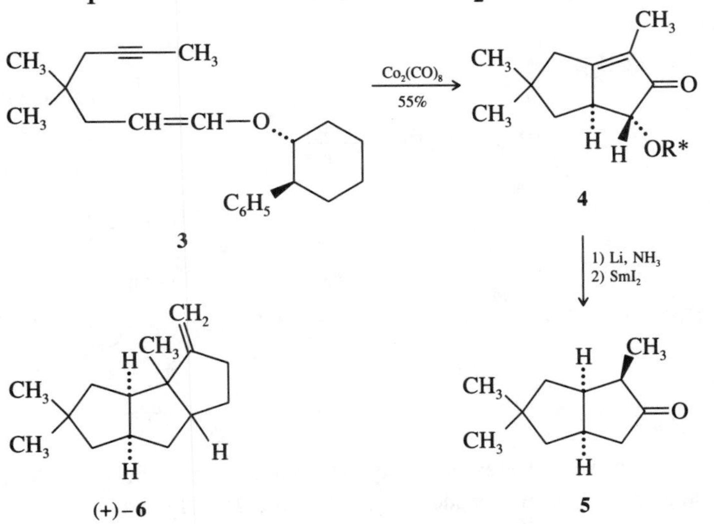

Asymmetric Pauson–Khand bicyclization.[1] Efficient asymmetric Pauson–Khand bicyclizations of enynes are possible using (1S,2R)-(+)-phenylcyclohexanol (**13**, 244; **14**, 128–129) as the chiral auxiliary. Thus the (E)-enol ether (**1**) derived from this alcohol when treated with $Co_2(CO)_8$ at 95° cyclizes mainly to the bicyclic [3.3.0]octenone **2** with high diastereoselectivity (7:1). This asymmetric synthesis has been used to obtain an intermediate (**5**) to (+)-hirsutene (**6**). Note that the chiral auxiliary is reductively cleaved by SmI_2 with 91% recovery (**13**, 270; **14**, 221).

[**2+2+1]Cycloaddition** (**14**, 118).   The $Co_2(CO)_6$-complexed allyl propargyl ethers (**1**) undergo [2+2+1]cycloaddition when heated at 45-60° on an adsorbent

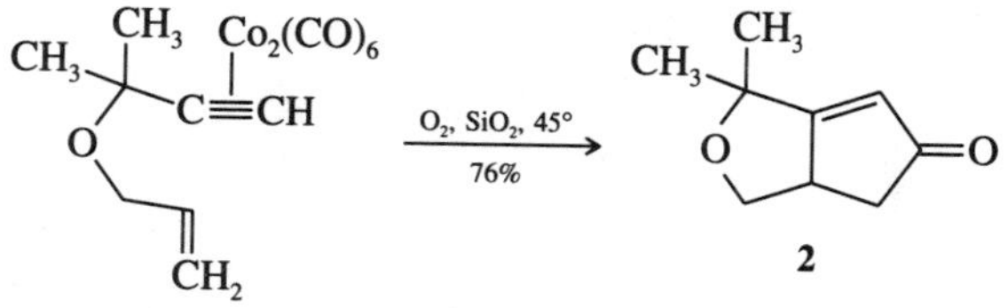

1

2

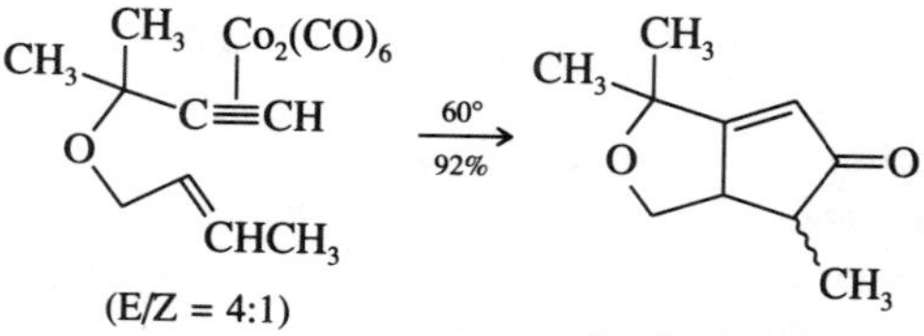

(E/Z = 4:1)

(*exo/endo* = 4:1)

such as SiO_2, Al_2O_3, or $MgO \cdot SiO_2$. The yield varies with the adsorbent, which should contain 10-15% H_2O, and decreases on addition of a solvent.[2]

Diastereoselective aldol coupling of alkynyl aldehydes.[3]   The $Co_2(CO)_6$ complexes (**2**) of alkynyl aldehydes react with silyl enol ethers to form aldols with

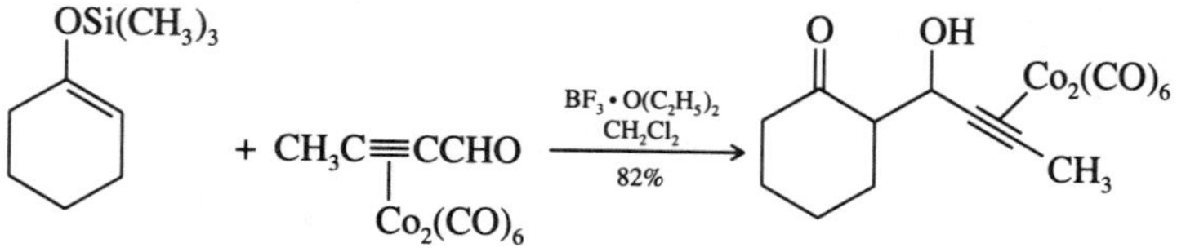

2

syn/anti = 1.7:1

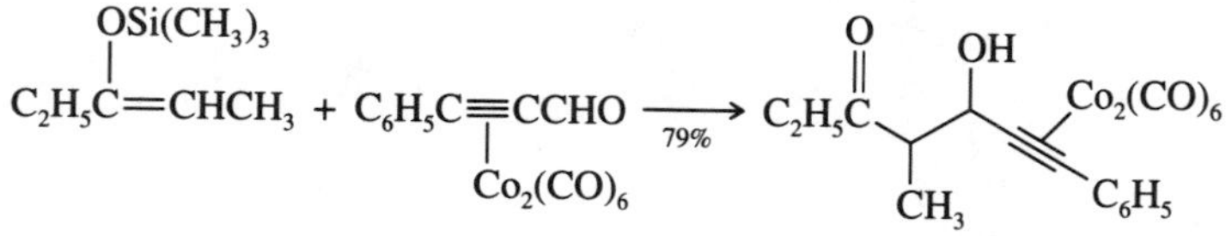

syn/anti = 32:1

moderate to high *syn*-diastereoselectivity, depending on the structure of the enol and on the temperature, but independent of the Lewis acid and the alkyne substituent.

Oxymethylation (**12**, 166).   Cyclic ethers are cleaved by carbon monoxide and a silane when catalyzed by $Co_2(CO)_8$.[4] The reactivity decreases in the order 4 > 3 > 5 > > 6 or 7-membered rings. Both electronic and steric factors affect the regioselectivity. The actual reagent is probably $R_3SiCo(CO)_4$.

Benzyl acetates react with trimethylsilane and CO in the presence of $Co_2(CO)_8$ as catalyst to give β-phenethyl alcohols by a one-carbon homologation. The active catalyst is assumed to be $(CH_3)_3SiCo(CO)_4$. The reaction proceeds under CO at atmospheric pressure at 25°. It fails with benzyl alcohol itself, but is successful with benzyl formate and benzyl methyl ether.[5]

[1] J. Castro, H. Sörenson, A. Riera, C. Morin, A. Moyano, M. A. Pericàs, and A. E. Greene, *Am. Soc.*, **112**, 9388 (1990).

[2] W. A. Smit, S. O. Simonyan, V. A. Tarasov, G. S. Mikaelian, A. S. Gybin, I. I. Ibragimov, R. Caple, D. Froen, and A. Kreager, *Synthesis*, 472 (1989).

[3] J. Ju, B. R. Reddy, M. Khan, and K. M. Nicholas, *J. Org.*, **54**, 5426 (1989).

[4] T. Murai, E. Yasui, S. Kato, Y. Hatayama, S. Suzuki, Y. Yamasaki, N. Sonoda, H. Kurosawa, Y. Kawasaki, and S. Murai, *Am. Soc.*, **111**, 7938 (1989).

[5] N. Chatani, T. Sano, K. Ohe, Y. Kawasaki, and S. Murai, *J. Org.*, **55**, 5923 (1990).

Dichloroacetic acid.

3-Chlorobutenolide annelation.[1]   Reaction of a cycloalkanone with dilithiodichloroacetate (prepared with LDA) in THF/HMPA at −84° results in a product that cyclizes to a β-lactone on treatment with benzenesulfonyl chloride in pyridine. Treatment of the β-lactone with freshly prepared magnesium bromide results in loss of HCl and rearrangement to a fused 3-chlorobutenolide. The overall yield of this annelated product increases with the ring size of the ketone, being 92% in the annelation of cyclopentadecanone.

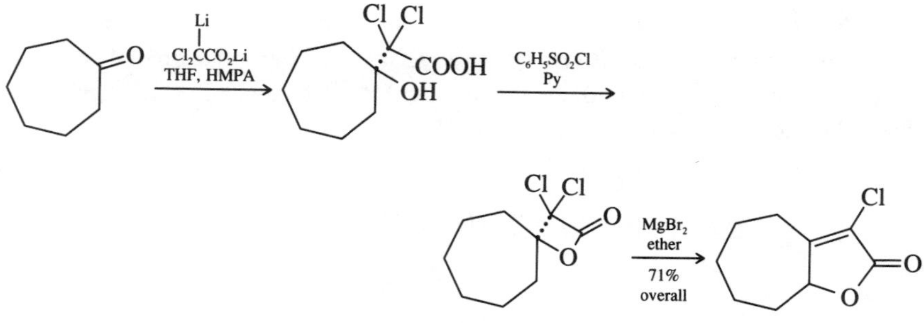

[1] T. H. Black and T. S. McDermott, *J.C.S. Chem. Comm.*, 184 (1991).

Dichlorobis(cyclopentadienyl)titanium, Cp_2TiCl_2.

Hydromagnesiation; butenolides.[1]   Reaction of optically pure γ-trimethylsilyl propargylic alcohols with isobutylmagnesium bromide catalyzed by Cp_2TiCl_2 (**14**,

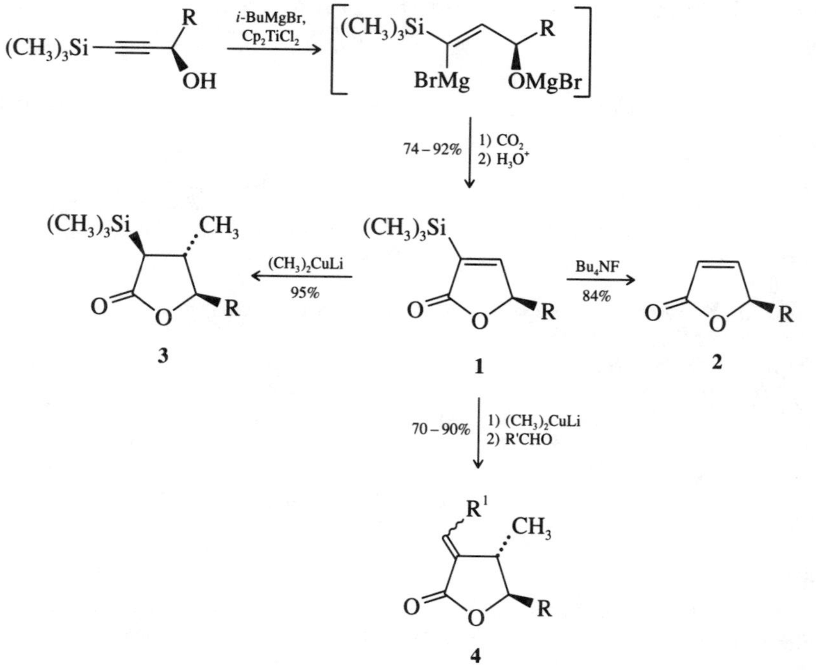

120) and then with CO_2 provides optically pure α-silyl-α,β-unsaturated butenolides (1) in 74–92% yield. The products are useful precursors to various butenolides such as 2–4.

[1] T. Ito, S. Okamoto, and F. Sato, *Tetrahedron Letters*, **31**, 6399 (1990).

Dichlorobis(cyclopentadienyl)titanium–Ethylaluminum dichloride, Cp_2TiCl_2–$C_2H_5AlCl_2$.

Cyclization of alkenes.[1]   This Ziegler–Natta polymerization catalyst can effect intramolecular cyclization of unactivated alkenes. Thus transmetallation of a Grignard reagent (1) with Cp_2TiCl_2 results in a titanium complex (2), which on

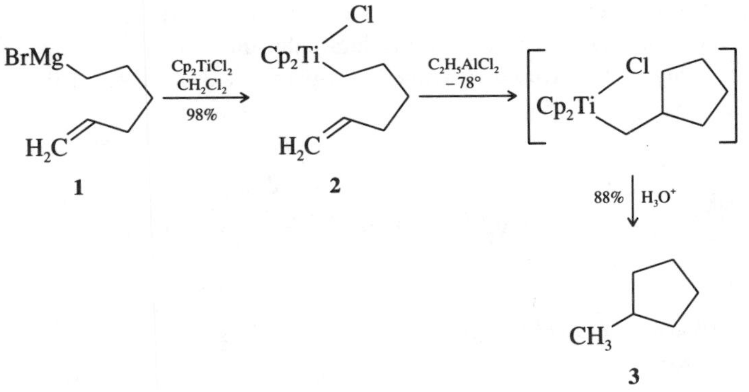

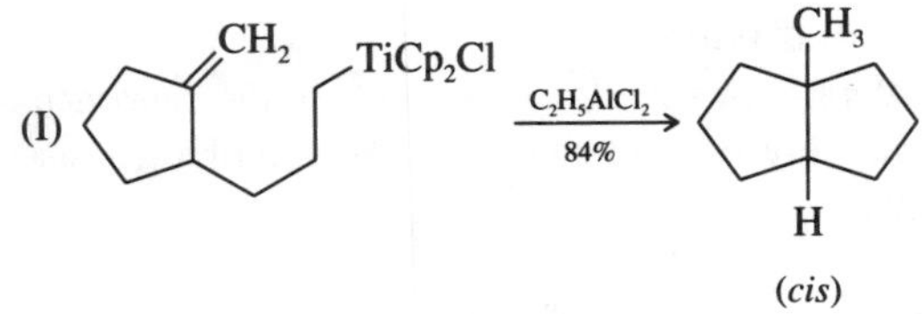

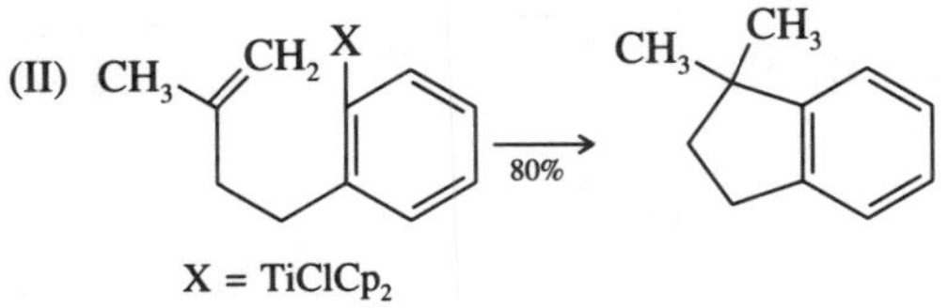

treatment with $C_2H_5AlCl_2$ followed by an acidic workup furnished methylcyclopentane (3) in 88% yield.

This cyclization is also useful for cyclization of various disubstituted alkenes, and can be used to obtain bicyclic systems (equations I and II). The most notable feature is the selective formation of cyclopentane rings, even when a quaternary center is formed. In this respect, Ti-induced cyclization is more useful than free-radical initiation (equation III).

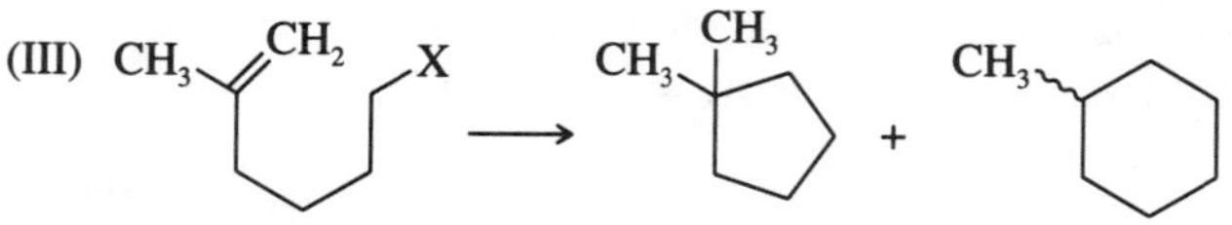

(III)				
4, X = TiClCp$_2$	93%	1%
5, X = Br (Bu$_3$SnH, AIBN)	40%	60%

[1] P. Rigollier, J. R. Young, L. A. Fowley, and J. R. Stille, *Am. Soc.*, **112**, 9441 (1990).

Dichlorobis(triphenylphosphine)nickel(II).

Coupling of dithioacetals with RMgX; butadienylsilanes.[1] This coupling is possible when catalyzed by Ni(II). Thus 2-(2-phenylethenyl)-1,3-dithiolane (1) couples with (trimethylsilyl)methylmagnesium chloride in THF in the presence of this nickel complex to form (E,E)-trimethyl(4-phenyl-1,3-butadienyl)silane (3) in 91% yield.

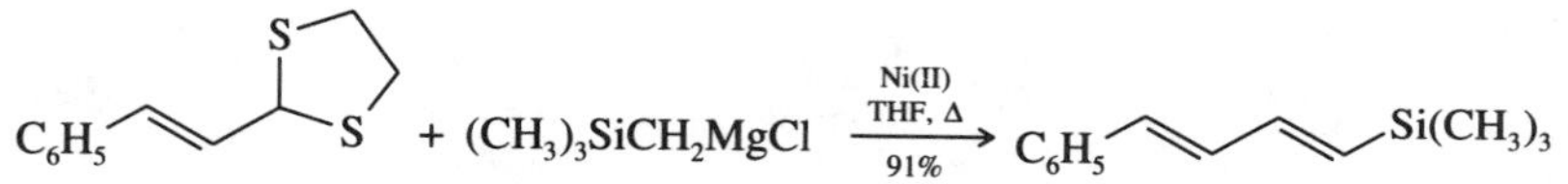

[1] Z.-J. Ni and T.-Y. Luh, *Org. Syn.*, submitted (1989).

Dichlorobis(triphenylphosphine)palladium(II).

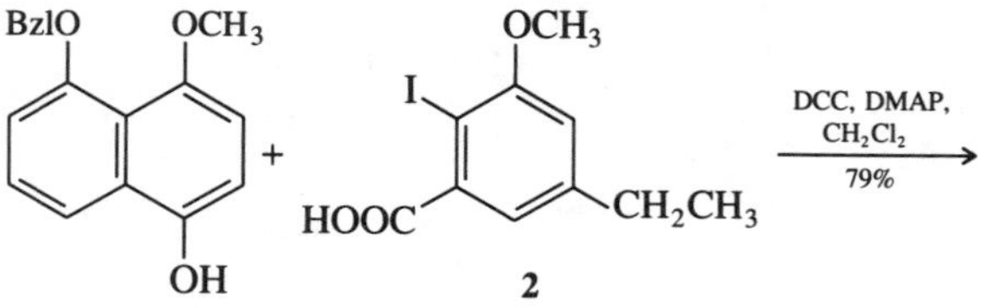

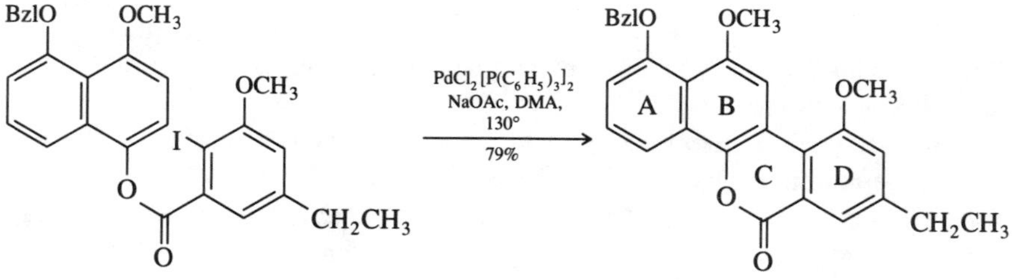

Intramolecular biaryl coupling.[1] A novel, concise route to the benzonaphtho-pyranone ring system (4) of the aglycones of the gilvocarcin antibiotics uses an intramolecular biaryl coupling to join ring B to ring D. Thus esterification of the naphthol **1** with the α-iodobenzoic acid **2** provides the precursor **3**. This product cyclizes to **4** when heated in N,N-dimethylacetamide at 130° in the presence of sodium acetate and $Cl_2Pd[P(C_6H_5)_3]_2$ in 79% yield. A related intramolecular coupling under these conditions has been used for the synthesis of naphthylisoquinolines.[2]

Reductive coupling of ArCOCl.[3] Aroyl halides substituted by an electron-withdrawing group couple to biphenyls when heated with a disilane such as **1** at 160° in the presence of a Pd catalyst.

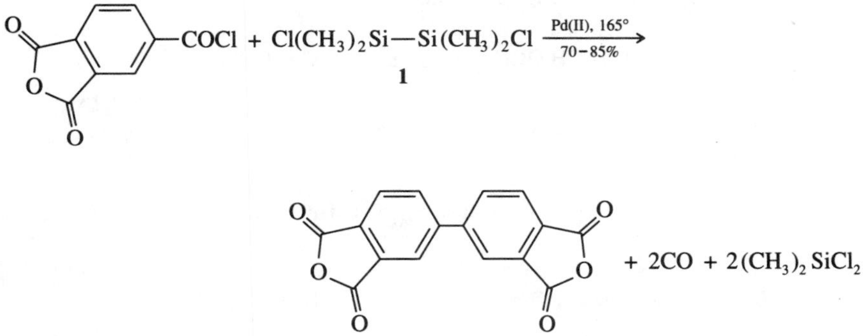

[1] P. P. Deshpande and O. R. Martin, *Tetrahedron Letters*, **31**, 6313 (1990).
[2] G. Bringmann, J. R. Jansen, and H.-P. Rink, *Angew. Chem. Int. Ed.*, **25**, 913 (1986).
[3] T. E. Krafft, J. D. Rich, and P. J. McDermott, *J. Org.*, **55**, 5430 (1990).

(−)-α,α-Dichlorocamphorsulfonyloxaziridine (1).

Preparation from (−)-camphorbenzenesulfonimine:

m.p. 93°, α_D–36°

m.p. 142°, α_D–29°

(–)-**1**, m.p. 122°
α_D–150°

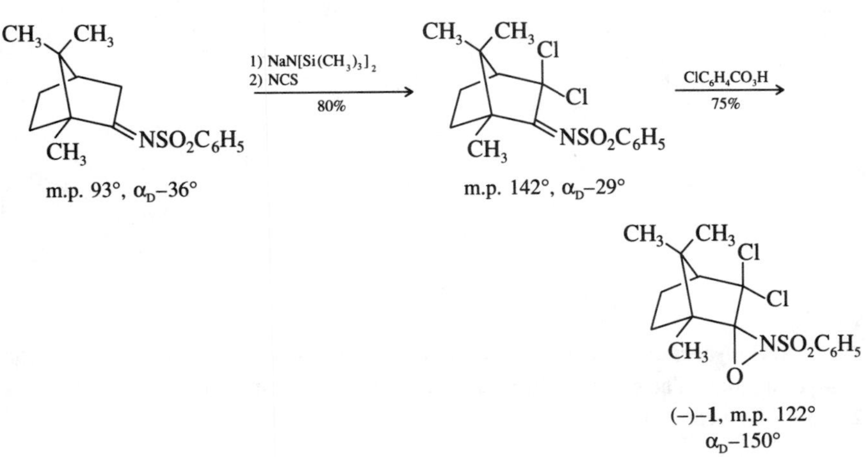

Chiral sulfoxides or selenoxides.[1]   This oxaziridine (1) is generally more effective than the modified Sharpless reagent of Kagan (13, 52) for enantioselective oxidation of alkyl aryl sulfides or selenides to the corresponding sulfoxides or selenoxides. The polar Cl groups of 1 improve both rate and the enantioselectivity.

[1] F. A. Davis, R. ThimmaReddy, and M. C. Weismiller, *Am. Soc.*, 111, 5964 (1989).

Dichlorodicyano-*o*-benzoquinone (DDQ).

Oxidation of allyl (or benzyl) methyl ethers.[1]   These ethers are oxidized by DDQ in refluxing toluene to carbonyl compounds.

$$C_6H_5CH_2OCH_3 \xrightarrow[75\%]{DDQ,\ C_6H_5CH_3,\ \Delta} C_6H_5CHO$$

$$C_6H_5CH{=}CHCH_2OCH_3 \xrightarrow[85\%]{} C_6H_5CH{=}CHCHO$$

$$\xrightarrow[50\%]{} C_6H_5OCH_3$$

[1] E. Lee-Ruff and F. J. Ablenas, *Can. J. Chem.*, 67, 699 (1989).

Dichloro(dicyclopentadienyl)hafnium–Silver perchlorate, Cp_2HfCl_2–$AgClO_4$.

Glycosidation of glycosyl fluoride.[1]   This reaction can be effected with a 1:1 ratio of these two reagents, but a 1:2 ratio is now favored. They should be premixed at 25° to ensure generation of the effective complex *in situ*, which is considered to be $Cp_2Hf(ClO_4)_2$. This glycoside synthesis does not require neighboring-group assistance and is β-selective (1.5–13:1).

(β/α = 13:1)

C-Aryl glycosides.[2]   This combination (1:2) also activates glycosyl fluoride for C-glycosidation. The glycosidation was used to obtain the anthracene β-C-olivosides 2 and 3 of vineomycin B_2.

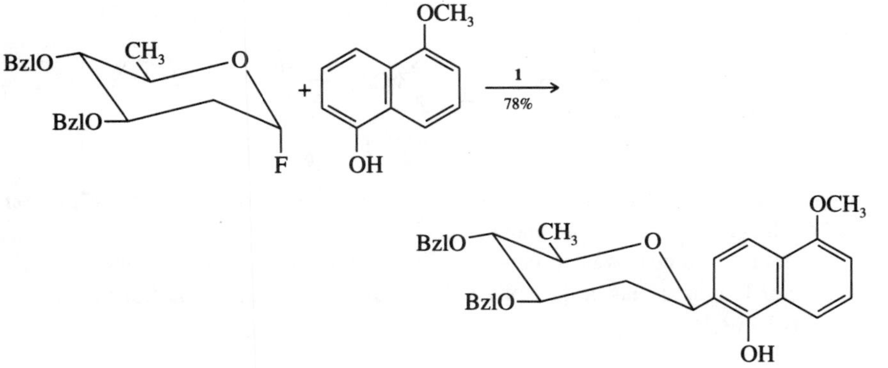

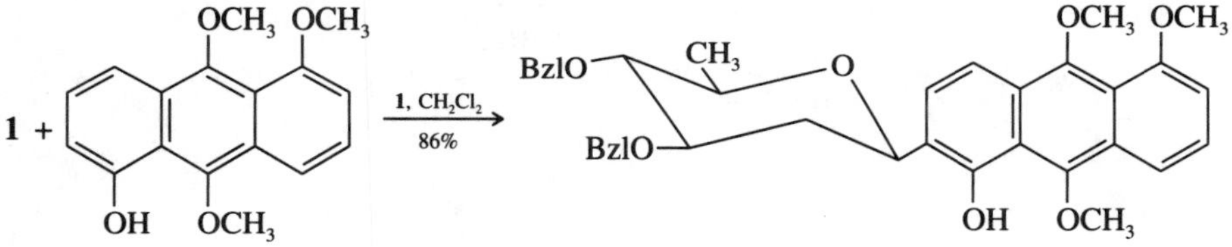

O-Glycosidation of phenols.[3]   O-Aryl glycosides can be obtained in high yield by reaction of glycosyl fluorides with phenols in the presence of 4-Å MS and this activator system. This hafnium complex is superior to Cp_2ZrCl_2, which is more useful for glycosidation of alcohols.

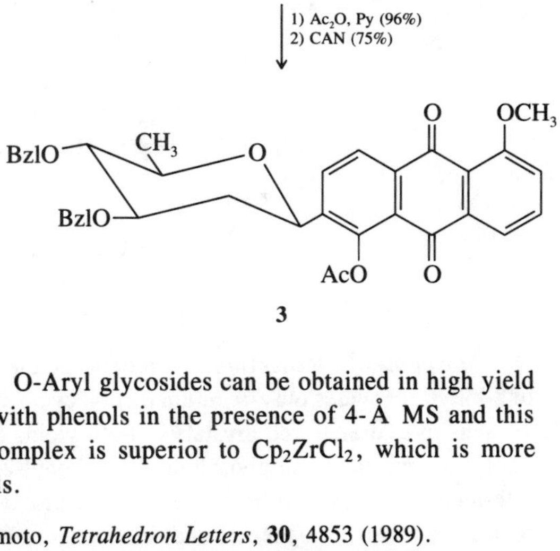

[1] K. Suzuki, H. Maeta, and T. Matsumoto, *Tetrahedron Letters*, **30**, 4853 (1989).
[2] T. Matsumoto, T. M. Katsuki, H. Jona, and K. Suzuki, *ibid.*, **30**, 6185 (1989).
[3] T. Matsumoto, M. Katsuki, and K. Suzuki, *Chem. Letters*, 437 (1989).

Dichlorodi(cyclopentadienyl)zirconium–Butyllithium, 14, 122–123.

"ZrCp₂."   Treatment of dichlorozirconocene with $HgCl_2$ (1 equiv.) and Mg (10 equiv.) produces a new reagent designated for convenience as "ZrCp₂."[1] This re-

Si(CH$_3$)$_3$ Si(CH$_3$)$_3$ Si(CH$_3$)$_3$

"ZrCp$_2$"   →   ZrCp$_2$ $\xrightarrow[\text{overall}]{\text{CO} \atop 55-65\%}$ =O

CH$_2$

agent promotes bicyclization of enynes and diynes to zirconabicycles, which on carbonylation furnish bicyclic enones.

A more convenient way to generate "ZrCp$_2$" involves reaction with an alkyllithium or Grignard reagent (2 equiv.). The reagent generated from BuLi is formulated as **1a** or **1b**.

C$_2$H$_5$ H
      ZrCp$_2$ ⟷   C$_2$H$_5$ H
H   H   ‖—ZrCp$_2$
                           CH$_2$

1a   **1b**

This reagent converts diynes separated by two- to five-carbon chains into (E,E)-exocyclic dienes containing four- to seven-membered rings in 40–89% isolated yields.

    —C≡CCH$_3$ CH$_3$
                                       CH
(CH$_2$)$_3$ $\xrightarrow[67\%]{1}$
                                       C
    —C≡CBu   Bu

Phospholes.[2] Reduction of zirconocene dichloride (**1**) by BuLi in the presence of 2-butyne provides the zirconium metallacycle **2**, which can be isolated if desired as air-sensitive, orange-red crystals in 85% yield. Reaction of **2** with dichlorophenylphosphine provides the phosphole **3** (1-phenyl-2,3,4,5-tetramethylphosphole). The intermediate **2** can also be converted into arsoles, stiboles, bismoles, siloles, germoles, stannoles, thiophenes, and selenophenes by the use of appropriate organometallics.

Cyclization of dienes.[3] "Cp$_2$Zr" also promotes cyclization of 1,6-dienes to *trans*-1,2-disubstituted cyclopentanes. Thus 1,6-heptadiene (**1**) on treatment with "Cp$_2$Zr" followed by bromination affords the *trans*-dibromide **2**. In contrast, use of a related reagent, Cp*ZrCl (Cp* = pentamethylcyclopentadienyl), effects cyclization to the isomeric *cis*-dibromide (equation I). Electrophiles other than bromine can be used: O_2, H^+, and $SeCl_2$. The "Cp_2Zr" cyclization fails for 1,5-hexadiene and 1,8-nonadiene, and cyclization of 1,7-octadiene results mainly in *cis*-products. But

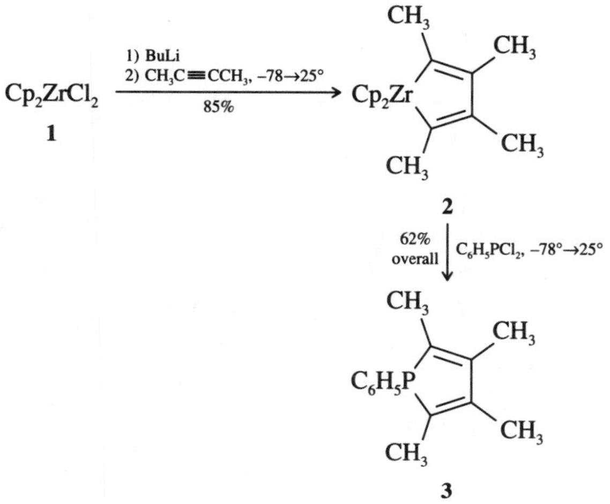

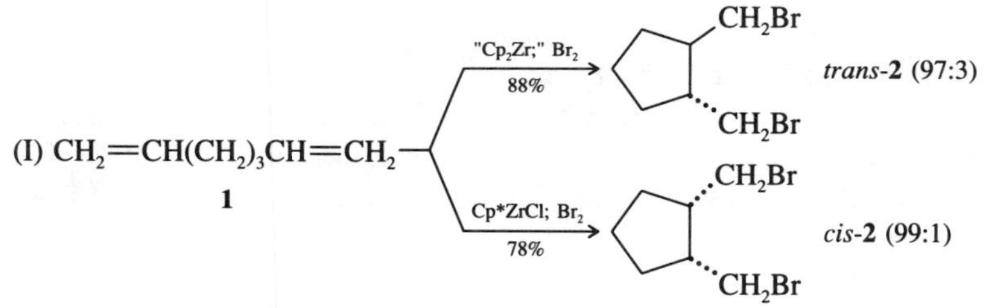

cyclization of substituted 1,6-heptadienes consistently affords *trans*-disubstituted cyclopentanes.

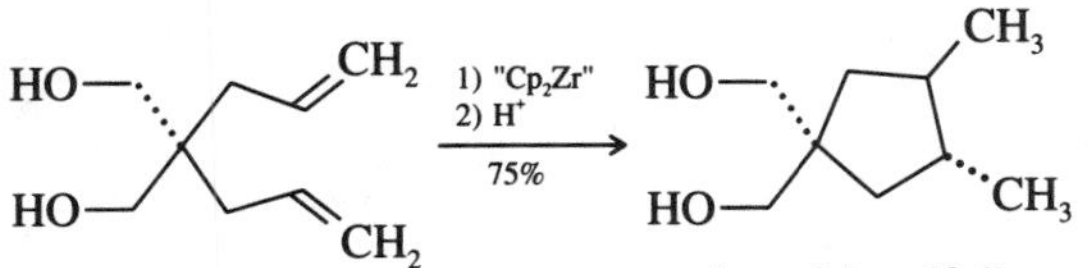

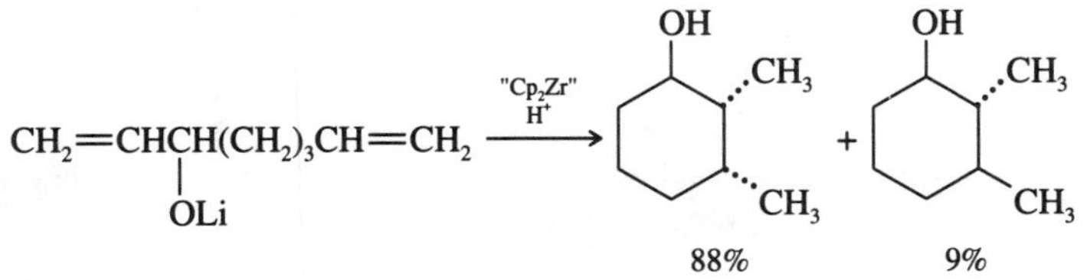

[1] E. Negishi, S. J. Holmes, J. M. Tour, J. A. Miller, F. E. Cederbaum, D. R. Swanson, and T. Takahashi, *Am. Soc.*, **111**, 3336 (1989).
[2] P. J. Fagan and W. A. Nugent, *Org. Syn.*, submitted (1989).
[3] W. A. Nugent and D. F. Taber, *Am. Soc.*, **111**, 6435 (1989).

Dichloro[(1,2-diphenylphosphino)ethane]nickel(II), NiCl₂(dppe).

Eliminative alkylation of dithioacetals.[1]   The reaction of CH_3MgI with allylic dithioacetals in the presence of this Ni-phosphine catalyst (3 mole %) effects *gem*-dimethylation. A substituent at C_2 favors formation of a 1,3-diene.

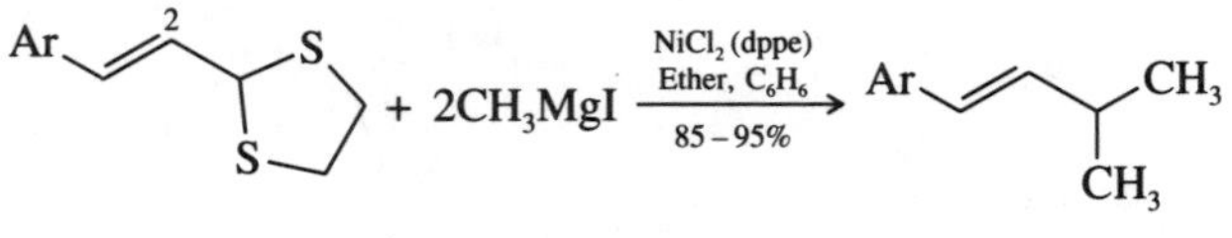

[1] P.-F. Yang, Z.-J. Ni, and T.-Y. Luh, *J. Org.*, **54**, 2261 (1989).

Dichloroketene.

α-Chloro-α,β-enones.[1]   The α,α-dichlorocyclobutanones (**1**), available by reaction of dichloroketene with alkenes, are converted into α-chloroenol acetates (**2**) by reaction with lithium dimethylcuprate or BuLi in acetic anhydride. When heated at about 90° for 24 hours, **2** rearranges to an α-chloroenone (**3**). This transformation can be extended to bicyclic dichlorobutanones, and in this case results in a two-

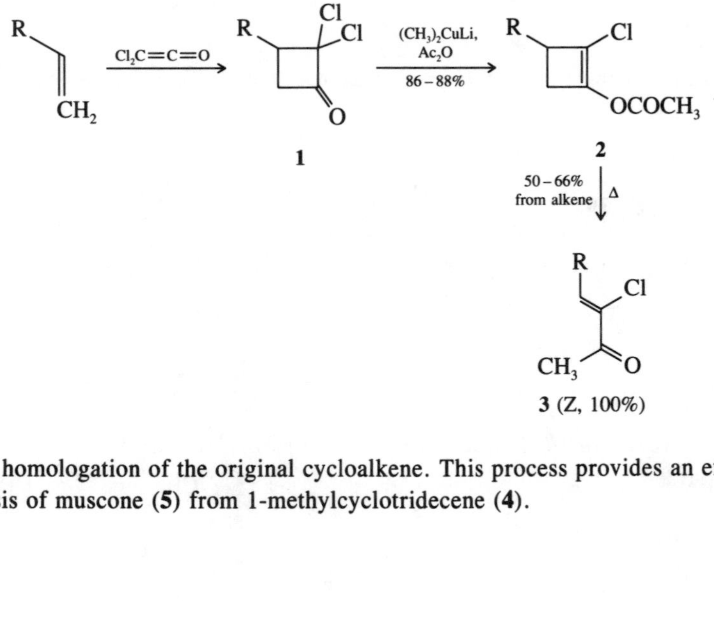

carbon homologation of the original cycloalkene. This process provides an efficient synthesis of muscone (**5**) from 1-methylcyclotridecene (**4**).

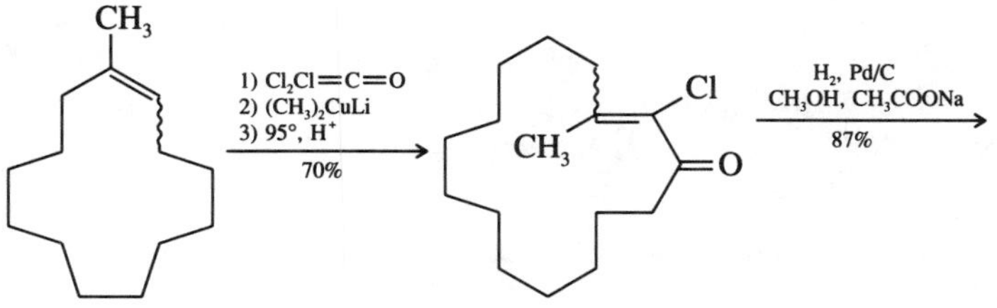

4

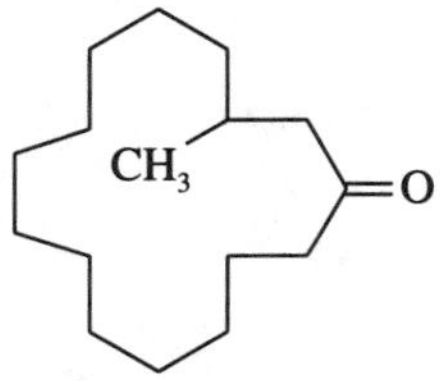

5

[1] J.-P. Després, B. Navarro, and A. E. Greene, *Tetrahedron*, **45**, 2989 (1989).

1,3-Dichloro-1,1,3,3-tetraisopropyldisiloxane, $\{[(CH_3)_2CH]_2Si(Cl)\}_2O$ **(1)**, b. p. 120°/15 mm. Supplier: Aldrich.

Protection of diols. The reaction of diols with **1** and pyridine provides protected derivatives, which are cleaved by aqueous HF or R_3NHF.[1] These disiloxane derivatives are exceedingly useful for preparation of *myo*-inositol phosphates.[2] Thus

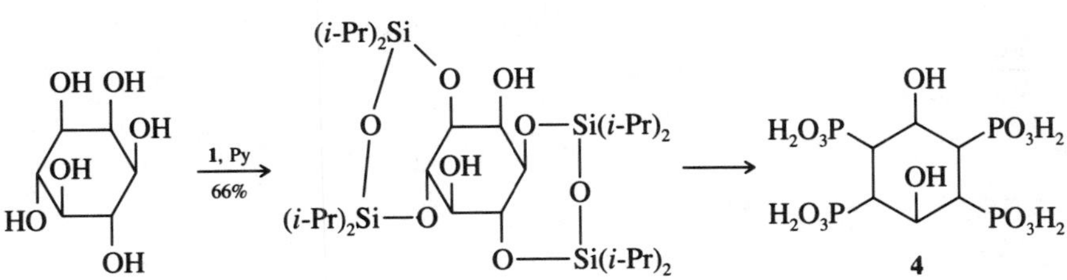

2   **3**   **4**

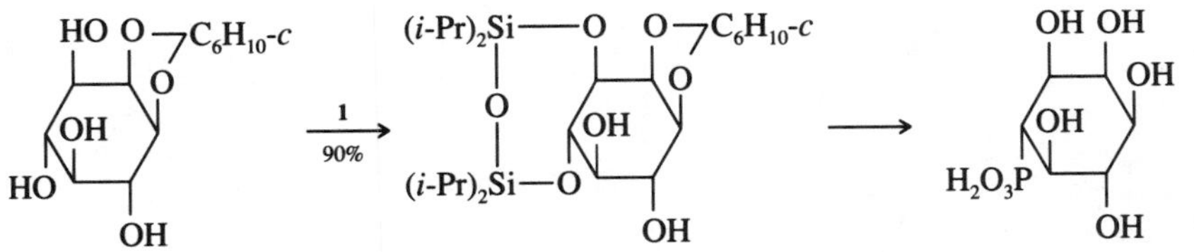

5   **6**   **7**

reaction of **1** with *myo*-inositol (**2**) forms the bisdisiloxane **3**, which has been used for the first synthesis of the *myo*-inositol tetrakisphosphate **4**. The reaction of **1** with 1,2-O-cyclohexylidene-*myo*-inositol (**5**) provides the disiloxane **6**, which was used for a synthesis of *myo*-inositol 4-phosphate (**7**) and *myo*-inositol 1,3,4-trisphosphate.

[1] *Aldrichim. Acta*, **15**, 11 (1982).
[2] Y. Watanabe, M. Mitani, T. Morita, and S. Ozaki, *J.C.S. Chem. Comm.*, 482 (1989).

Dichlorotris(triphenylphosphine)ruthenium(II), $RuCl_2[P(C_6H_5)_3]_3$.

α,β-*Enones*.[1] In the presence of this Ru(II) catalyst, primary alcohols and allylic acetates react to form enones. The reaction is best effected at 150° in the presence of carbon monoxide under pressure.

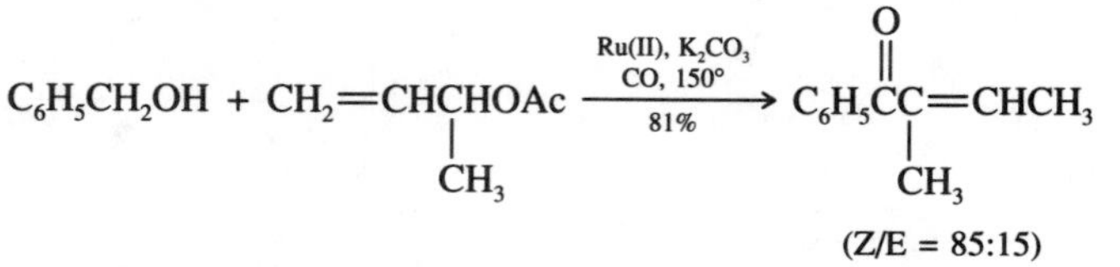

$$(Z/E = 85{:}15)$$

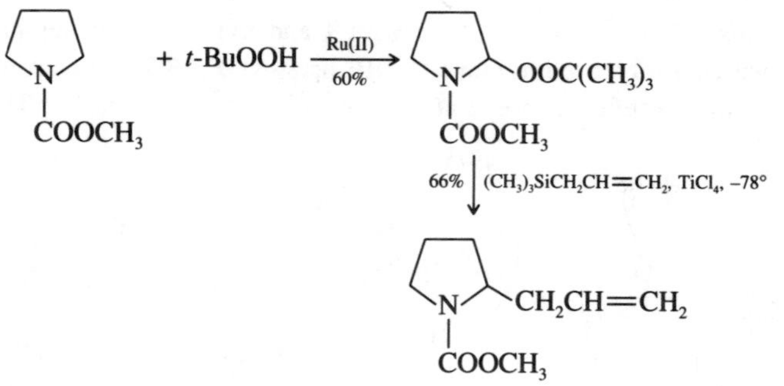

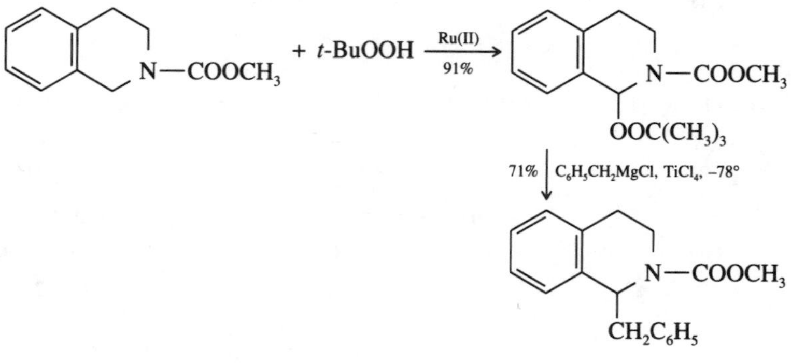

***Oxidation α to nitrogen*.**[2] In the presence of this Ru catalyst, *t*-BuOOH oxidizes amides to the α-(*t*-butyldioxy)amides in high yield, probably via an oxoruthenium(IV). This oxidation can be used to effect alkylation at the α-position of amides by displacement of the *t*-butyldioxy group by a nucleophile induced with $TiCl_4$.

Reaction of β-lactams with peracetic acid catalyzed by $Cl_2Ru[P(C_6H_5)_3]_3$ or Ru-on-carbon results in β-acetoxylation. Thus 2-azetidinones are converted into 4-acetoxy-2-azetidinones. This reaction can be effected with high diastereoselectivity.

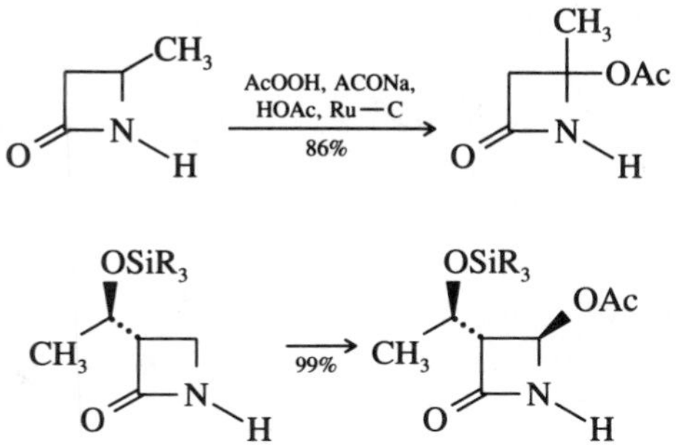

Reactions of 1,4-epiperoxides and Ru(II).[3]   1,4-Epiperoxides (endoperoxides) in the presence of this complex can undergo fragmentation, reduction, and disproportionation, reactions which differ from those induced by thermolysis or reaction with Fe(II) salts. Although several products are usually formed, fragmentation can lead to single products.

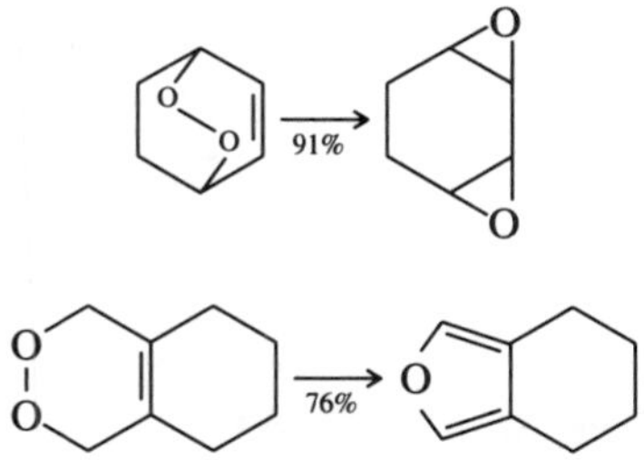

Isomerization of 2-ynols.[4]   This Ru complex in combination with a trialkylphosphine is an effective catalyst for isomerization of 2-ynols to α,β-enals in refluxing toluene.

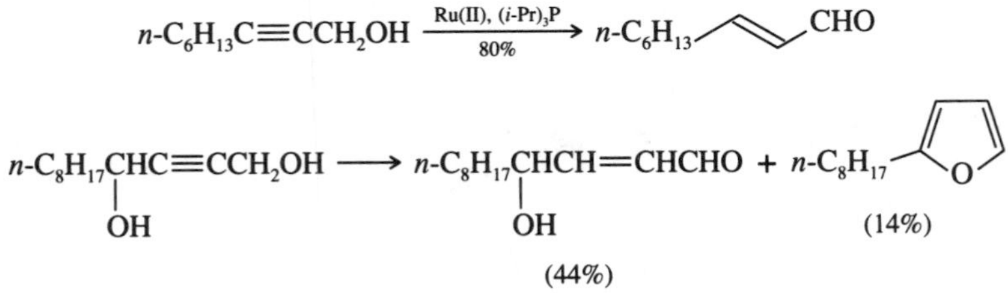

[1] T. Kondo, T. Mukai, and Y. Watanabe, *J. Org.*, **56**, 487 (1991).
[2] S.-I. Murahashi, T. Naota, T. Kuwabara, T. Saito, H. Kumobayashi, and S. Akutagawa, *Am. Soc.*, **112**, 7820 (1990).
[3] M. Suzuki, H. Ohtake, Y. Kameya, N. Hamanaka, and R. Noyori, *J. Org.*, **54**, 5292 (1989).
[4] D. Ma and X. Lu, *J.C.S. Chem. Comm.*, 890 (1989).

1,3-Dicyclohexylcarbodiimide, $C_6H_{11}N=C=NC_6H_{11}$ (DCC, **1**).

Solvent-free esterification. DCC esterifications are generally carried out in refluxing CH_2Cl_2. Yields can be poor in the case of angelate esters, which can isomerize to tiglate esters in slow esterifications. In the case of the khellactone **1** esterification with angelic acid in refluxing CH_2Cl_2 with DCC and 4-pyrrolidinopyri-

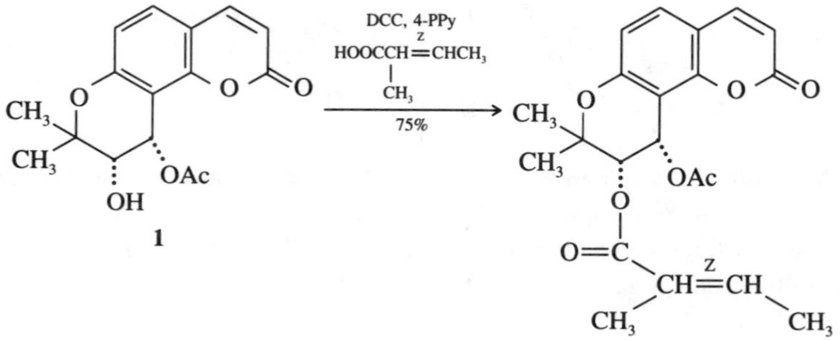

dine (4-PPy) gives a mixture of the angelate ester and the tiglate ester in a 1:11 ratio. However, in the absence of a solvent, either the angelate and tiglate esters can be prepared in high yield.

[1] S. Bal-Tembe, D. N. Bhedi, N. J. de Souza, and R. H. Rupp, *Heterocycles*, **29**, 1239 (1989).

Diethylaminosulfur trifluoride (DAST).

OH → F.[1]   Reaction of neat DAST with the optically active natural inositol quebrachitol (**1**) results in two products, each formed by replacement of one of the two axial hydroxyl groups of **1**. However, on deprotection (BBr_3) each provides (−)-fluoro-*myo*-inositol (**4**).

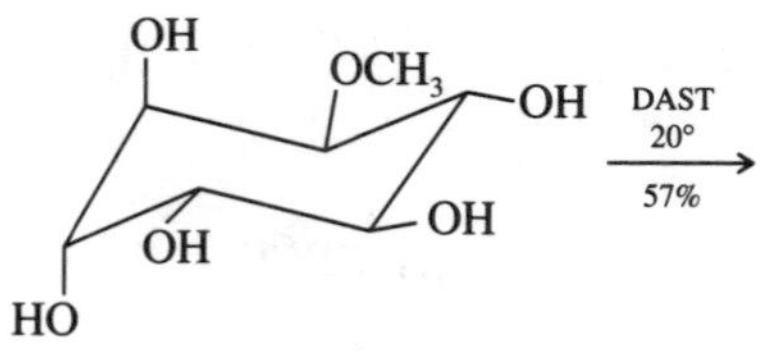

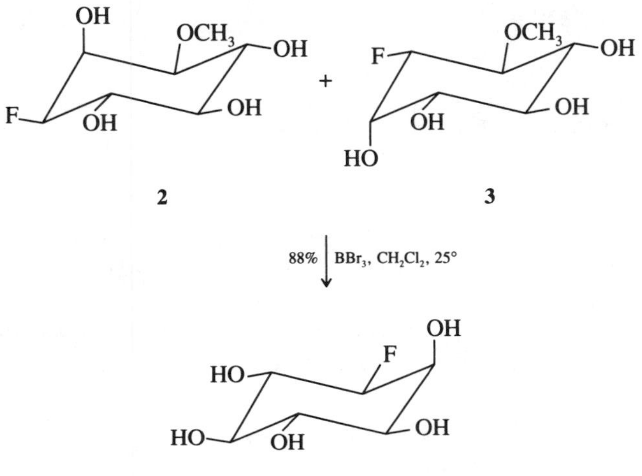

2   3

88% | BBr$_3$, CH$_2$Cl$_2$, 25°

(–)-4

Fluoromethylhomocysteine.[2] A protected derivative (**2**) of methionine can be obtained in 67% yield by reaction of DAST with the protected methionine sulfoxide (**1**). Deprotection of **2** is not possible because of the instability of a fluoroamino group.

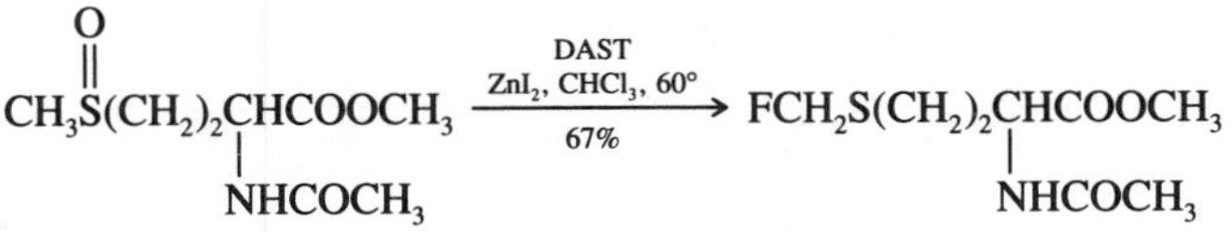

1   2

α,α-*Difluoroalkyl ethers*, RCF$_2$OR′. DAST does not react with esters, but it does convert thioesters into α,α-difluoroalkyl ethers.[3]

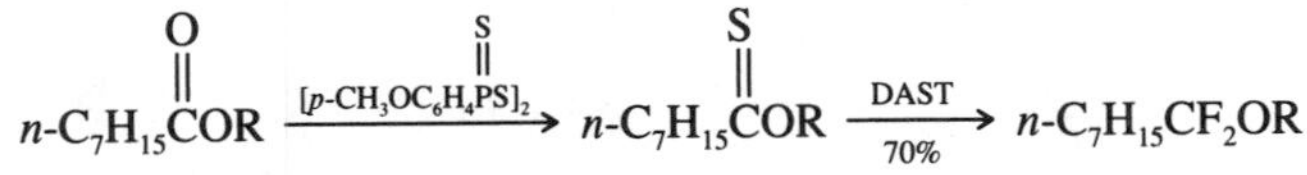

[1] A. P. Kozikowski, A. H. Fauq and J. M. Rusnak, *Tetrahedron Letters*, **30**, 3365 (1989).
[2] M. E. Houston, Jr. and J. F. Honek, *J.C.S. Chem. Comm.*, 761 (1989).
[3] W. H. Bunnelle, B. R. McKinnis, and B. A. Narayanan, *J. Org.*, **55**, 768 (1990).

Diethyl azodicarboxylate.

$RC\!\equiv\!CCH_2OH \rightarrow RCH\!=\!C\!=\!CH_2$.[1] This reaction can be effected by conversion of a 2-alkyne-1-ol (**1**) into an alkynylhydrazine (**2**), which on treatment with an oxidant rearranges, presumably via a diazene, with loss of N_2 to an allene (3). Slow oxidation is observed in air, but azo compounds such as DEAD are superior to air in terms of rapidity and yield.

This reaction can be used to obtain optically active allenes, as shown by conversion of 4 into 5, with complete retention of optical activity.

(R)-4 (76% ee)   (S)-5 (75% ee)

[1] A. G. Myers, N. S. Finney, and E. Y. Kuo, *Tetrahedron Letters*, **30**, 5747 (1989).

Difluoromethyldiphenylphosphine oxide, $(C_6H_5)_2\overset{\overset{\displaystyle O}{\|}}{P}CHF_2$ **(1).**

$$(C_6H_5)_2\overset{\overset{\displaystyle O}{\|}}{P}H \xrightarrow[52\%]{\begin{array}{l}1)\ BuLi\\2)\ CHClF_2\end{array}} \textbf{1, m.p. } 93-94°$$

1,1-*Difluoroalkenes*[1] (*cf.*, **11**, 180). The anion (LDA) of this reagent effects difluoromethylenation of aldehydes or ketones by a Wittig–Horner type reaction.

$$\textbf{1} \xrightarrow[47-81\%]{\begin{array}{l}1)\ LDA\\2)\ R^1COR^2\end{array}} R^1\overset{\overset{\displaystyle CF_2}{\|}}{C}R^2$$

[1] M. L. Edwards, D. M. Stemerick, E. T. Jarvi, D. P. Matthews, and J. R. McCarthy, *Tetrahedron Letters*, **31**, 5571 (1990).

Dihydridotetrakis(triphenylphosphine)ruthenium, 15, 135.

Dehydrogenation of **1,4-** *or* **1,5-*diols to lactones*.** This Ru-catalyzed transfer dehydrogenation can be effected with high selectivity when the 2-position of **1** is

(I) 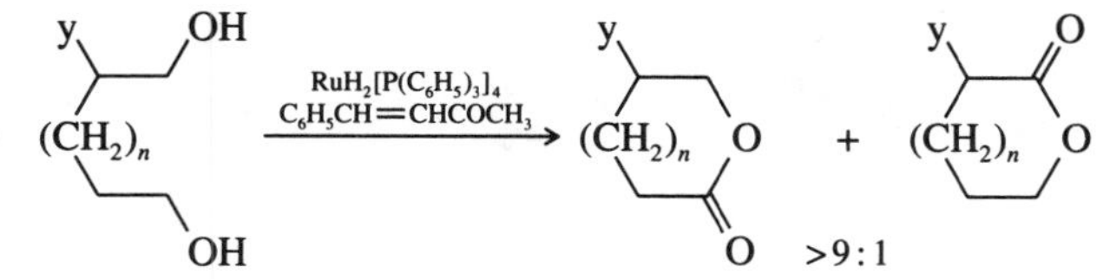 $1(n=0, 1,$
$y=R, RO, C_6H_5)$

substituted by an alkyl, alkoxy, or phenyl group because of steric effects.[1] This dehydrogenation also shows high selectivity in the conversion of 2,4-disubstituted 1,5-diols to δ-valerolactones (equation II). In the case of **2a**, both possible lactones are formed, but **2b** with the more bulky substituent is converted only into **3**.[2]

(II) 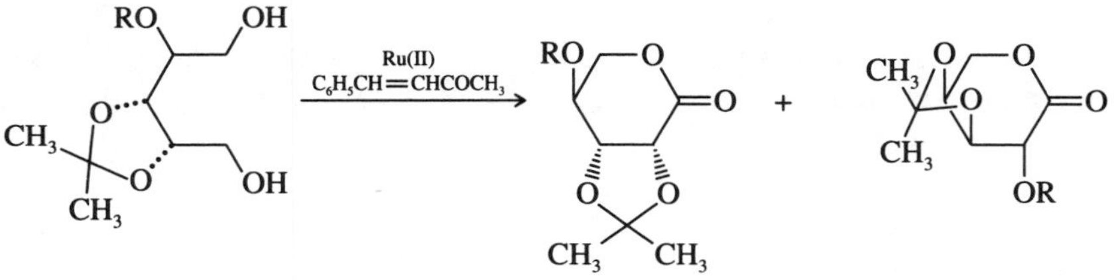 2a, R = Bzl quant. **3** 63:37 **4**
b, R = Si(CH$_3$)$_2$-t-Bu 84%

Aldol and Michael reactions of nitriles.[3] Activated nitriles such as ethyl cyanoacetate react with aldehydes or ketones in the presence of this ruthenium catalyst

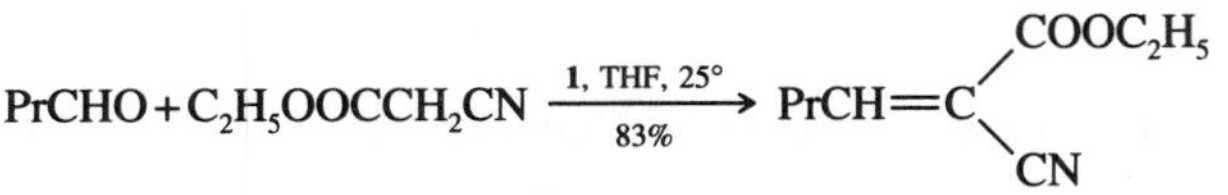

to form α,β-unsaturated nitriles by an aldol-type reaction. Under the same conditions nitriles react with α,β-enones or α,β-enals to afford only Michael adducts.

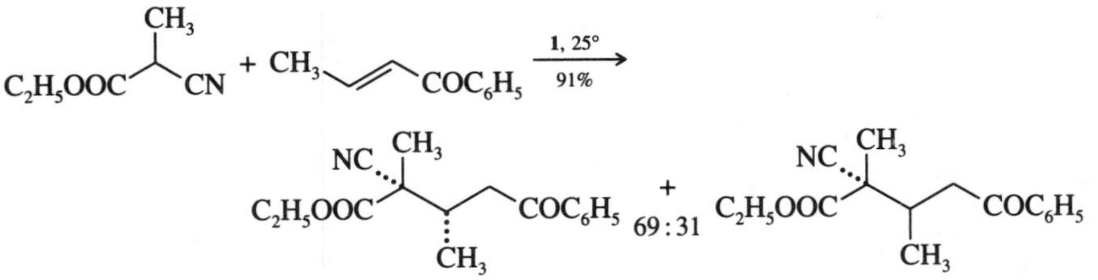

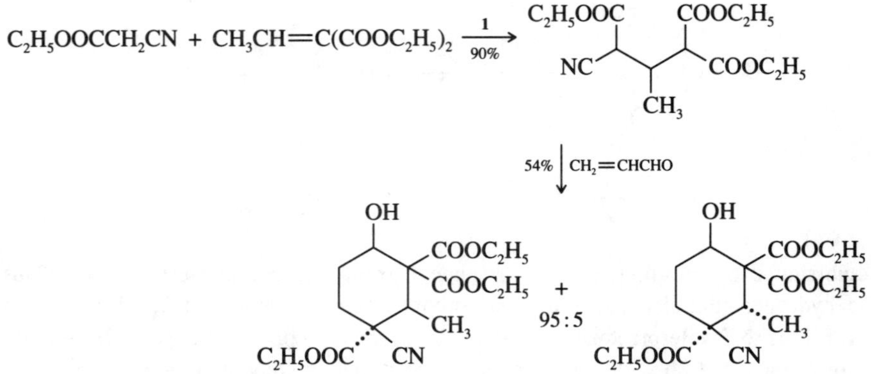
C2H5OOCCH2CN + CH3CH=C(COOC2H5)2
1
90%
C2H5OOC
COOC2H5
NC
COOC2H5
CH3
54%
CH2=CHCHO
OH
COOC2H5
COOC2H5
CH3
C2H5OOC
CN
+
95:5
OH
COOC2H5
COOC2H5
CH3
C2H5OOC
CN

Michael addition followed by an aldol cyclization can be used for stereoselective synthesis of cyclohexanes.

Isomerization of enynones to trienones.[4] Cyclic enones conjugated with a triple bond such as **1** isomerize to *trans*, *trans*-trienones **2** in the presence of this ruthenium catalyst.

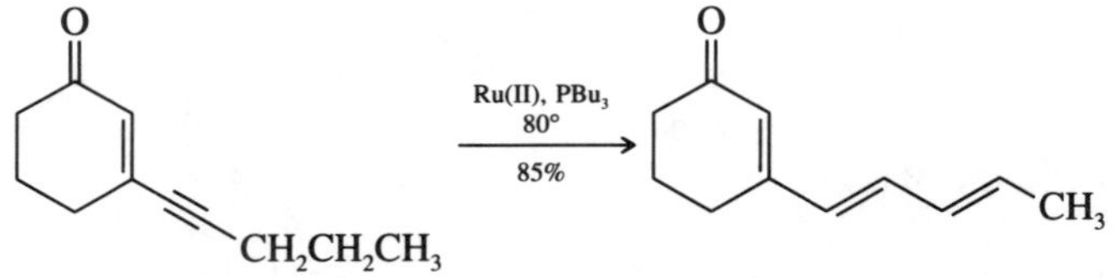
O
CH2CH2CH3
Ru(II), PBu3
80°
85%
O
CH3

1 **2**

[1] Y. Ishii, K. Osakada, T. Ikariya, M. Saburi, and S. Yoshikawa, *J. Org.*, **51**, 2034 (1986).
[2] M. Saburi, Y. Ishii, N. Kaji, T. Aoi, I. Sasaki, S. Yoshikawa, and Y. Uchida, *Chem. Letters*, 563 (1989).
[3] T. Naota, H. Taki, M. Mizuno, and S.-I. Murahashi, *Am. Soc.*, **111**, 5954 (1989).
[4] X. Lu, C. Guo, and D. Ma, *Synlett*, 357 (1990).

2,2'-Dihydroxy-1,1'-binaphthyl (BINOL).

Diastereoselective α-alkylation of arylacetic acids.[1] The binaphthyl ester (**1**) of phenylacetic acid undergoes highly diastereoselective methylation on treatment of

(I)

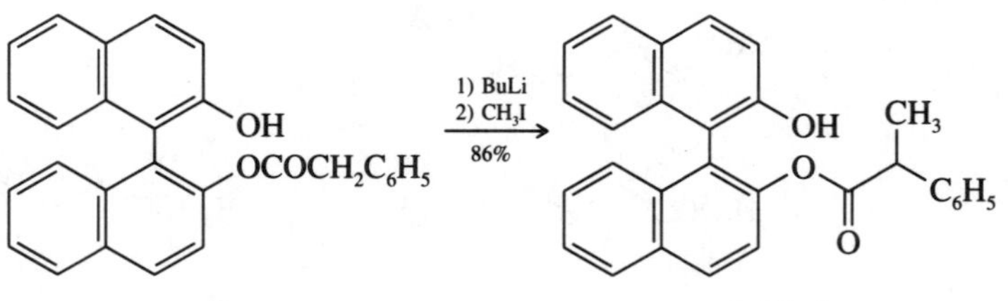
OH
OCOCH2C6H5
1) BuLi
2) CH3I
86%
OH
CH3
O
C6H5
O

1 **2** (96:4)

the enolate (BuLi) with CH_3I (equation I). This alkylation can be used for preparation of (S)-naproxen (**4**) by methylation of the (S)-binaphthyl ester (**3**) followed by hydrolysis (LiOH).

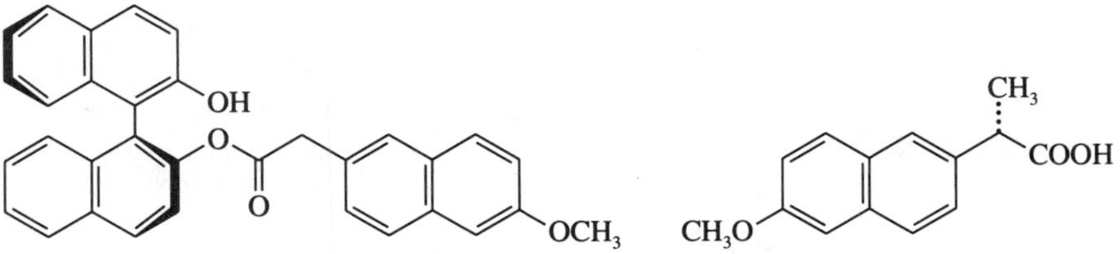

3

4 (82% ee)

[1] K. Fuji, M. Node, and F. Tanaka, *Tetrahedron Letters*, **31**, 6553 (1990).

(R)- and (S)-2,2'-Dihydroxy-1,1'-binaphthyl–Lithium aluminum hydride. (Noyori's reagent, BINAL-H), **9**, 169–170; **10**, 148–149; **12**, 190–191).

Enantioselective reduction of ketones.   Marshall *et al.*[1] report that more consistent results obtain when BINAL-H is prepared by refluxing a mixture of the binaphthol, $LiAlH_4$, and ethanol in THF for a short time before use. They also note that the expensive binaphthol can be recovered and reused.

The same group used (R)-(+)-BINAL-H for reduction of acylstannanes (**2**) to (S)-(α-alkoxyallyl)stannanes (**3**) in >95% ee. The epimeric (R)-stannanes (**3**) can be

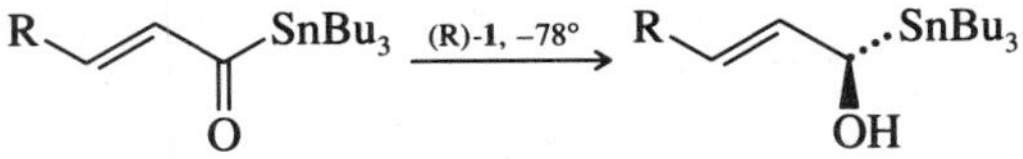

2

(S)-**3**, >95% ee

obtained in somewhat lower enantioselectivity by use of the commercially available (2S,3R)-(+)-4-dimethylamino-1,2-diphenyl-3-methyl-2-butanol (Chirald®, **5**, 231; **8**, 184) in combination with $LiAlH_4$.

[1] J. A. Marshall, G. S. Welmaker, and B. W. Gung, *Am. Soc.*, **113**, 647 (1991).

2,2'-Dihydroxy-3,3'-bis(triphenylsilyl)1,1'-binaphthyl–Trimethylaluminum (1), **14**, 46–47; **15**, 136–137.

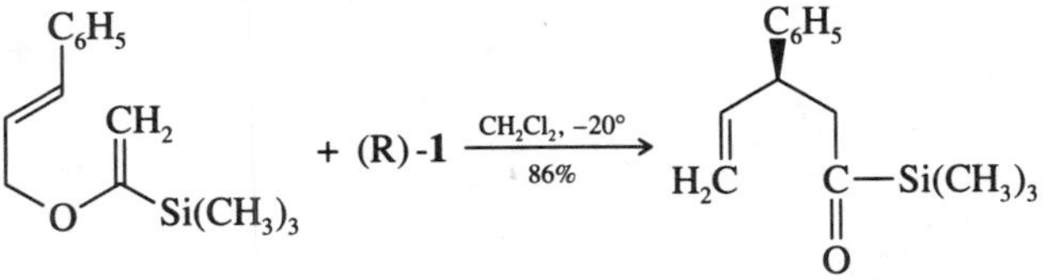

(80% ee)

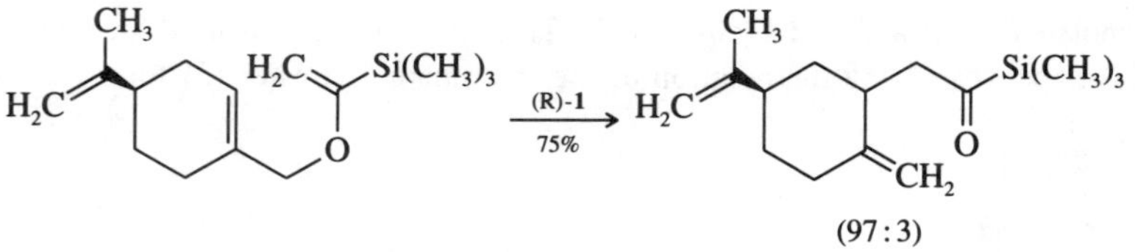

(97:3)

Claisen rearrangement.[1] Claisen rearrangement of simple allyl vinyl ethers effected with (R)- and (S)-1 shows low enantioselectivity, but the rearrangement of allyl vinylsilyl ethers results in acylsilanes in high optical yield (80–90%).

[1] K. Maruoka, H. Banno, and H. Yamamoto, *Am. Soc.*, **112**, 7791 (1990).

Diiodosilane, SiH_2I_2.

Acyl iodides.[1] This reagent converts acyl chlorides into acyl iodides at 25°. In combination with iodine (1:1) it also converts carboxylic acids, esters, and anhydrides into acyl iodides in generally high yield. This reaction in combination with an alcohol is a useful method for transesterification of hindered alcohols.

[1] E. Keinan and M. Sahai, *J. Org.*, **55**, 3922 (1990).

Diisobutylaluminum hydride.

Diastereoselective reduction of chiral β-keto sulfoxides[1] (**13**, 115–116). This reaction, which can be controlled to provide either diastereomer of a chiral β-hydroxy sulfoxide, has been used to obtain (R)- or (S)-4-hydroxy-2-cyclohexenones from the monoketal (**2**) of 1,4-cyclohexanedione. Sulfinylation of **2** with (S)-(−)-

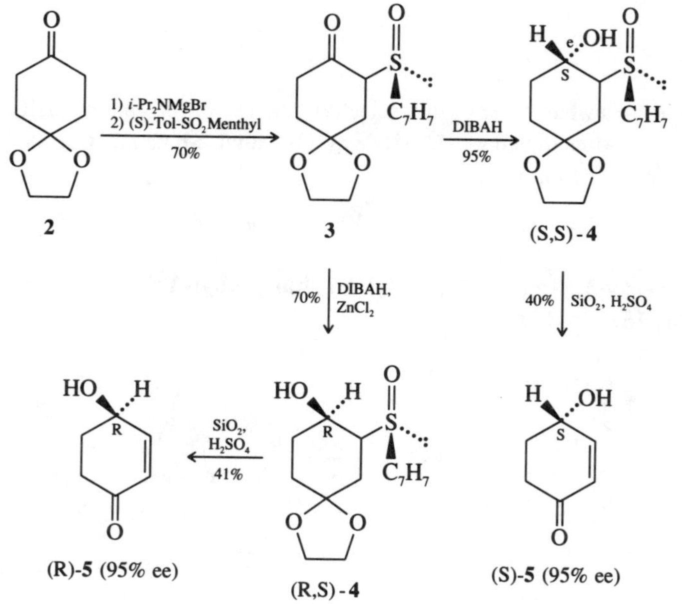

menthyl *p*-toluenesulfinate provides the chiral β-keto sulfoxide **3**. Reduction of **3** with DIBAH provides the equatorial β-hydroxy sulfoxide (S,S)-**4**, whereas reduction with DIBAH and $ZnCl_2$ provides mainly the axial β-hydroxy sulfoxide (R,S-**4**). Hydrolysis of the ketal group and sulfoxide elimination is affected in one operation with silica gel and H_2SO_4 to provide (R)- and (S)-**5**, both in high optical yield. The yield in the last step can be improved by prior acetylation or benzylation.

[1] M. Carmen Carreño, J. L. García-Ruano, M. Garrido, M. Pilar Ruiz, and G. Solladié, *Tetrahedron Letters*, **31**, 6653 (1990).

Diisopinocampheylborane, Ipc_2BH.

syn-*Aldols*. 1,4-Hydroboration of an (E)-α,β-enone (**1**) with Ipc_2BH provides a (Z)-vinyloxyborane (**2**), which reacts with an aldehyde to provide a *syn*-aldol (**3**) in high enantiomeric excess.

3 (90% ee)

[1] G. P. Boldrini, F. Mancini, E. Tagliavini, C. Trombini, and A. Umani-Ronchi, *J.C.S. Chem. Comm.*, 1680 (1990).

Diisopropyl Tartarate (E)-γ-[Cyclohexyloxy)dimethylsilyl]allylboronate,

(R,R)-**1**

Preparation.[1]

Optically active anti-1,2-*diols*.[2] *syn*-1,2-Diols are available by addition of various γ-alkoxyallylmetal reagents to aldehydes. In contrast, this reagent reacts with aldehydes to form *anti*-1,2-diol monoethers, which can be oxidized to the diol. In contrast, the reaction of **1** with some chiral aldehydes can show high enantioselectiv-

(69% ee)

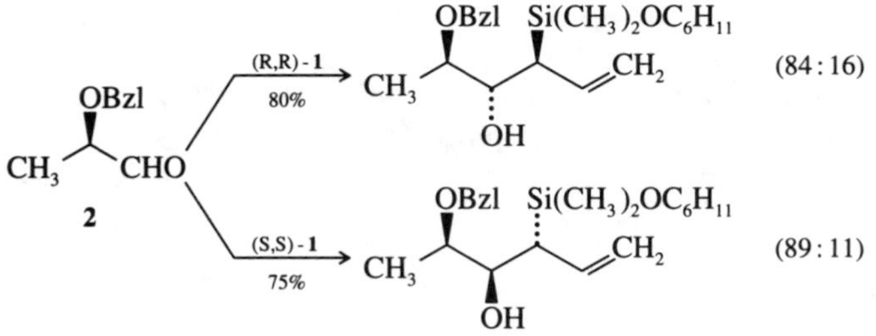

OBzl   Si(CH$_3$)$_2$OC$_6$H$_{11}$

(R,R)-1
80%   CH$_3$ CH$_2$ (84:16)

OBzl   OH

CH$_3$ CHO

2

OBzl   Si(CH$_3$)$_2$OC$_6$H$_{11}$

(S,S)-1
75%   CH$_3$ CH$_2$ (89:11)

OH

ity, owing to double or matched asymmetric reactions. Thus reactions of the aldehyde 2 with (R,R)-1 provides the major product as an 84:16 mixture, but the reaction with (S,S)-1 results in an 89:11 mixture (matched pair).

[1] W. R. Roush, P. T. Grover, and X. Lin, *Tetrahedron Letters*, **31**, 7563 (1990).
[2] W. R. Roush, K. Ando, D. B. Powers, R. L. Halterman, and A. D. Palkowitz, *Am. Soc.*, **112**, 6339 (1990).

1,2;5,6-Di-O-isopropylidene-α-D-glucofuranose (Diacetone-D-glucose, 1).

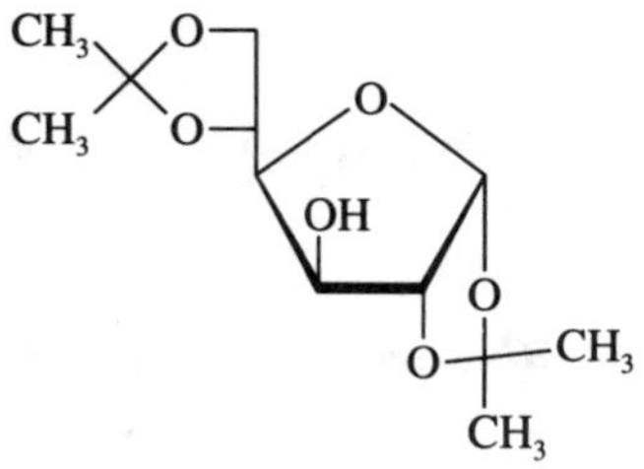

CH$_3$ O

CH$_3$ O   O

OH

O

O   CH$_3$

CH$_3$

1 (R*OH), α$_D$-18.5°, m.p. 105–108°

Chiral Ti-carbohydrate complex. Reaction of CpTiCl$_3$ with 1 and N(C$_2$H$_5$)$_3$ provides a yellow, crystalline solid (2) that can be stored in the absence of air and which is shown by X-ray analysis to consist of CpTiCl(OR*)$_2$.

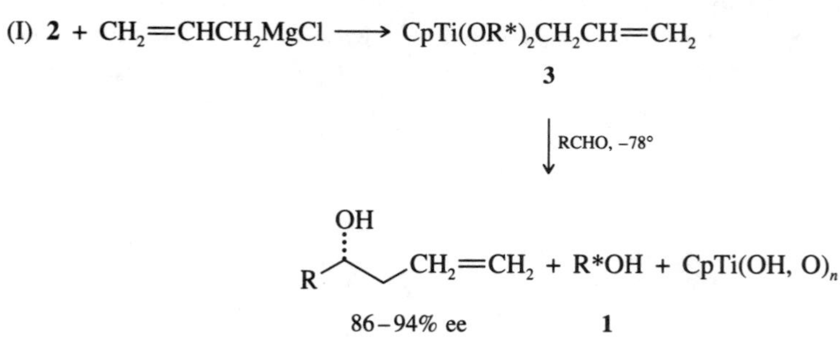

(I) 2 + CH$_2$=CHCH$_2$MgCl ⟶ CpTi(OR*)$_2$CH$_2$CH=CH$_2$

3

RCHO, −78°

OH

R   CH$_2$=CH$_2$ + R*OH + CpTi(OH, O)$_n$

86–94% ee   1

Enantioselective allylation.[1]   The reaction of **2** with allylmagnesium chloride in ether affords a chiral orange allyltitanium complex **3**. This complex reacts with an aldehyde at $-78°$ to afford homoallyl alcohols in 55–88% yield and in 86–94% ee and with release of **1** and a titanate that can be reconverted to $CpTiCl_3$ (equation I). Reaction of **3** with aryl ketones requires a temperature of $0°$, and the enantioselectivity is only about 50%.

Enantioselective aldol reaction; β-hydroxy esters.[2]   The lithium enolate **4** of *t*-butyl acetate reacts with an aldehyde in the presence of **2** to form β-hydroxy esters **6** in 90–96% ee.

$$
\begin{array}{ccc}
\overset{\displaystyle OLi}{\underset{|}{CH_2{=}C}}{-}OC(CH_3)_3 & + & CpTi(OR^*)_2Cl \longrightarrow & \overset{\displaystyle OTi(Cp)(OR^*)_2}{\underset{|}{CH_2{=}C}}{-}OC(CH_3)_3 \\
\mathbf{4} & & \mathbf{2} & \mathbf{5}
\end{array}
$$

$$50{-}87\% \downarrow RCHO,\ -78°$$

$$R \overset{\displaystyle OH}{\diagup\diagdown}COOC(CH_3)_3$$

6 (90–96% ee)

D-threo-β-Hydroxy-α-amino esters.[3]   An example of use of the complex **2** for synthesis of these uncommon amino esters is shown in equation (II).

(II)

$$\xrightarrow[\substack{2)\ \mathbf{2} \\ 3)\ CH_3CHO}]{1)\ LDA}$$

$$53\%\ \text{overall} \downarrow Boc_2O$$

(≥98% de)

[1] M. Riediker and R. O. Duthaler, *Angew. Chem. Int. Ed.*, **28**, 494 (1989); M. Riediker, A. Hafner, U. Piantini, G. Rihs, and A. Togni, *ibid.*, **28**, 499 (1989).

[2] R. O. Duthaler, P. Herold, W. Lottenbach, K. Oertle, and M. Riediker, *ibid.*, **28**, 495 (1989).
[3] G. Bold, R. O. Duthaler, and M. Riediker, *ibid.*, **28**, 497 (1989).

4,4-Dimethoxy-2-trimethylsilylmethyl-1-butene,

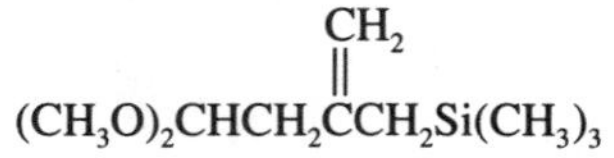

$$CH_2$$
$$\|$$
$$(CH_3O)_2CHCH_2CCH_2Si(CH_3)_3$$

Preparation.[1]

Spirocycles.[2] In the presence of trimethylsilyl triflate, the silyl enol ether of an aldehyde reacts selectively with the acetal function of **1** to form an intermediate with a carbonyl group that can then react with the initially inert allylsilane group of **1** to form a spirocyclic system. This methodology permits synthesis of spirocyclic sys-

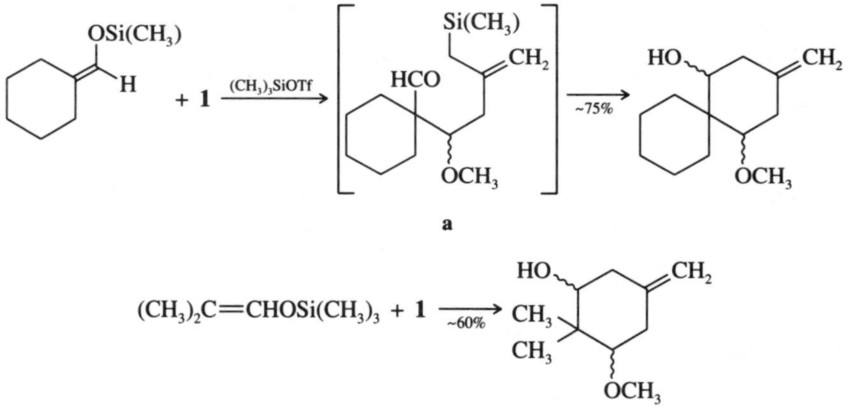

a

tems of various sizes as well as six-membered rings containing a quaternary center. The homolog of **1**, 5,5-dimethoxy-2-trimethylsilyl-1-pentene, is also available and provides a route to seven-membered rings, generally in yields lower than those observed with **1**.

[1] T. V. Lee, J. A. Channon, C. Cregg, J. R. Porter, F. S. Roden, and H. Yeoh, *Tetrahedron*, **45**, 5877 (1989).
[2] T. V. Lee and C. Cregg, *Synlett*, 317 (1990).

3,3-Dimethoxy-2-trimethylsilylmethyl-1-propene, $(CH_3)_3SiCH_2C(=CH_2)CH-(OCH_3)_2$ **(1)**.

Preparation (**15**, 72):

$$(CH_3O)_2CHCOOCH_3 + (CH_3)_3SiCH_2MgCl \xrightarrow[\substack{40\%}]{\substack{1)\ CeCl_3 \\ 2)\ SiO_2}} \mathbf{1}$$

Annelation-ring cleavage.[1] A bifunctional reagent with an acetal and an allylsilane group can be used for annulation of silyl enol ethers to six- and seven-membered carbocycles (equation I).[2] The reaction involves conjugate addition to give an adduct that undergoes intramolecular cyclization.

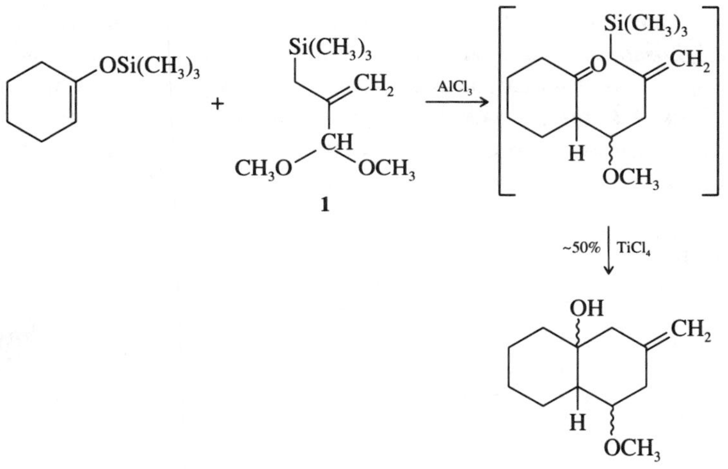

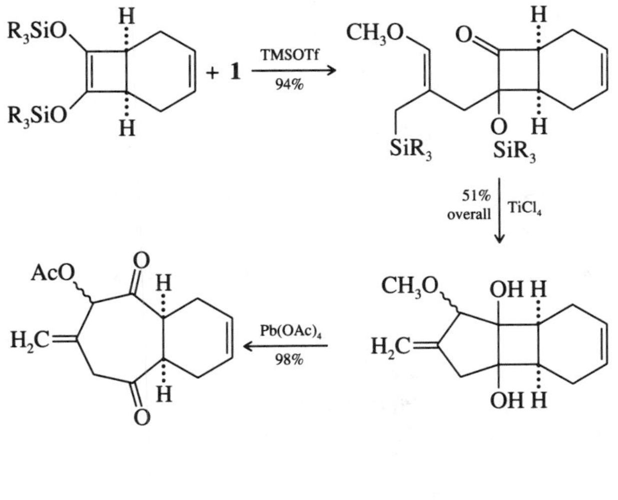

This annelation can be used as a route to carbocycles containing seven to nine members, which are difficult to obtain by usual methods. Thus reaction of the allylsilane-acetal **1** with a silylated enediol provides a bicyclic 1,2-diol, known to be oxidized to a ring-expanded diketone. Use of this annelation/ring cleavage route to a seven-membered ring system is formulated in equation II. The method was also used to prepare eight- and nine-membered rings.[3]

[1] T. V. Lee, J. Channon, C. Cregg, J. R. Porter, F. S. Roden, and H. Yeoh, *Tetrahedron*, **45**, 5877 (1989).
[2] T. V. Lee, R. J. Boucher, J. R. Porter, and C. J. M. Rockell, *ibid.*, **45**, 5887 (1989).
[3] T. V. Lee, J. R. Porter, and F. S. Roden, *ibid.*, **47**, 139 (1991).

1,3-Dimethoxy-1-(trimethylsilyloxy)butadiene (Brassard's diene, **12**, 196–197).

α,β-Unsaturated lactones.[1] These lactones (**2**) can be obtained stereoselectively by a Lewis-acid catalyzed reaction of **1** with an N-protected α-amino aldehyde. Ozonolysis of **2** provides optically active 4-amino-3-hydroxy carboxylic esters (**3**). The stereochemistry of the cyclization can be controlled by the N-protecting group as

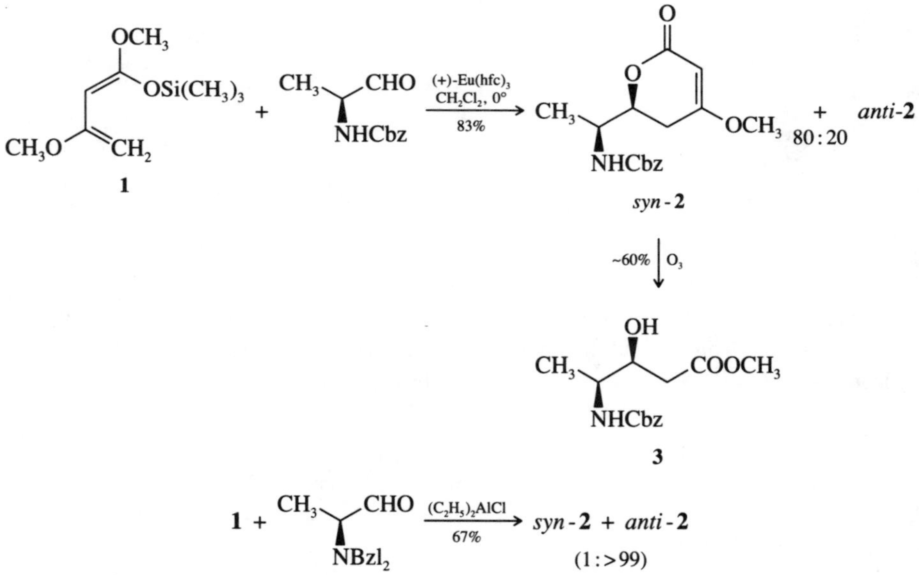

shown. These products are useful precursors to natural products containing a β-amino alcohol group.

[1] M. M. Midland and M. M. Afonso, *Am. Soc.*, **111**, 4368 (1989).

Dimethylaminodimethylchlorosilane, $(CH_3)_2NSi(CH_3)_2Cl$ (**1**).

Alkenyl or alkynyl silyl ethers. These can be prepared in high yield by reaction of an alkenyl- or alkynyllithium with **1**, followed by reaction of the product with an alcohol.

$$RC{\equiv}CLi \xrightarrow{\ 1\ } RC{\equiv}CSi(CH_3)_2N(CH_3)_2 \xrightarrow[\substack{85-95\% \\ overall}]{R'OH} RC{\equiv}CSiOR'$$

(with CH₃ groups above and below the Si in the product)

[1] G. Stork and P. F. Keitz, *Tetrahedron Letters*, **30**, 6981 (1989).

(−)-3-*exo*-(Dimethylamino)isoborneol,   (1).

Alkylation with R₂Zn.[1] After pursuing several clues, Noyori *et al.* found that this chiral, cyclic β-dimethylamino alcohol (1) is an effective catalyst for enantiose-lective alkylation of carbonyl compounds with R₂Zn. Thus in the presence of 2 mole % of (−) 1, diethylzinc and C₆H₅CHO react to form the (S)-alcohol (2) in 98% ee.

$$C_6H_5CHO + (C_2H_5)_2Zn \xrightarrow[97\%]{(-)-1}$$

(S)-2 (98% ee)

Since the original report, about 12 other reagents have been reported to catalyze this reaction, but this reagent of Noyori (1) and that of Oguni (15, 268) seem to be the most effective. In addition, both can effect chiral amplification, an increase of enan-tioselectivity over that of the catalyst. Noyori suggests that the alkyl transfer reaction involves a dinuclear Zn complex such as 3, whose structure has been established by

3

X-ray analysis. At the present time, only dialkylzinc compounds containing primary alkyl groups can be used for this alkylation because secondary and tertiary groups undergo alkene elimination.

[1] M. Kitamura, S. Suga, K. Kawai, and R. Noyori, *Am. Soc.*, **108**, 6071 (1986); M. Kitamura, S. Okada, S. Suga, and R. Noyori, *ibid.*, **111**, 4028 (1989); R. Noyori and M. Kitamura, *Angew. Chem. Int. Ed.*, **30**, 49 (1991).

(*p*-Dimethylaminophenyl)diphenylphosphine, p-$(CH_3)_2NC_6H_4P(C_6H_5)_2$ **(1).**

This phosphine has been recommended as an alternative to triphenylphosphine in Wittig[1] and Mitsunobu[2] reactions, because the corresponding phosphine oxide is readily removed by an acid wash, facilitating work-up of these reactions.

[1] S. Trippet and D. M. Walker, *J. Chem. Soc.*, 2130 (1961).
[2] M. von Itzstein and M. Mocerino, *Syn. Comm.*, **20**, 2049 (1990).

4-Dimethylaminopyridine (DMAP).

Dealkoxycarbonylation.[1] Enolizable β-keto esters undergo this reaction in 60–70% yield when refluxed in slightly aqueous toluene (90°) containing 1 equiv. of DMAP and buffered to pH 5–7. DABCO, N,N-dimethylaniline, and pyridine are not effective.

Acetoacetylation with DMAP.[2] Alcohols (even *tert*-ones) undergo acetoacetylation when treated with diketene in the presence of DMAP (55–100% yield). The reaction generally occurs at room temperature.

[1] D. F. Taber, J. C. Amedio, Jr., and F. Gulino, *J. Org.*, **54**, 3474 (1989).
[2] A. Nudelman, R. Kelner, N. Broida, and H. E. Gottlieb, *Synthesis*, 387 (1989).

Dimethyldioxirane.

Oxidation. Benzylic secondary amines are oxidized in high yield by dimethyldioxirane to nitrones,[1] probably via hydroxylamines.[2]

Epoxides of enol silyl ethers or acetates can be obtained in high yield by reactions with dimethyldioxirane at low temperatures.[3] Dimethyldioxirane is generally more efficient than alkaline hydrogen peroxide for epoxidation of α,β-enones and -enoates.[4]

Epoxidation of glycals can be effected in almost quantitative yield by dimethyldioxirane in acetone at 0°. In the case of nonparticipating protecting groups, the α-epoxide is formed almost exclusively. These α-1,2-anhydro sugars react with alcohols with clean inversion to form β-glycosides.[5]

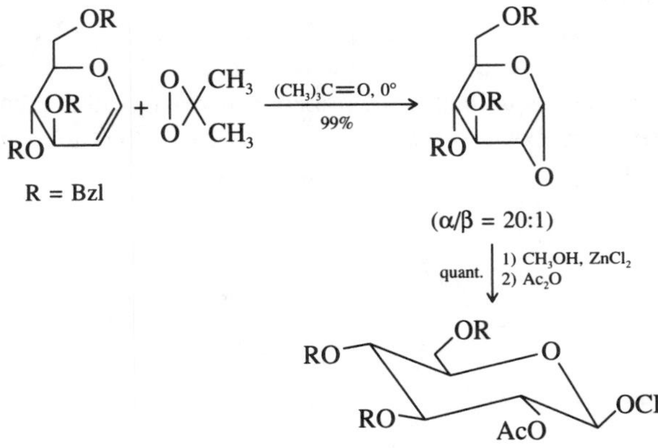

Epoxidation of exocyclic enol lactones.[6]   Peracids, even under buffered conditions, are not useful for this epoxidation because of rearrangement and decomposition. Dimethyldioxirane effects epoxidation of γ-methylene-γ-butyrolactones (**1**) in 94–96% yield in 2–3.5 hours. It is also effective for epoxidation of endocyclic enol lactones such as **3**.

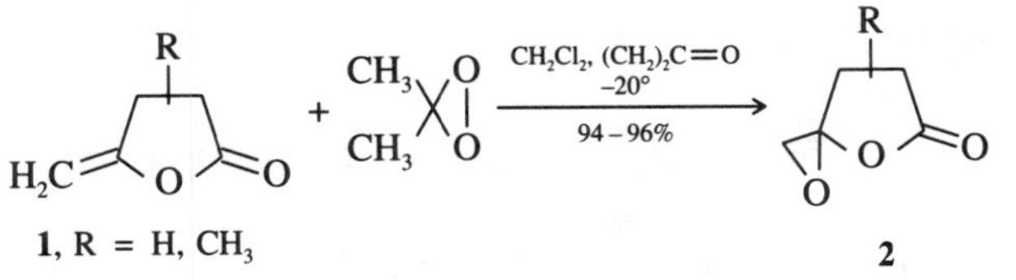

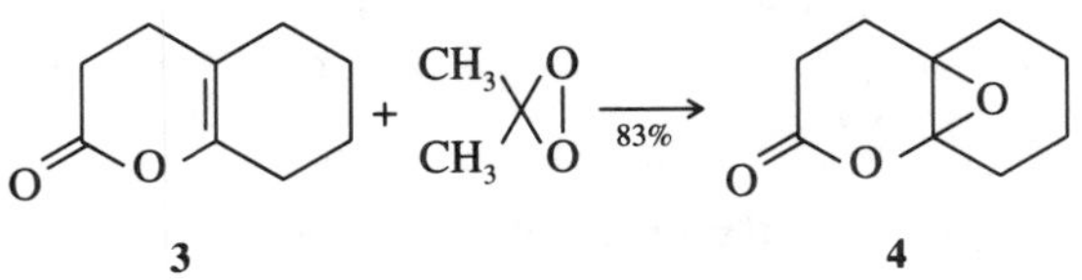

Reviews.[7,8]   The chemistry of dioxiranes, particularly of dimethyloxirane, has been reviewed by two active investigators in this field.

[1] R. W. Murray and M. Singh, *J. Org.*, **55**, 2954 (1990).
[2] M. D. Wittman, R. L. Halcomb, and S. J. Danishefsky, *ibid.*, **55**, 1981 (1990).

[3] W. Adam, L. Hadjiarapoglou, X. Wang, *Tetrahedron Letters*, **30**, 6497 (1989).
[4] W. Adam, L. Hadjiarapoglou, and B. Nestler, *ibid.*, **31**, 331 (1990).
[5] R. L. Halcomb and S. J. Danishefsky, *Am. Soc.*, **111**, 6661 (1989).
[6] W. Adam, L. Hadjiarapoglou, V. Jäger, and B. Seidel, *Tetrahedron Letters*, **30**, 4223 (1989).
[7] W. Adam, R. Curci, and J. O. Edwards, *Acc. Chem. Res.* **22**, 205 (1989).
[8] R. W. Murray, *Chem. Rev.* **89**, 1187 (1989).

Dimethylformamide.

Carboxamidation of RLi or RMgX.[1]   This reaction can be effected by reaction of RLi (or RMgX) with DMF to give a hemiaminal **a** followed by an Oppenauer oxidation. The second step requires the presence of a magnesium alkoxide such as magnesium 2-ethoxyethoxide **1**, $Mg(OCH_2CH_2OC_2H_5)_2$, either as a catalyst for the oxidation or for stabilization of **a**, possibly as a mixed cluster.

$$RLi + (CH_3)_2NCHO \xrightarrow{\text{THF}} \left[\begin{array}{c} R \diagdown \diagup OLi \\ \times \\ H \diagup \diagdown N(CH_3)_2 \end{array} \right]$$

$$\mathbf{a}$$

$$\mathbf{a} + (C_6H_5)_2C{=}O \xrightarrow[30-80\%]{1} RCN(CH_3)_2 + (C_6H_5)_2CHOH$$

Formylation.[2]   Dimethylformamide and trifluoromethanesulfonic anhydride form an iminium salt (**1**) that is more reactive than that formed from dimethylformamide and $POCl_3$, which is generally used for formylation (Vilsmeier reagent). Although the Vilsmeier reagent does not react with naphthalene or phenanthrene, reactions with **1** results in 1-naphthalenacarbaldehyde in 50% yield and in 3-phenanthrenecarbaldehyde in 25% yield.

$$(CH_3)_2\overset{+}{N}{=}CHOTf\ OTf^-\quad (1)$$

[1] C. G. Screttas and B. R. Steele, *J. Org.*, **53**, 5151 (1988).
[2] A. G. Martinez, R. M. Alvarez, J. O. Barcina, S. de la Moya Cerero, E. T. Vilar, A. G. Fraile, M. Hanack, and L. R. Subramanian, *J.C.S. Chem. Comm.*, 1571 (1990).

Dimethylgallium chloride; dimethylgallium triflate.

Glycosidation.[1]   These gallium compounds serve as efficient activators for glycosidation of glycopyranosyl fluorides in CH_2Cl_2 or toluene with moderate β-selectivity.

[1] S. Kobayashi, K. Koide, and M. Ohno, *Tetrahedron Letters*, **31**, 2435 (1990).

1,1-Dimethylhydrazine.

4-Alkyl-1,3-cycloalkanediones.[1] These products can be obtained from cyclo-hexane- or cyclopentane-1,3-diones by conversion to the monodimethylhydrazones

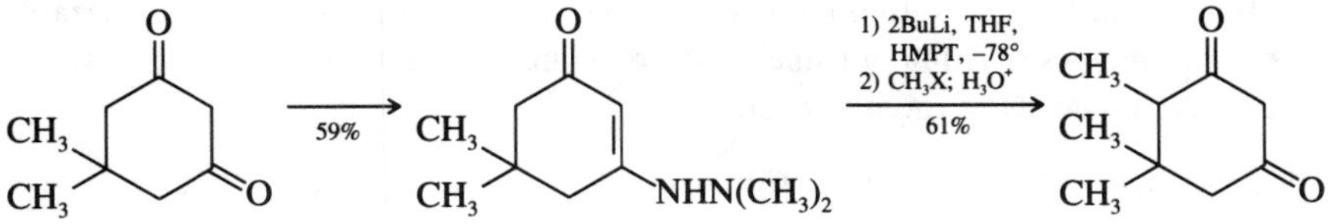

(7, 126–130), the dianions of which are alkylated at C_4 in 56–65% yield. In contrast, alkylation of the lithium enolates of cyclic 1,3-diones provide 6-alkyl derivatives.

[1] A. S. Demir and D. Enders, *Tetrahedron Letters*, **30**, 1705 (1989).

Dimethyl methylphosphonate, $\overset{\text{O}}{\overset{\|}{\text{CH}_3\text{P(OCH}_3)_2}}$, 3,117; 11,203.

Rearrangement of 2,2-dialkyl-1,3-cyclohexanediones.[1] Reaction of these sub-strates with the anion (LDA) of this phosphonate results in rearrangement to 3-alkyl-2-cyclohexenones in 60–93% yield. Addition of $ClSi(CH_3)_3$ improves the yield. The

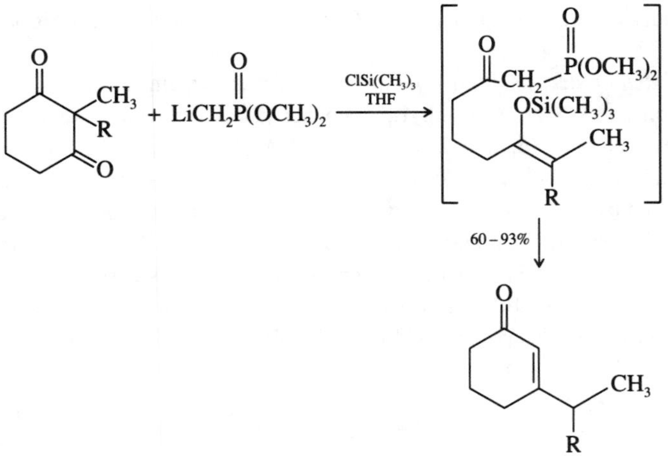

rearrangement is believed to involve a retro-aldol cleavage followed by a Wittig–Horner intramolecular reaction.

[1] Y. Yamamoto and T. Furuta, *J. Org.*, **55**, 3971 (1990).

$$\text{O} \atop \|$$

Dimethyloxosulfonium methylide, $(CH_3)_2\overset{\text{O}}{\underset{\|}{S}}{=}CH_2$ **(1).**

Preparation. The original procedure **(1,** 315–318) for the preparation employed reaction of trimethyloxosulfonium iodide with NaH in DMSO. A newer, less hazardous route involves reaction of trimethylsulfoxonium iodide (Aldrich) with potassium *t*-butoxide in DMSO at room temperature.[1]

[1] J. S. Ng, *Syn. Comm.*, **20,** 1193 (1990).

Dimethyl(phenyl)silyllithium.

Cyclic silyl enol ethers.[1] This reagent undergoes 1,2-addition to cyclic α,β-enones usually in high yield. Brook rearrangement of the adduct results in silyl enol ethers.

[1] M. Koreeda and S. Koo, *Tetrahedron Letters*, **31,** 831 (1990).

N,N'-Dimethyl-N,N'-propyleneurea (1,3-dimethyloxohydropyrimidine, DMPU, **11,** 207; **13,** 122).

Silyl ketene acetals from esters.[1] Ireland has examined various factors in the enolization and silylation of ethyl propionate **(1)** as a model system. As expected from previous work **(6,** 276–277), use of LDA (1 equiv.) in THF at $-78 \to 25°$ results mainly in (E)-**2,** formed from the (Z)-enolate. The stereoselectivity is markedly affected by the solvent. Addition of TMEDA results in a 60 : 40 ratio of (Z)- and (E)-**2** and lowers the yield significantly. Use of THF/23% HMPA provides (Z)- and (E)-**2** in the ratio of 85 : 15 with no decrease in yield. This system has been widely used for (E)-selective lithium enolate formation from esters and ketones. Highest stereoselectivity is observed by addition of DMPU, recently introduced as a noncar-

THF	90%	6:94
THF/25% TMEDA	50%	60:40
THF/23% HMPA	90%	85:15
THF/45% DMPU	90%	93:7

cinogenic substitute for HMPA. The higher efficiency of DMPU results from the fact that it can be used in higher concentrations at $-78°$ than HMPA.

A number of factors other than the solvent can affect the stereoselectivity of deprotonation of esters, such as the acid–base ratio and the nature of the base. But selective formation of (E)-silyl ketene acetals from esters remains a problem, particularly since they are more reactive than the (Z)-isomers.

The paper reports a highly efficient Claisen rearrangement of a (Z)-silyl ketene acetal (equation I) by use of DMPU to control the stereoselectivity. Use of HMPA lowers the diastereoselectivity from 96 to 84% de.

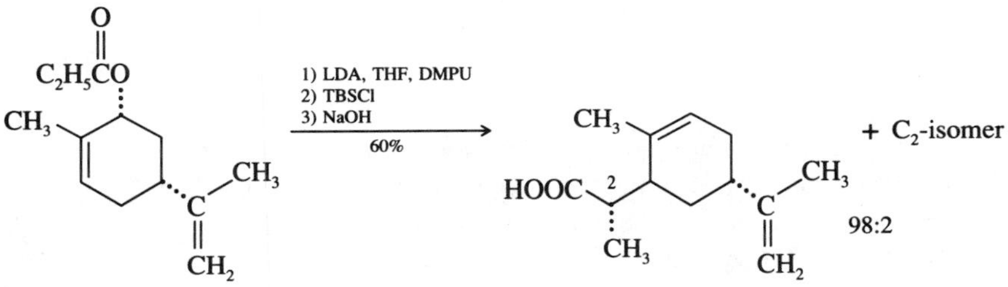

[1] R. E. Ireland, P. Wipf, and J. D. Armstrong, III, *J. Org.*, **56**, 650 (1991).

(S,S)- or (R,R)-2,5-*trans*-Dimethylpyrrolidine, (1)

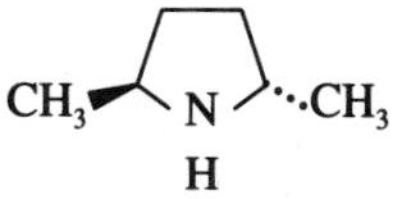

Preparation.[1]

Asymmetric radical reactions.   Several groups have reported asymmetric radical reactions observed with (S,S)- or (R,R)-**1** as the chiral auxiliary. Thus the iodide **2** in the presence of Bu$_3$SnH and AIBN cyclizes mainly to two diastereomeric *endo*-cyclic products **3** in the ratio 14:1.[2]

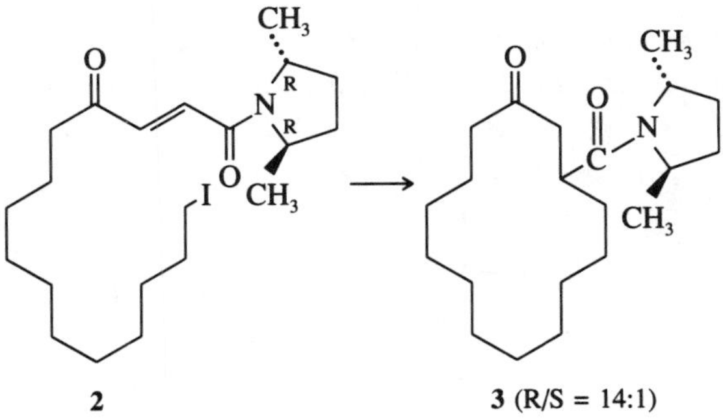

2   3 (R/S = 14:1)

Analogous results obtain in intermolecular addition of alkyl radicals to unsaturated amides. Thus addition of *n*-hexyl radical to the unsaturated amide **4** derived from (S,S)-**1** results in four products, two by addition α to the ketone and two by addition α to the amide enol. The former products are formed with slight selectivity, (60:40), but the latter products (**5**) are formed in ratio of 93:7.[3]

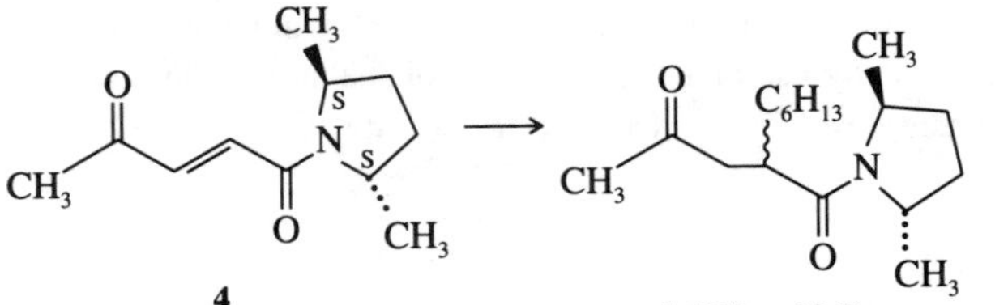

4

5 (S/R = 93:7)

Stereoinduction is also possible in addition of chiral radicals to ethyl acrylate.[4] Thus the radical formed from the bromo amide **6** derived from (R,R)-2,5-dimethylpyrrolidine reacts with ethyl acrylate to give the mono- and di-adducts **7** and **8**. The monoadduct is formed with 36:1 stereoselectivity at C_2.

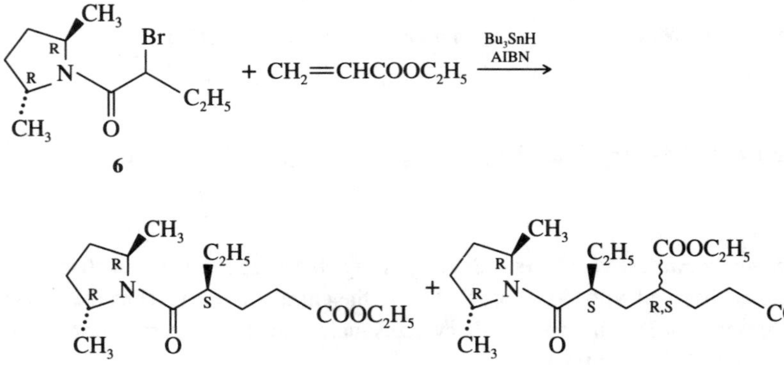

6

7 (35–50%, S/R = 36:1)

8 (15–25%, C_4 R/S = 1:1)

Asymmetric synthesis of cyclobutanones.[5] The ω-unsaturated keteniminium salts (**a**) formed from amides (**2**) derived from chiral R,R-**1** undergo intramolecular

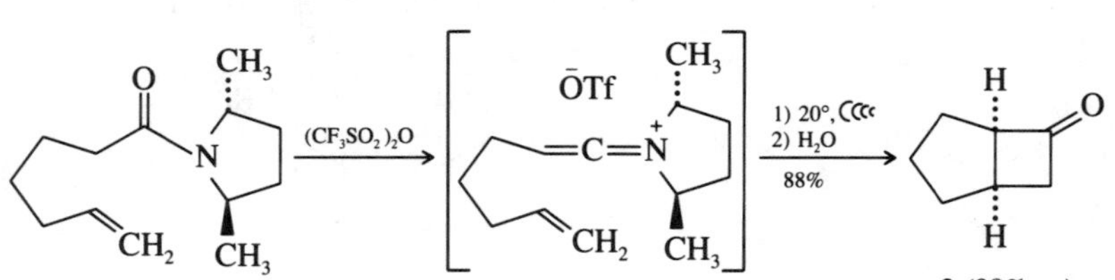

2

a

3 (98% ee)

[2 + 2]cycloaddition with high facial selectivity, particularly when carried out at 20° with ultrasonic activation.

[1] R. P. Short, R. M. Kennedy, S. Masamune, *J. Org.*, **54**, 1755 (1989).
[2] N. A. Porter, B. Lacher, V. H.-T. Chang, and D. R. Magnin, *Am. Soc.*, **111**, 8309 (1989).
[3] N. A. Porter, B. Giese, H. J. Lindner, *et al.*, *ibid.*, **111**, 8311 (1989).
[4] N. A. Porter, E. Swann, J. Nally and A. T. McPhail, *ibid.*, **112**, 6740 (1990); B. Giese, M. Zehnder, M. Roth, and H.-G. Zeitz, *ibid.*, **112**, 6741 (1990).
[5] L. Chen and L. Ghosez, *Tetrahedron Letters*, **31**, 4467 (1990).

Dimethyl sulfoxide.

Deprotection of acetals.[1]   Dialkyl acetals are cleaved to the carbonyl compound when refluxed in aqueous DMSO. Deprotection of cyclic acetals (1,3-dioxolanes, 1,3-dioxanes) requires more drastic conditions, such as refluxing aqueous DMSO in combination with a cosolvent such as ethanol or 2-butanone.

[1] T. Kametani, H. Kondoh, T. Honda, H. Ishizone, Y. Suzuki, and W. Mori, *Chem. Letters*, 901 (1989).

Dimethyl sulfoxide–Oxalyl chloride.

Oxidation of alkyl trimethyl- and triethylsilyl ethers, $ROSi(CH_3)_3$ or $ROSi$-$(C_2H_5)_3$.[1]   Silyl ethers of this type can be oxidized directly to carbonyl compounds by the Swern reagent. This oxidation provides an efficient route to the Corey aldehyde (**2**).

Dehydrogenation of diaziridines.[2]

Dehydrogenation of diaziridines.[2]   The Swern reagent is more effective than Ag_2O or NBS for oxidation of the diaziridines **1** to the diazirines **2**, of interest because they are photolyzed to reactive carbenes.

$$HN-NH \xrightarrow[71-86\%]{\underset{(COCl)_2}{DMSO}} N=N$$

$$RC_6H_4 \quad CF_3 \qquad\qquad RC_6H_4 \quad CF_3$$

$$\mathbf{1} \qquad\qquad\qquad \mathbf{2}$$

Review.[3]   Tidwell has reviewed the oxidation of alcohols by various versions of the Swern procedure, with particular emphasis on applications reported in the recent literature (1981–1989).

[1] G. A. Tolstikov, M. S. Miftakhov, M. E. Adler, N. G. Komissarova, O. M. Kuznetsov, and N. S. Vostrikov, *Synthesis*, 940 (1989).
[2] S. K. Richardson and R. J. Ife, *J.C.S. Perkin I*, 1172 (1989).
[3] T. T. Tidwell, *Synthesis*, 857 (1990).

Dimethyltitanocene, $Cp_2Ti(CH_3)_2$, **(1)** m.p. 97°. This reagent is prepared by reaction of titanocene dichloride, Cp_2TiCl_2, with CH_3Li (95% yield). It is stable for several months when stored in the dark in toluene or THF, but decomposes rapidly in the solid state.[1]

Methylenation.[2]   When heated in THF at 60–65°, this reagent effects methylenation of ketones, even readily enolizable ones, in 60–90% yield and of aldehydes (45–60% yield). It also converts esters and lactones into enol ethers, but this reaction is generally slower. It is thus an attractive alternative to the Tebbe and Grubbs reagent.

[1] V. K. Claus and H. Bestian, *Ann.*, **654**, 8 (1962).
[2] N. A. Petasis and E. I. Bzowej, *Am. Soc.*, **112**, 6392 (1990).

(S,S)-N,N'-Dineohexyl-2,2'-bipyrrolidine,

(S,S)−**1**

$$(CH_3)_3CCH_2CH_2 \qquad CH_2CH_2C(CH_3)_3$$

α_D-137°. Preparation.[1]

Enantioselective dihydroxylation.[2]   Highly enantioselective osmylation of *trans*-disubstituted and monosubstituted alkenes obtains with OsO_4 oxidations in the presence of 1 equiv. of this chiral 2,2'-bipyrrolidine ligand at −78°. Note, however, that the enantioselectivity for osmylation of *cis*-disubstituted alkenes is only ~65% ee and for osmylation of trisubstituted alkenes is less than 60% ee. X-ray analysis of

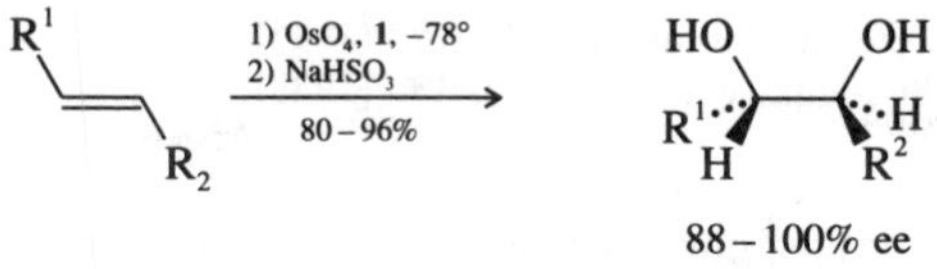

$$R^1 \underset{R_2}{\diagup\diagdown} \quad \xrightarrow[\substack{80-96\%}]{\substack{1)\ OsO_4,\ 1,\ -78° \\ 2)\ NaHSO_3}} \quad \underset{\substack{H \quad\quad R^2 \\ \\ 88-100\%\ ee}}{\overset{HO \quad\quad OH}{R^{1\cdots} \diagdown \diagup \cdots H}}$$

the complex (**2**) obtained from osmylation of stilbene shows that the diamine (**1**) is chelated to Os.

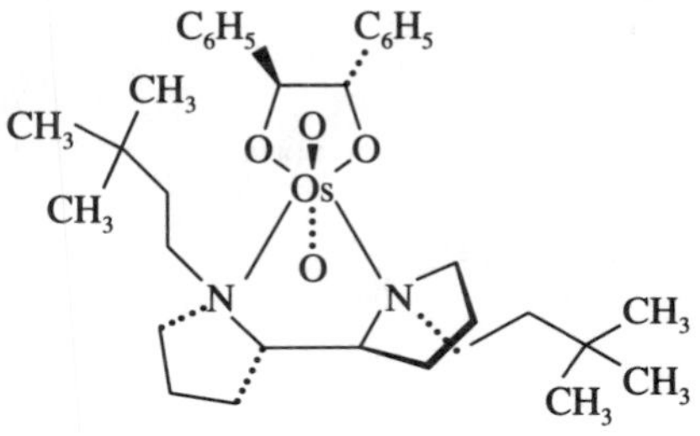

2

[1] M. Hirama, T. Oishi, and S. Ito, *J.C.S. Chem. Comm.*, 665 (1989).
[2] T. Oishi and M. Hirama, *J. Org.*, **54**, 5834 (1989).

trans-Dioxo(tetramesitylporphyrinato)ruthenium(VI) RuO$_2$(tmp), 1.

This porphyrin complex is obtained by chloroperbenzoic acid oxidation of Ru-(CO)(tmp).[1]

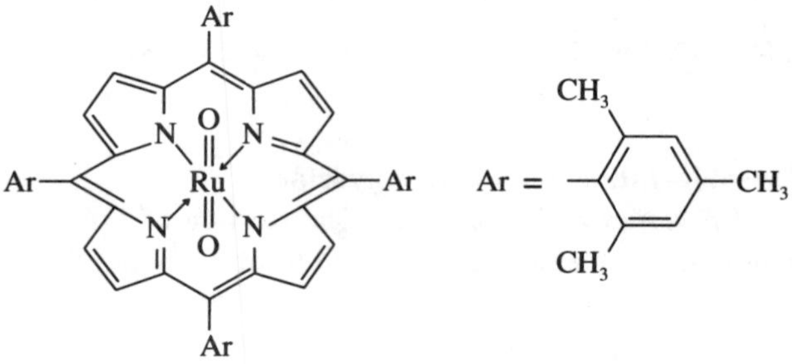

1

Epoxidation.[2]   The complex catalyzes air epoxidation of various alkenes at 25°. Thus air epoxidation of cholesteryl esters and various Δ^5-17-ketosteroids results in highly β-selective epoxidation of the 5,6-double bond in 72–95% yield with usually >99% β-selectivity.

[1] J. T. Groves and R. Quinn, *Am. Soc.*, **107**, 5790 (1985).

[2] J.-C. Marchon and R. Ramasseul, *Synthesis*, 389 (1989).

Diperoxohexamethylphosphoramidomolybdenum(VI), (MoOPH).

Vedejs hydroxylation of 2-azetidinone.[1] Some 3-alkyl-3-hydroxyazetidine-2-ones are known to have interesting biochemical properties, but have not been readily available. They are now available by alkylation of a 3-hydroxyazetidine-2-one derivative (**3**), which can be prepared by hydroxylation of (N-*t*-butyldimethylsilyl)azeti-

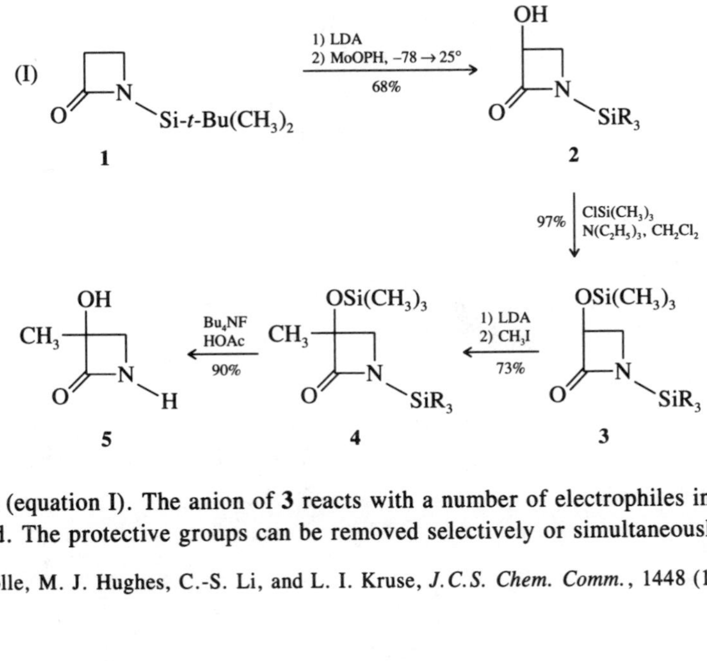

din-2-one (equation I). The anion of **3** reacts with a number of electrophiles in 50–85% yield. The protective groups can be removed selectively or simultaneously.

[1] R. E. Dolle, M. J. Hughes, C.-S. Li, and L. I. Kruse, *J.C.S. Chem. Comm.*, 1448 (1989).

Diphenyl diselenide–Lithium aluminum hydride.

Cleavage of oxetanes and oxolanes.[1] The reaction of LiAlH$_4$ with diphenyl diselenide in dioxane provides a particularly active form of LiSeC$_6$H$_5$, which cleaves

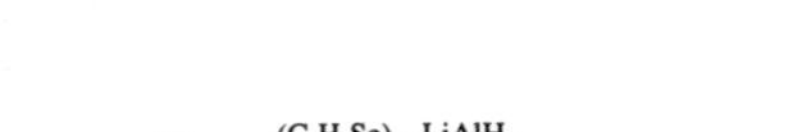

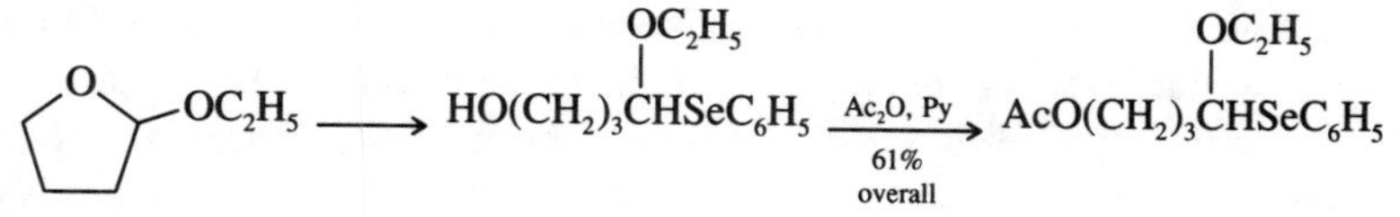

$$\text{(furanyl)}-OC_2H_5 \longrightarrow HO(CH_2)_3\overset{\displaystyle OC_2H_5}{\underset{\displaystyle |}{C}}HSeC_6H_5 \xrightarrow[\substack{61\% \\ \text{overall}}]{Ac_2O,\ Py} AcO(CH_2)_3\overset{\displaystyle OC_2H_5}{\underset{\displaystyle |}{C}}HSeC_6H_5$$

oxetanes and oxolanes to γ- and δ-phenylselenyl alcohols in generally high yield in contrast to the phenylselenide anion prepared from diphenyl diselenide with $NaBH_4$.

[1] K. Haraguchi, H. Tanaka, and T. Miyasaka, *Synthesis*, 434 (1989).

(R,R)- and (S,S)-1,2-Diphenyl-1,2-ethanediamine (1).
Preparation[1]:

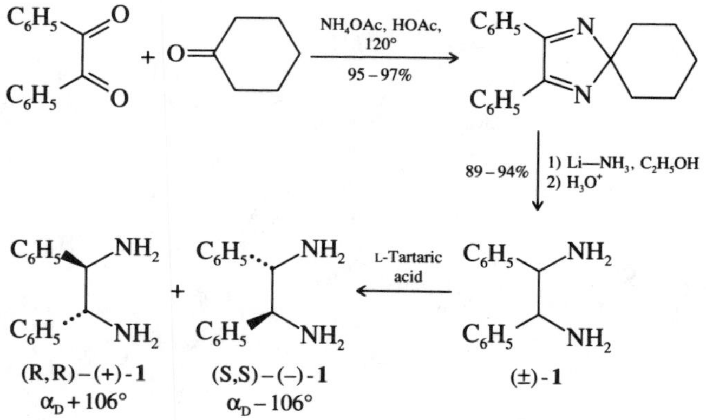

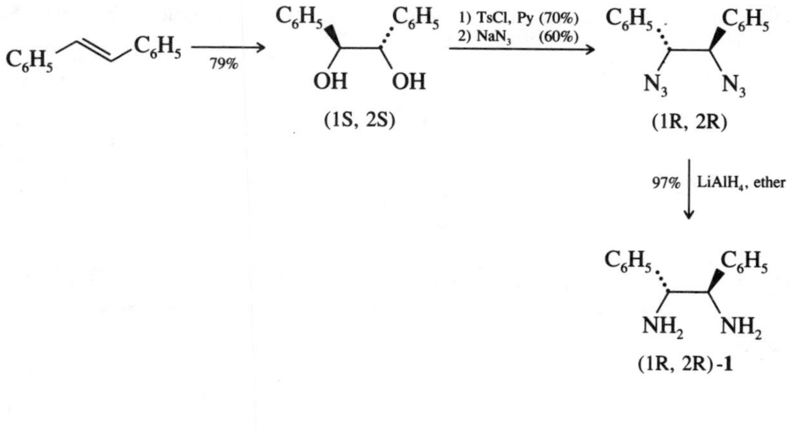

An alternate route to (+)- and (−)-1 involves asymmetric Sharpless dihydroxylation of *trans*-stilbene using dihydroquinine p-chlorobenzoate or dihydroquinidine p-chlorobenzoate as the chiral auxiliary for preparation of (−)- or (+)-1,2-diphenylethane-1,2-diol. Remaining steps involve conversion of the 1,2-diol via the ditosylate to the 1,2-diazide, which is reduced by LiAlH$_4$ to the diamine (**1**).[2]

These chiral C$_2$ symmetric 1,2-diamines have been used to prepare chiral reagents containing aluminum or boron, which have proved to be highly effective Lewis acid catalysts for several synthetic reactions.

Enantioselective Diels–Alder catalyst.[3] The aluminum reagent **2**, prepared by reaction of (CH$_3$)$_3$Al with the N,N-ditriflate of (S,S)-**1**, is an effective catalyst for asymmetric Diels–Alder reactions of acrylates with dienes. Particularly high enantioselectivity obtains in the reaction of 3-acrylyl-1,3-oxazolidinone-2 with cyclopentadiene. In this case, the *endo–exo* ratio is >50:1, and the optical yield is 90%. The stereoselectivity probably results from binding of the catalyst to the acrylyl carbon group.

(S, S)-**2**

(*endo/exo* = 96:4),
94% ee

Enantioselective aldol reactions.[3,4] A related borane reagent, (R,R)-**4**, prepared by reaction of BBr$_3$ with the N,N-bistosylsulfonamide of (R,R)-**1**, can effect highly enantioselective aldol reactions of ketones with aldehydes. Thus reaction of

(R,R)-**4**

diethyl ketone with (R,R)-**4** (1 equiv.) and a base (diisopropylethylamine) leads to a chiral boron enolate, which reacts with an aldehyde to afford *syn*-aldols in >95% ee (equation I).

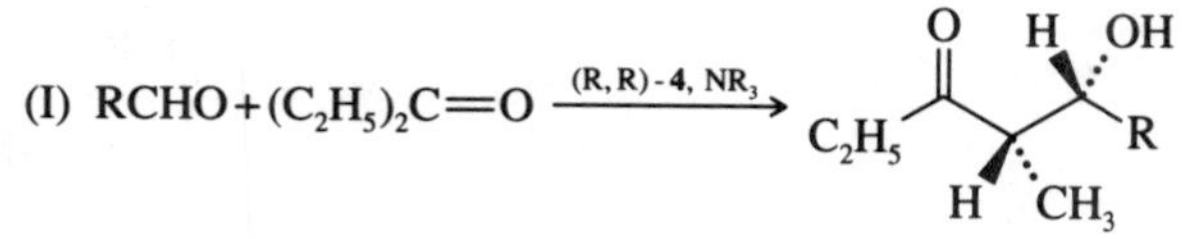

(I) $RCHO + (C_2H_5)_2C{=}O \xrightarrow{(R,R)\text{-}4,\ NR_3}$

$R = C_6H_5$	95%	97% ee, *syn/anti* = 94:6
$R = C_2H_5$	91%	98% ee, *syn/anti* = 98:2

This bromoborane can even effect enantioselective aldol reactions with an acetate ester and an aldehyde (equation II).

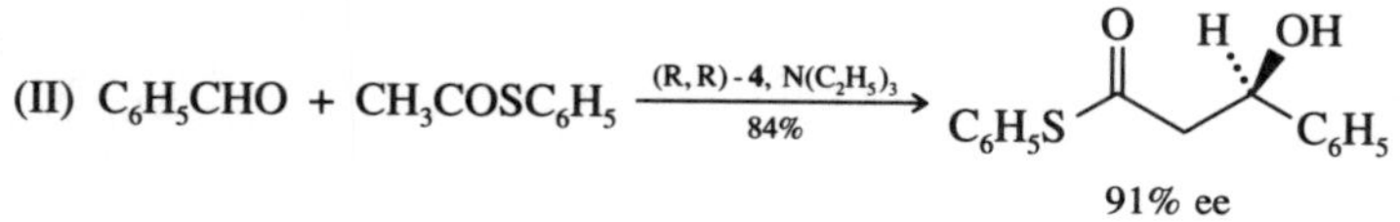

(II) $C_6H_5CHO + CH_3COSC_6H_5 \xrightarrow[84\%]{(R,R)\text{-}4,\ N(C_2H_5)_3}$

91% ee

Aldol reactions of propionate esters with aldehydes proceed more readily when promoted by the more Lewis acidic bromoborane (R,R)-5. This reagent is prepared

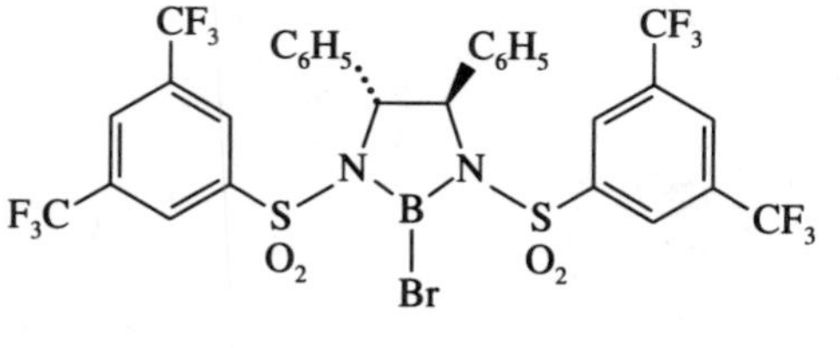

(R,R)-5

by conversion of (R,R)-1 to the bis(3,5-difluoromethyl)benzenesulfonamide followed by reaction with BBr₃. With this reagent phenylthiol esters are converted to *syn*-aldols (equation III), but *t*-butyl esters provide *anti*-aldols (equation IV), both

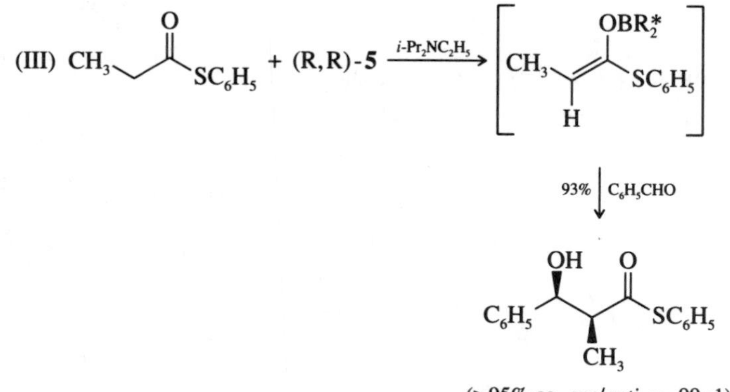

(III) $CH_3 \overset{O}{\underset{}{\|}} SC_6H_5$ + (R,R)-5 $\xrightarrow{i\text{-}Pr_2NC_2H_5}$

93% | C_6H_5CHO

(>95% ee, *syn/anti* = 99:1)

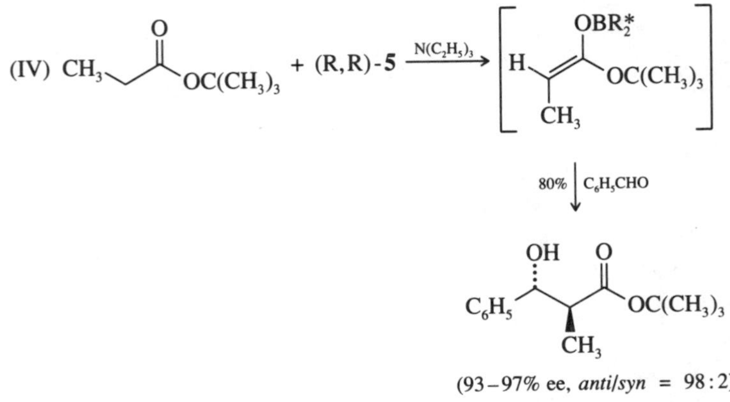

(IV) CH_3—C(O)—$OC(CH_3)_3$ + (R,R)-5 $\xrightarrow{N(C_2H_5)_3}$

$$\left[H—C=C(OBR_2^*)(OC(CH_3)_3) , CH_3 \right]$$

80% $\downarrow$ C_6H_5CHO

C_6H_5—CH(OH)—CH(CH_3)—C(O)—$OC(CH_3)_3$

(93–97% ee, *anti/syn* = 98:2)

with high enantio- and diastereoselectivity. Ethyldiisopropylamine is the preferred base for the *syn*-reaction, triethylamine for the *anti*-reaction.[4]

Allylation of aldehydes.[5] The (R,R)-allylborane **6**, obtained by reaction of the (R,R)-bromoborane **4** with allyltributyltin, adds to aldehydes to provide homoallylic

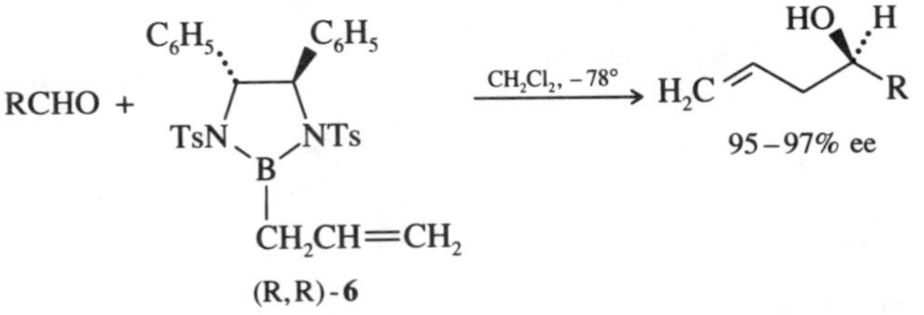

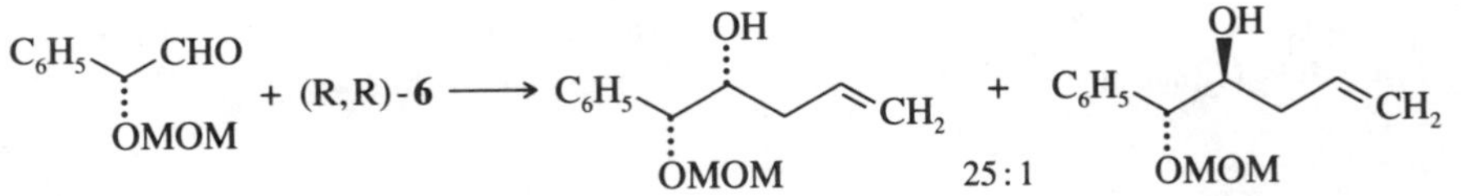

RCHO + [C_6H_5, C_6H_5 ring: TsN—B—NTs, $CH_2CH=CH_2$] $\xrightarrow{CH_2Cl_2, -78°}$ $H_2C=CH-CH_2-CH(OH)-R$

(R,R)-**6**   95–97% ee

C_6H_5—CH(OMOM)—CHO + (R,R)-**6** $\longrightarrow$ C_6H_5—CH(OMOM)—CH(OH)—CH_2—CH=CH_2 + C_6H_5—CH(OMOM)—CH(OH)—CH_2—CH=CH_2

25:1

alcohols in >90% yield and in 95–97% ee. A similar reaction with optically active aldehydes shows high diastereoselectivity (24–50:1). Similar addition of 2-bromo-

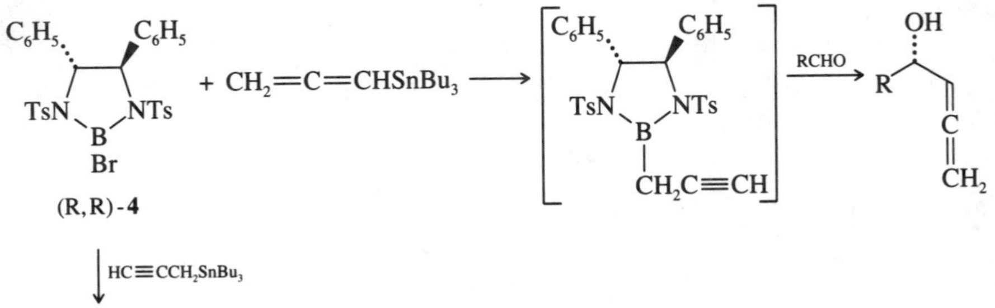

[C_6H_5, C_6H_5 ring: TsN—B—NTs, Br] + $CH_2=C=CHSnBu_3$ $\longrightarrow$ [C_6H_5, C_6H_5 ring: TsN—B—NTs, $CH_2C\equiv CH$] $\xrightarrow{RCHO}$ R—CH(OH)—C(=CH_2)—$C\equiv$

(R,R)-**4**

$\downarrow$ $HC\equiv CCH_2SnBu_3$

$$\left[\begin{array}{c} C_6H_5 \quad\quad C_6H_5 \\ TsN \quad NTs \\ B \\ CH{=}C{=}CH_2 \end{array}\right] \xrightarrow{RCHO} \begin{array}{c} OH \\ R \quad CH_2C{\equiv}CH \end{array}$$

and 2-chloroallyl groups is possible, but with lower enantioselectivity. Enantioselective synthesis of allenyl and propargyl carbinols by a similar process is also possible.[6]

Enantioselective dihydroxylation with OsO₄.[7] The (S,S)-diamine **6** [1,2-diphenyl-1,2-bis(2,4,6-trimethylbenzylamino)ethane] is prepared by reaction of (S,S)-**1** with mesitylaldehyde followed by NaBH₄ reduction. It markedly accelerates reac-

$$CH_3 \underset{CH_3}{\overset{C_6H_5 \quad C_6H_5}{\underset{\text{NH HN}}{\bigg|}}} CH_3 \quad CH_3 \quad\quad (S,S)\text{-}\mathbf{6}$$

tions of OsO₄ with alkenes. Rapid reactions even at $-90°$ are possible with terminal or (E)-1,2-disubstituted alkenes resulting in 1,2-diols with high enantioselectivity. The diamine is used in stoichiometric amounts, but the ligand and osmium can be recovered easily for recycling. The paper includes a rational explanation for the observed enantioselectivity based on a C_2-symmetric 1:1 complex of the diamine **6** and OsO₄.

$$(I) \quad \underset{R^1}{\overset{R^2}{\diagup}} + OsO_4 \xrightarrow[80-95\%]{\overset{(S,S)\text{-}\mathbf{6}}{CH_2Cl_2,\ -90°}} \underset{R^1 \quad R^2}{\overset{HO \quad OH}{\diagdown}}$$

82–98% ee

Asymmetric epoxidation of alkenes. Two groups have prepared chiral (salen)-manganese complexes such as **2** from (+)- or (−)-**1** and salicylaldehyde derivatives

$$\begin{array}{c} C_6H_5 \quad\quad C_6H_5 \\ {=}N \quad N{=} \\ Mn^+ \\ O \quad O \\ PF_6^- \text{ or} \\ R \quad AcO^- \quad R \end{array} \quad \mathbf{2}$$

and report that these complexes can serve as catalysts for enantioselective epoxidation of unfunctionalized alkenes with iodosylbenzene or -mesitylene as the oxygen atom source. The complex in which R is t-butyl[8] is more effective than that in which R is 1-phenylpropyl.[9] Preliminary results suggest that sodium hypochlorite can also be used as the oxidant. The highest enantioselectivity obtains with cis-disubstituted alkenes or with terminal alkenes substituted by a bulky group.

$C_6H_5CH{=}CHCH_3$ $C_6H_5CH{=}CH_2$

(cis, 84% ee)   (57% ee)
($trans$, 53% ee)

(78% ee)   (59% ee)

[1] S. Pikul and E. J. Corey, *Org. Syn.*, submitted (1990).
[2] D. Pini, A. Iuliano, C. Rosini, and P. Salvadori, *Synthesis*, 1023 (1990).
[3] E. J. Corey, R. Imwinkelried, S. Pikul, and Y. B. Xiang, *Am. Soc.*, 111, 5493 (1989); S. Pikul and E. J. Corey, *Org. Syn.*, submitted (1990).
[4] E. J. Corey and S. S. Kim, *Am. Soc.*, 112, 4976 (1990); *idem.*, *Tetrahedron Letters*, 31, 3715 (1990).
[5] E. J. Corey, C.-M. Yu, and S. S. Kim, *Am. Soc.*, 111, 5495 (1989).
[6] E. J. Corey, C.-M. Yu, and D.-H. Lee, *ibid.*, 112, 878 (1990).
[7] E. J. Corey, P. D. S. Jardine, S. Virgil, P.-W. Yuen, and R. D. Connell, *ibid.*, 111, 9243 (1989).
[8] W. Zhang, J. L. Loebach, S. R. Wilson, and E. N. Jacobsen, *Am. Soc.*, 112, 2801 (1990).
[9] R. Irie, K. Noda, Y. Ito, N. Matsumoto, and T. Katsuki, *Tetrahedron Letters*, 31, 7345 (1990).

(R,R)-1,2-Diphenylethane-1,2-diol dimethyl ether, C_6H_5 C_6H_5 (**1**, $\alpha_D - 14°$)

CH_3O OCH_3

Conjugate addition of RLi to an α,β-unsaturated aldimine.[1] This chiral diether is a catalyst for this reaction and also controls the enantioselectivity. It is more

=NC_6H_{11}-c

+ RLi   →

1) **1**, $C_6H_5CH_3$
2) H_3O^+
3) $NaBH_4$, CH_3OH

CH_2OH

R

R = Bu   80%   91% ee
R = C_6H_5 94%   82% ee

$CH_3CH{=}CH{-}CH{=}NC_6H_{11}$-$c$ + C_6H_5Li $\xrightarrow{48\%}$

CH_2OH

C_6H_5

CH_3

(S), >99% ee

effective than the dimethyl ether of (R,R)-butane-2,3-diol. The enantioselectivity is also high with acyclic α,β-unsaturated aldimines, but chemical yields are often only moderate.

[1] K. Tomioka, M. Shindo, and K. Koga, *Am. Soc.*, **111**, 8266 (1989).

(S,S)-1,2-Diphenyl-1,2-ethanedimethylamine,

$$\begin{array}{ccc} C_6H_5 & & C_6H_5 \\ & \diagdown \diagup & \\ CH_3NH & HNCH_3 \end{array}$$

(1)

Chiral aminals; 1,4-dihydropyridine-3-carboxaldehydes.[1] The chiral aminal **2** prepared from pyridine-3-carboxylaldehyde and (S,S)-**1**[2] reacts with organocopper reagents in the presence of methyl chloroformate to give almost exclusively products of 1,4-addition, as expected from reactions of the free aldehyde.[3] No products of 1,2-addition are formed, but 1,6-adducts are minor products in some cases. The 1,4-adducts are formed in 82–95% de (R-configuration). Addition of butyl and ethyl groups is best effected with lithium cuprates, but addition of methyl, vinyl, or aryl groups is best effected with organomagnesium cuprates. Under these conditions,

2		
$(CH_3)_2CuMgBr$	90%	95% ee
$(C_2H_5)_2CuLi$	90%	85% ee
$(CH_2{=}CH)_2CuMgCl$	90%	95% ee

addition of organocuprates can be effected with high regio- and stereoselectivity. Acyl chlorides can replace methyl chloroformate as activators.

[1] R. Gosmini, P. Mangeney, A. Alexakis, M. Commerçon, and J.-F. Normant, *Synlett*, 111 (1991).
[2] M. Commerçon, P. Mangeney, T. Tejero, and A. Alexakis, *Tetrahedron Asymmetry*, **1**, 287 (1990).
[3] D. L. Comins and A. H. Abdullah, *J. Org.*, **47**, 4315 (1982).

(S)-1,1-Diphenylpropane-1,2-diol.

$$\begin{array}{ccc} H_{\cdots} & CH_3 & C_6H_5 \\ & \diagup & \diagup \\ & C{-}\!\!{-}C{-}C_6H_5, \\ \diagup & & \diagdown \\ HO & & OH \end{array}$$

m.p. 93°, α_D–100° (1).

The diol is prepared by reaction of ethyl (S)-lactate with C_6H_5MgBr (75% yield).

t-Butyl alkyl sulfoxides.[1] The cyclic sulfite 2, prepared from 1, reacts with RMgBr or RLi to form chiral alkylsulfinates, 3 or 4, both convertible into chiral sulfoxides by reaction with a second organometallic. Particularly clean reactions of 2 are obtained with *t*-butylmagnesium chloride resulting from cleavage (with inversion) at the more-hindered site.

2

3

(R, 85% ee)

4 (90 : 10)

(R, 100% ee)

[1] F. Rebiere and H. B. Kagan, *Tetrahedron Letters*, **30**, 3659 (1989).

Diphenylsilane–Triethylborane.

Deoxygenation of primary and secondary alcohols.[1] This deoxygenation has been effected by reduction of the thiocarbonyl esters with tributyltin hydride and AIBN as the radial initiator (**11**, 550). A newer, milder method uses diphenylsilane in a radical chain reaction initiated by triethylborane and air. Even secondary thionocarbonates, particularly those derived from 4-fluorophenol, are deoxygenated at 25°.

80–96%

[1] D. H. R. Barton, D. O. Jang, and J. Cs. Jaszberenyi, *Tetrahedron Letters*, **31**, 4681 (1990).

Diphenylphosphoryl azide, $(C_6H_5O)_2P(O)N_3$.

Macrocyclic lactams.[1] These lactams have generally been prepared by reaction of dicarboxylic acid chlorides and diamines under high dilution. Yields are generally mediocre because linear polyamides are also formed. Actually, free dicarboxylic acids, when activated by diphenylphosphoryl azide, can condense with diamines to form macrocyclic lactams (*cf.*, **4**, 211). This new approach does not require high dilution and provides considerably higher yields. Thus the macrobicyclic lactam **3** is obtained by reaction of the dicarboxylic acid **1** and the HCl salt of the diamine **2** with diphenylphosphoryl azide (2.5 equiv.) and triethylamine (5 equiv.) in DMF at 25° in a yield of 82%.

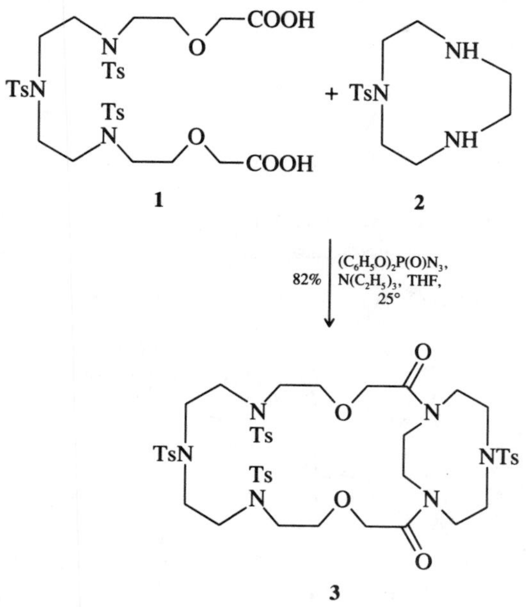

[1] L. Qian, Z. Sun, T. Deffo, and K. B. Mertes, *Tetrahedron Letters*, **31**, 6469 (1990).

1,3-Dithiane,

Review.[1]   A recent review of 1,3-dithianes and 2-lithio-1,3-dithianes covers the literature of 1977–1988 (126 references).

[1] P. C. B. Page, M. B. van Niel, and J. C. Prodger, *Tetrahedron*, **45**, 7643 (1989).

E

Ethanolamine.

Deprotection of 1-O-acyl sugars. Ethanolamine can selectively cleave the glycosyl ester bond of even peracylated aldoses.[1]

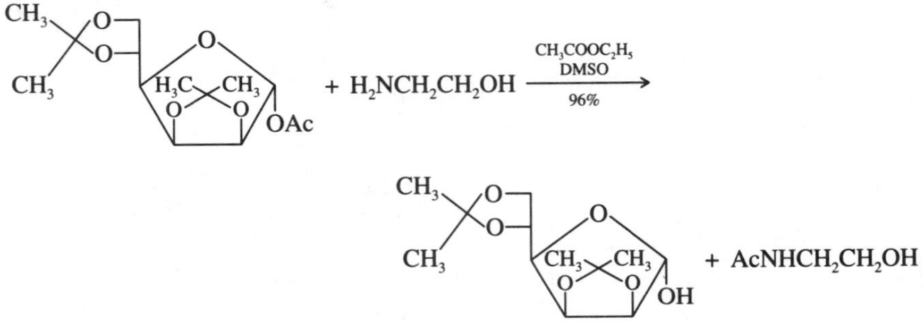

[1] G. Grynkiewicz, I. Fokt, W. Szeja, and H. Fitak, *J. Chem. Res. (S)*, 152 (1989).

3-Ethoxycarbonyl-2-methylallyltriphenylphosphonium bromide (1). Prepared by reaction of ethyl 4-bromo-3-methyl-2-butenoate with $P(C_6H_5)_3$.[1]

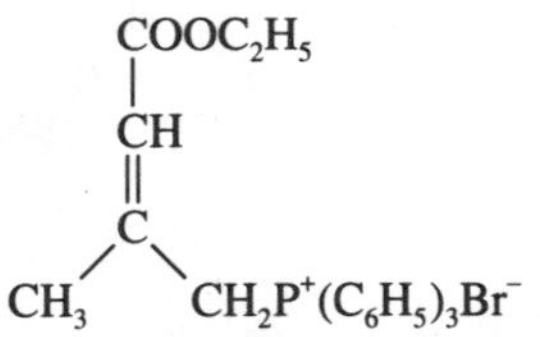

Cyclopentadienes.[2] This allylic triphenylphosphorane reacts with α-bromo ketones in the presence of $NaHCO_3$ at 25° to form cyclopentadienes by a [3 + 2]annelation. The reaction may involve intermediates such as **a** and **b**.

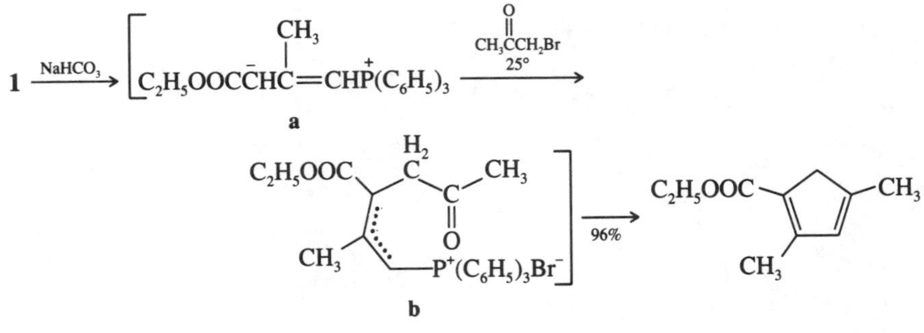

[1] E. J. Corey and B. W. Erickson, *J. Org.*, **39**, 821 (1974).
[2] M. Hatanaka, Y. Himeda, and I. Ueda, *J.C.S. Chem. Comm.*, 526 (1990).

Ethoxytrimethylsilylacetylene, $(CH_3)_3SiC\equiv COC_2H_5$ (1), **12**, 221.

Esters, lactones, and peptides.   The reagent serves as a dehydration agent for condensation of RCOOH with alcohols or amines to provide esters, lactones or lactams, or peptides, usually in the presence of a mercury catalyst (HgO).

$$R^1COOH + 1 \xrightarrow{Hg(II)} \left[\begin{array}{c} R^1COOC=CHSi(CH_3)_3 \\ | \\ OC_2H_5 \end{array} \right] \xrightarrow[98-100\%]{R^2OH} R^1COOR^2$$

$$C_6H_5CH_2NH(CH_2)_3COOH \xrightarrow{87\%}$$

[1] Y. Kita, S. Akai, M. Yamamoto, M. Taniguchi, and Y. Tamura, *Synthesis*, 334 (1989).

1-Ethoxy-3-(trimethylsilyloxy)-1,3-butadiene (1, Danishefsky's diene).

[4 + 2] *Cycloaddition to α-amino aldehydes.*[1] The diastereoselectivity of the $ZnBr_2$-catalyzed cycloaddition of 1 to D-alaninals (2) can be controlled by the N-protecting groups. Thus the reaction with NHBoc-2 is *syn*-selective, whereas NHTos-protected 2 reacts to give a 1:1 mixture of *anti*- and *syn*-3. In contrast,

	anti - 3	*syn* - 3
R^1 = H, R^2 = Boc 75%	25:75	
R^1, R^2 = Bzl 80%	90:10	
R^1 = Bzl, R^2 = Boc 85%	93:7	

diprotected D-alaninals react with high *anti*-selectivity and the diastereoselectivity is enhanced with increasing steric hindrance.

[1] J. Jurczak, A. Golebiowski, and J. Raczko, *J. Org.*, **54**, 2495 (1989).

Ethyl diazoacetate, $N_2=CHCOOC_2H_5$ (1).

RCHO → RCOCH₂CO₂C₂C₅.[1] This reaction can be effected with 1 in the presence of various Lewis catalysts, but $SnCl_2$, $GeCl_2$, and BF_3 are the most efficient, permitting reactions at $-25°$. Yields are generally 50–90%.

$$C_6H_5CH_2CH_2CHO \xrightarrow[\substack{86\%}]{\substack{1, SnCl_2, \\ CH_2Cl_2, -15°}} C_6H_5(CH_2)_2\overset{\overset{\displaystyle O}{\|}}{C}CH_2COOC_2H_5$$

[1] C. R. Holmquist and E. J. Roskamp, *J. Org.*, **54**, 3258 (1989).

Ethyl α-(hydroxymethyl)acrylate, $CH_2=C(CH_2OH)COOC_2H_5$. Preparation (**13,** 322).[1]

α-Methylene-γ-butyrolactones.[2]   These lactones have been obtained by reaction of the nickel carbonyl complex of 2-(bromomethyl)acrylate with carbonyl compounds (**6,** 93). They are also formed by allylation of carbonyl compounds with ethyl α-(hydroxymethyl)acrylate and SnCl₂ catalyzed by $PdCl_2(C_6H_5CN)_2$ in 1,3-dimethyl-imidazolidinone (DMI)/H_2O.

$$CH_2=C\substack{\diagup COOC_2H_5 \\ \diagdown CH_2OH} + C_6H_5CHO \xrightarrow[33\%]{\substack{Pd(II), \\ SnCl_2}} H_2C$$

This reaction shows high *syn*-selectivity when the hydroxymethyl group bears an alkyl substituent.

$$CH_2=C\substack{\diagup COOCH_3 \\ \diagdown CHOH \\ | \\ CH_3} + BuCHO \xrightarrow[40\%]{} H_2C$$

(*syn/anti* = 95:5)

[1] J. Villieras and M. Rambaud, *Org. Syn.*, **66**, 220 (1987).
[2] Y. Masuyama, Y. Nimura, and Y. Kurusu, *Tetrahedron Letters*, **32**, 225 (1991).

Ethyl isocyanoacetate, $CNCH_2COOC_2H_5$.

Pyrrole synthesis.   A new route to pyrroles[1,2] is based on a base-catalyzed Michael addition of an alkyl isocyanoacetate to a nitroalkene to give an intermediate that cyclizes to a pyrrole. The nitroalkene is generally obtained from a β-acetoxy nitroalkane (**1**), prepared by a nitro aldol reaction of an aldehyde with a nitroalkane. The synthesis of ethyl 3,4-diethylpyrrole-2-carboxylate (**2**) is typical.

Barton observed that this synthesis provides particularly useful precursors to porphyrins since the pyrroles are unsubstituted at C_5, and the 3- and 4-substituents

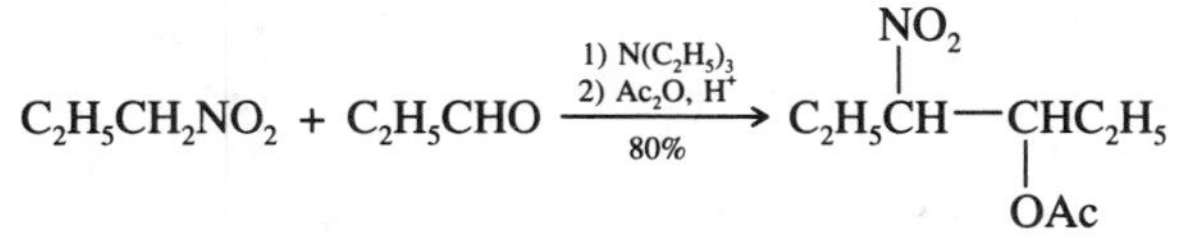

$$C_2H_5CH_2NO_2 + C_2H_5CHO \xrightarrow[80\%]{\substack{1)\ N(C_2H_5)_3 \\ 2)\ Ac_2O,\ H^+}} C_2H_5CH-CHC_2H_5$$

with the product bearing NO_2 and OAc groups

1

$$\mathbf{1} \xrightarrow{DBU} \left[\begin{array}{c} \underset{C_2H_5}{\overset{H}{\diagdown}}C=C\underset{C_2H_5}{\overset{NO_2}{\diagup}} \end{array} \right] \xrightarrow[DBU]{CNCH_2COOC_2H_5} \left[\begin{array}{c} ROOC{-}N \\ C_2H_5 {-}NO_2 \\ C_2H_5 \end{array} \right]$$

$$\xrightarrow[\text{from } \mathbf{1}]{\substack{86\% \\ -HNO_2}}$$

structure **2**: pyrrole with C_2H_5, C_2H_5 substituents and $COOC_2H_5$

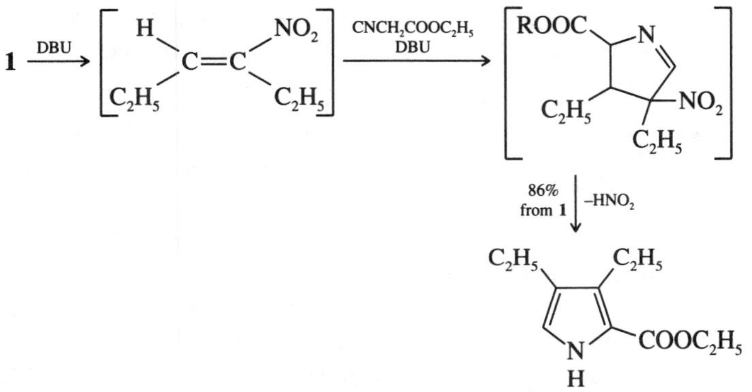

2

are determined by the structure of the nitroalkene, which is subject to wide variation. The pyrrole-2-carboxylates can be used directly for porphyrin synthesis by careful reduction with $LiAlH_4$ at $-20°$ to provide 2-hydroxymethylpyrroles. These undergo tetramerization directly to porphyrins in the presence of acid (TsOH) and an oxidant

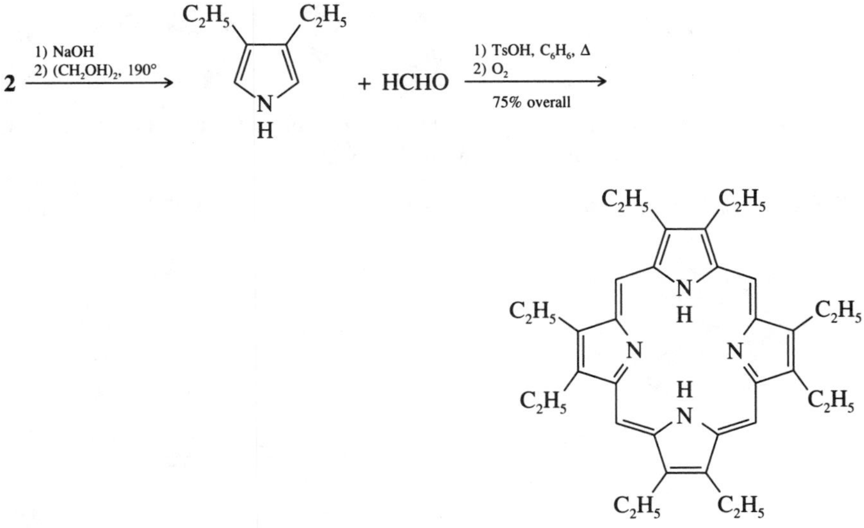

$$\mathbf{2} \xrightarrow[2)\ (CH_2OH)_2,\ 190°]{1)\ NaOH} \quad \text{(pyrrole with two } C_2H_5 \text{ groups)} \quad + \text{ HCHO } \xrightarrow[75\%\ \text{overall}]{\substack{1)\ TsOH,\ C_6H_6,\ \Delta \\ 2)\ O_2}}$$

porphyrin structure **3** (octaethylporphyrin)

3

(O_2 or chloranil). Yields are improved by addition of methylal as a source of formaldehyde. Under the best conditions, the pyrrole 2 was converted to octaethylporphyrin (3) in 55% yield.[2] This route to porphyrins is particularly useful for synthesis of porphyrins containing hindered aryl groups at C_3 of pyrrole.

Octaethylporphyrin can also be obtained from 2 by a more conventional approach.[3] The α-ester group is removed by saponification and decarboxylation, and the resulting 3,4-diethylpyrrole is condensed in benzene with aqueous formaldehyde in the presence of TsOH with removal of water and then oxidized with air (yield 75%).[3]

[1] D. H. R. Barton, J. Kervagoret, and S. Z. Zard, *Tetrahedron*, **46**, 7587 (1990).
[2] N. Ono, H. Kawamura, M. Bougauchi, and K. Maruyama, *ibid.*, **46**, 7483 (1990).
[3] J. L. Sessler, A. Mozaffari, and M. R. Johnson, *Org. Syn.*, **70**, 68 (1991).

Ethyl 3-oxo-4-pentenoate (1, Nazarov reagent).

This annelation reagent can be prepared most simply by condensation of the lithium enolate of ethyl acetate with acrolein to provide ethyl 3-hydroxy-4-pentenoate (2). Jones's oxidation provides the corresponding ketone (1) in 63–67% overall

yield. Methacrolein and crotonaldehyde can be used in place of acrolein, with only a slight decrease in the yield of the corresponding vinylketo esters.[1]

[1] R. Zibuck and J. M. Streiber, *Org. Syn.*, submitted (1991); *J. Org.*, **54**, 4717 (1989).

F

Ferric chloride.

Ring expansion of 1-*silyloxycycloalkanecarbaldehydes.*[1]   The method for one-carbon ring expansion of α-(silylmethyl)carbaldehydes with $Cl_2AlOC_6H_5$ (**14**, 119–120) can be extended to ring expansion of unsubstituted silyloxycycloalkanecarbaldehydes (equation I), but shows slight or no regioselectivity in expansion of 2-

(I) 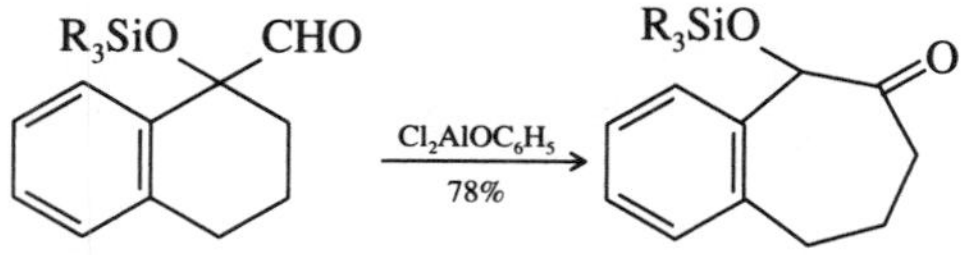 $\xrightarrow[78\%]{Cl_2AlOC_6H_5}$

substituted 1-silyloxycyclohexanecarbaldehydes. In this case $FeCl_3$ is the preferred Lewis acid (equations II, III).

(II) 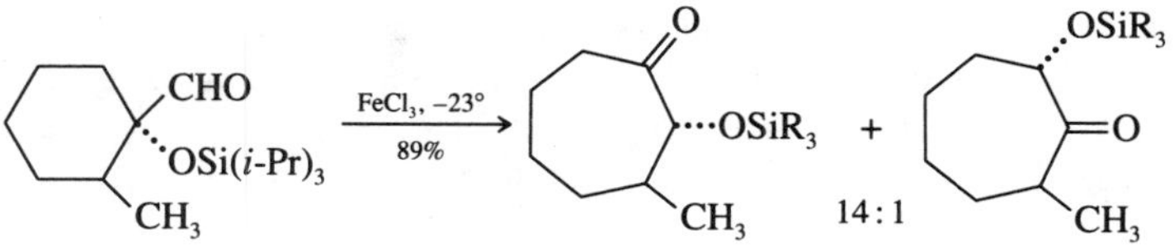 $\xrightarrow[89\%]{FeCl_3,\ -23°}$ ...   14:1

(III) 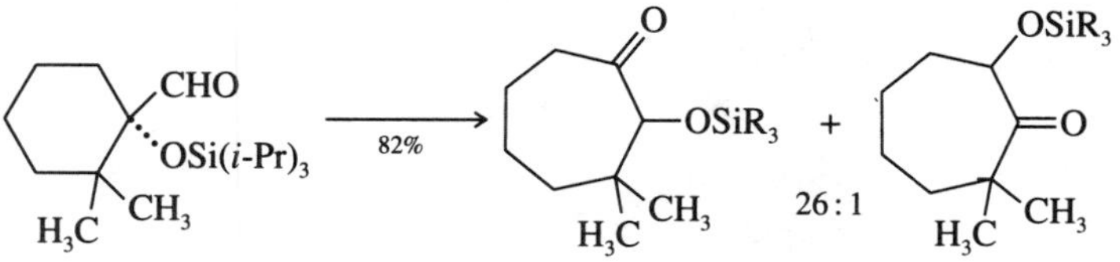 $\xrightarrow[82\%]{}$ ...   26:1

1,4-*Addition to acrylates.*[2] $FeCl_3$ is the most efficient Lewis-acid catalyst for 1,4-addition of primary or secondary amines to acrylates.

$$CH_2{=}CHCOOC_2H_5 + HN(C_2H_5)_2 \xrightarrow[96\%]{\substack{FeCl_3,\\ CH_2Cl_2,\ 25°}} (C_2H_5)_2NCH_2CH_2COOC_2H_5$$

$$CH_2{=}CHCOOC_2H_5 + BuNH_2 \xrightarrow[79\%]{} BuNHCH_2CH_2COOC_2H_5$$

Si-directed Nazarov cyclization (**14**, 164–165).   Denmark *et al.*[3] have examined the stereoselectivity of this cyclization with optically active β'-silyldivinyl ketones such as (+)- and (−)-**1**. In each case the cyclization occurs with essentially complete control by Si in the *anti*-S'_E sense.

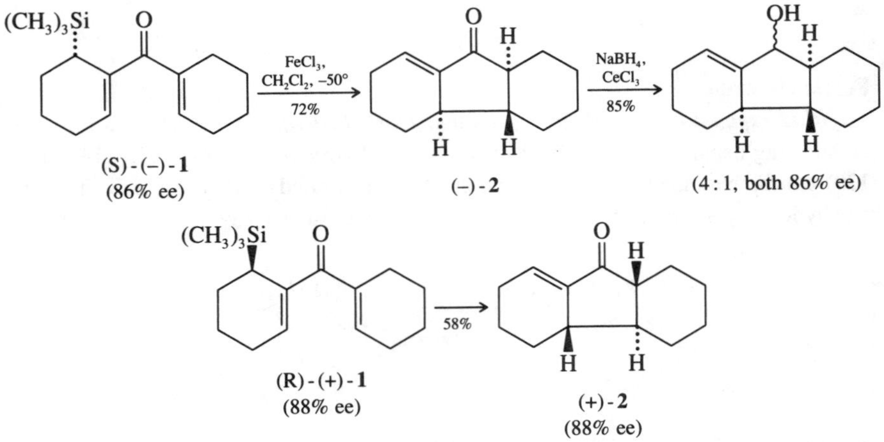

(S)-(−)-**1**
(86% ee)

(−)-**2**

(4:1, both 86% ee)

(R)-(+)-**1**
(88% ee)

(+)-**2**
(88% ee)

Stereoselective synthesis of perylenequinones.[4] Synthesis of the symmetrical perylenequinone phleichrome (**4**) has been effected by coupling of two identical naphthalene units to provide a binaphthol, which is then oxidized to a perylene-quinone. Thus the bromonaphthalene **1** on halo-lithium exchange (*t*-BuLi) followed by reaction of anhydrous FeCl₃ dimerizes to two optically active binaphthyls, (+)- and (−)-**2**, with 3:1 diastereoselectivity.

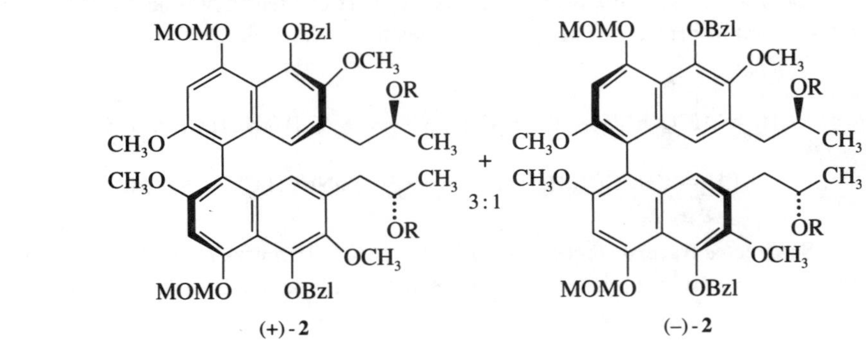

(S)-**1**, R = Si(C₆H₅)₂-*t*-Bu

(+)-**2**

(−)-**2**

The minor product of this dimerization, $(-)$-**2**, on desilylation and debenzylation provides $(-)$-**3**, which on treatment with $K_3Fe(CN)_6$ is converted into phleichrome (**4**) in 40% yield with retention of the axial chirality.

MOMO   OH
OCH$_3$
OH
CH$_3$O
CH$_3$
CH$_3$O
CH$_3$
OH
OCH$_3$
MOMO   OH

$(-)$-3

1) $K_3Fe(CN)_6$
2) H_3O^+
40%

HO   O
OCH$_3$
OH
CH$_3$O
CH$_3$
CH$_3$O
CH$_3$
OH
OCH$_3$
HO   O

4

Inversion of the stereocenter of (S)-**1** by the Mitsunobu reaction provides (R)-**1**. This bromonaphthalene is converted by the same sequence as above into the perylenequinone calphostin D (**5**). In this case the axial stereoselectivity in the dimerization step is opposite to that observed with (S)-**1**.

MOMO   OBzl
OCH$_3$
OR
CH$_3$O
CH$_3$
Br

(R)-1

1) t-BuLi
2) $FeCl_3$
3) $K_3Fe(CN)_6$

OH   O
OCH$_3$
OH
CH$_3$O
CH$_3$
CH$_3$O
CH$_3$
OH
OCH$_3$
OH   O

5

Of a number of metal salts tested, $CoCl_2$, $CuCl_2$, and $(C_6H_5)_2CuCl_2$ also effected coupling of model naphthyl bromides, but they are less effective than $FeCl_3$.

[1] T. Matsuda, K. Tanino, and I. Kuwajima, *Tetrahedron Letters*, **30**, 4267 (1989).
[2] J. Cabral, P. Laszlo, L. Mahé, M.-T. Montaufier, and S. L. Randriamahefa, *ibid.*, **30**, 3969 (1989).
[3] S. E. Denmark, M. A. Wallace, and C. B. Walker, Jr., *J. Org.*, **55**, 5543 (1990).
[4] C. A. Broka, *Tetrahedron Letters*, **32**, 859 (1991).

Ferrocenylphosphines, chiral, 11, 237–240; 14, 165–167.

Asymmetric cross-coupling.[1]   Japanese chemists have prepared a new chiral C_2-symmetric ferrocenylphosphine, which combines with $PdCl_2$ to form a highly active catalyst (1) for asymmetric cross-coupling of alkylzinc chlorides with vinyl bromides.

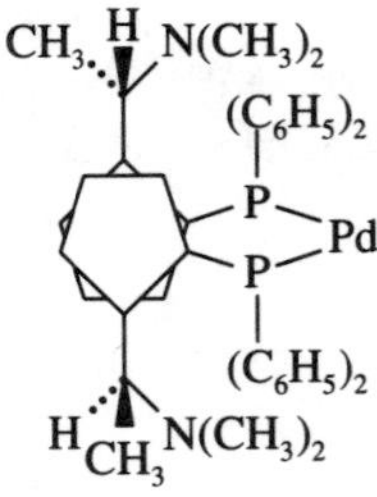

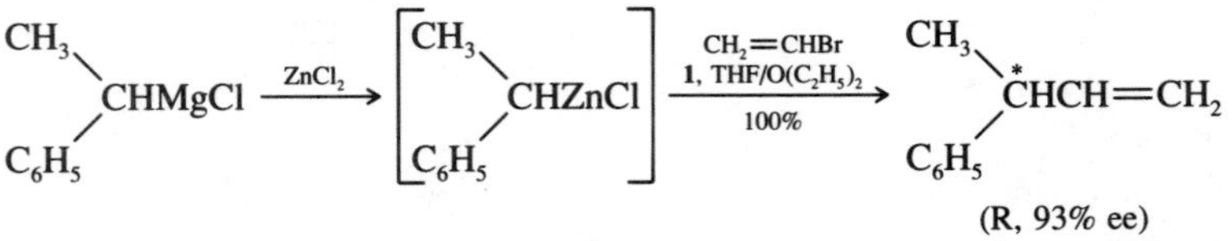

(R, 93% ee)

[1] T. Hayashi, A. Yamamoto, M. Hojo, and Y. Ito, *J.C.S. Chem. Comm.*, 495 (1989).

N-Fluoropyridinium trifluoromethanesulfonate.

Although a number of electrophilic fluorinating reagents are known, most are toxic, unstable, and difficult to handle. These new, stable N-fluoropyridinium salts are readily obtained in 60–80% yield by reaction of a pyridine with $F_2/N_2(1/9)$ in CH_3CN in the presence of sodium triflate.

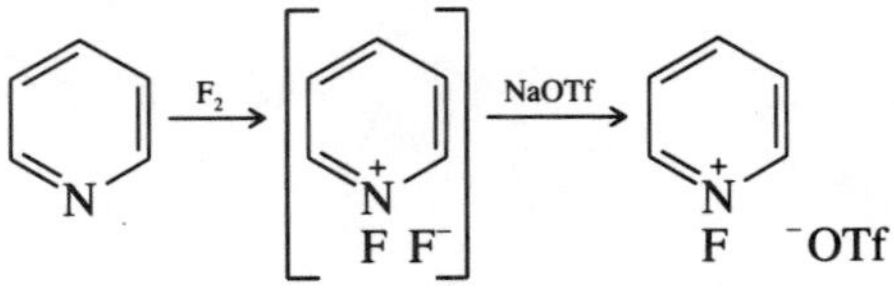

Fluorination.[1]   One advantage of these fluorinating reagents is that the reactivity can be controlled by variation of substituents on the pyridine ring. The 2,4,6-trimethylpyridinium triflate is less active than the parent reagent, whereas the pentachloro triflate is the most reactive. Because of this variation, a wide range of nucleophiles can be fluorinated, and selective fluorinations are possible. Benzene and naphthalene are fluorinated by the pentachloro salt in CH_2Cl_2, but phenol and naphthol are fluorinated by the parent unsubstituted salt. The trimethyl triflate converts alkyl or aryl Grignards into fluorides, but not the corresponding organolithium reagents. The trimethyl triflate can also fluorinate active methylene compounds when activated by $ZnCl_2$ or $AlCl_3$. Thus by variation of substituents on the ring, it is possible to fluorinate aromatics, carbanions, active methylene compounds, enol alkyl or silyl ethers, vinyl acetate, and even alkenes. These triflates can also be used for regioselective fluorination. Thus fluorination of the steroid **2** possessing three enol silyl ether groups with 1 equiv. of the parent N-fluoropyridinium triflate gives mainly the 9α-fluorosteroid **3**. No reaction was observed between **2** and the 2,4,6-trimethyl derivative of **1**. The *ortho/para* ratio of fluorination of phenols can be varied remarkably by change in the counter ion. In contrast to triflates, tetrafluoroborates of **1** favor *ortho*-substitution.

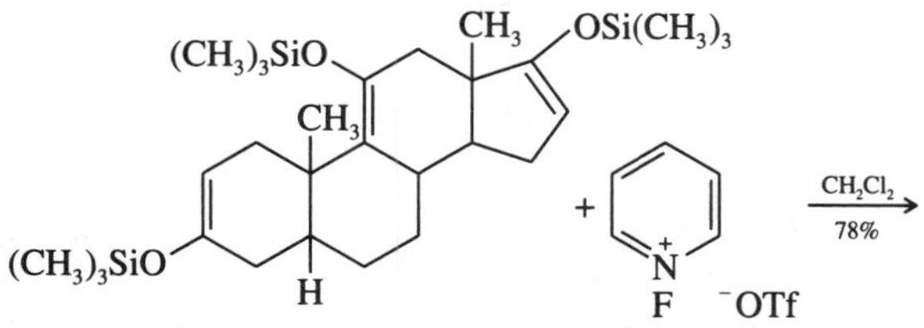

[1] T. Umemoto and K. Tomita, *Tetrahedron Letters*, **27**, 3271 (1986); T. Umemoto, S. Fukami, G. Tomizawa, K. Harasawa, K. Kawada, and K. Tomita, *Am. Soc.*, **112**, 8563 (1990).

Formaldehyde.

Hydroxymethylation of anthraquinones (Marschalk reaction). Krohn[1] has reviewed this reaction, particularly for the synthesis of anthracyclinones. It is particularly useful for preparation of optically active rhodomycinones by use of chiral aldehydes (166 references).

[1] K. Krohn, *Tetrahedron*, **46**, 291 (1990).

G

Grignard reagents.

Benzannelation of ketones.[1] This reaction can be effected by reaction of an allylic Grignard reagent such as methallylmagnesium chloride with the silyl ether of a 2-hydroxymethylene ketone as shown by conversion of **1** to **2** in 86% overall yield.

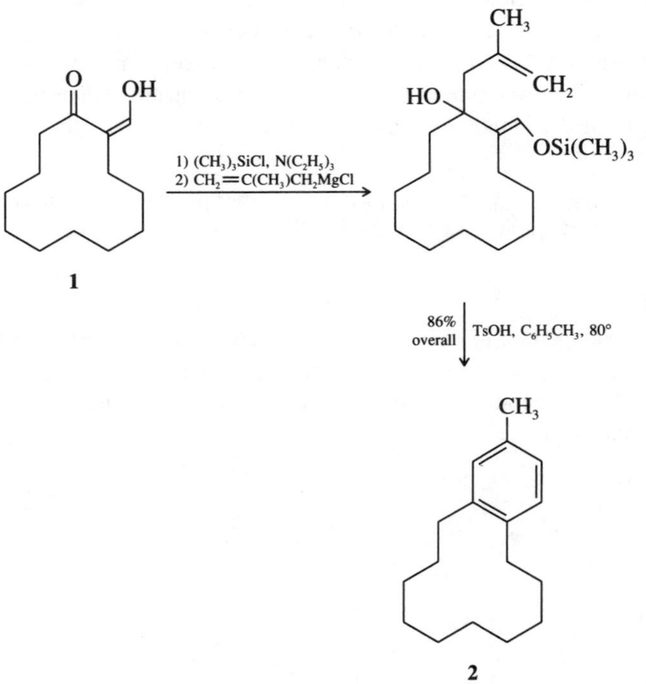

High yields require a cation-stabilizing group at C_2 of the Grignard reagent. Thus use of allylmagnesium chloride suffers from low yields, but methyl-unsubstituted benzannelation can be effected with 2-trimethylsilyl-2-propenylmagnesium chloride. Use of benzylmagnesium bromide results in annelation to a naphthalene derivative. Acyclic ketones as well as phenones can be precursors. Use of the latter results in biphenyls (**3** → **4**).

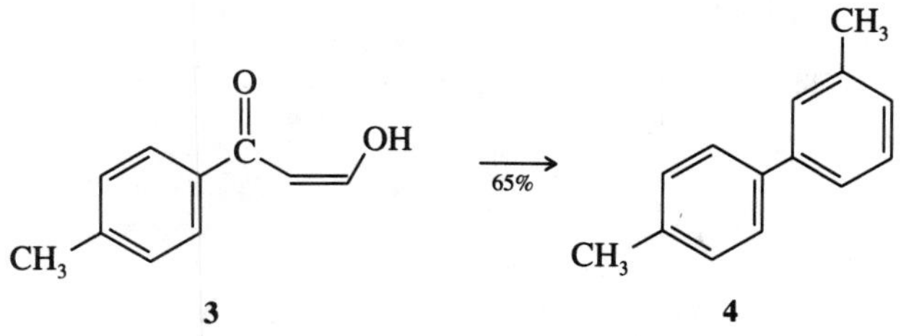

3   4

[1] M. A. Tius and G. S. K. Kannangara, *Org. Syn.*, submitted (1990).

H

Hexabutylditin.

Atom transfer cyclization (**14**, 173–174). This method of cyclization can be more useful than reductive tin hydride cyclization in the case of 6-substituted 5-hexynyl iodides such as **1** and has the further advantage of resulting in a vinyl iodide, which can be used for further functionalization.[1] The reaction is best conducted by photolysis in benzene at 80° with catalytic amounts of hexabutylditin.

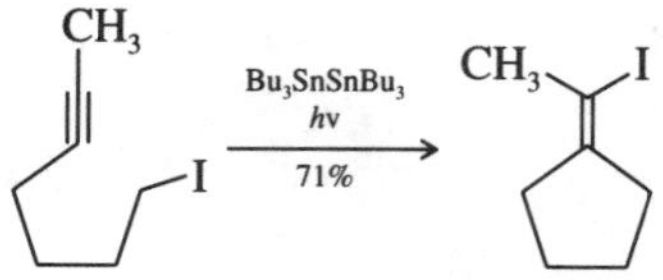

1 **2**

[1] D. P. Curran, M.-H. Chen, and D. Kim, *Am. Soc.*, **111**, 6265 (1989).

Hexamethylphosphoric triamide (HMPA).

Vinyl sulfides(selenides). Dialkyl disulfides or diselenides react in HMPA with vinyl bromides in the presence or absence of CuI to form vinyl sulfides or selenides

$$C_6H_5CH{=}CHBr + C_6H_5SeSeC_6H_5 \xrightarrow[82\%]{HMPA,\ \Delta} C_6H_5CH{=}CHSeC_6H_5 \quad (E/Z = 5:1)$$

$$CH_2{=}C(Br)CH_2C_6H_5 + C_6H_5SSC_6H_5 \xrightarrow[60\%]{HMPA\ (CuI)} C_6H_5CH{=}C(CH_3)SC_6H_5$$

as the major product. DMF and DMSO are less effective as the solvent, and THF and dioxane are ineffective.

[1] T. Ogawa, K. Hayami, and H. Suzuki, *Chem. Letters*, 769 (1989).

Hydrazine.

Azapeptides. Gante[1] has reviewed the synthesis and properties of azapeptides, substances in which the α-CH group of one or more amino acids is replaced by a

$$\text{(I)} \quad H_2NNHR + O{=}C{=}\overset{\overset{\displaystyle R^1}{|}}{N}CHCOOR^2 \longrightarrow H_2\overset{\overset{\displaystyle R}{|}}{N}NCO{-}\overset{\overset{\displaystyle R^1}{|}}{N}HCHCOOR^2$$

nitrogen atom. These peptides are of interest because of possible more potent drug activity or improved enzymic reactions. Most of the azaamino acids analogs of natural amino acids can be prepared by reaction of Boc-hydrazine with aldehydes or ketones to give the Boc-hydrazones, followed by catalytic hydrogenation. The amide linkage of azaamino acids can be formed by reaction of hydrazines with α-isocyanato esters (equation I). More frequently, the active esters usual to peptide synthesis are employed because of greater flexibility.

Wharton reaction[2] (1, 439–440).   This rearrangement of α,β-enones to allylic alcohols via an epoxy hydrazone has seen only occasional use outside of the sterol field because yields can be low. Some expedients have been shown by Dupuy and

Luche to improve the yield. When the intermediate epoxy hydrazone is isolable, addition of a strong base, potassium *t*-butoxide or potassium diisopropylamide (KDA), improves the yield. Triethylamine is the base of choice in the case of an unstable epoxy hydrazone (equation I).

[1] J. Gante, *Synthesis*, 405 (1989).
[2] C. Dupuy and J. L. Luche, *Tetrahedron*, **45**, 3437 (1989).

Hydridotris(triphenylphosphine)copper hexamer (1), **14**, 175; **15**, 166.

Reduction of alkynes to cis-*alkenes.*[1]   This relatively stable copper hydride (**1**) reduces terminal alkynes at 25°; internal alkynes are reduced at 80° to *cis*-alkenes exclusively. These reductions require 2 hydride equivalents of the complex, and 5–10 equiv. of H_2O is necessary for satisfactory yields. Under these conditions, a tertiary propargyl acetate was converted to an allene (equation I).

[1] J. F. Daeuble, C. McGettigan, and J. M. Stryker, *Tetrahedron Letters*, **31**, 2397 (1990).

Hydrofluoric acid–Boron trifluoride.

Cyclodehydration. The commercial preparation of an antihistamine (Lorata-dine, **3**) requires cyclodehydration of the ketone **1**. Use of the usual reagent for this reaction, polyphosphoric acid at 190°, leads to a mixture of two products in about 45% yield. The most useful and reasonably priced reagent is HF and BF_3 at −30°, which gives **2** in >90% yield.

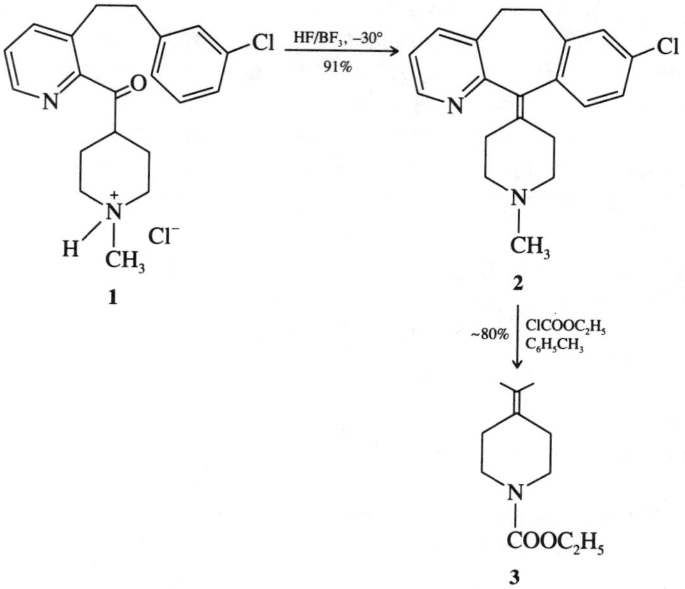

[1] D. P. Schumacher, B. L. Murphy, J. E. Clark, P. Tahbaz, and T. A. Mann, *J. Org.*, **54**, 2242 (1989).

Hydrogen azide–Diisopropylethylamine.

Selective cleavage of 2,3-epoxy esters.[1] The highly toxic reagent $i\text{-}Pr_2(C_2H_5)$-NHN_3 (**1**)[2], obtained by combination of hydrogen azide and the amine, cleaves 2,3-

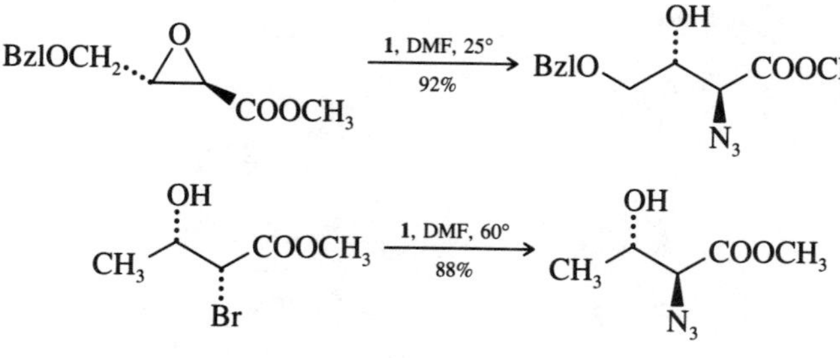

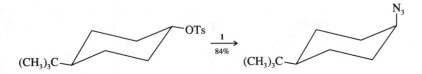

epoxy esters with marked preference for attack at C_2. Note that a similar tendency obtains in cleavage with dialkylamines (13, 312). The C_2-selectivity is consistently high (10–73:1) for chiral *trans*-2,3-epoxy esters, and depends on the steric bulk of the C_3-substituent. However, the C_2-selectivity for *cis*-2,3-epoxy ester is generally low (1.3–5.8:1). Even so, this reaction provides a useful route to β-hydroxy-α-amino acids. The reagent can also convert halides and sulfonates into azides.

[1] S. Saito, N. Takahashi, T. Ishikawa, and T. Moriwake, *Tetrahedron Letters*, **32**, 667 (1991).
[2] S. Saito, H. Yokoyama, T. Ishikawa, N. Niwa, and T. Moriwake, *ibid.*, **32**, 663 (1991).

Hydrogen peroxide–Areneseleninic acids.

Baeyer–Villiger oxidation.[1] Aromatic aldehydes can be converted to phenols in generally high yield by oxidation with hydrogen peroxide (30%) activated by an areneseleninic acid. Polymethoxyacetophenones are also oxidized to phenols, but in lower yields.

$$2,5\text{-}(CH_3O)_2C_6H_3CHO \xrightarrow[\substack{H_2O,\ CH_2Cl_2,\ 25°}]{H_2O_2,\ C_6H_5SeO_2H} \left[\begin{array}{c} O \\ \| \\ ArOCH \end{array}\right] \xrightarrow[93\%]{OH^-} 2,5\text{-}(CH_3O)_2C_6H_3OH$$

$$2,4\text{-}(CH_3O)_2C_6H_3COCH_3 \xrightarrow[\substack{[(NO_2)_2C_6H_3Se]_2 \\ 65\%}]{H_2O_2} 2,4\text{-}(CH_3O)_2C_6H_3OH$$

[1] L. Syper, *Synthesis*, 167 (1989).

Hydrogen peroxide–Dimethyl sulfoxide.

Hydration of nitriles. This conversion can be effected conveniently in 65–99% yield by reaction of nitriles with H_2O_2 (30%) in DMSO in the presence of K_2CO_3 as catalyst. The reaction is complete in 1–30 minutes and ester or urethane bonds are not affected.

[1] A. R. Katritzky, B. Pilarski, and L. Urogdi, *Synthesis*, 949 (1989).

Hydrogen peroxide–Formic acid–Polyphosphoric acid (PPA).

$>S \rightarrow >S{=}O$.[1] This oxidation can be carried out with H_2O_2 and excess formic or acetic acid in CH_2Cl_2.[1] If polyphosphoric acid (1 equiv.) is also present, this procedure is useful for oxidation of penicillin and cephalosporin esters, particularly diphenylmethyl esters (generally >90% yield).[2]

[1] R. G. Micetich, R. Singh, and S. N. Maiti, *Heterocycles*, **22**, 531 (1984).
[2] M. Tanaka, T. Konoike, and M. Yoshioka, *Synthesis*, 197 (1989).

Hydrogen peroxide–Tungstic acid, H_2O_2–H_2WO_4.

Oxidative cleavage of alkenes to carboxylic acids.[1]   Alkenes are oxidized to carboxylic acids by H_2O_2 (35%) catalyzed by H_2WO_4 in a weakly acidic medium (pH 4–5) maintained by addition of KOH. The oxidation probably involves initial oxidation to a 1,2-diol followed by dehydrogenation to an α-ketol, which is then cleaved to a mono- or dicarboxylic acid.

$$CH_3(CH_2)_5CH{=}CH_2 \longrightarrow CH_3(CH_2)_5COOH + CH_3(CH_2)_5CHOHCH_2OH$$

                                   (33%)   (16%)

[1] T. Oguchi, T. Ura, Y. Ishii, and M. Ogawa, *Chem. Letters*, 857 (1989).

(R)-3-Hydroxybutanoic acid.

Complete details are available[1] for depolymerization of the biopolymer of this acid, which is produced on an industrial scale and is available from Fluka and Aldrich.

[1] D. Seebach, A. K. Beck, R. Breitschuh, and K. Job, *Org. Syn.*, submitted (1990).

(2S,5R)-(−)-2-Hydroxydiphenylmethyl-5-(dimethylamino)methylpyrrolidine, 1. The pyrrolidine is prepared from ethyl (S)-N-(ethoxycarbonyl)pyroglutamate.

Enantioselective addition of $(C_2H_5)_2Zn$ to C_6H_5CHO.[1]   Guided by results of Noyori's group with a β-amino alcohol (**14**, 233–234) and of his own group with the oxazaborolidines, Corey et al.[1] have prepared the pyrrolidine **1** as a controller for enantioselective addition of R_2Zn to aldehydes. Reaction of **1** with $(C_2H_5)_2Zn$ forms a crystalline complex (**2**). This complex can function as an efficient catalyst for enantioselective addition of $(C_2H_5)_2Zn$ to benzaldehyde to form (S)-1-phenylpropanol in >95% yield and in 94% ee. An intermediate complex such as **a** may be involved.

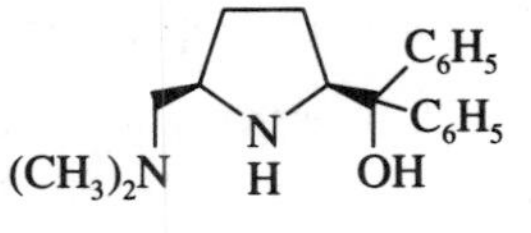

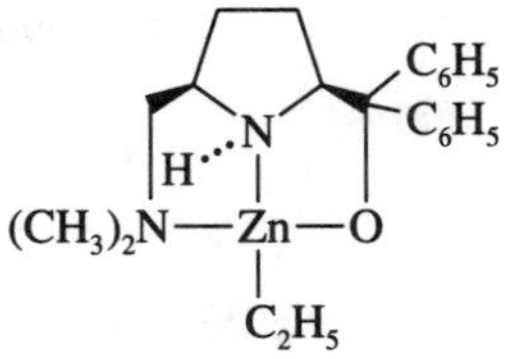

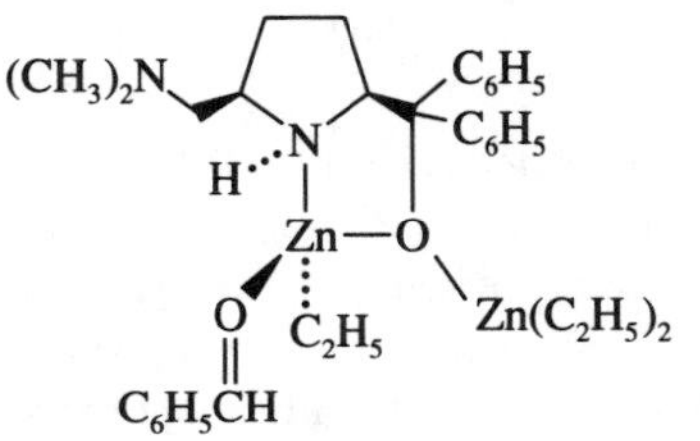

[1] E. J. Corey, P.-W. Yuen, F. J. Hannon, and D. A. Wierda, *J. Org.*, **55**, 784 (1990).

[Hydroxy(tosyloxy)iodo]benzene, $CH_3C_6H_4SO_3I(OH)C_6H_5$ (1), 14, 179–180.

Flavanones → Isoflavones. Although Koser's reagent (1) is known to effect α-tosyloxylation of ketones,[1] the reaction with flavanones (2) results in an oxidative 1,2-aryl shift to provide isoflavones (3) in 74–80% yield.[2] This conversion has been effected previously with thallium salts.

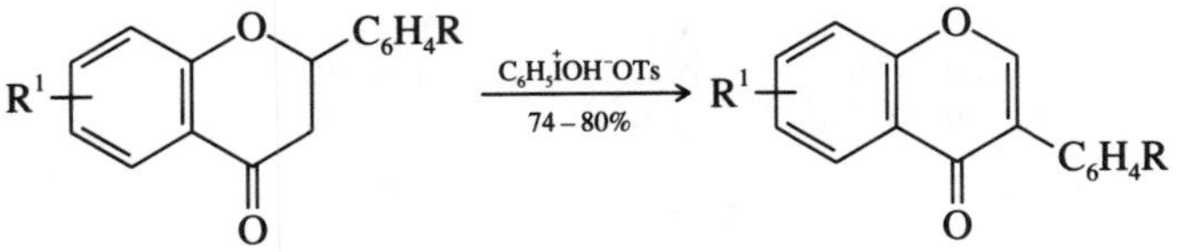

Review. Moriarty and Koser[3] have reviewed this hypervalent iodine(III) compound. It is particularly useful for *vic*, *cis*-ditosyloxylation of alkenes, α-tosylation of ketones, and conversion of primary amides to amines.

[1] G. F. Koser, A. G. Relenyi, A. N. Kalosa, L. Rebrovic, and R. H. Wettach, *J. Org.*, **47**, 2487 (1982).
[2] O. Prakash, S. Pahuja, S. Goyal, S. N. Sawhney, and R. M. Moriarty, *Synlett*, 337 (1990).
[3] R. M. Moriarty, R. K. Vaid, and G. F. Koser, *Synlett*, 365 (1990).

(R)-(+)-2-Hydroxy-1,2,2-triphenylethyl acetate (1). Preparation from methyl (R)-(−)-mandelate[1]:

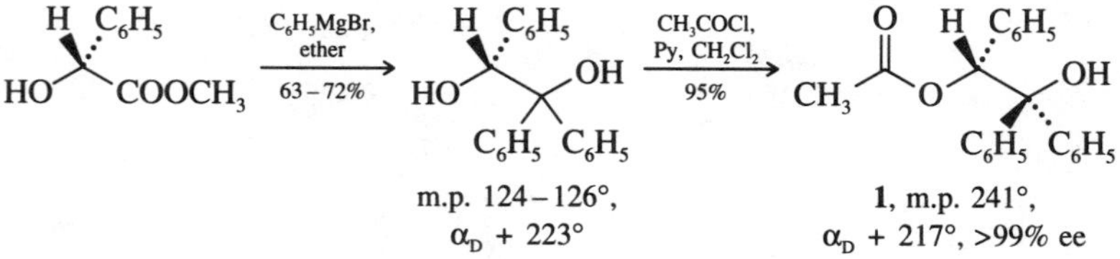

β-*Hydroxy carboxylic acids* (**12**, 3).[2]   This acetate on double deprotonation with LDA undergoes diastereoselective aldol reactions with aldehydes. The adducts are easily hydrolyzed to optically active β-hydroxycarboxylic acids with release of (R)-(+)-1,1,2-triphenyl-1,2-ethanediol, the precursor to **1**. Optically pure acids can be obtained by crystallization of the salt with an optically active amine such as (S)-(−)-1-phenylethylamine.

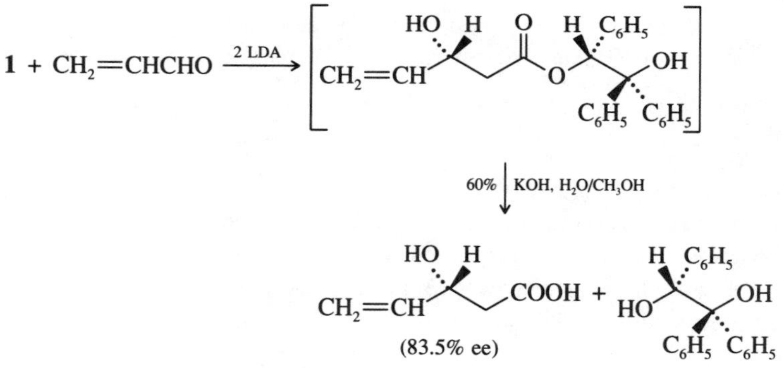

[1] M. Braun, S. Schneider, and S. Houben, *Org. Syn.*, submitted (1990).
[2] M. Braun and S. Schneider, *ibid.*, submitted (1990).

I

Indium, In.

Cyclopropanation.[1]   Electron-deficient alkenes react with active methylene dibromides and indium to form cyclopropanes in 35–95% yield. Reaction of aldehydes with dibromomalononitrile, lithium iodide, and indium also provides epoxides.

$$CH_2{=}CHCOCH_3 + Br_2C(CN)_2 \xrightarrow[94\%]{\substack{In,\ LiI,\\ DMF}}$$

(product: cyclopropane bearing two CN groups and $CH_3C(=O)$)

$$C_2H_5CHO + Br_2C\begin{smallmatrix}CN\\COOC_2H_5\end{smallmatrix} \xrightarrow{62\%}$$

(product: epoxide C_2H_5, H, CN, $COOC_2H_5$)

[1] S. Araki and Y. Butsugan, *J.C.S. Chem. Comm.*, 1286 (1989).

Indium(III) chloride, InCl₃.

Catalyst for reactions of O-trimethylsilyl monothioacetals with nucleophiles.[1]   Indium(III) chloride in combination with $ClSi(CH_3)_3$ is an effective catalyst

$$C_6H_5\overset{OSi(CH_3)_3}{\underset{SC_2H_5}{\overset{|}{C}}}CH_3 + (C_2H_5)_3SiH \xrightarrow[97\%]{\substack{ClSi(CH_3)_3,\ InCl_3\\ CH_2Cl_2,\ 25^\circ}} C_6H_5\overset{CH_3}{\underset{}{C}}{-}SC_2H_5$$

for conversion of O-trimethylsilyl monothioacetals to sulfides by reduction with $(C_2H_5)_3SiH$.

$$C_6H_5CH\begin{smallmatrix}OSi(CH_3)_3\\SC_2H_5\end{smallmatrix} + C_6H_5C\begin{smallmatrix}OSi(CH_3)_3\\CHCH_3\end{smallmatrix} \xrightarrow[83\%]{ClSi(CH_3)_3,\ InCl_3} C_6H_5\overset{O}{\overset{\|}{C}}CH\underset{CH_3}{\overset{SC_2H_5}{\overset{|}{C}}}HC_6H_5$$

$$60\%\ \Big\downarrow\ \substack{CH_2{=}CHCH_2Si(CH_3)_3\\ ClSi(CH_3)_3,\ InCl_3}$$

$$C_6H_5\overset{SC_2H_5}{\overset{|}{C}}HCH_2CH{=}CH_2$$

This system also promotes reaction of these acetals with silyl enol ethers or allyltrimethylsilanes.

[1] T. Mukaiyama, T. Ohno, T. Nishimura, J. S. Han, and S. Kobayashi, *Chem. Letters*, 2239 (1990).

Iodine.

Iodolactamization (**13**, 149).   Full details are available for this conversion of γ,δ-unsaturated amides into γ-lactams.[1] The report includes nine examples.

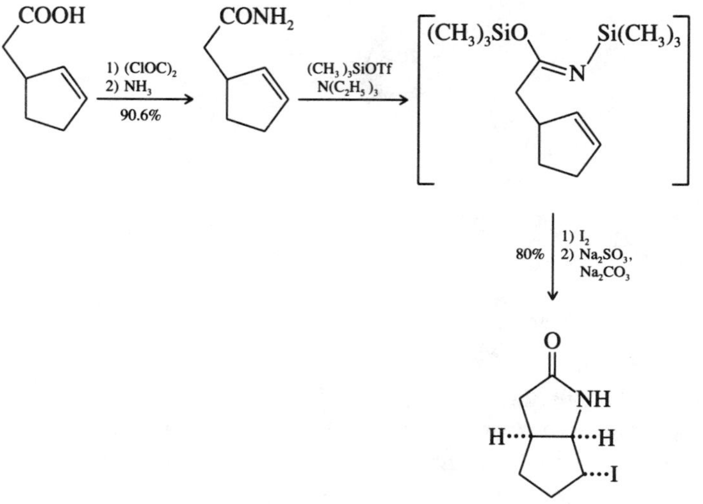

[1] S. Knapp and F. S. Gibson, *Org. Syn.*, **70**, 101 (1991).

Iodine–Borane: N,N-diethylaniline.

ROH → RI.[1]   Reaction of I_2 (1 equiv.) with BH_3: $C_6H_5N(C_2H_5)_2$ provides a diiodoborane: amine complex *in situ* that converts alcohols into alkyl iodides in 62–86% yield. It also effects reductive iodination of carbonyl compounds.

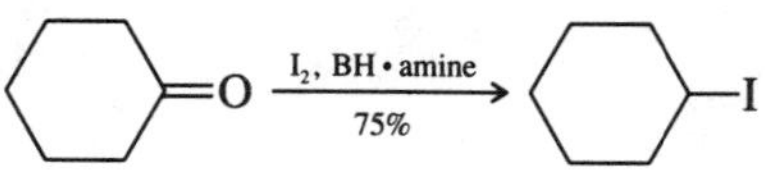

BI₃: amine complex.[2]   Reaction of BH_3: $C_6H_5N(C_2H_5)_2$ with 1.5 equiv. of iodine provides an amine complex of BI_3. This complex when combined with acetic acid liberates hydroiodic acid, which undergoes Markovnikov addition to alkenes and alkynes in good yield.

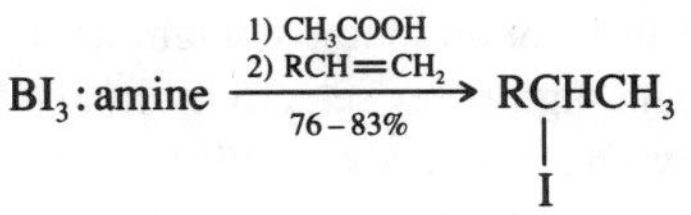

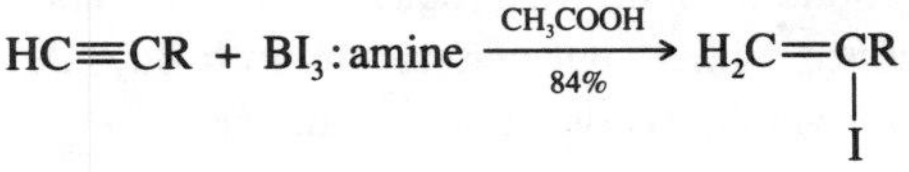

This BI_3 complex cleaves aryl methyl ethers to phenols at room temperature in good yield. Dialkyl ethers are converted into iodinated products. It also converts terminal *gem*-diacetates into aldehydes.[3]

[1] C. K. Reddy and M. Periasamy, *Tetrahedron Letters*, **30**, 5663 (1989).
[2] *Idem, ibid.*, **31**, 1919 (1990).
[3] C. Narayana, S. Padmanabhan, and G. W. Kabalka, *ibid.*, **31**, 6977 (1990).

Iodine–Silver nitrite, I_2–$AgNO_2$.

Iodination of arenes.[1] These two reagents generate nitryl iodide, which can iodinate arenes, particularly methylarenes, at 28°. The yield increases with the number of methyl substituents, being 90–95% with tri- and tetramethylbenzenes.[1]

[1] W.-W. Sy and B. A. Lodge, *Tetrahedron Letters*, **30**, 3769 (1989).

Iodomalononitriles, $RCI(CN)_2$. These reagents are prepared by reaction of $RCH(CN)_2$ with NaH in the presence of N-iodosuccinimide (NIS).

Radical reactions.[1] Unlike iodomalonic esters, iodomalononitriles can add to di- and trisubstituted alkenes. Thus the propargyliodomalononitrile **1** when heated at

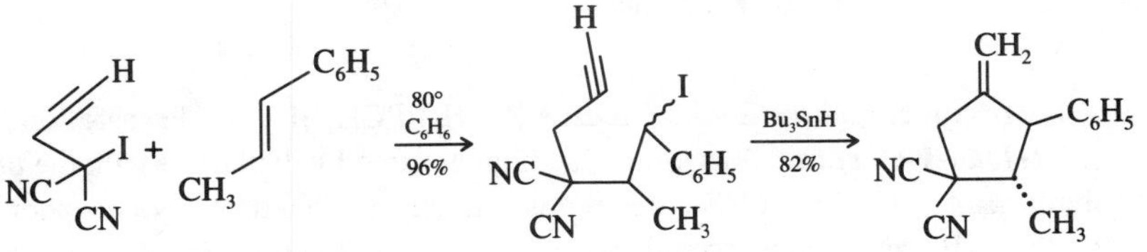

1 **2** **3** **4**

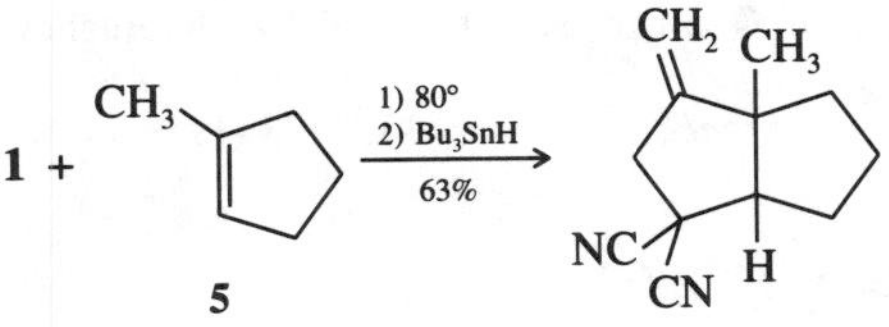

5 **6**

80° adds to an alkene (**2**) to form an iodine-containing adduct that cyclizes to **4** on treatment with Bu₃SnH. The high *trans* selectivity in a radical reaction is noteworthy. Similar reaction of **1** with a cyclic alkene (**5**) is best effected by a one-pot group transfer and cyclization with Bu₃SnH.

Reaction of crotyliodomalononitrile (equation I) with alkenes leads to adducts similar to those obtained with **1**, but cyclization with Bu₃SnH results in two products, the major one of which involves nitrile transfer.

(I)

20:1

Macrocyclization can also be effected with an iodomalononitrile such as **7**. To facilitate isolation, the crude atom-transfer product was reduced with Bu₃SnH. The resulting product, surprisingly, is the mononitrile **8**.

7   **8**

[1] D. P. Curran and C. M. Seong, *Am. Soc.*, **112**, 9401 (1990).

Iodomethyltriphenylphosphonium iodide [(C₆H₅)₃P̄CH₂I]I⁻ (**1**). Preparation.[1]

(*Z*)-**1-*Iodo*-1-alkenes.**   Reaction of sodium hexamethyldisilazane with the phosphonium salt **1** in THF generates iodomethylenetriphenylphosphorane, (C₆H₅)₃P=CHI, which converts aldehydes into (Z)-vinyl iodides (15–62 : 1) in reactions conducted at −78° in THF/HMPT in 61–91% yield.[2]

[1] D. Seyferth, J. K. Heeren, G. Singh, S. O. Grim, and W. R. Hughes, *J. Organometal Chem.*, **5**, 267 (1966).
[2] G. Stork and K. Zhao, *Tetrahedron Letters*, **30**, 2173 (1989).

Iodomethylzinc iodide, ICH₂ZnI.

Organocopper–zinc reagents.[1]   This Simmons–Smith reagent is obtained from CH₂I₂ and zinc in THF. It can react with some organocopper nucleophiles to give

$$ICH_2ZnI + RSCu \xrightarrow{-50°} [RSCH_2Cu \cdot ZnI_2]$$

$$(R = n\text{-}C_{10}H_{21}) \qquad \qquad \textbf{a}$$

$$75\% \downarrow \quad \overset{BrCH_2C=CH_2,\ 0°}{\underset{CH_3}{|}}$$

$$\overset{CH_3}{\underset{|}{}}$$
$$RSCH_2CH_2C=CH_2$$

insertion products formulated as **a**, which can then react with an activated electrophile such as an allylic halide to form a product in which the nucleophile is linked
to the electrophile by a methylene group.

This procedure is apparently limited to cyanomethylcopper derivatives, copper
amides, heteroarylcopper compounds, and copper thiolates. Use of bis(iodomethyl)-
zinc, $(ICH_2)_2Zn$, results in products from double insertion.

[1] P. Knochel, N. Jeong, M. J. Rozema, and M. C. P. Yeh, *Am. Soc.*, **111**, 6474 (1989).

N-Iodosuccinimide, NIS.

β-Lactams.[1]   A new synthesis of β-lactams involves oxidative coupling of dianions of acyclic amides such as **1**, prepared as shown from *p*-anisidine (equation I).
The amide **1** is converted into the dianion by BuLi (2 equiv.) and DABCO or by *t*-

(I)   1) $BrCH_2COOC(CH_3)_3$, $N(C_2H_5)_3$
      2) RCH_2COCl, $N(C_2H_5)_3$
      68–93%

1

BuLi. On treatment with an oxidant, NIS or $Cu(OCOR)_2$, the dianion undergoes
coupling to *cis*- and *trans*-β-lactams (**2**). Use of NIS favors formation of *cis*-lactams
regardless of the R substituent, whereas Cu(II) is slightly *trans*-selective.

1 (R = C_2H_5)   1) *t*-BuLi
                   2) NIS, −78°
                   51%   *cis*-**2**   + *trans*-**2**
                                                           10:1

1 (R=Bzl$_2$N) $\xrightarrow{48\%}$ *cis*-**2** + *trans*-**2**
                                            7:1

Use of (R)-(+)-1-phenylethylamine as the starting material furnishes an optically active amide (4), which was used for an asymmetric synthesis of a natural β-lactam.

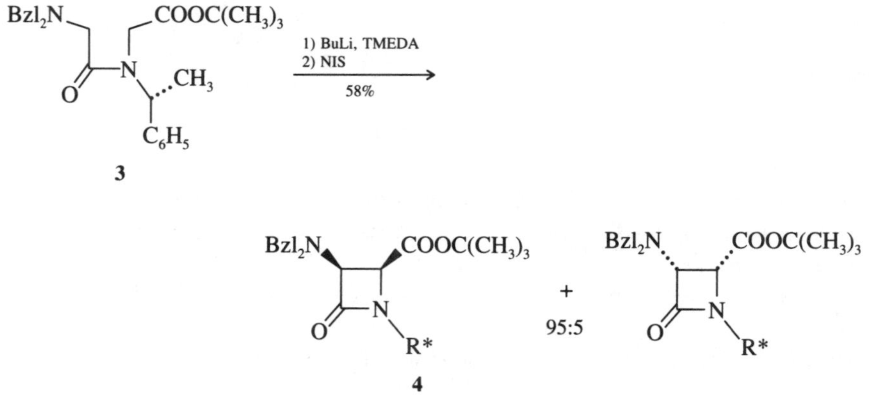

3

4

[1] T. Kawabata, K. Sumi, and T. Hiyama, *Am. Soc.*, **111**, 6843 (1989).

Iodosylbenzene, $C_6H_5I=O$.

Oxidation of 2-(trimethylsilyloxy)furan.[1]   Oxidation of 2-(trimethylsilyloxy)-furan with iodosylbenzene–BF_3 etherate can afford 5-substituted 2(5H)-furanones.

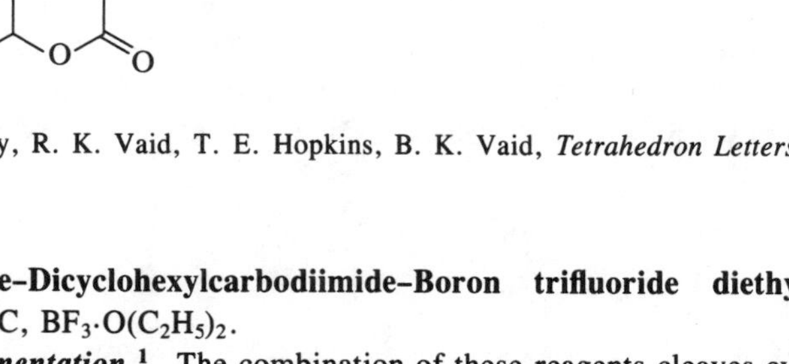

[1] R. M. Moriarty, R. K. Vaid, T. E. Hopkins, B. K. Vaid, *Tetrahedron Letters*, **30**, 3019 (1989).

Iodosylbenzene–Dicyclohexylcarbodiimide–Boron   trifluoride   diethyl   ether,
$C_6H_5I=O$, DCC, $BF_3 \cdot O(C_2H_5)_2$.

Grob fragmentation.[1]   The combination of these reagents cleaves cyclic γ-tri-butylstannyl alcohols at 25° in CH_2Cl_2 to unsaturated aldehydes and ketones.[2] The actual reagent is considered to be **1**, a modified Pfitzner–Moffatt reagent, which converts the substrate into an iodine(III) intermediate.

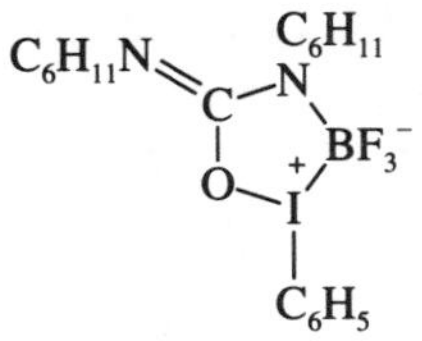

$$C_6H_{11}N=\underset{|}{\overset{C_6H_{11}}{C}}-\underset{|}{\overset{}{N}} \quad BF_3^- \quad (1)$$

$$\longrightarrow CH_2=CH(CH_2)_3CHO$$

$$+ C_6H_5I + (C_6H_{11}NH)_2C=O$$

$$\xrightarrow[81\%]{} CH_2=CH(CH_2)_3\overset{O}{\overset{\|}{C}}C_6H_5$$

$$\xrightarrow[67\%]{} CH_3\overset{O}{\overset{\|}{C}}(CH_2)_3CH\overset{(E)}{=}CHC_2H_5$$

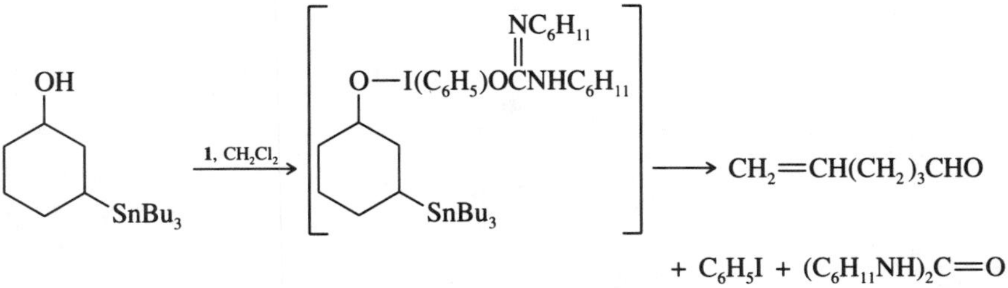

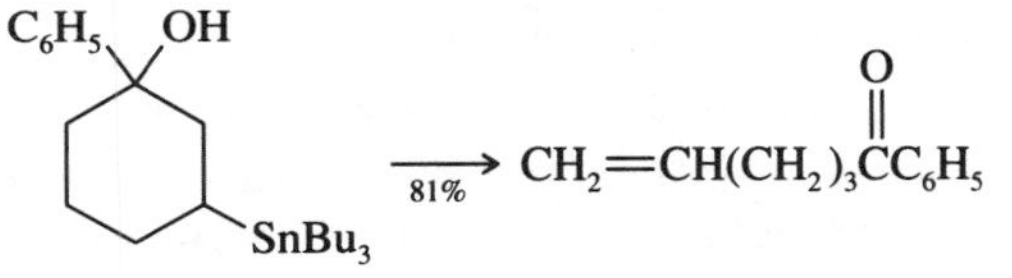

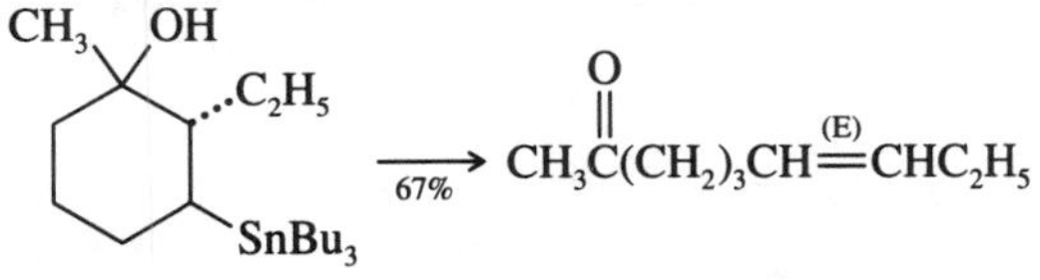

(89 : 11)

[1] C. A. Grob, *Angew. Chem. Int. Ed.*, **8**, 535 (1969).
[2] M. Ochiai, T. Ukita, S. Iwaki, Y. Nagao, and E. Fujita, *J. Org.*, **54**, 4832 (1989).

Iodosylbenzene tetrafloroborates, $(C_6H_5I^+)_2O\cdot2BF_4^-$ or $C_6H_5IO\cdot HBF_4$. These reagents are prepared by addition of HBF_4 to $C_6H_5I(OAc)_2$ in $CHCl_3$ or by reaction of
C_6H_5IO with $HBF_4\cdot O(CH_3)_2$ in CH_2Cl_2.

   α-Ketomethyl aryliodonium salts, $C_6H_5I^+CH_2COC_6H_5BF_4^-$ (2).[1] This salt (2)
can be generated by reaction of $C_6H_5IO\cdot HBF_4$ with the silyl enol ether of acetophe-
none, $C_6H_5COCH_3$. Such salts are not formed by reaction of $C_6H_5IO\cdot HBF_4$ with
aliphatic ketones, but the salt 2 couples with enol silyl ethers of aliphatic or aromatic

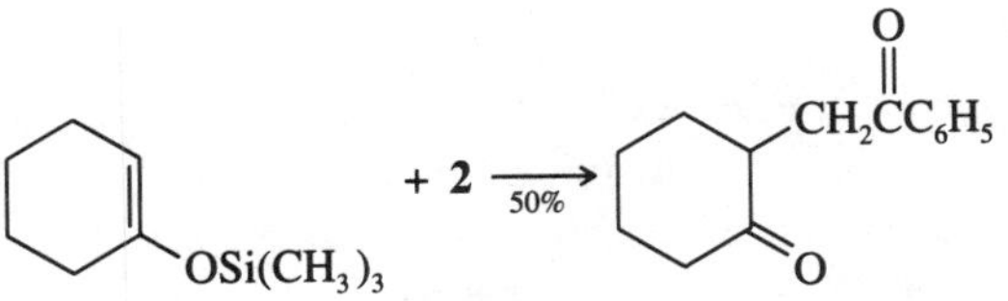

ketones in about 50% yield to form 1,4-butanediones (equation I) or with alkenes (equation II and III).

(II)

(III) $(CH_3)_2C=C(CH_3)_2 + 2 \xrightarrow{80\%}$

$(C_6H_5I^+)_2O\cdot2BF_4^-$ **(1).** This salt (yellow) is insoluble in water or $CHCl_3$ but soluble in polar organic solvents (DMSO, CH_3CN). The corresponding hexafluoroantimonate and hexafluorophosphate salts have similar properties. The reactivity of these salts can be enhanced by addition of $HBF_4\cdot O(CH_3)_2$. The principal difference between these salts and $C_6H_5IO\cdot HBF_4$ is their enhanced stability. Typical reactions of **1** with alkenes, silyl enol ethers, and alkynes are formulated (equations IV–VI).

(IV)

(*cis*)

(V) $C_6H_5\underset{\underset{OSi(CH_3)_3}{|}}{C}=CH_2 \xrightarrow[94\%]{\overset{1}{CH_2Cl_2}} C_6H_5\underset{\underset{O}{\|}}{C}CH_2CH_2\underset{\underset{O}{\|}}{C}C_6H_5$

(VI) $PrC\equiv CH \xrightarrow[42\%]{\overset{1}{CH_2Cl_2}} PrC\equiv CI^+C_6H_5BF_4^-$

[1] V. V. Zhdankin, M. Mulliken, R. Tykwinski, B. Berglund, R. Caple, N. S. Zefirov, and A. S. Koz'min, *J. Org.*, **54**, 2605, 2609 (1989).

Iodotrimethylsilane.

Deoxygenation of 1,4-endooxides.[1] The transformation of a 1,4-endooxide **(1)** to the corresponding aromatic hydrocarbon **(2)** can be effected in high yield with iodotrimethylsilane. The deoxygenation was used for a novel synthesis of 1,4-dimethylphenanthrene **(3)** as shown in equation (I).

1 2

(I)

3

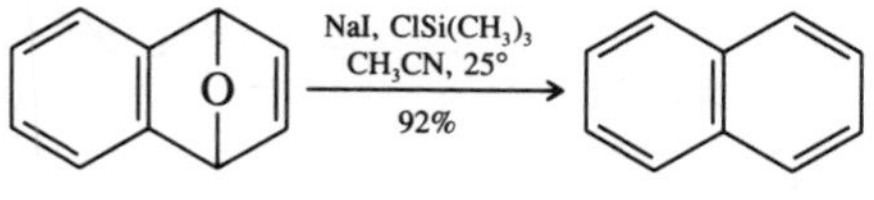

[1] K.-Y. Jung and M. Koreeda, *J. Org.*, **54**, 5667 (1989).

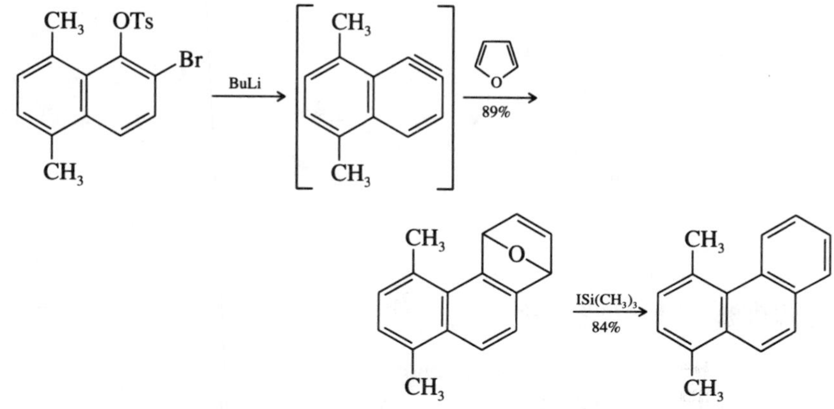

Iron(0).

Cyclization of trienes.[1] In combination with 2,2-bipyridyl (bpy), an iron(0) species prepared by reduction of iron(III) 2,4-pentanedionate with $Al(C_2H_5)_3$ (3 equiv.) promotes cyclization of trienes (**1**) in which a 1,3-diene unit is tethered to an allylic or homoallylic ether group. *vic*-Disubstituted cyclopentanes or cyclohexanes are formed, and the *cis/trans* disposition of the substituents is controlled by the geometry of the allylic double bond. Thus (Z,E-)**1** is cyclized to *trans*-**2**, whereas (E,E)-**1** is cyclized to *cis*-**2**.

(Z,E)-**1**, n = 1,2 *trans*-**2** *trans*-**3**

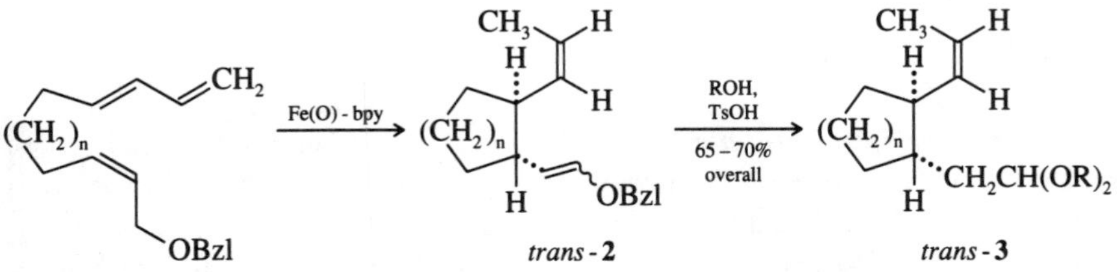

The presence of an oxygen atom in the linking chain lowers the yield of cyclization markedly, but cyclization is facilitated by a nitrogen atom in the chain. Thus this cyclization is a useful route to N-acylpiperidines (equation I).[2]

[1] J. M. Takacs, L. G. Anderson, M. W. Creswell, and B. E. Takacs, *Tetrahedron Letters*, **28**, 5627 (1987); J. M. Takacs and L. G. Anderson, *Am. Soc.*, **109**, 2200 (1987).
[2] B. E. Takacs and J. M. Takacs, *Tetrahedron Letters*, **31**, 2865 (1990).

Iron(III) chloride.

Oxidative phenolic coupling.[1] This coupling with $FeCl_3 \cdot 6H_2O$ (1) is more efficient when conducted in the solid state than in solution and can be accelerated by ultrasound in the former case. A catalytic amount of $FeCl_3$ is sometimes effective. Thus a mixture of β-naphthol and finely powdered $FeCl_3 \cdot 6H_2O$ when heated to 50° in a test tube is converted to 2 in 95% yield. The same reaction in refluxing CH_3OH-H_2O provides 2 in 60% yield. The water molecules of 1 are not involved because coupling to 2 can also be effected with $[Fe(DMF)_3Cl_2]^+FeCl_4^-$ in the solid state in 79% yield. The conversion of β-naphthol to 2 in 89% yield can also be effected with 0.2 molar amount of $FeCl_3 \cdot 6H_2O$ when irradiated with ultrasound at 50°.

Asymmetric intramolecular ene cyclization.[2] 1,7-Dienes, particularly those substituted by two electron-withdrawing groups on the enophile, undergo ene cyclization to *trans*-disubstituted cyclohexanes when heated or in the presence of Lewis acids (13, 349). This reaction can proceed with high diastereoselectivity when applied to the chiral 1,7-diene 1, obtained by a Knoevenagel reaction of (R)-citronellal with dimethyl malonate. The highest induced diastereoselectivity obtains in reactions at −78° → 20° catalyzed by $FeCl_3$ adsorbed on alumina. Unlike the more common Lewis acids such as $ZnBr_2$, only 0.1 equiv. of $FeCl_3$ is required for cyclization at −78°.

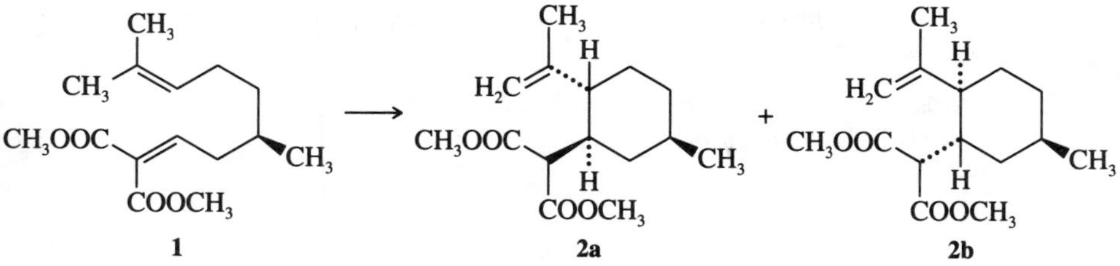

180°,	75%	89.7 : 10.3
ZnBr$_2$, 25°	86%	96.6 : 3.4
FeCl$_3$/Al$_2$O$_3$, −78°	77%	98.8 : 1.2

[1] F. Toda, K. Tanaka, and S. Iwata, *J. Org.*, **54**, 3007 (1989).
[2] L. F. Tietze, U. Beifuss, and M. Ruther, *ibid.*, **54**, 3120 (1989); L. F. Tietze and U. Beifuss, *Org. Syn.*, submitted (1990).

Iron(III) sulfate.

Dehydration of alcohols.[1]   This iron salt, as well as CuSO$_4$ and NaHSO$_4$, when supported on silica gel effects dehydration of alcohols in various solvents at 100–125°. The order of reactivity of alcohols is tertiary > secondary > primary. Silica gel is essential for dehydration; other solid supports are not effective, and the rate of dehydration is increased by increasing amounts of SiO$_2$ and then becomes constant.

[1] T. Nishiguchi and C. Kamio, *J.C.S. Perkin I*, 707 (1989).

Isopropylidenetriphenylphosphorane, $(C_6H_5)_3P=C(CH_3)_2$ (1). Preparation.[1]

Stereoselective cyclopropanation.[2]   Reaction of the chiral oxazolidine **2**, prepared from (1R,2S)-norephedrine, reacts with **1** (3 equiv.) in benzene–hexane at

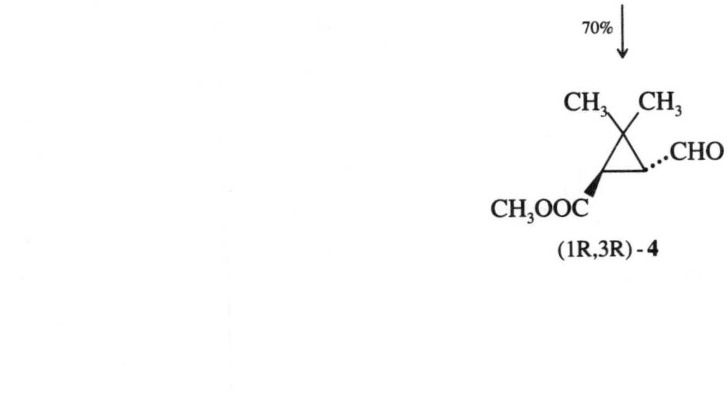

25° to afford the cyclopropane **3** in 60% yield. This product is a precursor to hemicaronic aldehyde (**4**).

[1] P. A. Grieco and R. S. Finkelhor, *Tetrahedron Letters*, 3781 (1972).
[2] A. Bernardi, C. Scolastico, and R. Villa, *ibid.*, **30**, 3733 (1989).

L

Lead tetraacetate.

1-*Alkynyllead triacetates* (*cf.*, **14**, 188).   These organometallics (**1**) can be prepared more directly and in higher yield by reaction of 1-alkynyllithiums with Pb(OAc)$_4$.[1] They are useful for preparation of α-alkenyl ketones (**4**) from β-keto benzyl esters (**2**). Thus the adduct (**3**), from reaction of **1** with **2**, on partial hydrogenation (Lindlar) and reductive debenzyloxycarbonylation provides the (Z)-α-alkenyl ketone **4**.

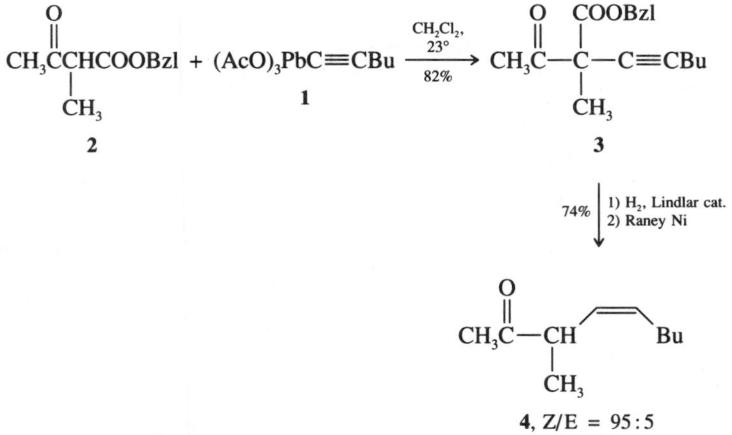

4, Z/E = 95:5

Oxidative ring expansion of enamides.[2]   A general route to tetrahydro-3-benzazepine-2-ones involves oxidative ring expansion of enamides of dihydroisoquinolines, which are easily formed by reaction with anhydrides or acid chlorides. The

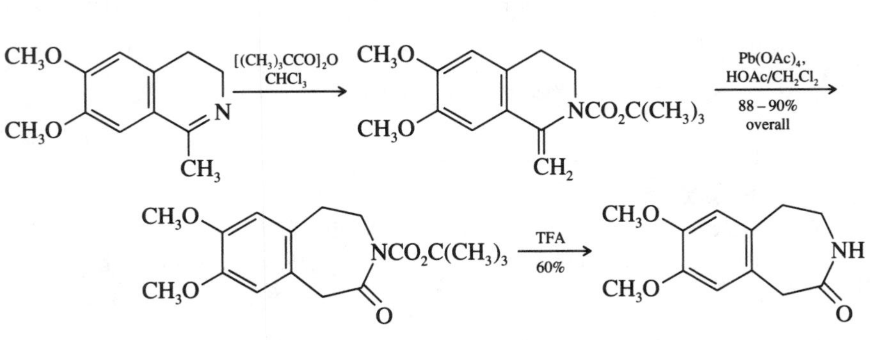

193

ring expansion is general with respect to substitution on the aromatic system, but fails if the exocyclic methylene group is disubstituted.

[1] S. Hashimoto, Y. Miyazaki, T. Shinoda, and S. Ikegami, *J.C.S. Chem. Comm.*, 1100 (1990).
[2] G. R. Lenz and C. Costanza, *J. Org.*, **53**, 1176 (1988); G. R. Lenz and R. A. Lessor, *Org. Syn.*, submitted (1989).

Lithium N-benzyltrimethylsilylamide (LSA), LiN
$$\text{LiN}\begin{array}{c}\text{Bzl}\\ \diagdown\\ \text{Si(CH}_3)_3\end{array}$$
(1).

Addition to methyl crotonate.   LSA adds to methyl crotonate to give exclusively the (Z)-enolate (**a**), which reacts with aldehydes to form the *anti*, *syn*-aldol **2** as the major product. The lithium enolate (Z)-**a** can be converted into the (E)-isomer by

(Z)-**a**

anti, syn-**2**   82:18   syn, syn-**2**

(E)-**a**   syn, anti-**2**   90:10   anti, anti-**2**

reaction with LDA and then ClSi(CH$_3$)$_3$. This (E)-enolate reacts with aldehydes to form mainly the *syn, anti*-aldol.

[1] T. Uyehara, N. Asao, and Y. Yamamoto, *J.C.S. Chem. Comm.*, 753 (1989).

Lithium chloride–Dimethylformamide.

Cleavage of ROAr.[1]   LiCl (3 equiv.) in refluxing DMF can effect dealkylation of these ethers if the aryl group is substituted by *o*- or *p*-electron-withdrawing substituents (NO$_2$, X, COOR).

[1] A. M. Bernard, M. R. Ghiani, P. P. Piras, A. Rivoldini, *Synthesis*, 287 (1989).

Lithium di-*t*-butylbiphenylide (LDBB), **7**, 200; **10**, 240.

2-Lithiotetrahydropyrans.[1] LDBB is superior to LDMAN (**12**, 279–280) for reduction of 2-thiophenyltetrahydropyrans at −78° to axial 2-lithiotetrahydropyrans, which isomerize to the equatorial epimers at −20° or higher. Thus lithiation of **1** and trapping provides mainly axial **2a**, whereas lithiation at −78° followed by trapping at 20° provides mainly **2b**. This methodology is widely applicable to 2-(phenylthio)tetrahydropyrans.

1 **2a** 96.8 : 3.2 **2b**

$$\mathbf{1} \xrightarrow[\text{2) } (CH_3)_2C=O,\ 20°]{\text{1) LDBB, THF, } -78°} \mathbf{2a} + \mathbf{2b}$$

4.8 : 95.2

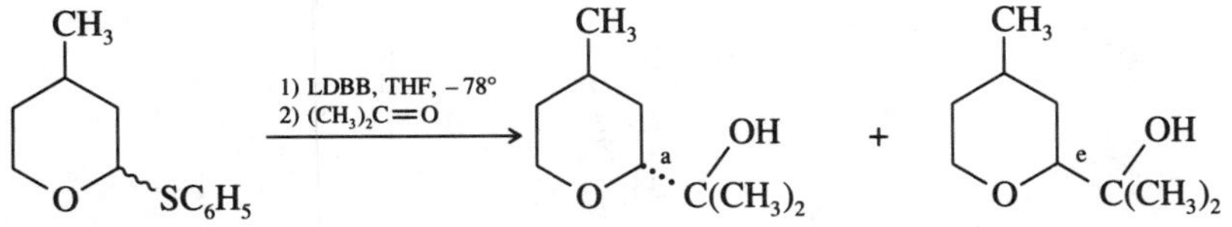

Reductive lithiation of oxetanes.[2] Oxetanes are cleaved by LDBB at 0° in THF to γ-lithioalkoxides (**a**) which are trapped by aldehydes or ketones to give 1,4-diols.

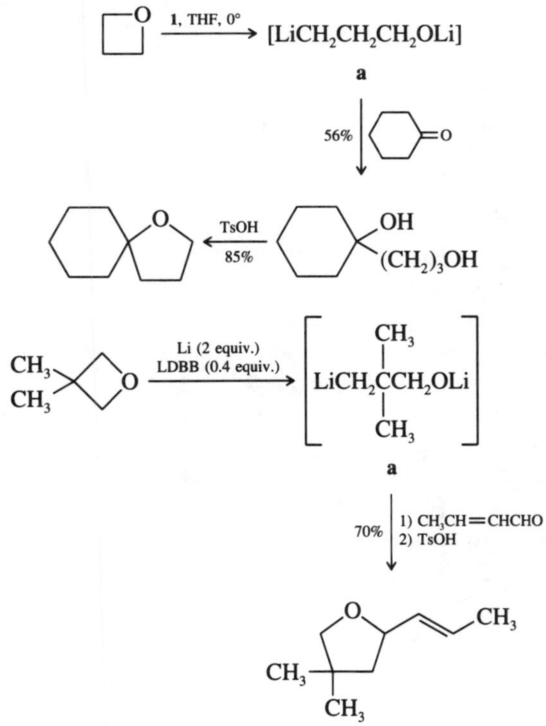

These diols can be cyclized to tetrahydrofurans. Oxetanes are also cleaved in the same way by lithium and a catalytic amount of LDBB.

Cycloalkenyllithiums.[3] Alkenyllithiums are usually prepared by reductive lithiation of trisylhydrazones of ketones with butyllithium, but this method fails with the hydrazones of cyclic ketones. However, the cycloalkenyl sulfides, prepared by reaction of cyclic ketones with thiophenol, can be reductively lithiated with LDBB at −78°. This lithiation fails in the case of cyclopentenyl sulfides, but is useful in the case of the vinyl sulfides obtained from 6-, 7-, and 8-membered cycloalkanones.

[1] S. D. Rychnovsky and D. E. Mickus, *Tetrahedron Letters*, **30**, 3011 (1989).
[2] B. Mudryk and T. Cohen, *J. Org.*, **54**, 5657 (1989).
[3] T. Cohen and M. D. Doubleday, *ibid.*, **55**, 4784 (1990).

Lithium diisopropylamide.

C-Alkylation of polypeptides.   Seebach et al.[1] report that selective C-alkylation of a sarcosine (N-methylglycine) unit in a tri- or hexapeptide with an N-protected terminal group is possible if polylithiation is effected with excess LDA in THF with addition of LiCl (5-6 equiv.) to provide a homogeneous mixture. In some cases, addition of an aprotic dipolar solvent (DMPU) or of BuLi can improve the yield. Under these conditions, epimerization and N-alkylation are slight. The new center introduced by alkylation of Sar tends to have the (R)-configuration when the configuration at the other centers is (S). In the methylation of the tripeptide formulated in equation (I), the yield is particularly high because the Sar unit is flanked by an N-methylamino acid. In this case, aldehydes can also serve as electrophiles. Alkylation and benzylation are also possible in the presence of added DMPU.

The results of this investigation suggest that C-alkylation may provide a useful route to novel polypeptides.

(I) Boc — (structure)

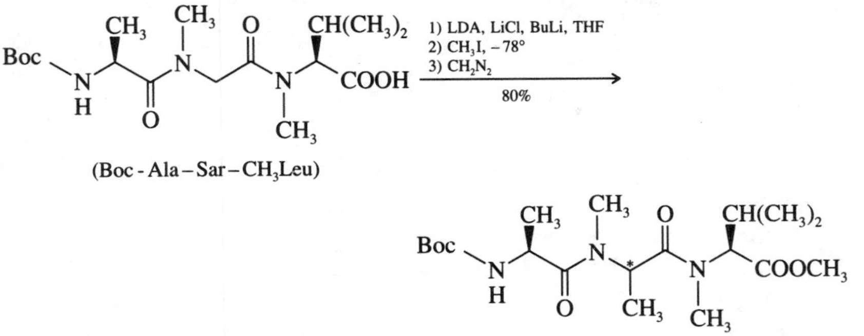

(Boc - Ala – Sar – CH₃Leu)

1) LDA, LiCl, BuLi, THF
2) CH₃I, –78°
3) CH₂N₂
⟶
80%

(structure)

R/S = 3.7 : 1

¹ D. Seebach, H. Bossler, H. Gründler, S. Shoda, and R. Wenger, *Helv.*, **74**, 1974 (1991).

Lithium tri-*t*-butoxyaluminum hydride–Lithium iodide (1).

anti-1,3-Polyols.[1] Reduction of the chiral β-alkoxy β'-hydroxy ketone **2** with this hydride catalyzed by LiI provides the *anti*-1,3-diol with high stereoselectivity.

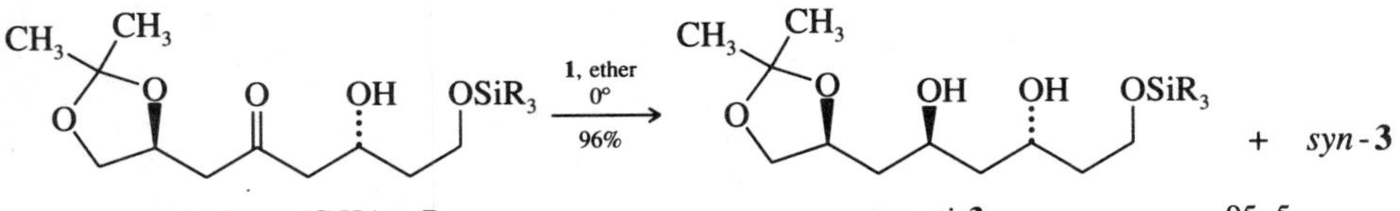

(reaction scheme)

1, ether
0°
⟶
96%

+   *syn* - **3**

2, α_D –9°; R₃ = (C₆H₅)₂-*t*-Bu

anti -**3**   95 : 5

This step was used for a general route to *anti*-1,3-polyols containing a 1,3-*syn*-3,5-*anti*-triol unit as in **4**.

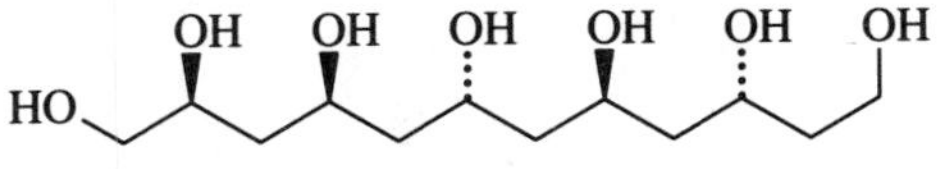

(structure)

4

¹ Y. Mori and M. Suzuki, *Tetrahedron Letters*, **30**, 4383, 4387 (1989).

M

Magnesium, Mg.

Complexes with **1,3-***dienes.* The reaction of activated Mg* (**11**, 307) in THF at 25° with (E,E)-1,4-diphenyl-1,3-butadiene results in the deep red complex (1,4-diphenyl-1,2-butene-1,4-diylmagnesium). This halide-free bismagnesium reagent (**1**) forms a 1,4-adduct with α,ω-dibromoalkanes, which undergoes intramolecular

$$C_6H_5CH=CHCH=CHC_6H_5 \xrightarrow[25°]{Mg*,\ THF} \quad C_6H_5 \overset{\displaystyle \diagup\!\!\!\diagdown}{\underset{Mg}{\diagdown\!\!\!\diagup}} C_6H_5$$

1

alkylation to afford disubstituted cycloalkanes. Products from R_2SiCl_2 are stable 1,4-adducts (equations I and II). Stepwise electrophilic addition can be effected in the case of adduct **2** of Mg* with 2,3-dimethyl-1,3-butadiene (equation III).

(I) $\mathbf{1}$ + Br(CH$_2$)$_3$Br $\xrightarrow[66\%]{\text{THF} \\ -78°}$ C$_6$H$_5$CH=CH— (cyclopentane with C$_6$H$_5$)

(II) $\mathbf{1}$ + (CH$_3$)$_2$SiCl$_2$ $\xrightarrow{60\%}$ (silacyclopentene ring with C$_6$H$_5$, H, CH$_3$ substituents)

(III) (structure **2**: 3,4-dimethyl ring with Mg) $\xrightarrow[62\%]{1)\ Br(CH_2)_4Br,\ -78° \\ 2)\ CH_3COCl,\ 0°}$ (product with CH$_3$, CH$_3$, (CH$_2$)$_4$Br, H$_2$C, O=, CH$_3$)

RCHO → ***RC≡CH***. A new two-step method for conversion of RCHO to RC≡CH involves conversion first to a 1,1-dibromoalkene, $RCH=CBr_2$, followed by debromination. The most convenient reagent for this second step is magnesium in refluxing THF.

$$\text{RCHO} \xrightarrow[\substack{80-90\%}]{\substack{P(C_6H_5)_3,\ CBr_4 \\ CH_2Cl_2}} \text{RCH}{=}\text{CBr}_2 \xrightarrow[75-95\%]{\text{Mg, THF}} \text{RC}{\equiv}\text{CH}$$

[1] H. Xiong and R. D. Rieke, *J. Org.*, **54**, 3247 (1989).
[2] L. Van Hijfte, M. Kolb, and P. Witz, *Tetrahedron Letters*, **30**, 3655 (1989).

Magnesium bromide, $MgBr_2$.

Thioacetals. Aldehydes, ketones, or acetals, both cyclic and acyclic, can be converted into thioacetals by reaction with a thiol or dithiol and $MgBr_2$ (2.1 equiv.) in ether at 25°. The difference in reactivity between acetals and ketones permits selective conversion of acetals into thioacetals without acetalization of a ketone.

$$C_6H_5CH(OCH_3)_2 + C_6H_5SH \xrightarrow[93\%]{MgBr_2} C_6H_5CH(SC_6H_5)_2$$

$$C_6H_5CH(OCH_3)_2 + RCOCH_3 + HSCH_2CH_2SH \xrightarrow[93\%]{MgBr_2} C_6H_5CH\underset{S}{\overset{S}{\diagdown}}$$

[1] J. H. Park and S. Kim, *Chem. Letters*, 629 (1989).

Magnesium iodide, MgI_2.

Deoxygenation of oxiranes.[1] MgI_2, prepared *in situ* from magnesium turnings and I_2 in refluxing ether, effects deoxygenation of oxiranes to the corresponding alkenes via an iodohydrin with retention of configuration in 85–90% yield.

[1] P. K. Chowdhury, *J. Chem. Res. (S)*, 192 (1990).

Magnesium monoperoxyphthalate (MMPP, 1), **14**, 197.

Alkene epoxidation.[1] In the presence of a Mn-porphinate, particularly 5,10,15,20-tetra-2,6-dichlorophenylporphinatomanganese(III) acetate,[2] MMPP can epoxidize alkenes in the presence of a phase-transfer catalyst in high yield. Addition of a base (pyridine) improves the rate.

Oxidation of furans to enediones.[3] Oxidation of 2,5-dialkylfurans to *cis*-enediones can be effected with MMPP rapidly at 25° in C_2H_5OH/H_2O in 90–99% yield. *m*-Chloroperbenzoic acid is less efficient for this oxidation.

Oxidative cleavage of N,N-dialkylhydrazones.[4] These hydrazones are cleaved to the parent ketone by treatment with this oxidant in aqueous methanol at 0° in yields of 76–91%. No racemization occurs in the case of SAMP or RAMP hydrazones.

[1] C. Querci and M. Ricci, *J.C.S. Chem. Comm.*, 889 (1989).
[2] S. Banfi, F. Montanari, and S. Quici, *J. Org.*, **53**, 2863 (1988).
[3] C. Dominguez, A. G. Csáky, and J. Plumet, *Tetrahedron Letters*, **31**, 7669 (1990).
[4] D. Enders and A. Plant, *Synlett*, 725 (1990).

Magnesium oxide, MgO.

α,β-*Enones* → allylic alcohols.[1] This reduction can be effected in 50–90% yield by hydrogen transfer from propanol-2 catalyzed by a finely powdered magnesium oxide obtained by thermolysis of $Mg(OH)_2$ at 350°. No other metal oxide is satisfactory. These conditions also reduce β,γ-enones to homoallylic alcohols.

Alkoxycarbonylation of malonates.[2] This inexpensive base is particularly efficient for reaction of alkyl chloroformate with dialkyl malonates to give methanetricarboxylic esters. Yields are typically 55–70%, but this method fails with methyl cyanoacetate.

[1] J. Kaspar, A. Trovarelli, M. Lenarda, and M. Graziani, *Tetrahedron Letters*, **30**, 2705 (1989).
[2] J. Skarzewski, *Tetrahedron*, **45**, 4593 (1989).

Manganese(III) acetate, $Mn_3O(OAc)_7$ (1).

Cyclization of allylic β-diesters.[1] The Mn(III) cyclization of allylic β-keto acids (12, 292–293) has been extended to lactonization of allylic β-diesters. In general, yields are improved by use of Mn(III) acetate (2 equiv.), $Cu(OAc)_2$ (1 equiv.), and NaOAc (1 equiv.).

Lactams.[2] N,N-Dialkenyl-β-oxoamides are oxidized by $Mn(OAc)_3$ to lactams and spirolactams via a radical cyclization. Ethanol is a better solvent than acetic acid or acetonitrile.

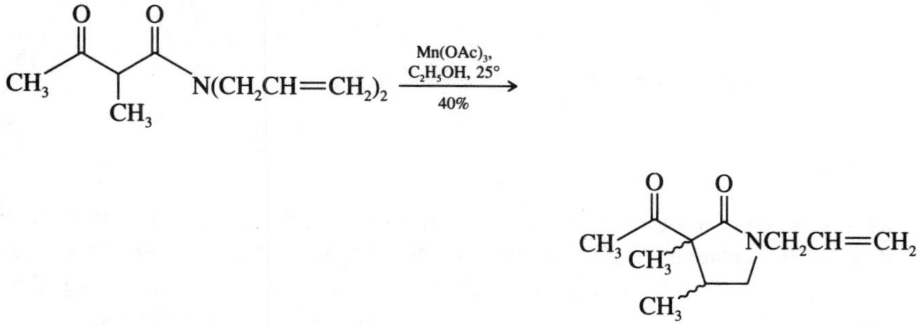

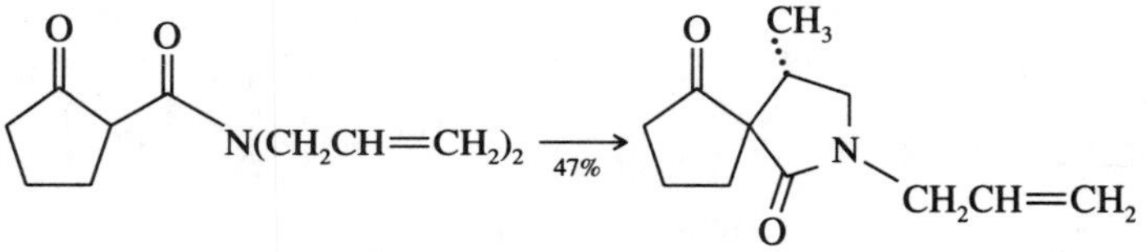

α'-Acyloxylation of enones.[3]   This oxidation can be effected by reaction of the enone with Mn(III) acetate (6 equiv.) and either 12 equiv. of a carboxylic acid or 6 equiv. of a Mn(II) carboxylate in benzene at reflux temperature for 6–18 hours.

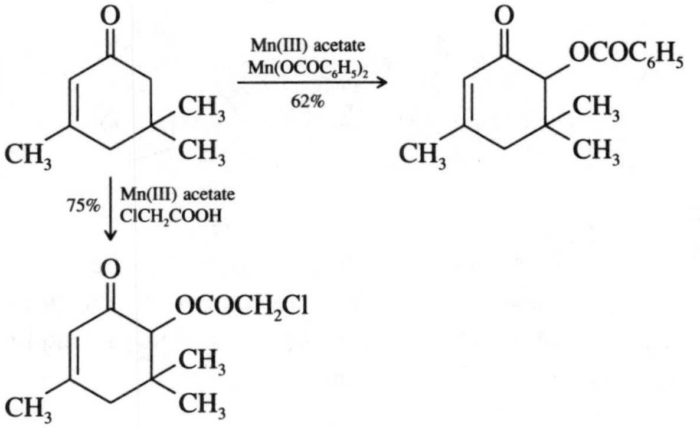

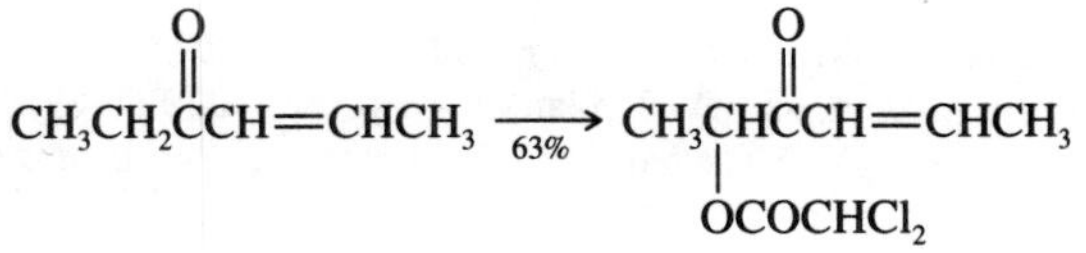

Addition of ·CH₂COOH to alkenes (2, 263–265; 6, 355–356).[4]   The conversion of alkenes to γ-lactones is generally effected with acetic acid and 1–2 equiv. of manganese(III) acetate, but this reagent is expensive and rather unstable. It can be

$$C_6H_5CH=CHC_6H_5 \xrightarrow[\substack{Mn(OAc)_2 - Cu(OAc)_2 \\ HOAc, Ac_2O \\ 61\%}]{-e}$$

generated in the presence of the alkene by anodic oxidation of the inexpensive manganese(II) acetate by $Cu(OAc)_2$. In this indirect process, the $Mn(OAc)_2$ is used in catalytic amounts with $Cu(OAc)_2$ as the stoichiometric agent. In general, the yields obtained by this anodic oxidation are significantly higher than those obtained by the conventional procedure.

$$(CH_3)_3CCH_2\overset{\overset{\displaystyle CH_3}{|}}{C}=CH_2 + CH_3COCH_2COOC_2H_5 \xrightarrow[82\%]{Mn(OAc)_2,}$$

$$\xrightarrow[78\%]{\substack{-e \\ Mn(OAc)_2,}}$$

Anodic oxidation of $Mn(OAc)_2$ (catalytic amounts) in the presence of nonactivated alkenes and ethyl acetoacetate provides a route to dihydrofurans (cf., 6, 356). This electrooxidation of $Mn(OAc)_2$ has been extended to coupling of activated methylene compounds with alkenes and dienes.

[1] H. Oumar-Mahamat, C. Moustrou, J.-M. Surzur, and M. P. Bertrand, J. Org., 54, 5684 (1989).
[2] J. Cossy and C. Leblanc, Tetrahedron Letters, 30, 4531 (1989).
[3] A. S. Demir, A. Jeganathan, and D. S. Watt, J. Org., 54, 4020 (1989).
[4] R. Shundo, I. Nishiguchi, Y. Matsubara, and T. Hirashima, Tetrahedron, 47, 831 (1991).

Manganese dioxide–Chlorotrimethylsilane.

α-*Chloro ketones.*[1] This combination (1:4) is believed to generate $MnCl_4$, which converts ketones in α-chloro ketones by a radical process. A methylene group is attacked in preference to a methyl group.

$$\underset{C_6H_5\overset{\displaystyle O}{\overset{\|}{C}}CH_2CH_3}{} \xrightarrow[\underset{90\%}{}]{\underset{HOAc}{MnO_2,\ ClSi(CH_3)_3}} \underset{C_6H_5\overset{\displaystyle O}{\overset{\|}{C}}\overset{\displaystyle}{\underset{\displaystyle Cl}{\overset{|}{C}H}}CH_3}{}$$

[1] F. Bellesia, F. Ghelfi, U. M. Pagnoni, and A. Pinetti, *J. Chem. Res. (S)*, 188 (1990).

Manganese(II) tetramethylheptane-3,5-dionate,

$$\begin{matrix} C(CH_3)_3 \\ \Big\backslash \\ O \cdot \cdot \\ \Big/ \Big) Mn\ (1) \\ O \\ C(CH_3)_3 \end{matrix}_2$$

α-Hydroxy carboxylic esters.[1]   This complex is an efficient catalyst for peroxygenation of α,β-unsaturated carboxylic esters. Reduction *in situ* with $C_6H_5SiH_3$ converts the hydroperoxide to an α-hydroxy ester. 2-Propanol is the solvent of choice. $Mn(acac)_2$ shows only moderate activity in this reaction. α-Hydroxy esters are obtained generally in high yield from esters with only one substituent at the β-position or even from α-substituted esters.

$$CH_3CH{=}CHCO_2CH_2C_6H_5 \xrightarrow[91\%]{\underset{1}{O_2,\ C_6H_5SiH_3,}} CH_3CH_2\overset{\displaystyle OH}{\overset{|}{C}H}CO_2CH_2C_6H_5$$

[1] S. Inoki, K. Kato, S. Isayama, and T. Mukaiyama, *Chem. Letters*, 1869 (1990).

Menthol.

Optically active α-hydroxy carboxylic acids.   The organozinc reagents prepared *in situ* from RMgX and $ZnCl_2$ or $ZnBr_2$ add selectively to the keto group of (−)-menthyl phenylglyoxalate (1). The adducts are hydrolyzed to optically active α-hydroxy carboxylic acids (3).[1]

$$\underset{1}{C_6H_5COCOOMen(-)} \xrightarrow{RMgX\ +\ ZnX_2} \underset{OH}{\underset{2}{C_6H_5\overset{\displaystyle R}{\overset{|}{C}}{*}COOMen(-)}} \xrightarrow{OH^-} \underset{OH}{\underset{(R)\text{-}3}{C_6H_5\overset{\displaystyle R}{\overset{|}{C}}{*}COOH}}$$

$R = C_2H_5$	88%	75% de	80%
$R = n\text{-}C_6H_{13}$	92%	88% de	83%

[1] G. Boireau, A. Deberly, and D. Abenhaim, *Tetrahedron*, **45**, 5837 (1989).

Menthone.

Deracemization of meso-1,3-diols[1] (**14**, 202–203).   The spiroacetals (**2**) formed from *meso*-1,3-diols (**1**) and *l*-menthone undergo highly selective cleavage of the equatorial C–O bond on treatment with allyltrimethylsilane and TiCl$_4$. The resulting free hydroxyl group of the diol can then be functionalized before removal of the chiral auxiliary to give chiral derivatives of the diol in high optical yields.

This transformation to chiral products is also applicable to 2-alkyl-1,3-propane-diols and *meso*-1,2-diols.

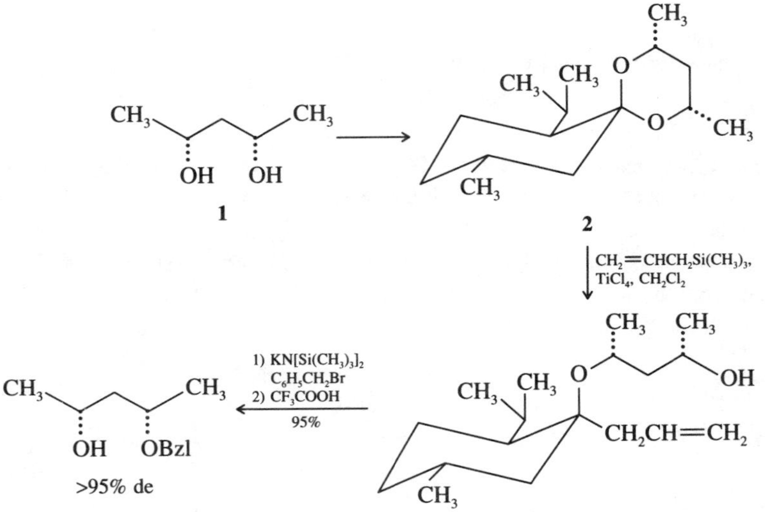

[1] T. Harada, Y. Ikemura, H. Nakajima, T. Ohnishi, and A. Oku, *Chem. Letters*, 1441 (1990).

(S)- or (R)-Menthyl p-toluenesulfinate (**1**), **14**, 203.

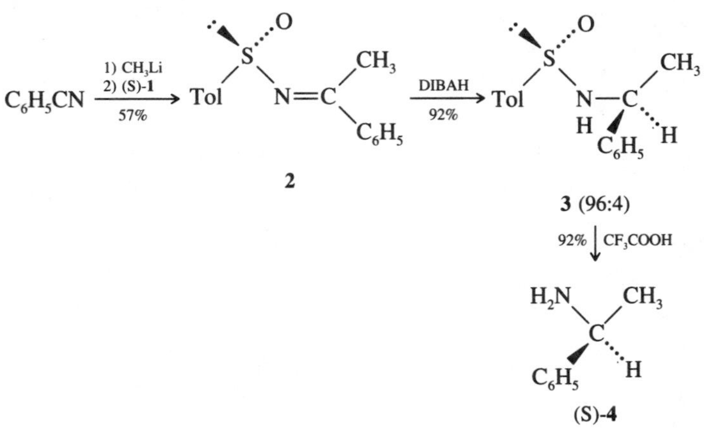

Chiral N-benzylidene-p-toluenesulfinamides; β- and γ-amino acids.[1]   Reaction of benzonitrile with CH$_3$Li and then with (S)-**1** at 0° provides sulfinamide **2**, which is reduced by DIBAH at 0° to the (S)-amine **3**. Chromatography and hydrolysis provide an optically pure (S)-amine (**4**).

Surprisingly, allylmagnesium bromide adds to **2** to give a single diastereomer, (R)-**5**, in 98% yield, which is hydrolyzed as above to the amine **6**.

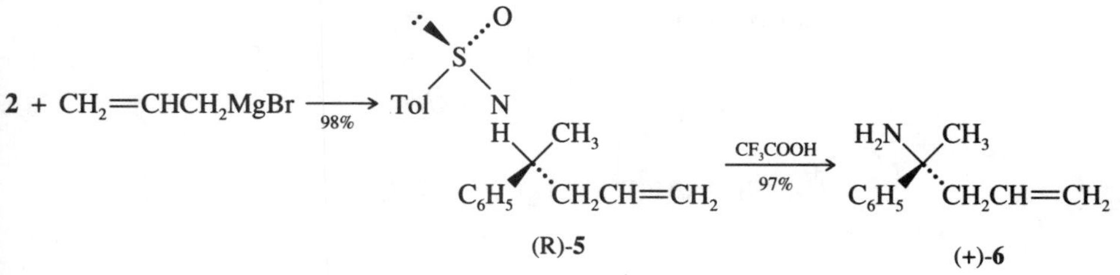

$$2 + CH_2{=}CHCH_2MgBr \xrightarrow{98\%}$$

(R)-**5**   (+)-**6**

Amine (+)-**6** can be converted into optically active β-amino acids (**7**) by acetylation, ozonolysis followed by oxidation (AgNO₃–KOH), and deacetylation. γ-Amino acids (**8**) are obtained by hydroboration (BH₃) of the acetate followed by oxidation (PCC; AgNO₃–KOH), and hydrolysis.

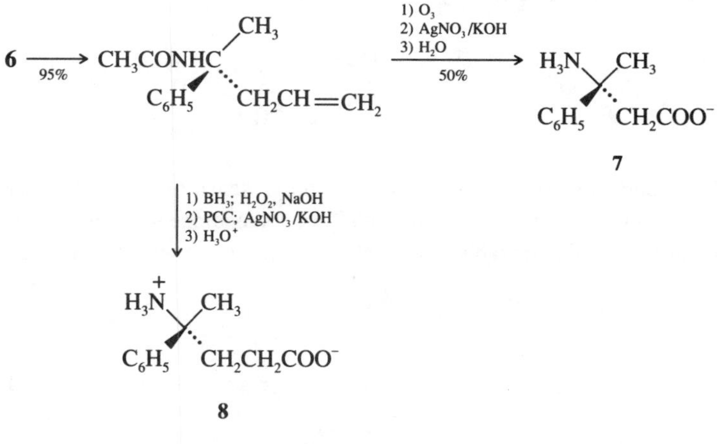

7

1) BH₃; H₂O₂, NaOH
2) PCC; AgNO₃/KOH
3) H₃O⁺

8

¹ D. H. Hua, S. W. Miao, J. S. Chen, and S. Iguchi, *J. Org.*, **56**, 4 (1991).

Mercury.

*Hydrodimerization of olefins.*¹ In addition to dehydrodimerization of alkanes (**15**, 198), hydrodimerization of alkenes can be effected by mercury-photosensitization, and has the advantage that it is applicable to a wide range of unsaturated substrates: alcohols and derivatives, ketones, and others. Since the hydrogen adds to the alkene to give the most stable intermediate (*tert* > *sec* > primary), this dimerization can be regioselective. The last example shows that cross-dimerization is possible. In this case the hydrodimer of both components is also formed, but in lower yield.

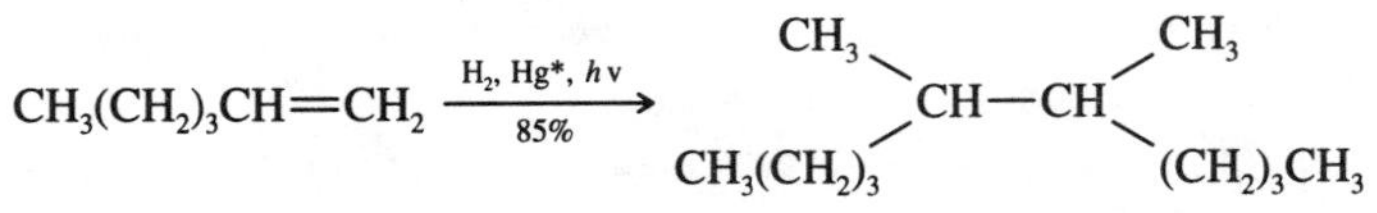

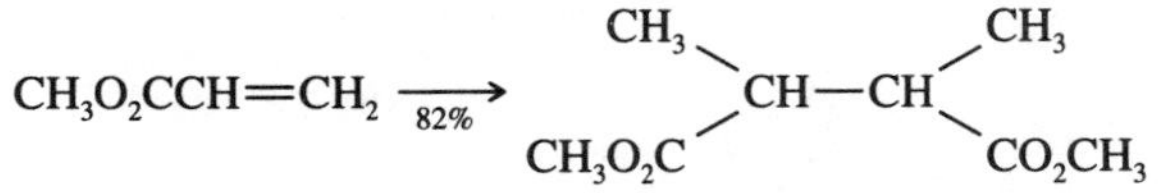

$$CH_3(CH_2)_3CH{=}CH_2 \xrightarrow[85\%]{H_2,\ Hg^*,\ h\nu} \begin{array}{c} CH_3 \qquad\qquad CH_3 \\ \diagdown \qquad\quad \diagup \\ CH{-}CH \\ \diagup \qquad\quad \diagdown \\ CH_3(CH_2)_3 \qquad (CH_2)_3CH_3 \end{array}$$

$$CH_3O_2CCH{=}CH_2 \xrightarrow[82\%]{} \begin{array}{c} CH_3 \qquad\qquad CH_3 \\ \diagdown \qquad\quad \diagup \\ CH{-}CH \\ \diagup \qquad\quad \diagdown \\ CH_3O_2C \qquad\quad CO_2CH_3 \end{array}$$

(D,L/*meso* ~ 1:1)

+ $CH_2{=}CHCH_2Si(CH_3)_3 \longrightarrow CH_3CH_2CH_2Si(CH_3)_3$

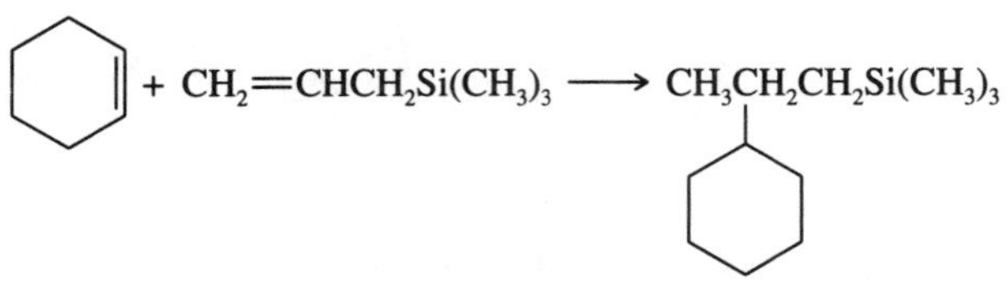

[1] C. A. Muedas, R. R. Ferguson, and R. H. Crabtree, *Tetrahedron Letters*, **30**, 3389 (1989).

1-Methoxy-1,3-bis(trimethylsilyloxy)-1,3-butadiene,

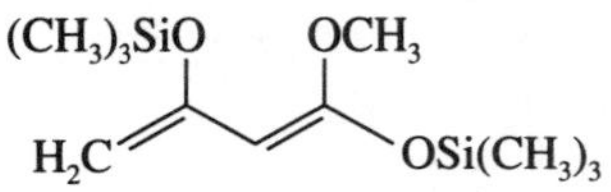

$$\begin{array}{cc} (CH_3)_3SiO & OCH_3 \\ \diagup & \diagup \\ H_2C\diagdown\quad\diagup\diagdown & OSi(CH_3)_3 \end{array} \qquad (1).$$

[3+4] *and* [3+5]*Cycloadditions*.[1]   In the presence of $TiCl_4$, **1** undergoes cycloadditions with 2,4-hexanedione or with 2,5-heptanedione to form bicyclic ethers **2** or **3**, respectively. Comparable results obtain in reactions with 1,4- or 1,5-oxoaldehydes.

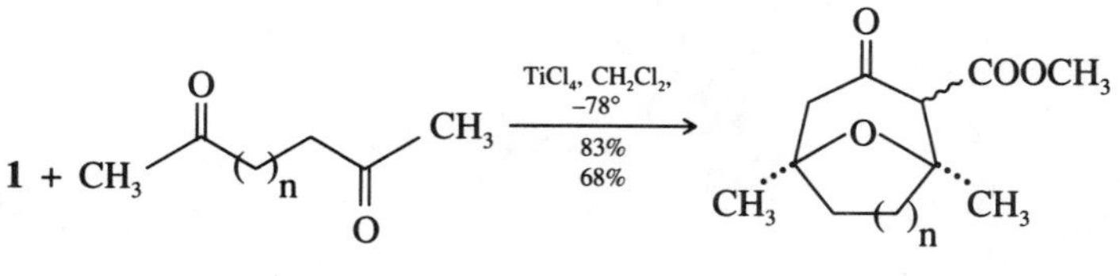

n = 1
n = 2

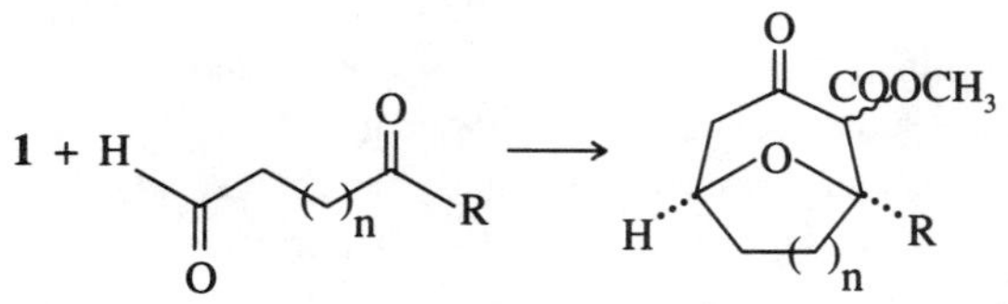

n = 1, R = CH₃   66%
n = 2, R = CH₃   77%

[1] G. A. Molander and S. W. Andrews, *Tetrahedron Letters*, **30**, 2351 (1989).

(Methoxymethoxy)allene, $CH_3OCH_2OCH{=}C{=}CH_2$ (**1**, **12**, 310; **13**, 177; **15**, 201).

Hydroxyquinone annelation. The lithio anion (**2**) of **1** adds to the vinylogous silyl ether **3** to provide the adduct **4**. Conversion to the aldehyde and epoxidation results in **6**, which undergoes cyclization in base to a hydroxyquinone, best isolated as the methyl ether **7**. This quinone annelation fails when applied to an acyclic vinylogous silyl ether.

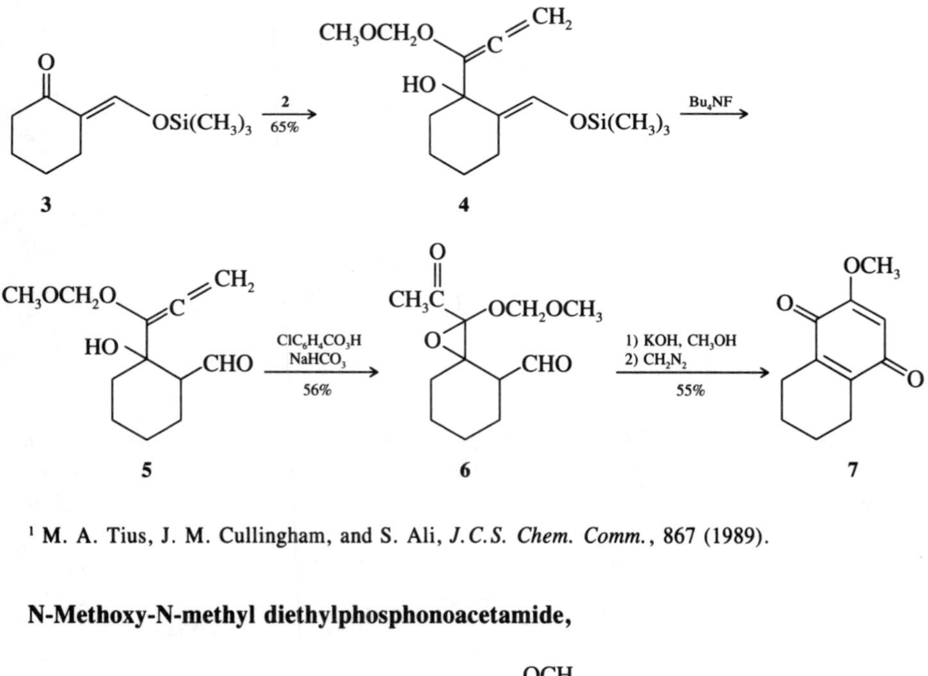

[1] M. A. Tius, J. M. Cullingham, and S. Ali, *J.C.S. Chem. Comm.*, 867 (1989).

N-Methoxy-N-methyl diethylphosphonoacetamide,

$$(C_2H_5O)_2P(O)CH_2CON\begin{matrix}OCH_3\\CH_3\end{matrix} \quad \textbf{(1)}.$$

This Wittig-Horner reagent is prepared from O,N-dimethylhydroxylamine and chloroacetyl chloride and $N(C_2H_5)_3$ in CH_2Cl_2 followed by reaction with triethylphosphite. The anion (BuLi) of **1** reacts with aldehydes or ketones to form hydroxamates, which are reduced to aldehydes by $LiAlH_4$ (**11**, 201–202).[1]

$$\mathbf{1} \xrightarrow[74\%]{1)\ BuLi\quad 2)\ C_6H_5CHO} C_6H_5CH{=}CHCON\begin{matrix}OCH_3\\OCH_3\end{matrix} \xrightarrow[84\%]{LiAlH_4} C_6H_5CH{=}CHCHO$$

(E/Z = 95:5)

[1] J.-M. Nuzillard, A. Boumendjel, and G. Massiot, *Tetrahedron Letters*, **30**, 3779 (1989).

2-(Methoxymethyl)pyrrolidine (1).

Enantioselective conjugate addition.[1] This asymmetric reaction can be effected by use of (S)-1 as a chiral leaving group. Thus the cyclohexenone 2, prepared by reaction of 2-(nitromethyl)-2-cylohexen-1-one with 1, undergoes conjugate addition with cuprates, followed by hydrolytic elimination of the pyrrolidine, to give 3-alkyl-2-*exo*-methylenecyclohexanones (3) in high enantiomeric excess.

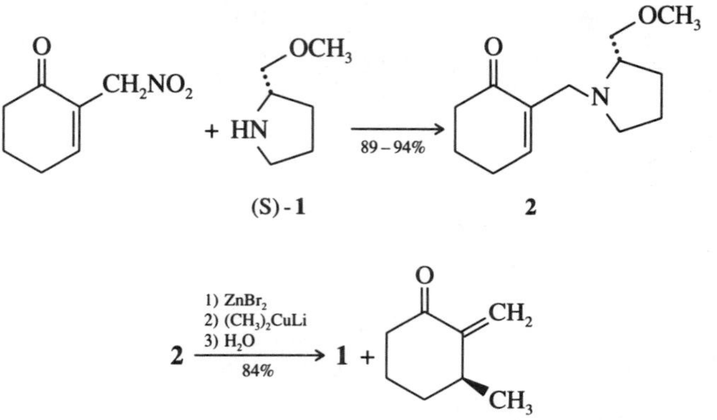

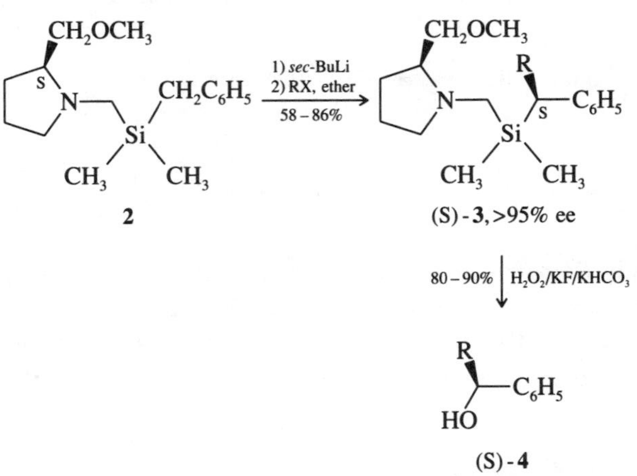

Diastereoselective alkylation of organosilanes.[2] Use of chiral organosilanes in which Si is the chiral center for enantioselective reactions has not been promising, but high diastereoselectivity has been obtained when a chiral auxiliary is attached by a methylene group to the silicon atom of an organosilane. Thus the anion (*sec*-BuLi) of **2**, prepared from (S)-2-(methoxymethyl)pyrrolidine and benzyl(chloromethyl)dimethylsilane, is alkylated in >95% de to give (S)-**3** in 58–86% yield. Hydrogen peroxide oxidation cleaves **3** to (S)-phenylcarbinols (**4**) with retention of configuration.

[1] R. Tamura, K. Watabe, H. Katayama, H. Suzuki, and Y. Yamamoto, *J. Org.*, **55**, 408 (1990).
[2] T. H. Chan and P. Pellon, *Am. Soc.*, **111**, 8737 (1989).

Methoxy(phenyldimethylsilyl)methyllithium (1).

α-Hydroxy aldehydes.[1] A new preparation of **1** is shown in equation (I). This anion reacts with a wide variety of aldehydes or ketones to furnish adducts (**2**)

$$\text{(I)} \quad C_6H_5(CH_3)_2SiCH_2Br \xrightarrow[\substack{2)\ \textit{sec}\text{-BuLi, TMEDA}\\50\%}]{1)\ CH_3OH,\ AgNO_3} C_6H_5(CH_3)_2SiCHLiOCH_3$$

1

$$\text{(II)} \quad \mathbf{1} + R^1COR^2 \longrightarrow$$

generally in 75–95% yield and usually in a 1:1 diastereomeric ratio. The products are converted into α-hydroxy aldehydes by oxidation of the acetate with H_2O_2 in the presence of Br_2 (equation II).

[1] D. J. Ager, J. E. Gano, and S. I. Parekh, *J.C.S. Chem. Comm.*, 1256 (1989).

Methylaluminum bis(4-bromo-2,6-di-*t*-butylphenoxide),

1

Claisen rearrangement.[1] Allyl vinyl ethers such as **3** undergo Claisen rearrangement reluctantly and in low yield when treated with methylaluminum bis(2,6-di-*t*-butylphenoxide), MAD; but this dibromo derivative, **1**, effects this rearrangement readily at −78° with high (Z)-selectivity. Evidently the bulky *t*-butyl groups control the stereoselectivity, for use of methylaluminum bis(2,6-diphenylphenoxide), **2**, results in (E)-**4** almost exclusively. In the latter case, electronic effects may be involved as well as the lower steric requirements. Optically active substrates are rearranged with conservation of chirality by both **1** and **2**.

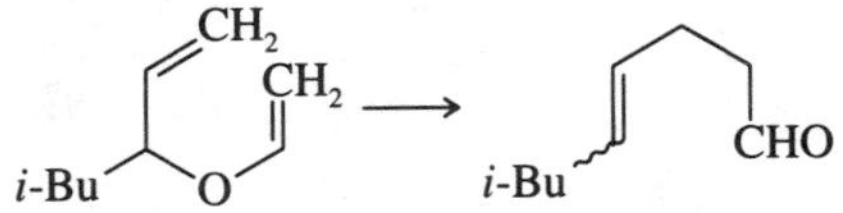

	3	**4**
+ **1**, −78°, 64%		E/Z = 7:93
+ **2**, −20°, 85%		E/Z = 97:3

Moreover, rearrangement of a bisallyl vinyl ether such as **5** involves the more substituted allylic system to provide (E)- and (Z)-**6**.

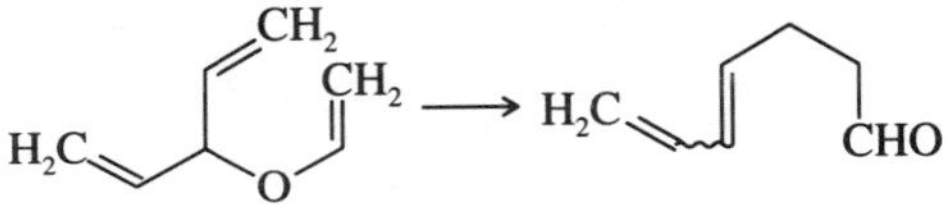

	5	**6**
1	97%	E/Z = 24:76
2	91%	E/Z = 90:10

Rearrangement of allyl phenyl ethers with **1** results mainly in *para*-substituted phenols rather than the *ortho*-substituted phenols formed by thermal rearrangements.[2]

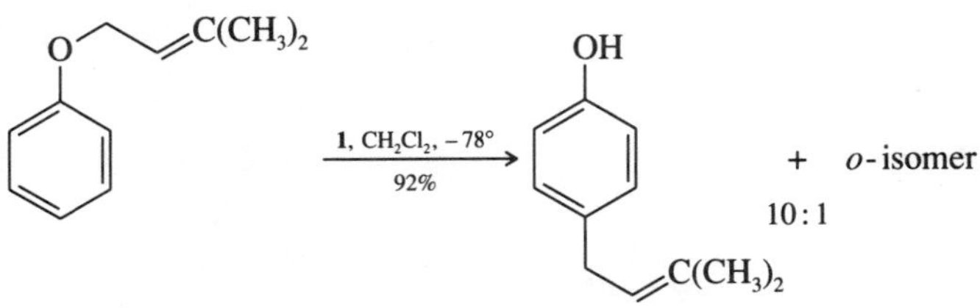

$$\xrightarrow[92\%]{1, \ CH_2Cl_2, \ -78°}$$

+   *o*-isomer

10:1

Intramolecular ene reactions.[3]   Ene reactions of δ,ε-enals with an α-methyl substituent (**2**) when promoted by dimethylaluminum chloride result in *cis*-methyl-

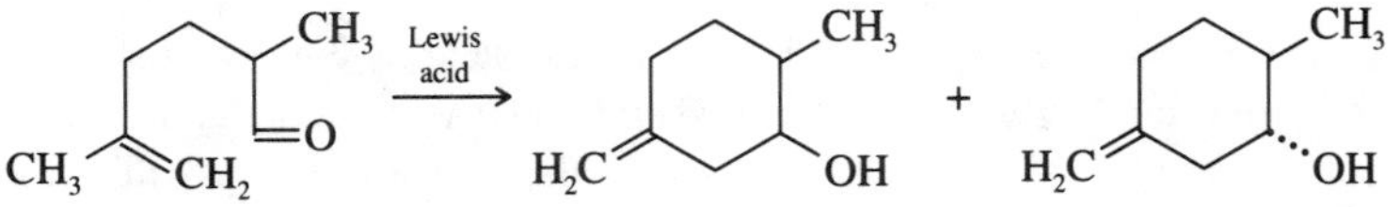

	2	*cis*-**3**	*trans*-**3**
(CH₃)₂AlCl	65%	9:1	
1	85%	1:32	

enecyclohexanols (**3**) with high *cis*-selectivity. The same *cis*-selectivity obtains in reactions promoted by the more traditional Lewis acids SnCl$_4$ or BF$_3$ etherate. In contrast, use of the bulky organoaluminum reagent **1** results in high *trans*-selectivity. This *trans*-selectivity extends to ene reactions with a trisubstituted double bond (equation I) and to rigid cyclic substrates (equation II).

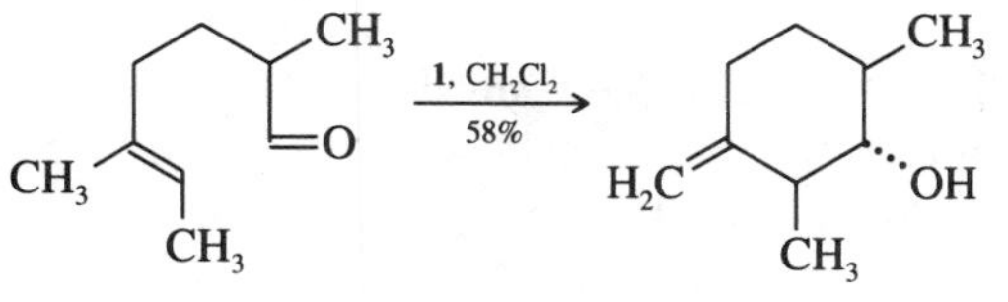

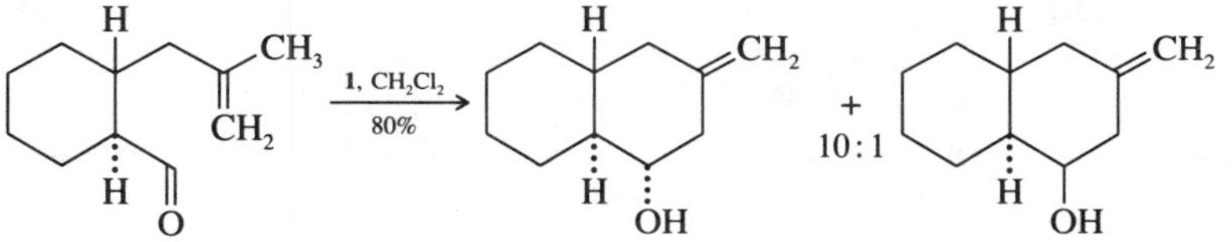

Epoxide rearrangement.[4] This reagent promotes rearrangement of chiral α,β-epoxy silyl ethers to β-silyloxy aldehydes with *anti*-migration of the —CH$_2$OSiR$_3$ group to the epoxide with no loss of the optical purity.[5] The rearrangement is retarded by related aluminum reagents (such as MAD, **13**, 203) lacking the *p*-bromo groups.

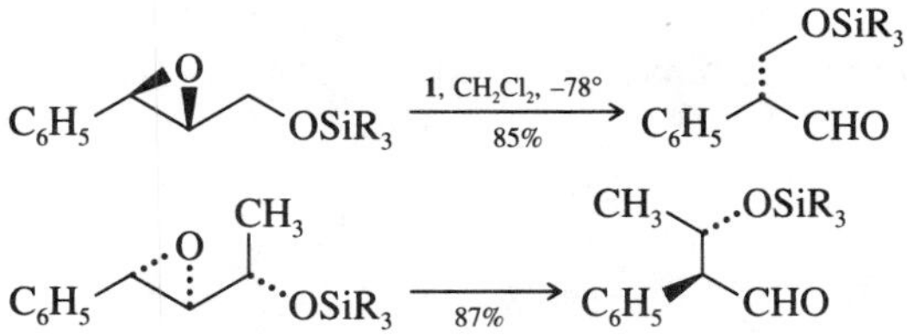

This aluminum reagent is also an effective catalyst for rearrangement of epoxides to aldehydes or ketones at −20° to −78°.[6] The related reagent in which the *t*-butyl groups are replaced by diisopropyl groups is generally ineffective.

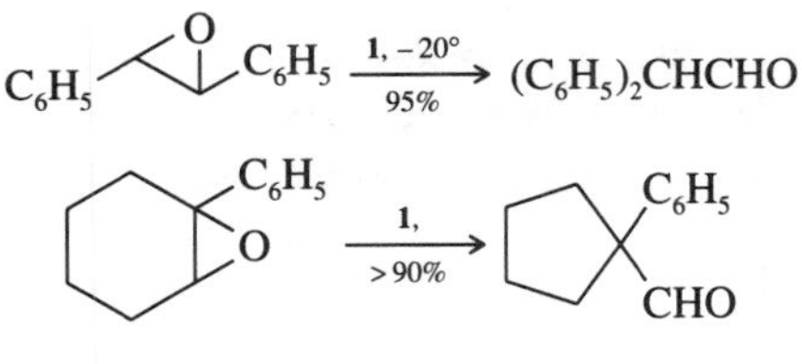

[1] K. Nonoshita, H. Banno, K. Maruoka, and H. Yamamoto, *ibid.*, **112**, 316 (1990).
[2] K. Maruoka, J. Sato, H. Banno, and H. Yamamoto, *Tetrahedron Letters*, **31**, 377 (1990).
[3] M. I. Johnston, J. A. Kwass, R. B. Beal, and B. B. Snider, *J. Org.*, **52**, 5419 (1987).
[4] K. Maruoka, T. Ooi, and H. Yamamoto, *Am. Soc.*, **112**, 9011 (1990).
[5] K. Maruoka, T. Ooi, and H. Yamamoto, *Am. Soc.*, **111**, 6431 (1989).
[6] K. Maruoka, S. Nagahara, T. Ooi, and H. Yamamoto, *Tetrahedron Letters*, **30**, 5607 (1989).

Methylaluminium bis(2,6-di-*t*-butyl-4-methylphenoxide) (MAD), **13**, 203.

Selective complexation of ethers.[1] This aluminum reagent shows remarkable selectivity in formation of complexes with ethers. Thus it effects virtually complete complexation of alkyl methyl ethers without effect on alkyl ethyl ethers. In general, ethers with less-hindered alkyl substituents form complexes more easily with MAD than their more bulky counterparts and the more basic etheral oxygens coordinate more readily to MAD than the less basic oxygen. The two bulky phenoxy groups are essential for this selective complexation, since methylaluminium bis(2,6-diisopropylphenoxide) does not form complexes with ethers under similar conditions. This selective complexation can be used to separate ethers by chromatography with MAD as the stationary phase.

1,4-Addition to quinone monoketals and quinol ethers.[2] Complexation of quinone monoketals or quinol ethers with MAD permits 1,4-addition of organolithium and Grignard reagents. Highest yields obtain with aryl, vinyl, and acetylenic organometallics.

[1] K. Maruoka, S. Nagahara, and H. Yamamoto, *Am. Soc.*, **112**, 6115 (1990).
[2] A. J. Stern, J. J. Rohde, and J. S. Swenton, *J. Org.*, **54**, 4413 (1989).

Methylaluminum bis(2,6-diphenylphenoxide), MAPH (1).

Formaldehyde–MAPH complexes.[1] Formaldehyde (gas) readily undergoes self-polymerization and is usually generated by thermal depolymerizaton of paraformaldehyde, a linear trimer. Treatment of trioxane in CH_2Cl_2 at 0° with MAPH generates a complex, $CH_2{=}O{\cdot}MAPH$, which is stable at 0° for 5 hours, but which decomposes slowly at 25°. The complex can be used for aldol and ene reactions of the aldehyde itself.

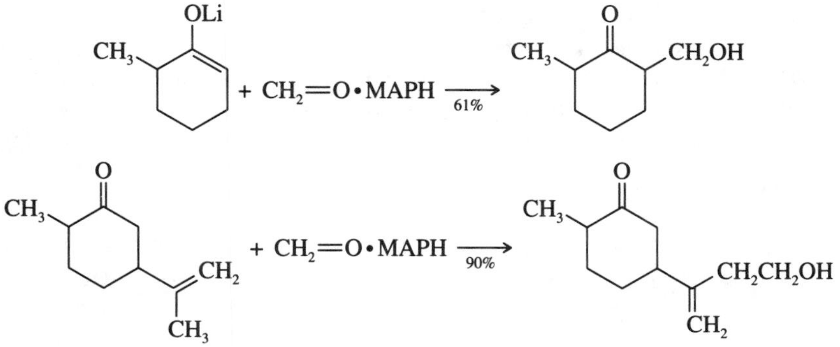

[1] K. Maruoka, A. B. Concepcion, N. Hirayama, and H. Yamamoto, *Am. Soc.*, **112**, 7422 (1990).

o-(Methylamino)benzenethiol (1).

Protection of carbonyls.[1]   The reagent converts carbonyl groups into 3-methyl-benzothiazolines in 80–90% yield. The products can be hydrolyzed in good yields

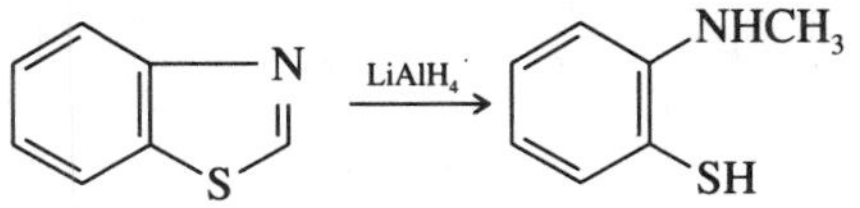

1

under mild and neutral conditions with $AgNO_3$, $HgCl_2$, NBS, chloramine-T, or by methylation (CH_3OSO_2F) followed by basic hydrolysis. Benzothiazolination is also useful for selective protection of a carbonyl group in the presence of another one. The relative order is cyclic one > linear one > cyclic enone > linear enone > > 2,4-dimethyl-3-pentanone (hindered ketone).

[1] H. Chikashita, S. Komazawa, N. Ishimoto, K. Inoue, and K. Itoh, *Bull. Chem. Soc. Japan*, **62**, 1215 (1989).

Methyl azodicarboxylate, ℗ $-CH_2OCON=NCOOCH_3$ (1).

Mitsunobu reactions.[1]   The polystyrene-supported reagent offers several advantages for Mitsunobu reaction; it is easily obtained, is nonexplosive, and provides comparable yields.

[1] L. D. Arnold, H. I. Assil, and J. C. Vederas, *Am. Soc.*, **111**, 3973 (1989).

Methyl benzenesulfenate, $C_6H_5SOCH_3$.

Polyene cyclization.[1]   The reaction of $C_6H_5SOCH_3$ and BF_3 with the diene **1** in CH_3NO_2 initiates cyclization to the tricyclic ring system **2** via a sulfenium ion. This product was converted to the diterpene nimbidiol (**3**) by reductive desulfuration (lithium naphthalenide) and oxidative decyanation ($SnCl_2$), and demethylation (BBr_3).

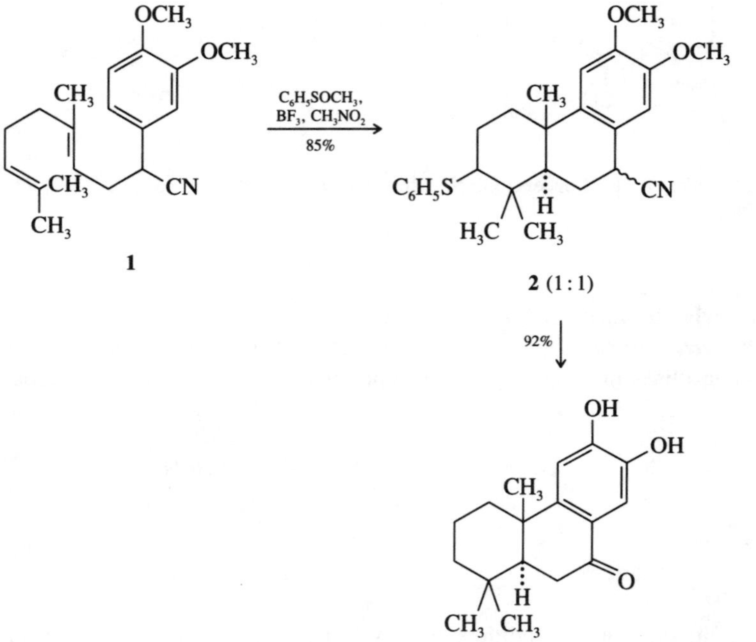

2 (1 : 1)

3

[1] S. R. Harring and T. Livinghouse, *Tetrahedron Letters*, **30**, 1499 (1989).

Methyl bis(1-naphthyl)bismuthinate, Np_2BiOCH_3.

The reagent is obtained in 80% yield by reaction of tris(1-naphthyl)bismuthine, Np_3Bi, with chloramine-T in methanol.[1]

N-Acetylation.[2]   Amides, thioamides, ureas, and thioureas are N-acetylated at 25° by acetic acid in CH_2Cl_2 in the presence of this reagent. It can serve as a catalyst, but stoichiometric amounts are required for high yields.

[1] T. Ogawa, T. Murafuji, K. Iwata, and H. Suzuki, *Chem. Letters*, 2021 (1988).
[2] T. Ogawa, K. Miyazaki, and H. Suzuki, *ibid.*, 1651 (1990).

N-Methyl-N,O-bis(trimethylsilyl)hydroxylamine, $CH_3N\begin{matrix}OSi(CH_3)_3\\Si(CH_3)_3\end{matrix}$ (1), b. p.

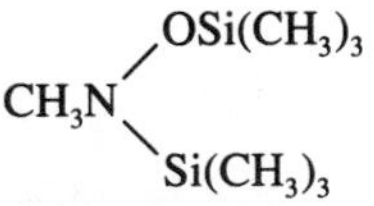

40–41°/10 mm. The reagent can be prepared in 52% yield by reaction of N-methyl-hydroxylamine hydrochloride and $N(C_2H_5)_3$ (3 equiv.) with $ClSi(CH_3)_3$ (2 equiv.) at 25° with stirring for three days.

N-Methylnitrones. The reagent converts aldehydes and aliphatic ketones into N-methylnitrones in high yield via a hemiaminal intermediate (**a**), which can be isolated in some cases.

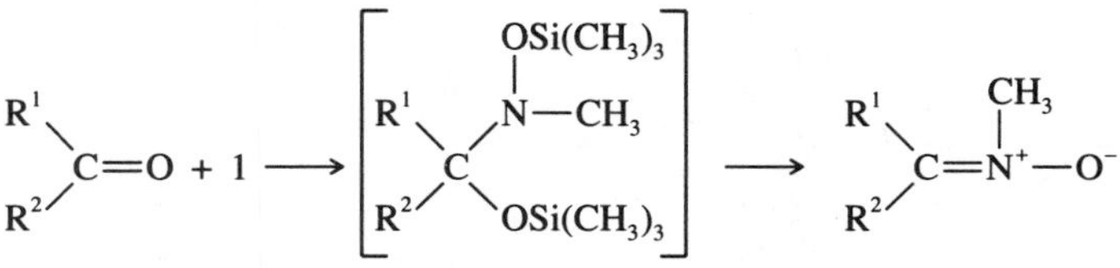

a   62–98%

[1] J. R. Hwu, J. A. Robl, N. Wang, D. A. Anderson, J. Ku, and E. Chen, *J.C.S. Perkin I*, 1823 (1989).

(2S,4S)-N-Methylcarbamoyl-4-dicyclohexylphosphino-2[diphenylphosphino)-methyl]pyrrolidine (MCCPM, 1) 15, 52–53.

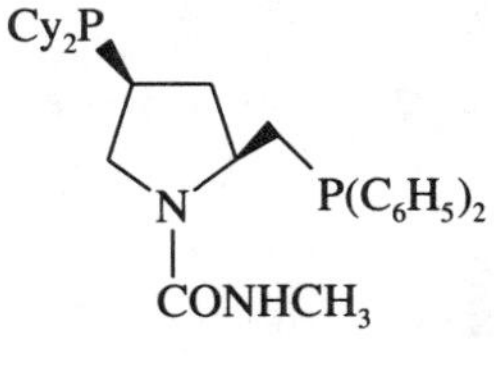

Asymmetric hydrogenation of α-amino ketones.[1] A catalyst prepared from $[Rh(COD)Cl]_2$ and (2S,4S)-1 effects highly enantioselective hydrogenation of α-amino ketones. This reaction provides a ready route to (S)-propranolol (3) from a 3-aryloxy-2-oxo-1-propylamine (2). A number of related α-amino ketones are hydrogenated under these conditions to the corresponding (2S)-alcohols in 85–96% ee, but β-amino ketones are reduced with lower enantioselectivity.

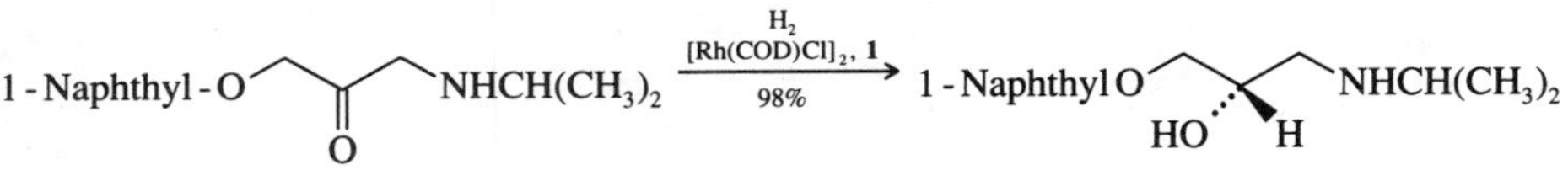

[1] H. Takahashi, S. Sakuraba, H. Takeda, and K. Achiwa, *Am. Soc.*, **112**, 5876 (1990).

α-Methylcinnamyl alcohol, $C_6H_5CH{=}CHCH(CH_3)OH$, b.p. 80–85°/0.5 mm.

The alcohol is obtained by reaction of CH_3MgBr with cinnamyl aldehyde in THF at 0° (95% yield).

Protection of carboxylic acids.[1] MEC esters (**1**) are stable to weak acid or base, but are cleaved selectively by reaction with the thiostannane $(CH_3)_2Sn(SCH_3)_2$ (**2**) and $BF_3{\cdot}O(C_2H_5)_2$ (1 equiv.) at 0°.

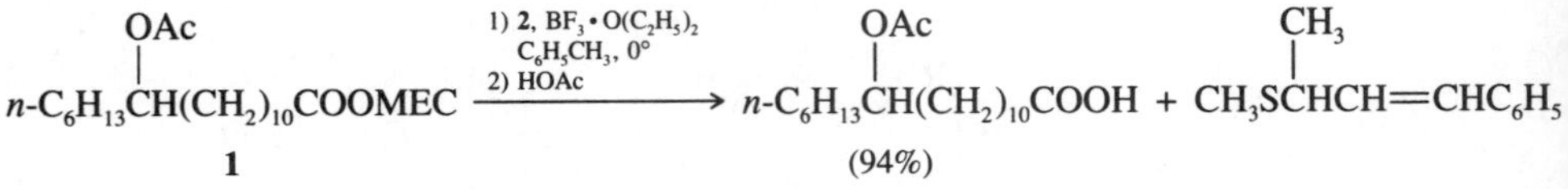

$$C_6H_5COOMEC + CH_3C_6H_4COOC_2H_5 \longrightarrow \underset{(100\%)}{C_6H_5COOH} + \underset{(100\%)}{CH_3C_6H_4COOC_2H_5}$$

An added advantage of these esters is the ready transformation into other esters by alkylation of the intermediate organotin carboxylates in the above reaction.

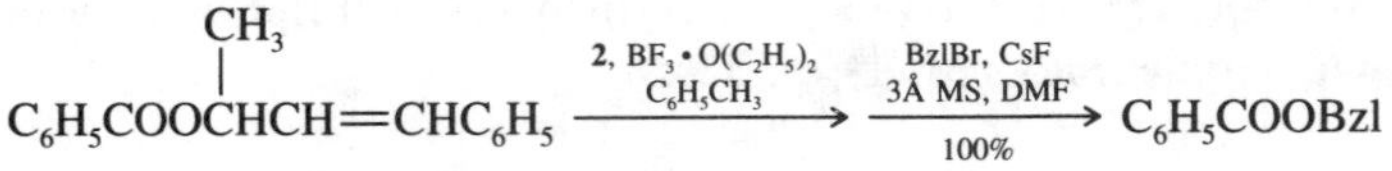

[1] T. Sato, J. Otera, and H. Nozaki, *Tetrahedron Letters*, **30**, 2959 (1989).

Methylcopper–Diisobutylaluminum hydride.

1,4-Reduction of enones.[1] Reduction of CH_3Cu, prepared *in situ* from CuI and CH_3Li in THF, with DIBAH in the presence of HMPA results in a form of copper hydride that effects efficient and selective 1,4-reduction of enals, enones, and enoates, and 1,6-reduction of dienones and dienoates. The reagent does not reduce isolated carbonyl groups or double bonds.[2] It can also be used for regiospecific preparation of enol silyl ethers from an enone.[2]

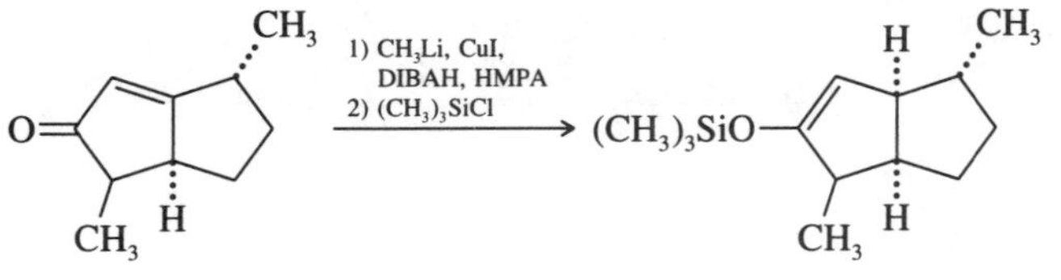

[1] T. Tsuda, T. Hayashi, H. Satomi, T. Kawamoto, and T. Saegusa, *J. Org.*, **51**, 537 (1986).
[2] K. R. Dahmke and L. A. Paquette, *Org. Syn.*, submitted (1990).

Methyl diazoacetate, $N_2CHCOOCH_3$.

Methoxycarbonylmethylation.[1]   The reaction of silyl enol ethers of aldehydes or ketones with methyl diazoacetate [$Rh_2(OAc)_4$ or $Cu(acac)_2$] forms silyloxycyclopropanecarboxylates, which are opened by $N(C_2H_5)_3 \cdot HF$ (aldehydes) or HCl (ketones) to form α-methoxycarbonylmethylated aldehydes or ketones.

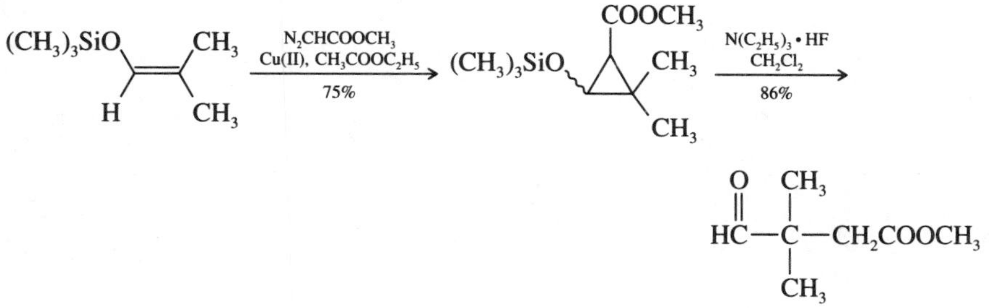

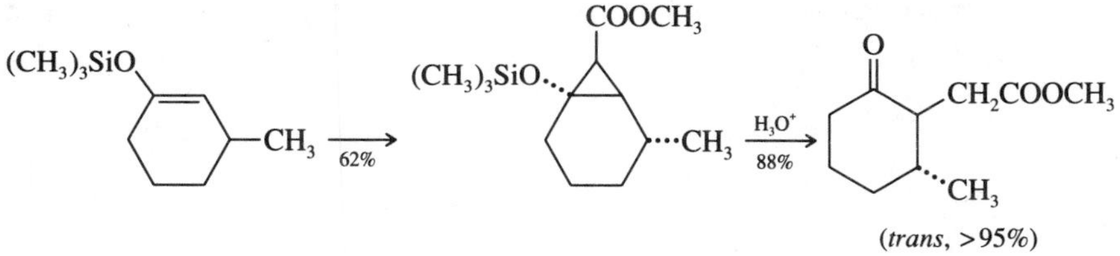

[1] H.-U. Reissig, I. Reichelt, and T. Kunz, *Org. Syn.*, submitted (1989).

Methyl 2,2-dimethyl-1-cyclopropenecarboxylate (1).
Preparation:

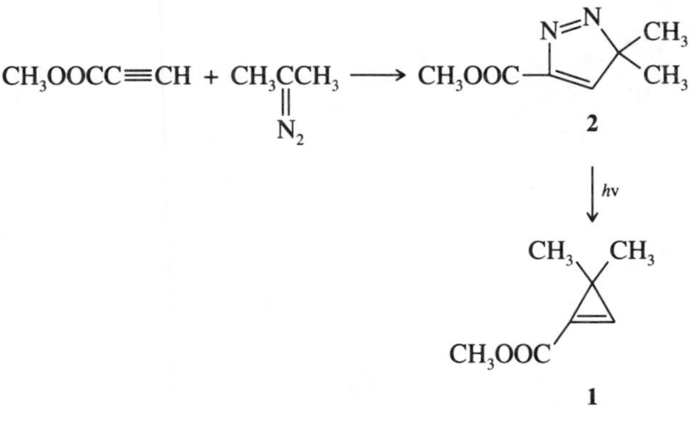

Cycloadditions. This activated *gem*-dimethylcyclopropene undergoes Diels–Alder reactions at pressures of 6–10 kbar. The reaction with 1,3-dienes shows high *exo*-selectivity as shown in equation (I). In contrast, reaction of the precursor (2) of 1 with the same diene followed by extrusion of N_2 results in the same products, but with opposite diastereoselectivity.

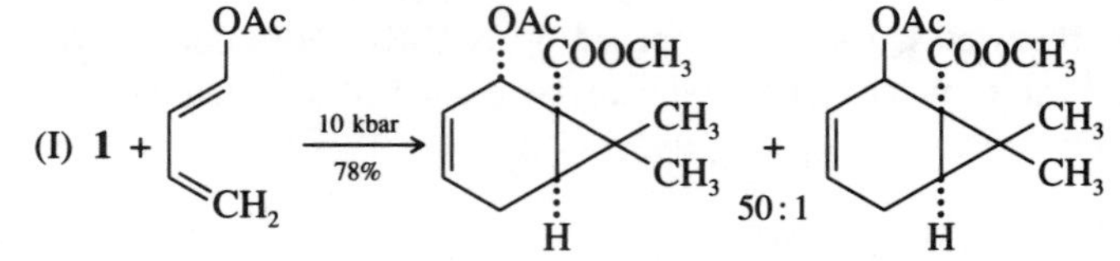

exo - **3**   *endo* - **3**

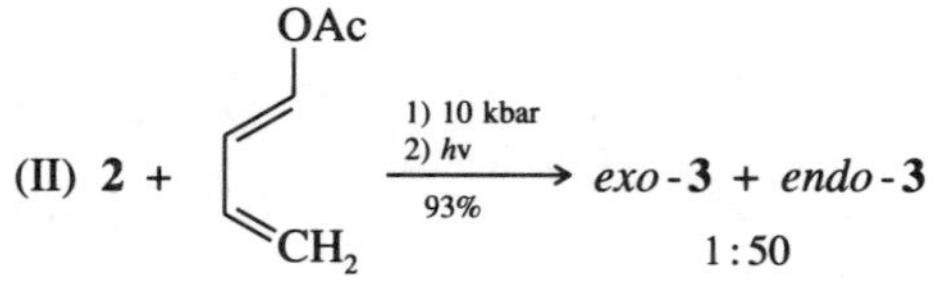

(II) **2** + [diene] → *exo* - **3** + *endo* - **3**
                        1 : 50

This reaction was used to generate a model of the tricyclic ring system **4** present in some terpenes.

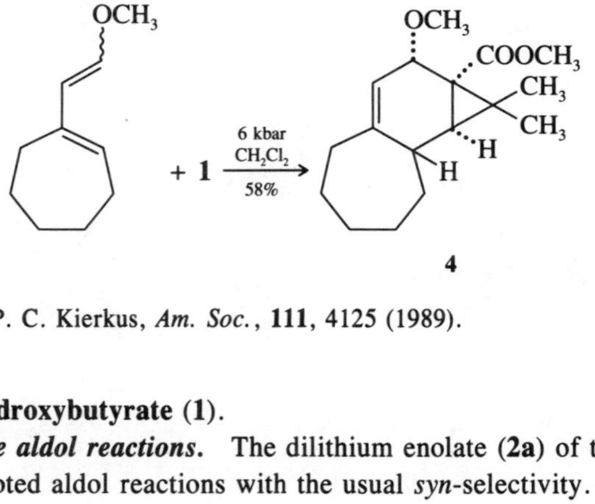

4

[1] J. H. Rigby and P. C. Kierkus, *Am. Soc.*, **111**, 4125 (1989).

Methyl (R)-3-hydroxybutyrate (1).

anti-*Selective aldol reactions.* The dilithium enolate (**2a**) of this ester undergoes $TiCl_4$-promoted aldol reactions with the usual *syn*-selectivity. In contrast, the

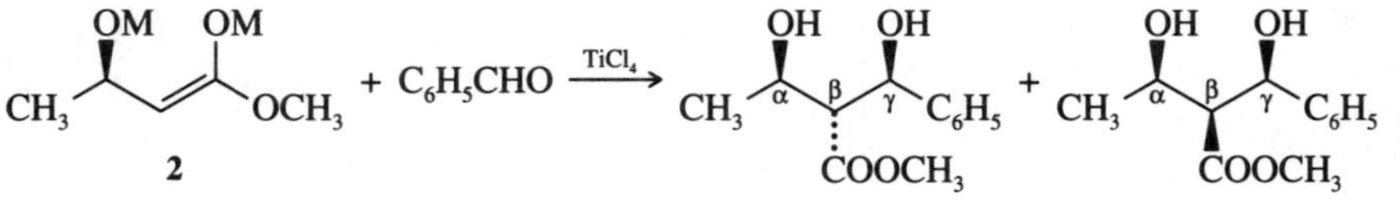

3 (*anti,anti*)   **4** (*syn,syn*)

a) M = Li	86%	22:78
b) M = Si(C₂H₅)₃	75%	93:7

silyl ketene acetal (**2b**), obtained by reaction of **2a** with $(C_2H_5)_3SiCl$, shows double *anti*-diastereofacial selectivity in aldol reactions with a variety of aldehydes.

[1] F. Shirai, J.-H. Gu, and T. Nakai, *Chem. Letters*, 1931 (1990).

Methyllithium–Chlorotrimethylsilane.

Methyl ketones from esters (**12**, 126).[1] The reaction of a carboxylic acid or ester with CH_3Li and $ClSi(CH_3)_3$ to provide a methyl ketone has been used to obtain a silyl protected (S)-3-hydroxy-2-butanone (**2**) from ethyl lactate (**1**). The addition of

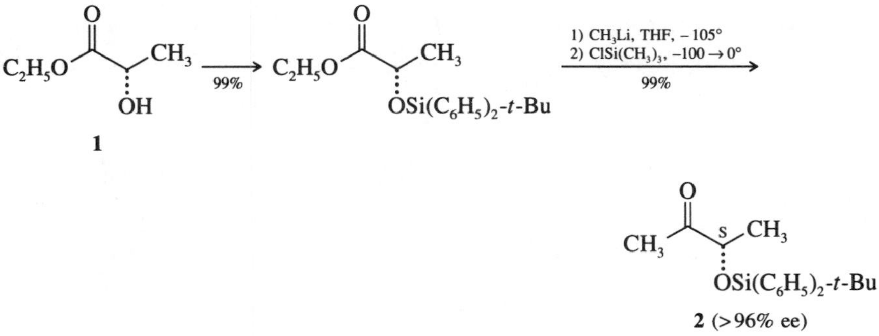

$ClSi(CH_3)_3$ is essential; use of NH_4Cl quench results in large amounts of a tertiary alcohol. This route to ketones is apparently limited to methyl ketones, since reactions with butyl- or phenyllithium promoted with chlorotrimethylsilane give significant amounts of a tertiary alcohol and starting material.

[1] L. E. Overman and G. M. Rishton, *Org. Syn.*, submitted (1991).

Methyl (R)- and (S)-mandelate, $C_6H_5\overset{*}{C}H(OH)COOCH_3$. This hydroxy acid is a useful chiral auxiliary because it is available commercially in both enantiomeric forms and is removable by hydrogenolysis.

Asymmetric Diels–Alder reactions.[1] Both the acrylate and the fumarate of methyl (R)-mandelate undergo cycloaddition with α-hydroxy-*o*-quinonedimethane to give 1-hydroxy-1,2,3,4-tetrahydronaphthalenes in >95% de.

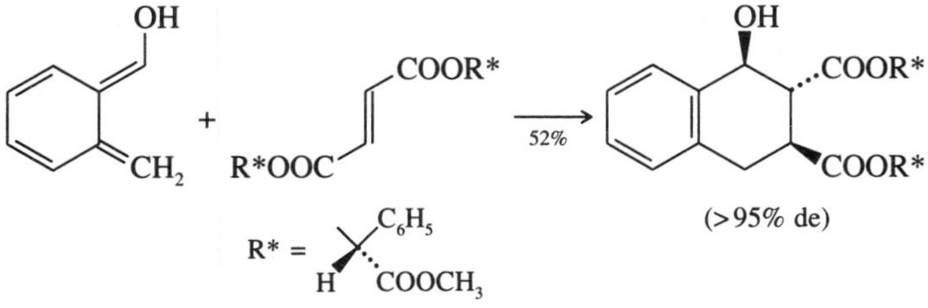

[1] J. L. Charlton and K. Koh, *Synlett*, 333 (1990).

(S)-1-Methyl-2-[(N-1-naphthylamino)methyl]pyrrolidine,

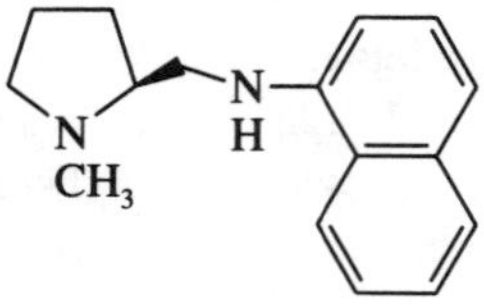

(1)

Asymmetric aldol reactions.[1]   This diamine (1) when coordinated with tin(II) triflate and dibutyltin diacetate promotes highly stereoselective aldol-type reactions between silyl enol ethers and aldehydes.

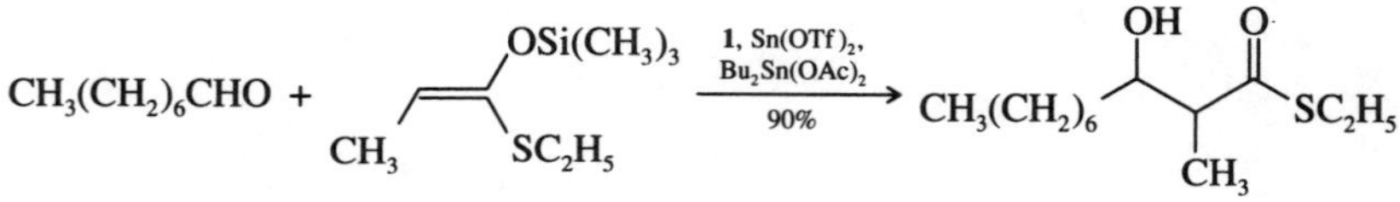

syn, >98% ee

[1] T. Mukaiyama, H. Uchiro, and S. Kobayashi, *Chem. Letters*, 1757 (1989); T. Mukaiyama, S. Kobayashi, H. Uchiro, and I. Shiina, *ibid.*, 129 (1990).

(S)-(−)-Methyl 1-naphthyl sulfoxide (1). This sulfoxide is obtained in optically pure form by oxidation of the complex of methyl 1-naphthyl sulfide and β-cyclodextrin with peracetic acid followed by crystallization; m.p. 58°, α_D-460°.

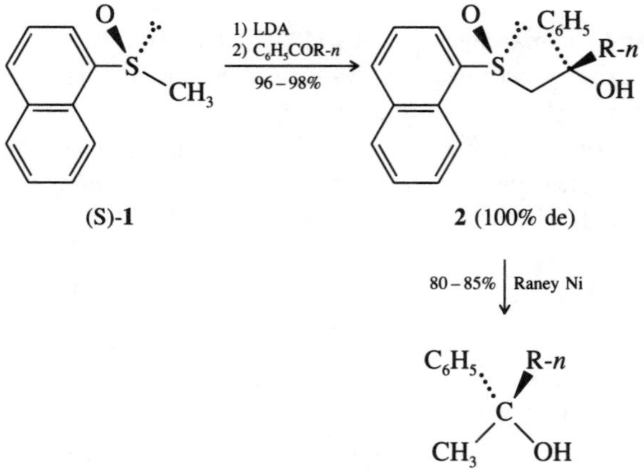

(S)-1   →   2 (100% de)

80−85% | Raney Ni

(S)-3

The anion (LDA) of (S)-**1** adds to *n*-alkyl phenyl ketones to give the (S,S)-adduct (**2**) in essentially 100% de. Desulfurization of the adducts provides optically pure (S)-tertiary alcohols (**3**).[1] Reaction of the anion of **1** with aliphatic ketones shows almost no diastereoselectivity.

[1] H. Sakuraba and S. Ushiki, *Tetrahedron Letters*, **31**, 5349 (1990).

10-Methyl-9-oxa-10-borabicyclo[3.3.2]decane (1).

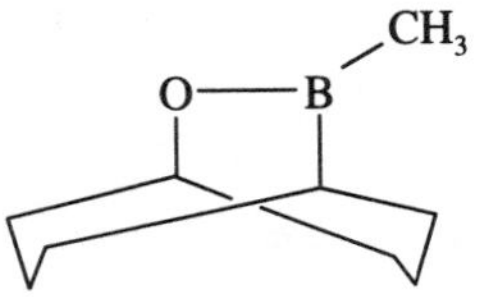

The borane is prepared by reaction of $B-CH_3O$-9-BBN with CH_3Li and then trimethylamine oxide.

Coupling with vinyl, alkenyl, and aryl bromides.[1] This reagent couples with these bromides under Suzuki conditions (**14**, 124–125; cat. Pd, base) to give the corresponding methylated products. 10-CH_3-9-BBN is not useful for this coupling because it is spontaneously flammable in air.

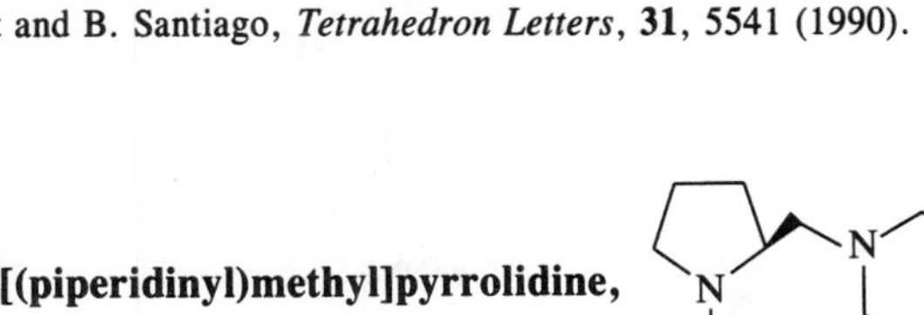

[1] J. A. Soderquist and B. Santiago, *Tetrahedron Letters*, **31**, 5541 (1990).

(S)-1-Methyl-2-[(piperidinyl)methyl]pyrrolidine,

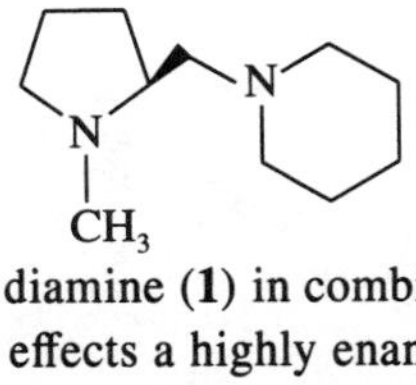

(**1**)

Asymmetric aldol-type reactions.[1] This chiral diamine (**1**) in combination with tin(II) triflate and tributyltin fluoride (**15**, 314–315) effects a highly enantioselective aldol-type reaction between ketene silyl acetals and aldehydes. A tentative structure (**2**) has been suggested for the promotor.

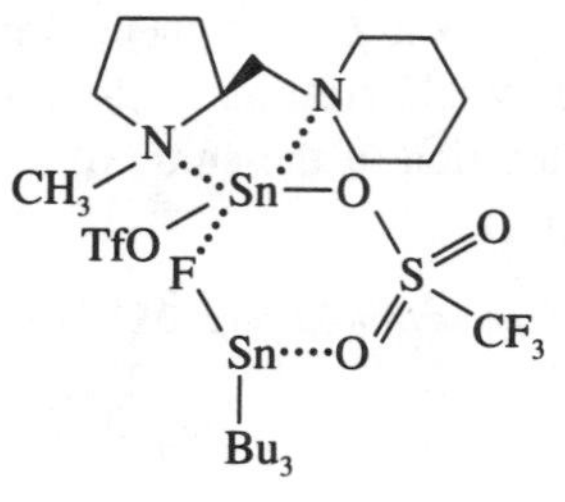

$$RCHO + CH_2=C \overset{OSi(CH_3)_3}{\underset{OBzl}{\big<}} \xrightarrow[50-80\%]{2} \quad BzlO \overset{O \quad OH}{\diagup\!\!\!\diagdown\!\!\!R}$$

(S), 89–98% ee

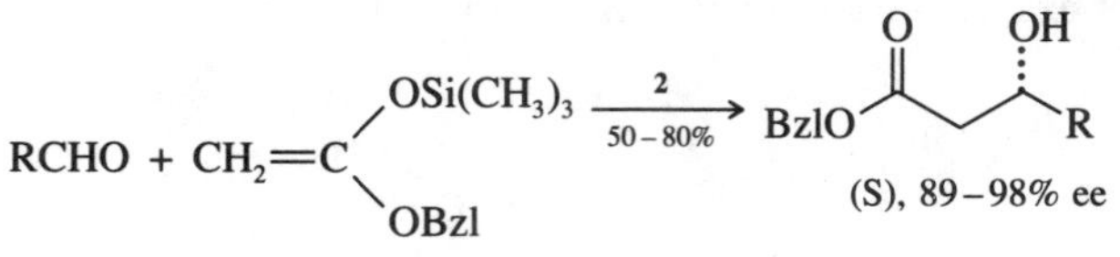

[1] T. Mukaiyama, S. Kobayashi, and T. Sano, *Tetrahedron*, **46**, 4653 (1990).

2-Methyl-β-propiolactone,   (**1**).

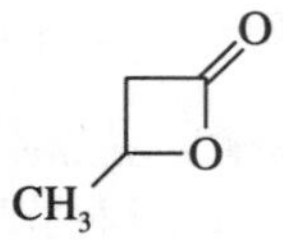

α,β-*Enones.*   This β-propiolactone reacts with phosphorus ylides in toluene at 40° to form ketophosphoranes (**3**). When heated these products eliminate triphenylphosphine oxide with formation of α,β-enones, possibly via an oxaphosphinine intermediate (**a**).

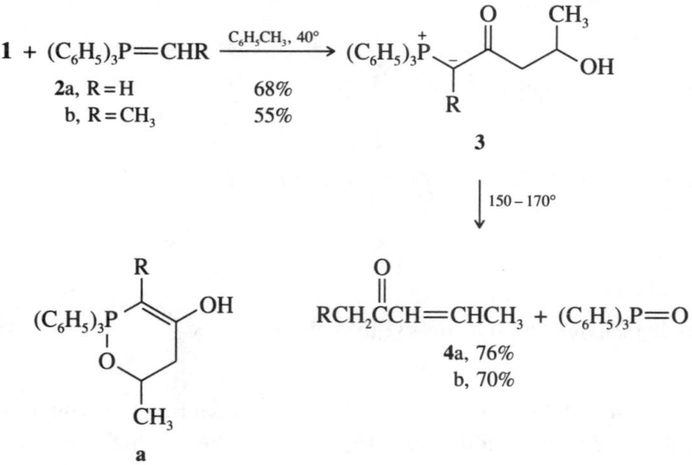

1 + $(C_6H_5)_3P=CHR$ $\xrightarrow{C_6H_5CH_3, \ 40°}$ $(C_6H_5)_3\overset{+}{P}\underset{R}{\diagdown}\overset{O \quad CH_3}{\diagup\!\!\!\diagdown\!\!\!\diagup\!\!\!\diagdown OH}$

2a, R=H	68%
b, R=CH_3	55%

3

↓ 150–170°

$(C_6H_5)_3P\underset{O}{\overset{R}{\diagup\!\!\!\diagdown}}\!\!\!OH$ ... CH_3

a

$$RCH_2\overset{O}{\overset{\|}{C}}CH=CHCH_3 + (C_6H_5)_3P=O$$

4a, 76%
b, 70%

[1] J. Le Roux and M. Le Corre, *J.C.S. Chem. Comm.*, 1464 (1989).

4-Methyl-1,2,4-triazoline-3,5-dione,

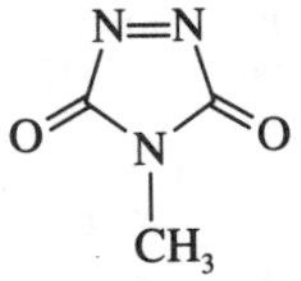

Oxidation of hydrazines to diazines. Propargylic hydrazines are oxidized rapidly and efficiently in CH_3OH at 0° by 4-methyl-1,2,4-triazoline-3,5-dione (MTAD) or diethyl azodicarboxylate (DEAD) with evolution of nitrogen to provide the corresponding allenes in 50–70% yield. The reaction occurs with high stereospecificity, and can be used to obtain optically active allenes (equation I).[1]

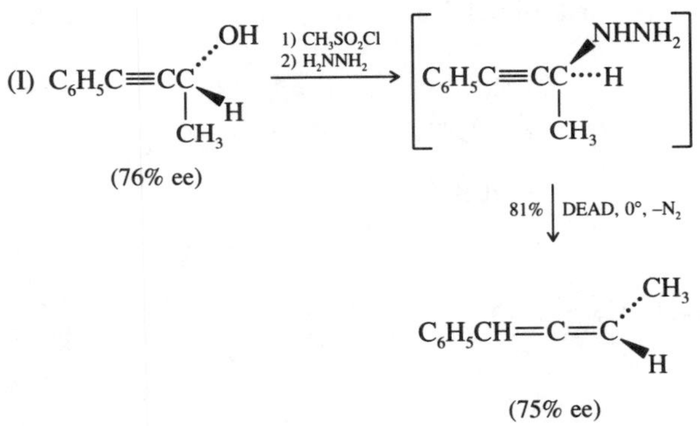

Of equal significance, the oxidation with MTAD can proceed at such low temperatures ($-95°$) that intermediates can be identified by NMR spectroscopy. This work suggests that the oxidation of (3-phenyl-2-propynyl)hydrazine (**1**) involves formation of both the (E)- and (Z)-diazine (**2**). (Z)-**2** rearranges rapidly to **3**, but (E)-**2** rear-

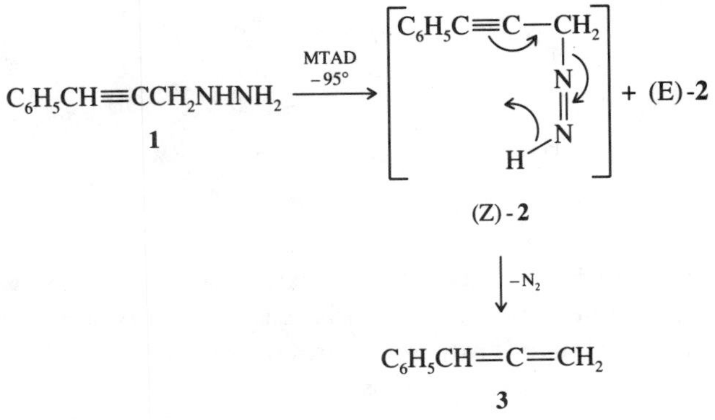

ranges more slowly at $-70°$, possibly catalyzed by the 4-methylurazole formed on oxidations with MTAD.[2]

[1] A. G. Myers, N. S. Finney, and E. Y. Kuo, *Tetrahedron Letters*, **30**, 5747 (1989).
[2] A. G. Myers and N. S. Finney, *Am. Soc.*, **112**, 9641 (1990).

Methyl(trifluoromethyl)dioxirane, (1), **15**, 212.

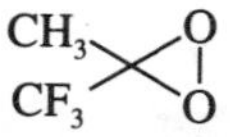

Oxidation of alkanes to alcohols and/or ketones.[1] This dioxirane oxidizes hydrocarbons in $CH_2Cl_2/1,1,1$-trifluoro-2-propanone (TFP) at -22 to $0°$ to alcohols or further oxidation products in high yield. Tertiary C–H bonds are attacked more rapidly than secondary ones, and primary C–H bonds are scarcely affected. The oxidation apparently involves insertion of O-atom. Oxidations can be stereospecific, as in the case of *cis*- and *trans*-1,2-dimethylcyclohexane.

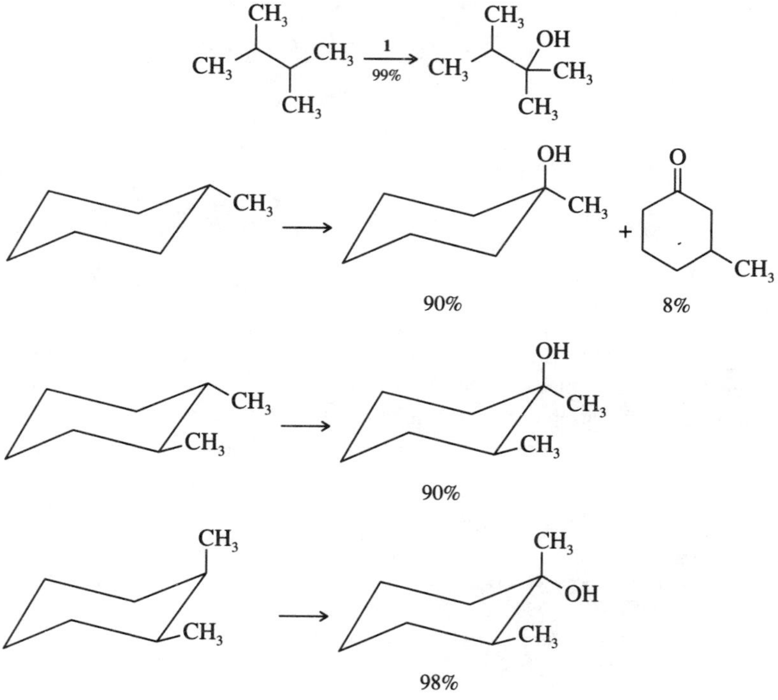

Oxidation of alcohols.[2] This dioxirane oxidizes secondary alcohols to the ketones very rapidly even at low temperatures in essentially quantitative yield. Primary alcohols are oxidized more slowly to mixtures of the aldehyde and the carboxylic acid. The oxidation presumably proceeds via a hemiacetal.

[1] R. Mello, M. Fiorentino, C. Fusco, and R. Curci, *Am. Soc.*, **111**, 6749 (1989).
[2] R. Mello, L. Cassidei, M. Fiorentino, C. Fusco, W. Hümmer, V. Jäger, and R. Curci, *Am. Soc.*, **113**, 2205 (1991).

N-Methyl trifluoromethanesulfonamide.

Methylation of alcohols[1] (**6**, 406).   The sensitive aldol product **1** was converted to the corresponding methyl ether in 83% yield by reaction with methyl triflate (15 equiv.) and 2,6-di-*t*-butylpyridine (30 equiv.) in CHCl$_3$ at 80° without retro-aldol cleavage or epimerization.

1

−OH → −NTfCH$_3$.   Primary or secondary alcohols are converted to protected secondary amines by this triflamide under Mitsunobu conditions (triphenylphosphine, diethyl azodicarboxylate) in 70–86% yield. The reaction proceeds with inversion, and is useful for preparation of optically active secondary amines.

[1] D. A. Evans and G. S. Sheppard, *J. Org.*, **55**, 5192 (1990).
[2] M. L. Edwards, D. M. Stemerick, and J. R. McCarthy, *Tetrahedron Letters*, **31**, 3417 (1990).

Molybdenum carbonyl.

Chiral molybdenum complexes of 2H-pyran.[1]   Enantiomerically pure Mo-complexes, (S)- and (R)-**1**, of 2*H*-pyran have been prepared by known methods (**13**, 194–195) from D- and L-arabinose, respectively. They react with a wide range of nucleophiles at an allylic position with 96% ee. The resulting complex can react with a second nucleophile at the other allylic position to form *cis*-disubstituted complexes, also with high enantioselectivity. The sequence can be used to obtain chiral *cis*-2,6-disubstituted tetrahydropyrans such as **2**, a component of the scent gland of the civer cat.

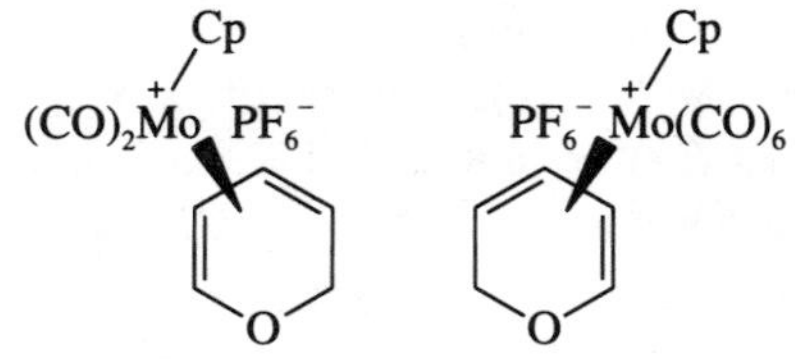

(S)-**1**   (R)-**1**

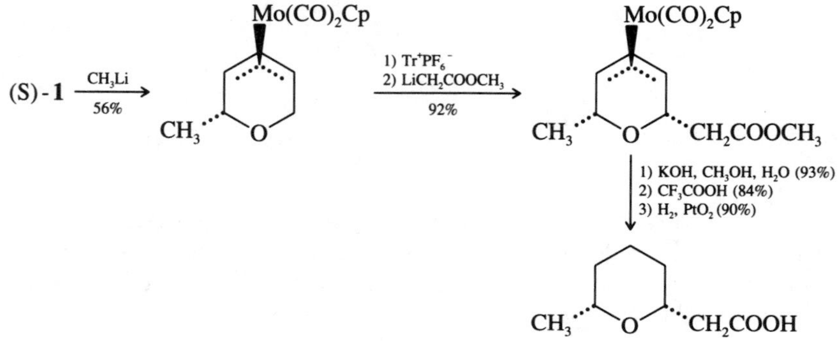

(R,R)-**2**, >90% ee

These complexes can also be used to prepare optically pure *cis*-2,5-disubstituted 5,6-dihydro-2*H*-pyrans such as **3** by using a molybdenum nitrosyl allyl complex as an intermediate.

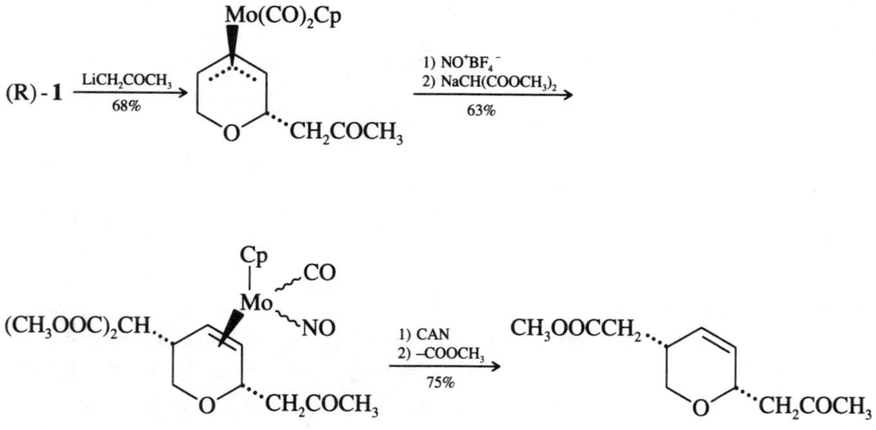

(−)-**3**, >90% ee

[1] S. Hansson, J. F. Miller, and L. S. Liebeskind, *Am. Soc.*, **112**, 9660 (1990).

Molybdenum(VI) oxide–Bis(tributyltin) oxide, MoO_3–$(Bu_3Sn)_2O$.

Epoxidation. The two reagents (premixed in $CHCl_3$) can serve as a catalyst for epoxidation of alkenes in H_2O by H_2O_2 (60% yield). In some cases (hexene, α-pinene) addition of trimethylamine is essential for high yields, 57–87%. Surprisingly, addition of a phase-transfer catalyst depresses the yield.

[1] T. Kamiyama, M. Inoue, and S. Enomoto, *Chem. Letters*, 1129 (1989).

2-Morpholinobutadienes,

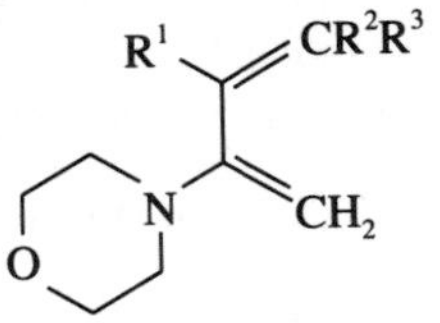

(1).

Butadienes of this type are readily available by addition of morpholine to 3-alkene-1-ynes in the presence of $Hg(OAc)_2$ (45–65% yield).[1]

Divinyl ketones.[2] These butadienes react with aromatic aldehydes in the presence of $MgBr_2 \cdot O(C_2H_5)_2$ or $ZnCl_2$ to furnish divinyl ketones (Nazarov reagents).

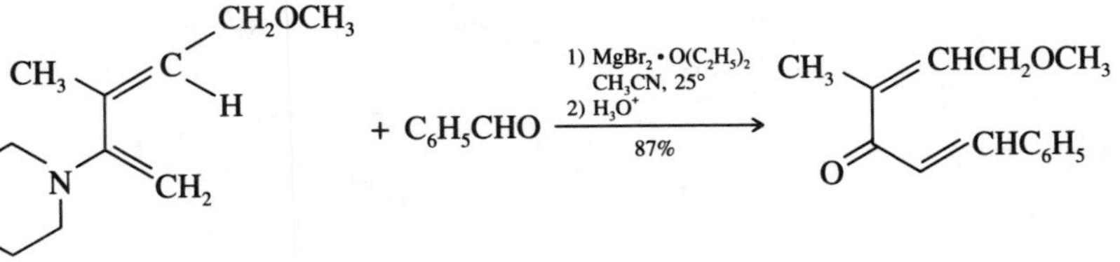

[1] J. Barluenga, F. Aznar, R. Liz, and M.-P. Cabal, *J.C.S. Chem. Comm.*, 1375 (1985).
[2] J. Barluenga, F. Aznar, M.-P. Cabal, and C. Valdes, *Tetrahedron Letters*, **30**, 1413 (1989).

N

Nickel boride, Ni$_2$B.

ArNO$_2$ → ArNH$_2$.[1] Aryl nitro compounds are reduced to arylamines by Ni$_2$B at 40° in $3N$ HCl or $15N$ NH$_4$OH in 80–96% yield without effect on alkene, keto, nitrile, amide, carboxyl, or ester functional groups. Nitroso-, azoxy-, azo-, and hydrazobenzene are reduced to amines under the same conditions.

[1] A. Nose and T. Kudo, *Chem. Pharm. Bull.*, **37**, 816 (1989).

Nickel cyanide, Ni(CN)$_2$·4H$_2$O.

Carbonylation of CH$_2$=CHX and halodienes.[1] Ni(CN)$_2$ is an active catalyst for carbonylation of unsaturated halides under phase-transfer conditions. The effective species is probably a cyanotricarbonylnickelate. Thus in the presence of cetyltri-

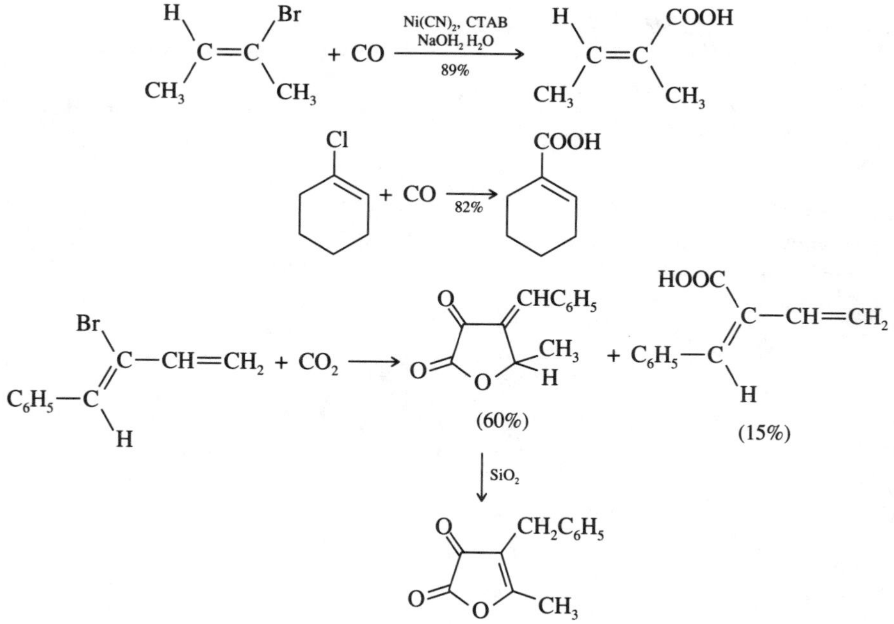

methylammonium bromide (CTAB) and Ni(CN)$_2$, carbon monoxide converts vinyl halides to α,β-unsaturated acids.

[1] H. Alper, I. Amer, and G. Vasapollo, *Tetrahedron Letters*, **30**, 2615, 2617 (1989).

Niobium(III) chloride–Dimethoxyethane, 14, 213–214.

Coupling of alkynes with 1,2-aryl dialdehydes.[1] The complexes of Nb-$Cl_3 \cdot$DME (1) with alkynes react with phthalic dicarboxaldehyde to give, after work-up with aqueous KOH, 2,3-disubstituted naphthols. In the case of unsymmetrical

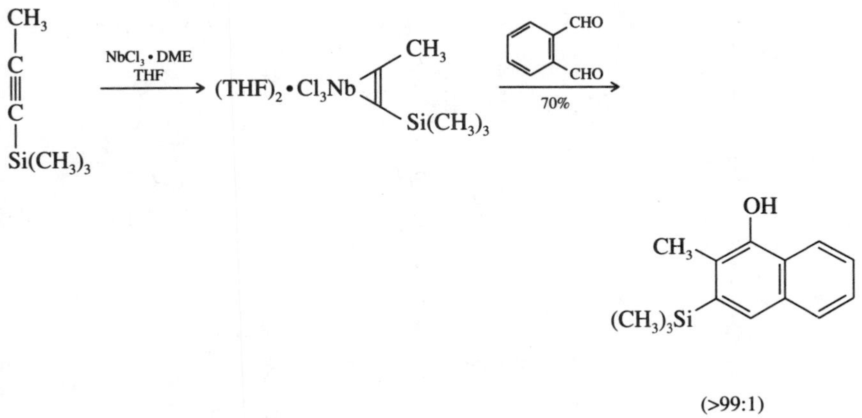

(>99:1)

alkynes, two isomeric naphthols are formed, but use of trialkylsilyl-substituted alkynes results in high yield of a single regioisomer.

Pyrrole synthesis.[2] This reagent effects coupling of α,β-unsaturated imines with an ester or DMF to form N-substituted pyrroles in 34–78% yield.

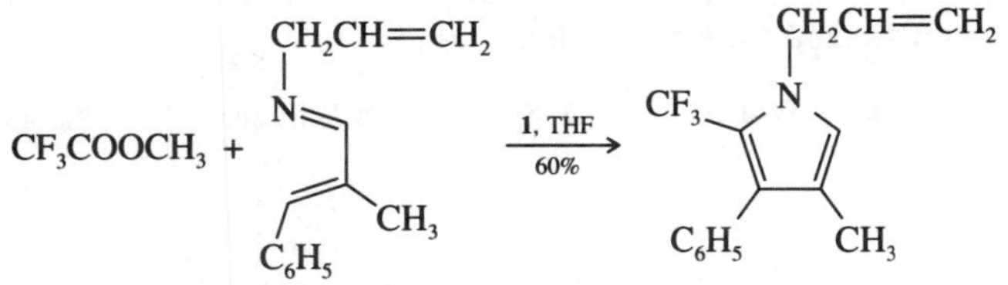

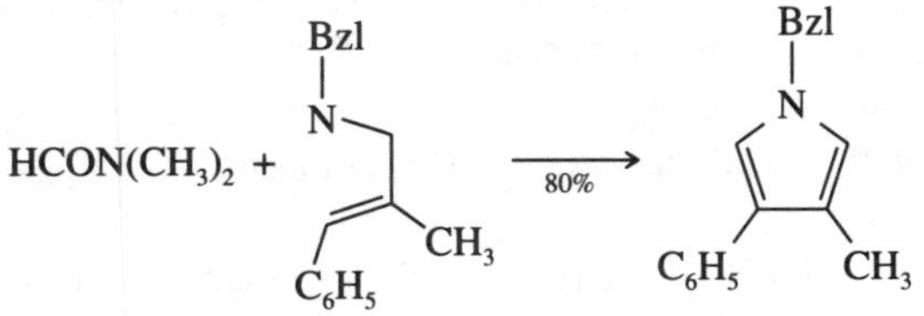

[1] J. B. Hartung, Jr., and S. F. Pedersen, *Am. Soc.*, **111**, 5468 (1989).
[2] E. J. Roskamp, P. S. Dragovich, J. B. Hartung, Jr., and S. F. Pedersen, *J. Org.*, **54**, 4736 (1989).

Nitrosobenzene, $C_6H_5N=O$.

Diels–Alder reactions.[1] Nitrosoarenes undergo Diels–Alder reactions at 25° with *cis*- and/or *trans*-hexadienals **2** to give unstable adducts that can be identified by IR and ¹H-NMR as **3** or the hemiacetals **4**. On standing or warming to 40° these primary products rearrange to pyridinium betaines (**5**) or pyrroloindoles (**6**) as the

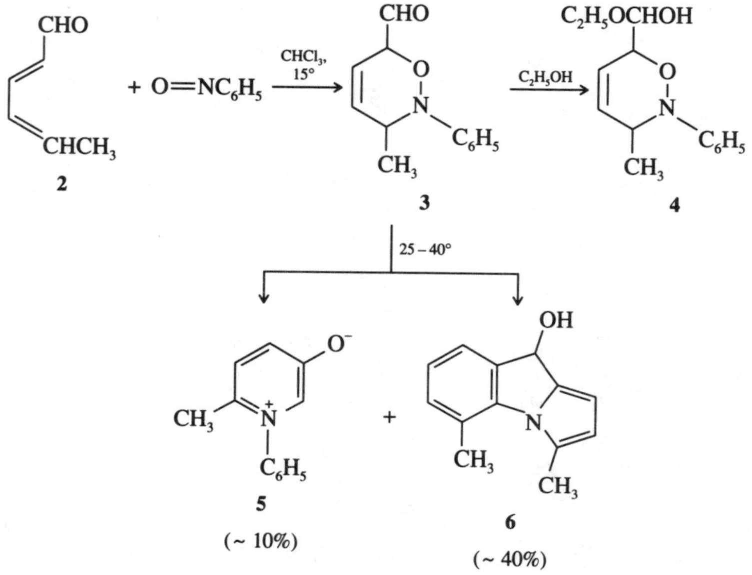

major products. The latter products are particularly interesting because they are related to antibiotic and antitumor mitomycins.

[1] A. Defoin, G. Geffroy, D. Le Nouen, D. Spileers, and J. Streith, *Helv.*, **72**, 1199 (1989).

Norephedrine.

γ-Hydroxy ketones.[1] An asymmetric synthesis of γ-hydroxy ketones employs selective addition of a dialkylzinc to γ-keto aldehydes catalyzed by (1S,2R)-(−)- or (1R,2S)-(+)-N,N-dibutylnorephedrine (**1**).

Conjugate addition of R_2Zn to enones.[2] A nickel catalyst (**2**) consisting of (1S,2R)-(−)-N,N-dibutylnorephedrine (**1**), $Ni(acac)_2$, and 2,2′-bipyridyl promotes

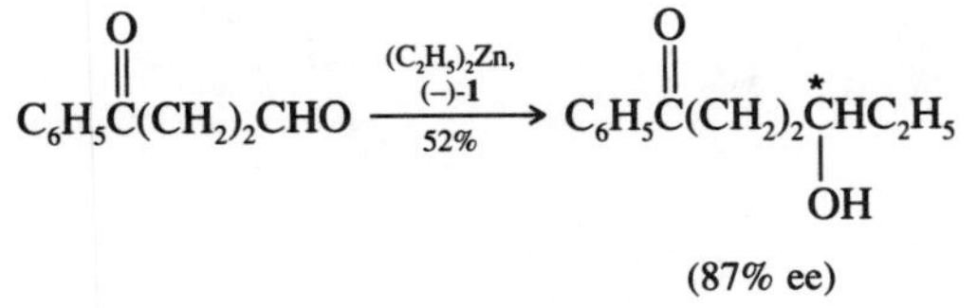

$$C_6H_5\overset{\displaystyle O}{\overset{\|}{C}}(CH_2)_2CHO \xrightarrow[52\%]{\substack{(C_2H_5)_2Zn, \\ (-)\text{-}1}} C_6H_5\overset{\displaystyle O}{\overset{\|}{C}}(CH_2)_2\overset{*}{\underset{\underset{\displaystyle OH}{|}}{C}}HC_2H_5$$

(87% ee)

asymmetric conjugate addition of dialkylzincs to α,β-enones in CH_3CN/toluene. Both $Ni(acac)_2$ and CH_3CN as well as **1** are essential for enantioselectivity. 2,2'-Bipyridyl can be replaced by piperazine or 1,10-phenanthroline without significant loss of enantioselectivity.

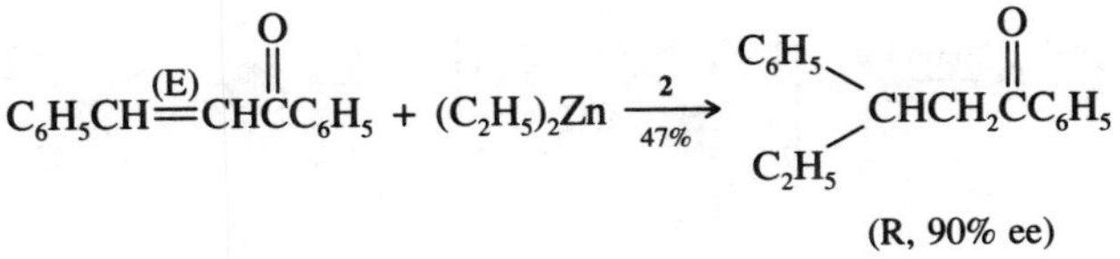

$$C_6H_5CH\overset{(E)}{=\!\!=}CH\overset{\displaystyle O}{\overset{\|}{C}}C_6H_5 + (C_2H_5)_2Zn \xrightarrow[47\%]{2} \underset{C_2H_5}{\overset{C_6H_5}{>}}CHCH_2\overset{\displaystyle O}{\overset{\|}{C}}C_6H_5$$

(R, 90% ee)

[1] K. Soai, M. Watanabe, and M. Koyano, *J.C.S. Chem. Comm.*, 534 (1989).
[2] K. Soai, T. Hayasaka, and S. Ugajin, *ibid.*, 516 (1989).

O

Organocerium reagents.

Spiroketals and oxaspirolactones.[1] An improved route to spiroketals from lactones involves use of cerium 3- or 4-cerioalkoxides such as **1**, prepared from 3,3-
dimethyloxetane by reductive cleavage with lithium di-*t*-butylbiphenylide (LDBB)
followed by transmetallation. Use of the intermediate lithium 3-lithioalkoxide as the

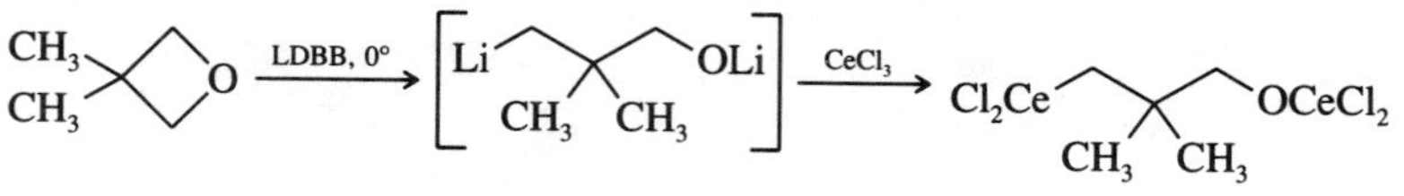

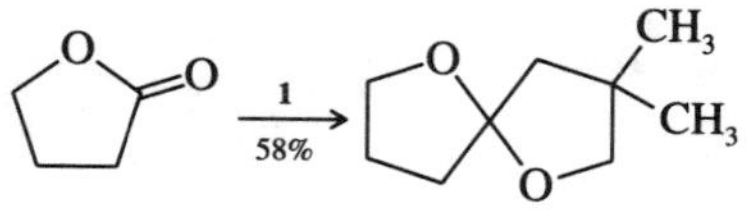

reagent results in low yields of the desired spiroketal because of double attack of the
lactone to give diols. A cerium reagent such as **1** also undergoes monoaddition to
anhydrides to provide an oxaspirolactone such as **2**.

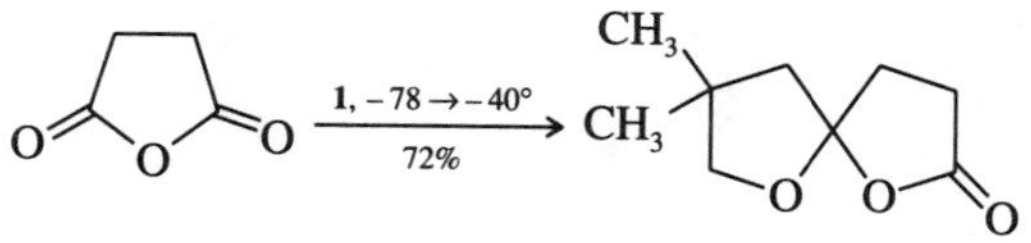

2

[1] B. Mudryk, C. A. Shook, and T. Cohen, *Am. Soc.*, **112**, 6389 (1990).

Organocopper reagents.

Trimethylstannylcuprates; $(CH_3)_3Sn(Bu)Cu(CN)Li_2$.[1] These cuprates are
most conveniently obtained by *in situ* generation of $(CH_3)_3Sn\text{-}Si(CH_3)_3$ followed by
reaction with $Bu_2Cu(CN)Li_2$.

These cuprates deliver the $(CH_3)_3Sn$ group to organic substrates by conjugate
addition or substitution of halo or triflate groups. Stannylcupration is also possible.

Chiral amino acids.[2] The key step in a new synthesis of chiral β-amino acids
involves displacement of mesyloxy groups by lithium dialkylcuprates. Thus (S)-

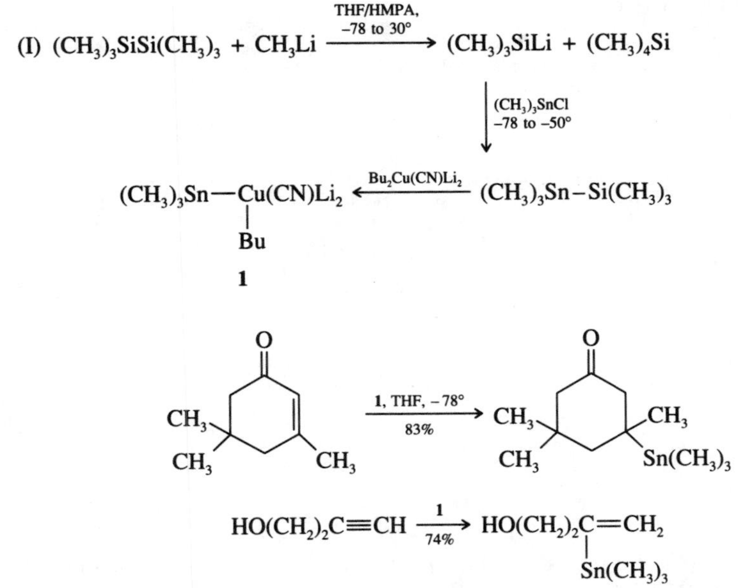

(I) $(CH_3)_3SiSi(CH_3)_3$ + CH_3Li $\xrightarrow[\text{−78 to 30°}]{\text{THF/HMPA,}}$ $(CH_3)_3SiLi$ + $(CH_3)_4Si$

$\downarrow$ $(CH_3)_3SnCl$
$\quad$ −78 to −50°

$(CH_3)_3Sn$—$\underset{\underset{\textbf{1}}{\overset{|}{Bu}}}{Cu(CN)Li_2}$ $\xleftarrow{Bu_2Cu(CN)Li_2}$ $(CH_3)_3Sn$—$Si(CH_3)_3$

$\xrightarrow[\textbf{74\%}]{\textbf{1}}$

$HO(CH_2)_2C\equiv CH$ $\xrightarrow[74\%]{1}$ $HO(CH_2)_2\underset{\underset{Sn(CH_3)_3}{|}}{C}=CH_2$

N,N′-dibenzylasparagine (**1**) is converted by standard reactions into **2**, an activated β-homoserine equivalent. Reaction of **2** with R_2CuLi gives **3** in 50–70% yield. The final step to a β-amino acid involves hydrolysis of the nitrile group and deprotection of nitrogen.

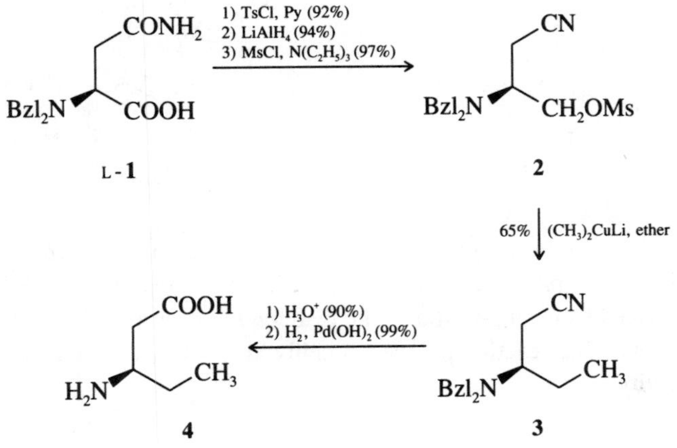

The ring opening of *t*-butyl (S)-N-tosylaziridine-2-carboxylate (**5**) with organocuprates provides products (**6** and **7**) that are precursors to optically active α- or β-amino acids.[3]

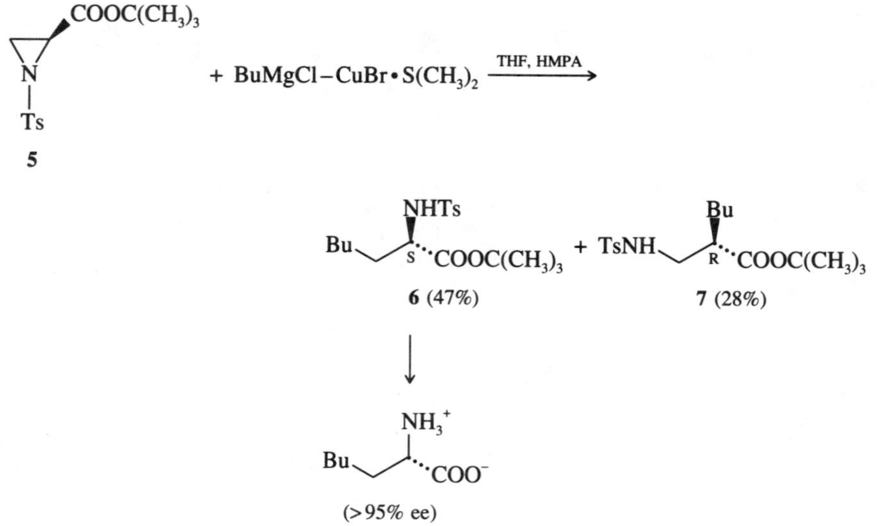

$$\text{5} \quad + \quad \text{BuMgCl–CuBr•S(CH}_3)_2 \xrightarrow{\text{THF, HMPA}}$$

6 (47%)   +   7 (28%)

(>95% ee)

(CH₃)₂Cu(CN)Li₂·BF₃ (1), $(CH_3)_2Cu(CN)Li_2 \cdot BF_3$ **(1)**,[4] prepared by addition of $BF_3 \cdot O(C_2H_5)_2$ to $(CH_3)_2Cu(CN)Li_2$ at $-78 \rightarrow 0°$ (aging is important), undergoes a highly diastereoselective reaction with γ-mesyloxy α,β-enoates to provide (E)-α-methyl-β,γ-enoates, regardless of the geometry of the starting material, and with highly diastereostereoselective 1,3-chirality transfer (>99:1). A γ-mesyloxy or a γ-tosyloxy group and THF or THF/ether as solvent are essential for high chirality transfer. This reaction provides chiral products from L-threonine.

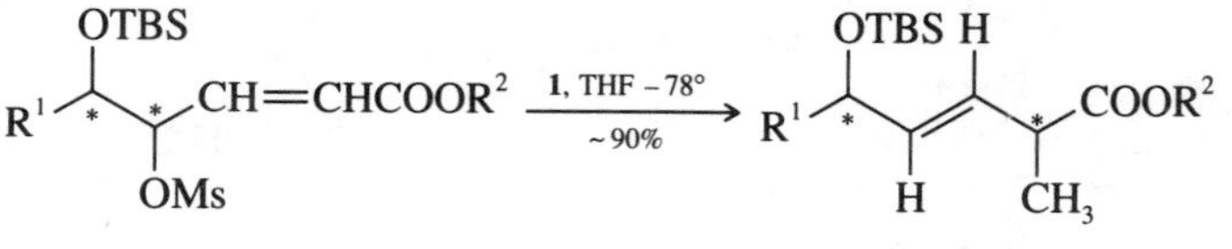

(>99:1 de at C_2)

Chiral, unsymmetrical divinylmethanols can be prepared by reaction of $CH_3Cu(CN)Li \cdot BF_3$, prepared from CH_3Li, with a monoprotected (E,E)-dienoate such as **1**, derived from an L-tartrate. The reaction involves an S_N2' reaction with the mesyloxy group. The reaction proceeds readily at $-78°$ with high regio- and (E)-stereoselectivity.[5]

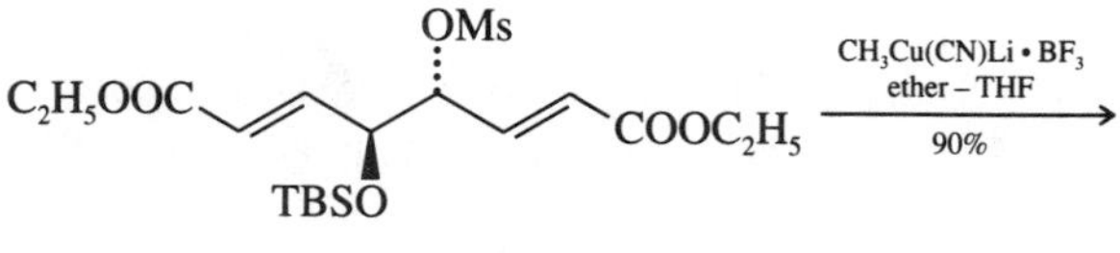

1 [TBS = t-Bu(CH₃)₂Si]

$$C_2H_5OOC\diagdown\diagup\diagdown\diagup\diagdown COOC_2H_5$$

(99% de)

Addition to allylic mesylates.[6] Conjugate addition of organocyanocuprates to acyclic allylic mesylates substituted at the β-position with a chiral sulfoxide group involves an S_N2'-substitution with high Z/E stereoselectivity and high asymmetric induction. This reaction provides a route to chiral trisubstituted vinyl sulfoxides.

$$C_2H_5 \diagdown \overset{OMs}{\underset{C_6H_5}{\diagup}}\overset{O}{\underset{}{S}}C_7H_7 + CH_3CuCNLi \xrightarrow{81\%} $$

94:6

$$C_2H_5 \diagdown \overset{OMs}{\underset{C_6H_5}{\diagup}}\overset{O}{\underset{}{S}}C_7H_7 + CH_3CuCNLi \xrightarrow{80\%} $$

72:28

Reaction of R₂CuLi with sec-tosylates.[7] Displacement of tosyloxy groups is facilitated by the presence of S or O atoms in the vicinity. In the case of methoxy-methyl ethers or (methylthio)methyl ethers, optimum conditions obtain when the S or O atoms are in a vicinal position. This reaction is useful because it involves complete inversion of configuration. As shown by the last entry in equation (I), the yield of the substitution is low if a heteroatom is absent. A sulfur atom is more effective than oxygen for coordination to copper.

(I) $R^1CH_2\overset{OTs}{\underset{}{C}}HCH_2CH_2R^2 + (CH_3)_2CuLi \longrightarrow R^1CH_2\overset{CH_3}{\underset{}{C}}HCH_2CH_2R^2$

$$R^1 = -CH_2SC_6H_5 \qquad 90\%$$
$$= -OCH_2SCH_3 \qquad 83\%$$
$$= -OCH_2OCH_3 \qquad 64\%$$
$$= -(CH_2)_2CH_3 \qquad 40\%$$

Allylic cyanocuprates.[8] Reagents of this type can be prepared in essentially quantitative yield by reaction of allyltributyltins with $(CH_3)_2Cu(CN)Li_2$ (equation I). Unlike allylic cuprates, these new allylic cyanocuprates are highly reactive even at −78°. Thus they can displace unactivated chlorides and cleave epoxides without rearrangement (equation II).

(I) $2 CH_2{=}CHCH_2SnBu_3 + (CH_3)_2Cu(CN)Li_2 \xrightarrow[-2CH_3SnBu_3]{THF,\ 0°} (CH_2{=}CHCH_2)_2Cu(CN)Li_2$

$$\mathbf{1}$$

$$86\% \downarrow C_6H_5O(CH_2)_3Br,\ -78°$$

$$C_6H_5O(CH_2)_4CH{=}CH_2$$

(II) 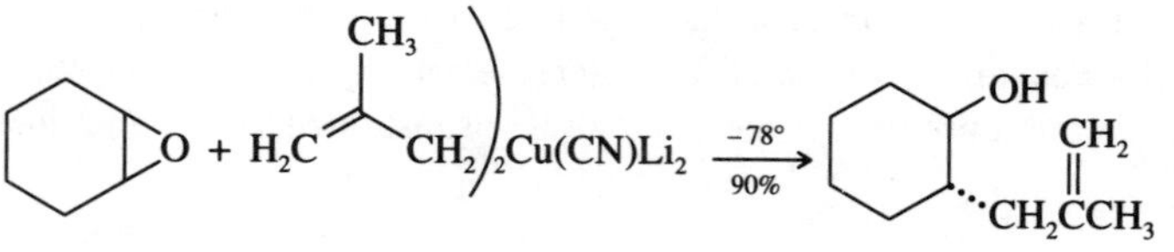 $+ \ H_2C{\Large\diagdown}^{CH_3}{\Big(}CH_2{\Big)}_2Cu(CN)Li_2 \xrightarrow[90\%]{-78°}$

Cyclic allylic cyanocuprates can also be prepared from cyclic allylic stannanes by reaction with CH_3Li followed by CuCN solubilized with LiCl. These cyanocuprates couple with enones, primary and vinylic halides and epoxides (equation III).[9]

(III)

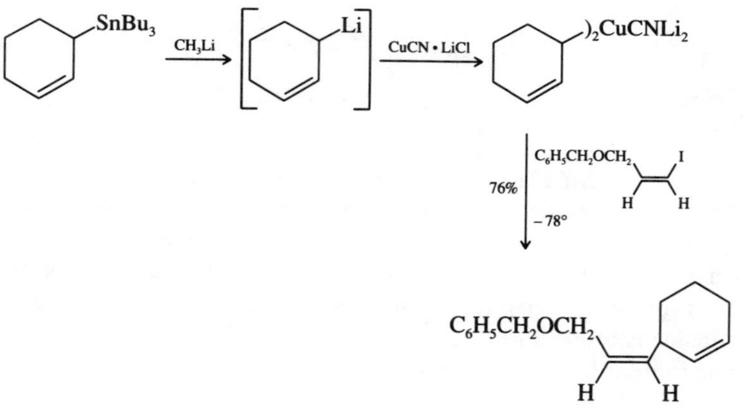

$$76\% \downarrow \ \begin{matrix} C_6H_5CH_2OCH_2 \\ \end{matrix} \ \ -78°$$

Dilithium diallylic cyanocuprates couple readily at $-78°$ in THF with vinyl triflates to form 1,4-dienes (equation IV).[10]

(IV) $C_6H_5CH_2CH{=}CHOTf + [(CH_3)_2C{=}CHCH_2]_2Cu(CN)Li_2 \xrightarrow{79\%}$

(E/Z = 8:1)

$$C_6H_5CH_2CH{=}CHCH_2CH{=}C(CH_3)_2$$

(E/Z = 8:1)

Allylcoppers.[11] In general, lithium diallylcuprates are not useful for conjugate addition to enones. Surprisingly, allylcoppers, prepared from allyltributyltins, methyllithium, and CuI·LiCl in THF, add to enones even at $-78°$, and this Michael reaction is markedly facilitated by the presence of $ClSi(CH_3)_3$. In the case of prenyl- and crotylcoppers, coupling occurs mainly at the α-position.

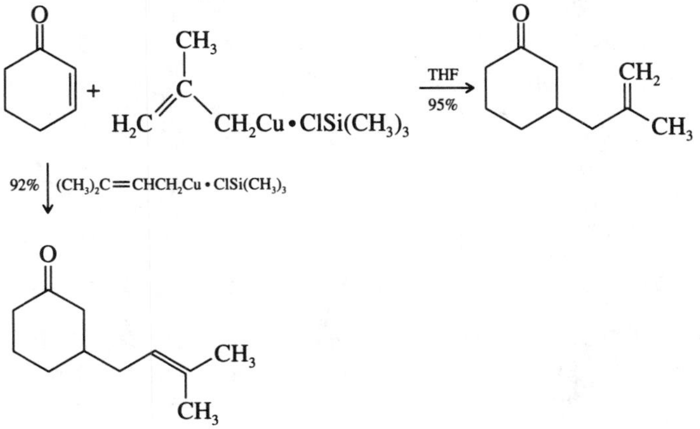

C-Glycosides. Transmetallation of glycosylstannanes with BuLi provides glycosyllithium reagents that undergo 1,2-addition to carbonyl compounds to provide C-glycosides.[12] Reaction of the glycosyllithium reagents with epoxides, however, even when promoted by a Lewis acid, proceeds in poor yield. Coupling with epoxides can be markedly improved by conversion of the glycosyllithium to a cuprate by reaction with 2-thienyl(cyano)copper lithium (**1, 14,** 226; **15,** 228). This cuprate reacts effi-

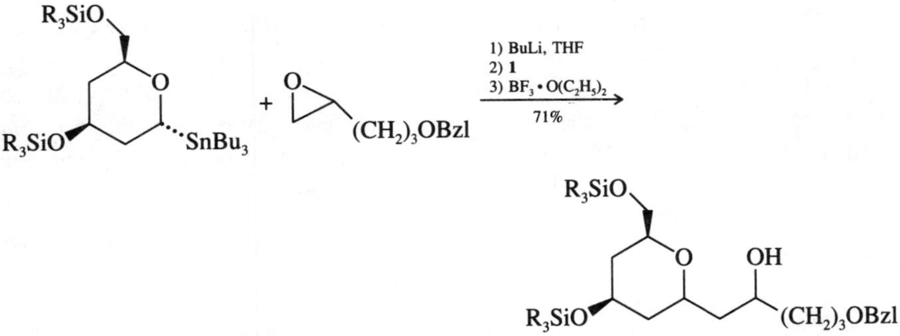

ciently with epoxides in the presence of BF_3 etherate to give C-glycosides in good yield.[13]

Organobis(cuprates), **14,** 225. A detailed report[14] of the spiroannelation of enones with these cuprates includes 11 examples. In all cases yields of 56–96% are obtainable.

[1] B. H. Lipshutz, S. Sharma, and D. C. Reuter, *Tetrahedron Letters*, **31**, 7253 (1990).

[2] P. Gmeiner, *Tetrahedron Letters*, **31**, 5717 (1990).

[3] J. E. Baldwin, R. M. Adlington, I. A. O'Neil, C. Schofield, A. C. Spivey, and J. B. Sweeney, *J.C.S. Chem. Comm.*, 1852 (1989).

[4] T. Ibuka, M. Tanaka, S. Nishii, and Y. Yamamoto, *Am. Soc.*, **111**, 4864 (1989).

[5] T. Ibuka, M. Tanaka, and Y. Yamamoto, *J.C.S. Chem. Comm.*, 967 (1989).

[6] J. P. Marino, A. Viso, R. Fernandez de la Pradilla, and P. Fernandez, *J. Org.*, **56**, 1349 (1991).

[7] S. Hanessian, B. Thavonekham, and B. DeHoff, *J. Org.*, **54**, 5831 (1989).

[8] B. H. Lipschutz, R. Crow, S. H. Dimock, E. L. Ellsworth, R. A. J. Smith, and J. R. Behling, *Am. Soc.*, **112**, 4063 (1990).

[9] B. H. Lipshutz, C. Ung, T. R. Elworthy, and D. C. Reuter, *Tetrahedron Letters*, **31**, 4539 (1990).

[10] B. H. Lipshutz and T. R. Elworthy, *J. Org.*, **55**, 1695 (1990).

[11] B. H. Lipshutz, E. L. Ellsworth, S. H. Dimock, and R. A. J. Smith, *Am. Soc.*, **112**, 4404 (1990).

[12] P. Lesimple, J.-M. Beau, G. Jaurand, and P. Sinaÿ, *Tetrahedron Letters*, **27**, 6201 (1986).

[13] J. Prandi, C. Audin, and J.-M. Beau, *ibid.*, **32**, 769 (1991).

[14] P. A. Wender, A. W. White, and F. E. McDonald, *Org. Syn.*, submitted (1989).

Organocopper/zinc reagents.

RCu(CN)ZnI. Organozinc iodides (or bromides) are readily formed by reaction of alkyl or aryl iodides with zinc foil or dust. This insertion reaction is compatible with a wide variety of functional groups: ester, keto, nitrile, acetoxy, amino, thioether. However, these reagents lack the high reactivity of organocopper compounds. In contrast, it is difficult to prepare organocopper compounds containing reactive constituents because they are usually obtained by reaction of Grignard or lithium reagents with a copper(I) species.

Organozinc halides, RZnI, undergo transmetallation to RCu(CN)ZnI on reaction with CuCN·2LiCl, prepared by reaction of CuCN with anhydrous LiCl in THF. The resulting dimetallic reagents are comparable to organocopper reagents in reactions with electrophiles and in addition to unsaturated substrates.[1]

This transmetallation has been used to obtain novel α-acetoxyalkylcopper/zinc reagents such as **1** from α-bromoalkyl acetates, obtained by addition of acetyl bromide to aldehydes.[2]

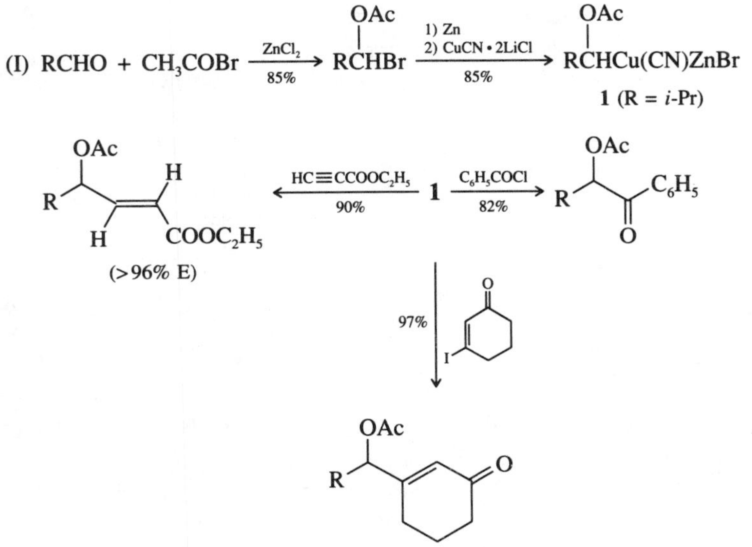

(I) $RCHO + CH_3COBr \xrightarrow[85\%]{ZnCl_2}$ RCHBr (OAc) $\xrightarrow[85\%]{\text{1) Zn} \atop \text{2) CuCN·2LiCl}}$ RCHCu(CN)ZnBr (OAc)

1 (R = *i*-Pr)

(>96% E)

$\xleftarrow[90\%]{HC\equiv CCOOC_2H_5}$ **1** $\xrightarrow[82\%]{C_6H_5COCl}$

97%

Allylcopper/zinc reagents can be prepared directly by reaction of vinyl copper reagents with (iodomethyl)zinc iodide, the Simmons–Smith reagent. These allylcopper/zinc reagents do not couple with an alkyl iodide or benzyl bromide, but react readily with electrophiles such as aldehydes, ketones, or imines.[3] This approach to organodimetallic reagents is apparently limited (see Iodomethylzinc iodide, this volume).

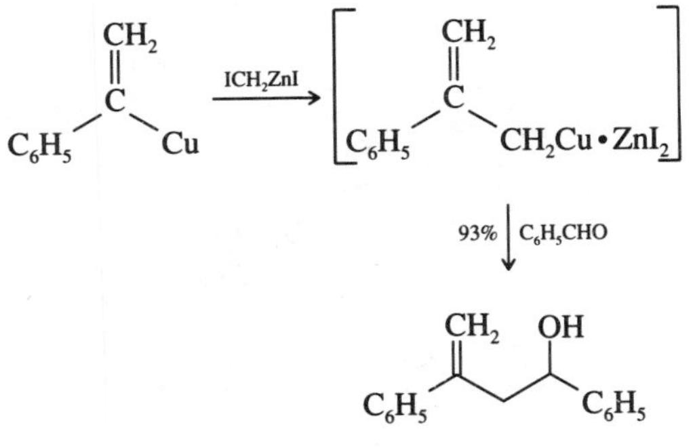

A similar insertion of zinc with aryl iodides requires N,N-dimethylformamide or -acetamide as solvent. These arylzinc iodides react with CuCN·2LiCl to form mixed zinc and copper organometallics, ArCu(CN)ZnI, which react with a variety of electrophiles to give functionalized aromatics.[4]

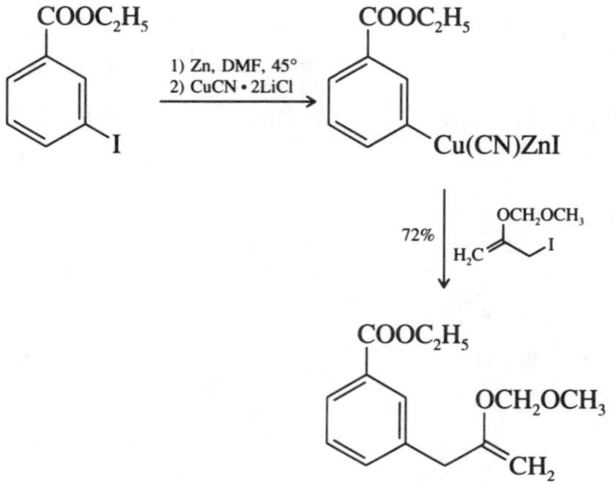

Sulfur-stabilized derivatives.[5] α-Chloroalkyl phenyl sulfides undergo a very rapid reaction with zinc to form the corresponding organozinc chloride, which reacts with CuCN·2LiCl to form an α-phenylthiocopper/zinc reagent such as **1**. Reagents of this type show enhanced reactivity with various electrophiles. Thus **1**, prepared as shown in equation (I), undergoes the usual reactions with electrophiles in generally high yield.

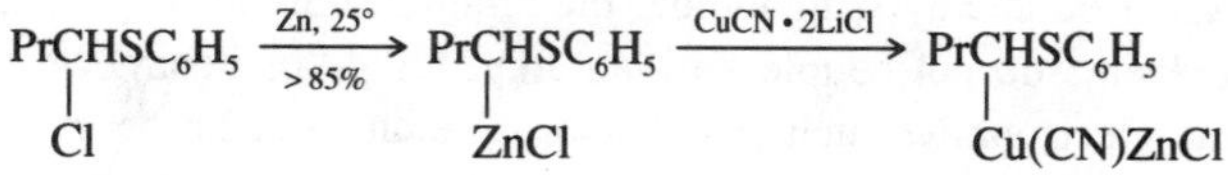

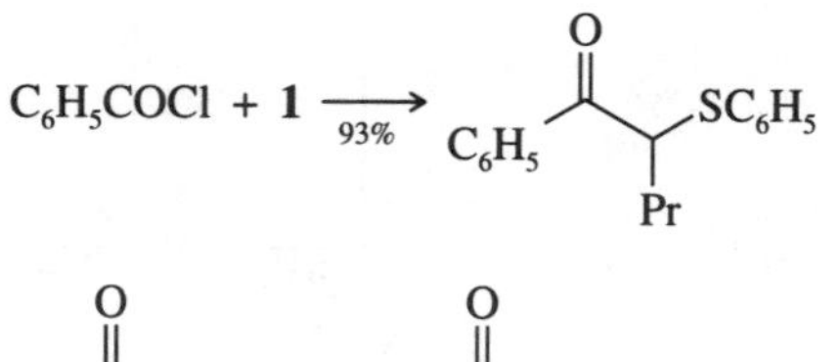

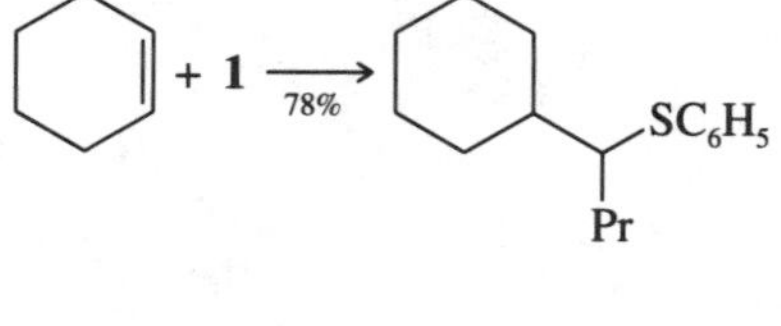

Organocopper/zinc reagents substituted by a phenylthio group at the γ-position can be prepared, and also show enhanced reactivity with electrophiles (equation II).

$$\text{(II)} \quad I(CH_2)_3SC_6H_5 \xrightarrow[\text{2) CuCN} \cdot \text{2LiCl}]{\text{1) Zn}} \underset{\mathbf{2}}{C_6H_5S(CH_2)_3Cu(CN)ZnI}$$

$$C_6H_5COCl + \mathbf{2} \xrightarrow[80\%]{} C_6H_5S(CH_2)_3COC_6H_5$$

Reaction with β-alkylthio nitro olefins.[6]   Nitro alkenes bearing an alkylthio or phenylsulfonyl group at the β-position undergo addition–elimination reactions with organocopper/zinc reagents to provide functionalized (E)-nitro alkenes.

$$NC(CH_2)_3Cu(CN)ZnI + (CH_3S)_2C{=}CHNO_2 \xrightarrow[86\%]{\substack{\text{THF, } -30° \\ \text{4 hr.}}} [NC(CH_2)_3]_2C{=}CHNO_2$$

[1] M. C. P. Yeh, H. G. Chen, and P. Knochel, *Org. Syn.*, submitted (1990).
[2] T.-S. Chou and P. Knochel, *J. Org.*, **55**, 5791 (1990).
[3] P. Knochel and S. A. Rao, *Am. Soc.*, **112**, 6146 (1990).
[4] T. N. Majid and P. Knochel, *Tetrahedron Letters*, **31**, 4413 (1990).
[5] S. A. Rao, C. E. Tucker, and P. Knochel, *ibid.*, **31**, 7575 (1990).
[6] C. Retherford and P. Knochel, *ibid.*, **32**, 441 (1991).

Organoiron reagents.

Alkyltetracarbonylferrates, $K^+[RCOFe(CO)_3]^-$.[1]   The reaction of these ferrates with alkyl vinyl ketones provides 1,4-diketones when carried out in dimethylacetamide. Addition of 18-crown-6 improves yields.

$$n\text{-}C_6H_{13}Br + K_2Fe(CO)_4 \longrightarrow K^+[n\text{-}C_6H_{13}COFe(CO)_3]^- \xrightarrow[84\%]{\overset{\overset{\text{O}}{\|}}{CH_3CCH{=}CH_2}}$$

$$n\text{-}C_6H_{13}\overset{\overset{\text{O}}{\|}}{C}CH_2CH_2\overset{\overset{\text{O}}{\|}}{C}CH_3$$

[1] M. Yamashita, H. Tashika, and R. Suemitsu, *Chem. Letters*, 691 (1989).

Organolead reagents.

***α-Alkoxy organolead compounds;* 1,2-*diols*.**[1] α-Methoxy organolead compounds (**1**) can be prepared by transmetallation of α-methoxy organotin reagents with BuLi followed by trapping with Bu_3PbBr. They react with aldehydes in the presence of Lewis acids to form 1,2 diols, with the stereochemistry controlled by the Lewis acid; use of $TiCl_4$ results in *syn*-diols, whereas *anti*-diols are favored by BF_3 etherate.

(I)

+ $TiCl_4$	84%	99:1
+ $BF_3 \cdot O(C_2H_5)_2$	95%	39:61

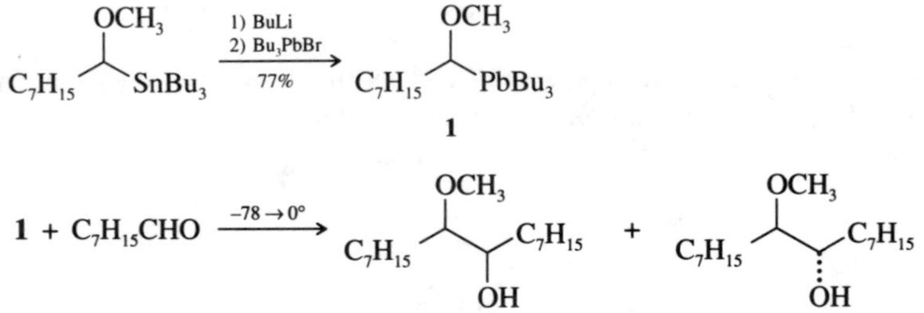

The reaction has been carried out on an optically active α-methoxy organolead reagent and shown to proceed with retention (equation II).

(II)

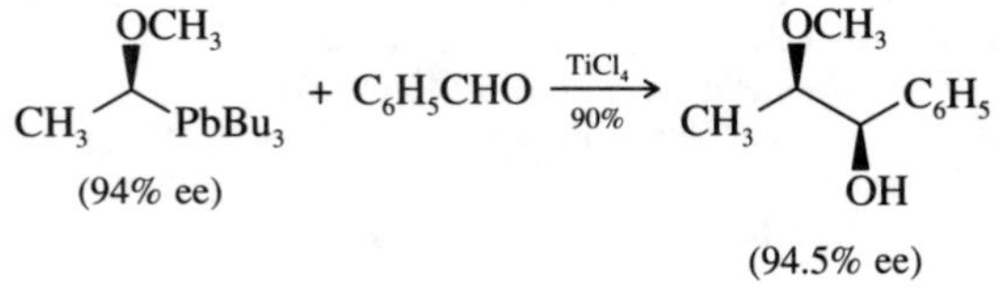

[1] J. Yamada, H. Abe, and Y. Yamamoto, *Am. Soc.*, **112**, 6118 (1990).

Organolithium compounds.

Reaction with epoxy silanes.[1] A variety of organolithiums containing aryl, alkenyl, alkynyl, amido, and cyano groups react with (E)- and (Z)-epoxysilanes to form adducts that on treatment with base (KH) are converted stereoselectivity into (E)- and (Z)-alkenes, respectively. Organocopper reagents can be preferable to the corresponding organolithium when the carbanion is an alkenyl or aryl group.

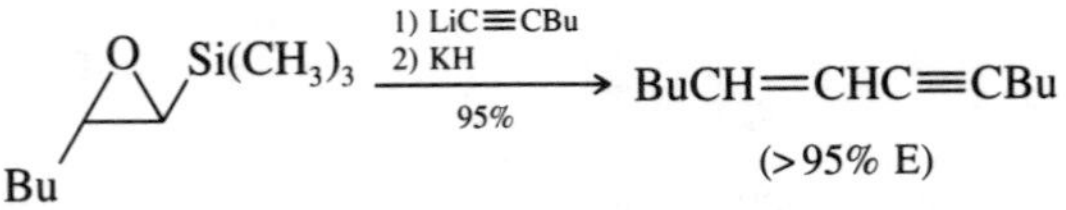

[1] Y. Zhang, J. A. Miller, and E. Negishi, *J. Org.*, **54**, 2043 (1989).

Organomanganese halides.

RMnBr.[1] Organomanganese iodides are very useful for acylation, but expensive since they are prepared from the expensive MnI_2. Organomanganese bromides have been difficult to prepare because $MnBr_2$ is insoluble in ether, but can be obtained by use of $MnBr_2$ and LiBr, which is soluble in ether at 20°, probably because $MnBr_3Li$ is formed. This ate complex reacts readily with RMgX to form RMnBr (+ MgXBr, LiBr). By addition of commercial LiBr to RMgX in ether followed by addition of $MnBr_2$, RMnBr can be obtained in 80–98% yield. The R group can be alkyl, alkenyl, alkynyl, or aryl. The RMnBr reagents are similar in reactivity to RMnI, but less expensive and they do not liberate I_2, which can be a problem.

1,4-*Addition to enones*.[2] In the presence of CuCl (5%), alkylmanganese chlorides add to α,β-enones in THF at 0° to give high yields of 1,4-adducts (88–98%). The nature of the alkyl group has slight effect on the yield. In general, RMnCl reagents are superior to R_2CuLi reagents, even when the latter are activated by

$$+ \text{CuCl (5\%)} \quad 95\%$$

$$+ \text{CuCl (1\%)} \qquad\qquad 67\%$$
$$+ \text{CuCl (1\%)} + MnCl_2 (10\%) \qquad 88\%$$

chlorotrimethylsilane. Use of both CuCl and $MnCl_2$ as catalysts can further increase the yield of 1,4-adducts of RMnCl to enones.

$$(\text{I}) \quad CH_3CH = C(COOC_2H_5)_2 + \text{BuMnCl} \xrightarrow[87\%]{\text{THF, } -30°}$$

Conjugate addition of RMnCl to alkylidenemalonic esters proceeds in generally good yields (80–87%, equation I), which are generally higher than those obtained by use of RLi or RMgCl. A new synthesis of citronellol (**1**) is based on this reaction.[3]

1

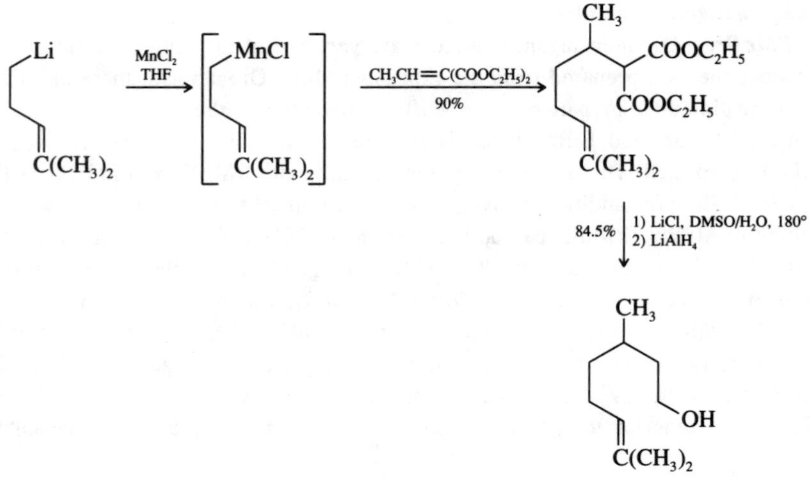

[1] G. Cahiez and B. Laboue, *Tetrahedron Letters*, **30**, 3545 (1989).
[2] G. Cahiez and M. Alami, *ibid.*, **30**, 3541 (1989).
[3] *Idem*, *Tetrahedron*, **45**, 4163 (1989).

Organomolybdenum reagents.

CpMo(NO)Cl(Crotyl)BF$_4$ **(1)**.[1] This complex is obtained as a single product by reaction of $CpMo(CO)_2$(crotyl) with $NO^+BF_4^-$ followed by addition of LiCl. This complex reacts with C_6H_5CHO to form homoallylic alcohols with high regio- and diastereoselectivity (equation I).

(I) **1** + C_6H_5CHO ⟶

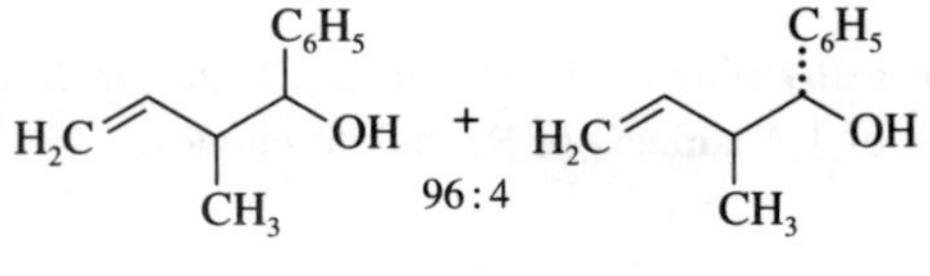

The neomenthylcyclopentadienyl analog (S-**2**) of (**1**) undergoes the same reaction to give (+)-(R,R)-2-methyl-1-phenyl-3-butene-1-ol (**3**) in >98% ee (equation II).

(II) (S)-**2** + 2C_6H_5CHO ⟶

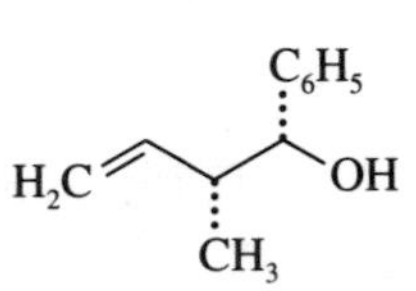

R,R-**3** (>98% ee, 92% de)

[1] J. W. Faller, J. A. John, and M. R. Mazzieri, *Tetrahedron Letters*, **30**, 1769 (1989).

Organopalladium reagents.

1,2-*Addition to* CH₂=CHY ($Y \neq C$).   Daves and Hallberg have reviewed the Heck-type coupling of organopalladium reagents to vinylsilanes, enol ethers, enol carboxylates, and related C–C double bonds, which proceed by 1,2-addition followed by elimination of palladium. Both steric and electronic factors can determine the regiochemistry. In general, triphenylphosphine ligands and coordinating solvents (CH₃CN) favor α-arylation. Also, the X group of ArPdX affects the regiocontrol.

$$p\text{-}CH_3OC_6H_4I + CH_2{=}CHOCH_3 \xrightarrow[55\%]{Pd(II)} p\text{-}CH_3OC_6H_4\overset{\overset{\displaystyle CH_2}{\|}}{C}OCH_3$$

(α-arylation)

$$p\text{-}O_2NC_6H_4COCl + CH_2{=}CHOBu \longrightarrow p\text{-}O_2NC_6H_4CH{=}CHOBu \quad + \quad p\text{-}O_2NC_6H_4\overset{\overset{\displaystyle CH_2}{\|}}{C}OBu$$

(β-arylation)   10 : 1   (α-arylation)

Aryl triflates favor α-arylation. In general vinylsilanes show opposite selectivity to that of enol ethers.

[1] G. D. Daves, Jr. and A. Hallberg, *Chem. Rev.*, **89**, 1433 (1989).

Organotitanium reagents.

Ti-carbohydrate complexes.   The Ciba-Geigy group has prepared a crystalline complex (**1**) from the reaction of cyclopentadienyltitanium(IV) trichloride, CpTiCl₃, and 2 equiv. of commercially available 1,2;5,6-di-O-isopropylidene-α-D-gluco-furanose (diacetone glucose) and shown to correspond to Cp(OR*)₂TiCl. Reaction of **1** with allylmagnesium bromide provides the complex **2**, which reacts with aldehydes to give homoallylic alcohols **3** with 85–91% ee.[1]

Reaction of the complex **1** with the lithium enolate of *t*-butyl acetate provides a complex **4**, which reacts with aldehydes to form β-hydroxy esters (**5**) in 90–96% ee.[2]

$$RCHO + 2 \xrightarrow[57-81\%]{ether, -30°}$$

3, 85–94% ee

The main disadvantage of these carbohydrate complexes is that the corresponding complexes from L-glucose are not readily available.

$$CpTi(OR^*)_2OCOC(CH_3)_3 \xrightarrow[57-81\%]{RCHO} R \overset{OH}{\wedge} COOC(CH_3)_3$$

4   **5**, 90–96% ee

[1] M. Riediker and R. O. Duthaler, *Angew. Chem. Int. Ed.*, **28**, 494 (1989); R. O. Duthaler, P. Herold, W. Lottenbach, K. Oertle, and M. Riediker, *ibid.*, **28**, 495 (1989).
[2] G. Bold, R. O. Duthaler, and M. Riediker, *ibid.*, **28**, 497 (1989); M. Riediker, A. Hafner, U. Piantini, G. Rihs, and A. Togni, *ibid.*, **28**, 499 (1989).

Organozinc reagents.

γ-Keto amino acids.[1]   The organozinc reagent (**2**), prepared from the β-iodo alanine derivative **1** by sonication with Zn/Cu, couples with acid chlorides in the presence of Pd(II) catalysts to give enantiomerically pure protected L-γ-keto α-amino acids (**3**).

1   **2**

3

Propargylation of acylsilanes.[2]   Attempted propargylation of aldehydes with propargylic Grignard or zinc reagents results in both α- and γ-adducts. However, if acylsilanes are used as the electrophiles, homopropargylic alcohols can be obtained in high yield after desilylation. Higher stereocontrol is possible by use of the triiso-propylsilyl group.

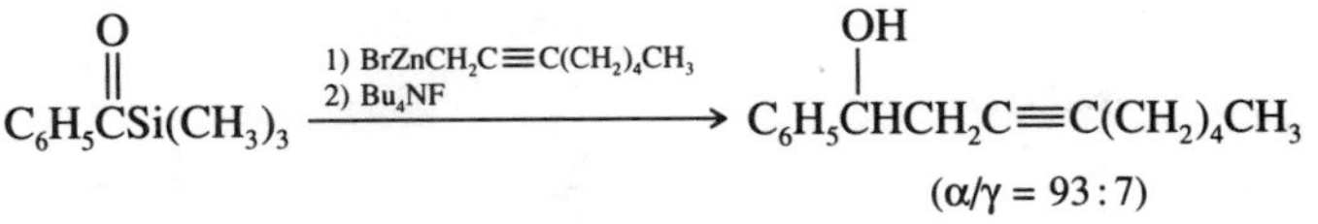

$$C_6H_5\overset{\overset{\displaystyle O}{\|}}{C}Si(CH_3)_3 \xrightarrow[\text{2) Bu}_4\text{NF}]{\text{1) BrZnCH}_2\text{C}\equiv\text{C(CH}_2)_4\text{CH}_3} C_6H_5\overset{\overset{\displaystyle OH}{|}}{C}HCH_2C\equiv C(CH_2)_4CH_3$$

$$(\alpha/\gamma = 93:7)$$

2-Carboalkoxycyclopentenones.[3] The zinc homoenolate **1**, prepared as shown (**13**, 349–350), can undergo a formal [3+2]cycloaddition to acetylenic esters in the presence of CuBr·S(CH₃)₂, ClSi(CH₃)₃, and HMPA to give 2-carboalkoxycyclopentenones. The reaction probably involves conjugate addition to give an allenolate followed by intramolecular cyclization.

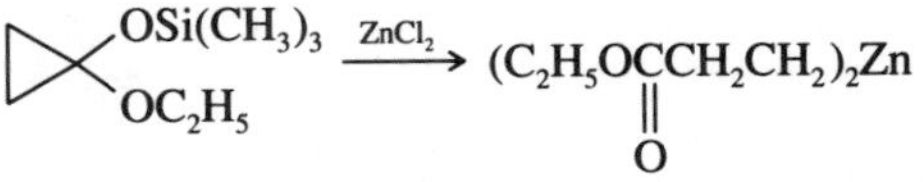

$$\text{1 + } \xrightarrow[\text{50 - 70\%}]{\substack{\text{CuBr · S(CH}_3)_2\text{, HMPA} \\ \text{ClSi(CH}_3)_3\text{, }))))}} $$

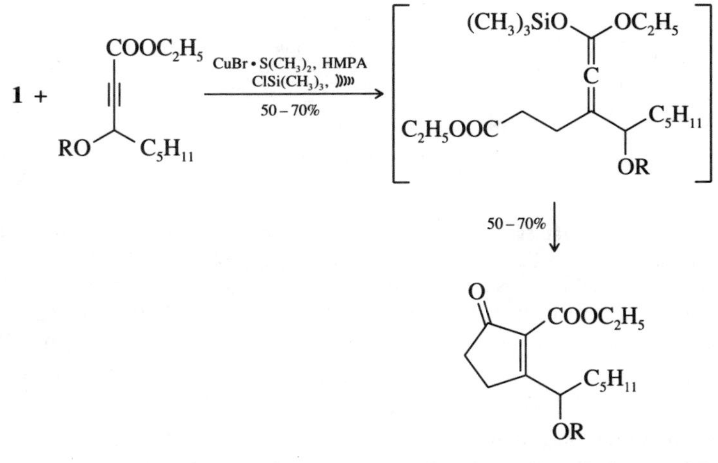

Nucleophilic displacement of anomeric sulfones.[4] 2-Benzenesulfonyl cyclic ethers undergo nucleophilic substitution with various organozinc reagents, particularly with those formed by reaction of ZnBr₂ with Grignard reagents, rather than an alkyllithium.

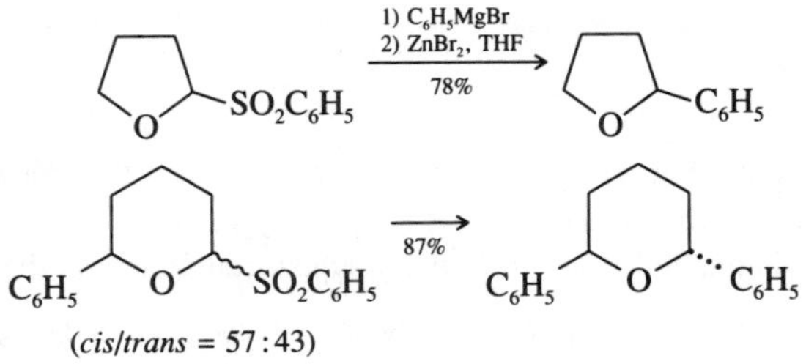

(*cis/trans* = 57:43)

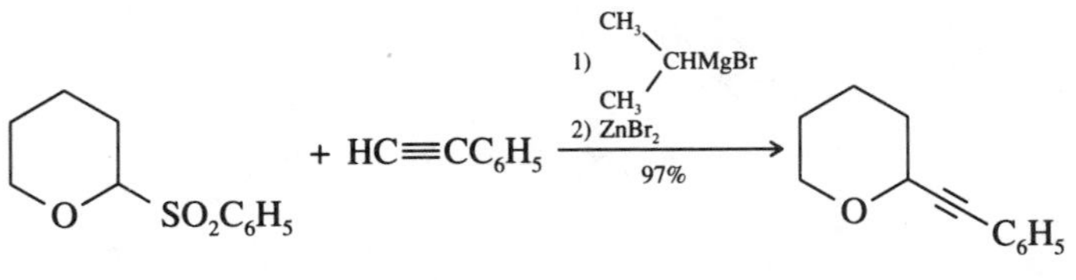

Zincate carbenoids.[5] Reaction of lithium trialkylzincates with a 1,1-dibro-moalkane results in a double insertion into the C–Br bonds to provide a secondary zincate carbenoid.[2] These products undergo Pd-catalyzed coupling with acid chlorides or vinyl bromides.

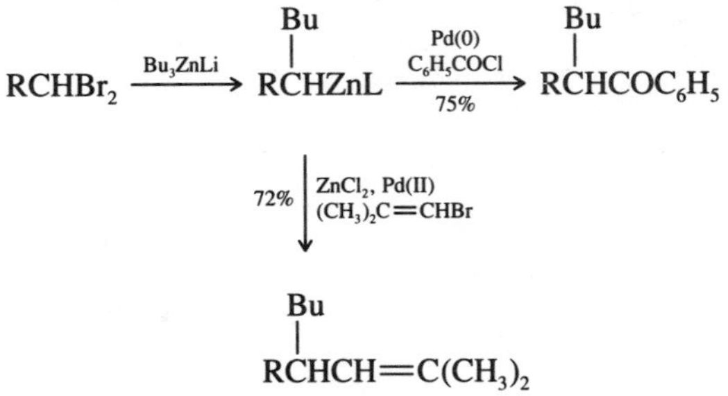

[1] R. F. W. Jackson, K. James, M. J. Wythes, and A. Wood, *J.C.S. Chem. Comm.*, 644 (1989).
[2] A. Yanagisawa, S. Habaue, and H. Yamamoto, *J. Org.*, **54**, 5198 (1989).
[3] M. T. Crimmins and P. G. Nantermet, *ibid.*, **55**, 4235 (1990).
[4] D. S. Brown, M. Bruno, R. J. Davenport, and S. V. Ley, *Tetrahedron*, **45**, 4293 (1989).
[5] T. Harada, Y. Kotani, T. Katsuhira, and A. Oku, *Tetrahedron Letters*, **32**, 1573 (1991).

Organoytterbium reagents.

Addition to carbonyls. Reaction of an alkyllithium or a Grignard reagent with tris(trifluoromethanesulfonate)ytterbium, $Yb(OTf)_3$, provides a species of organo-

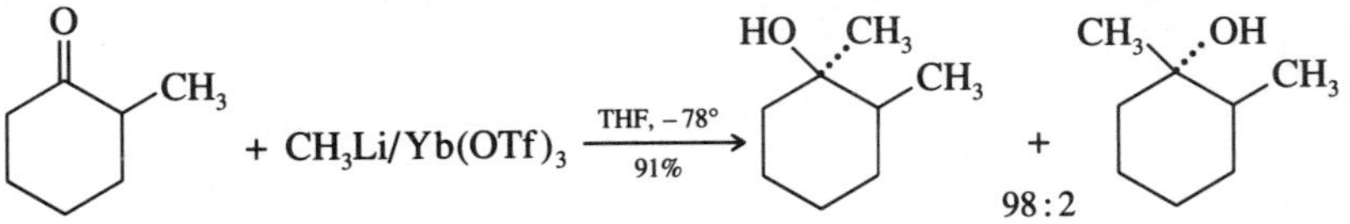

ytterbium reagents that adds to aldehydes or ketones with high stereoselectivity. Thus the methylytterbium reagent adds to 2-methylcyclohexanone to provide almost entirely the axial alcohol from preferential equatorial attack on the carbonyl group.

[1] G. A. Molander, E. R. Burkhardt, and P. Weinig, *J. Org.*, **55**, 4990 (1990).

Osmium tetroxide.

Catalytic asymmetric dihydroxylation (**14**, 237–239; **15**, 240–241).   Complete details are now available for this reaction with a solid substrate, *trans*-stilbene, in acetone/water (3:1, v/v) with dihydroquinidine 4-chlorobenzoate as catalyst.[1-4]

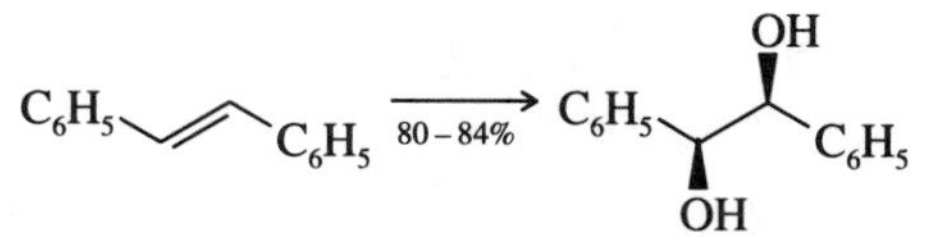

(R,R = 90% ee)

A new asymmetric synthesis of anthracycline antibiotics such as (+)-4-demeth-oxydaunomycinone (**5**) is based on dihydroxylation of **1** with OsO$_4$ in the presence of the diamine (−)-**2**, which provides the diol **3** in 82% ee.[5] The product is converted into (+)-**5** by four known reactions.

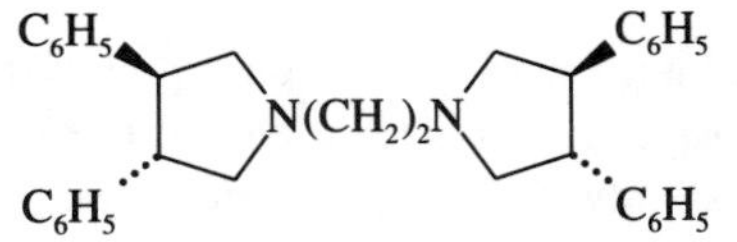

(−)-2

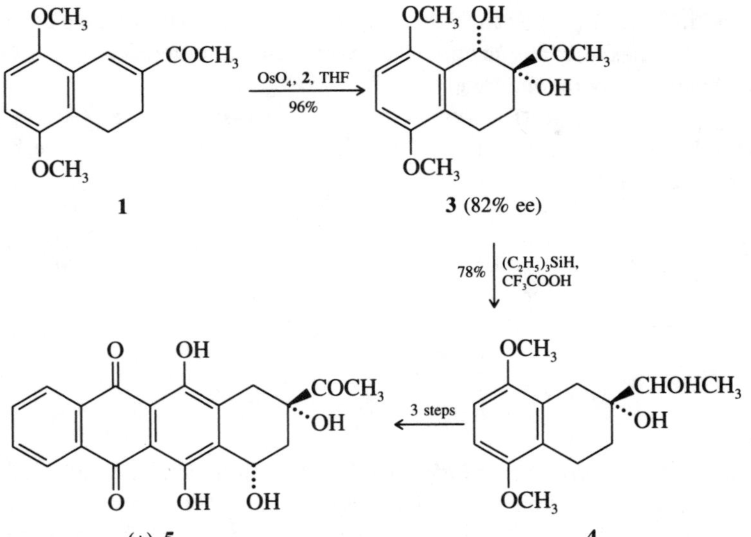

1

3 (82% ee)

(+)-5

4

Stereoselective osmylation.   Ochiai *et al.*[6] have reported an example of highly selective osmylation resulting from the conformational preferences of *cis*-1,3 substituents. Thus the cyclohexene **1** undergoes osmylation to provide a mixture of two *vic*-glycols (**2**) in approximately equal amounts. However, if one methyl group on Sn is replaced by chlorine by oxidation with iodosylbenzene and quenching with NH_4Cl, the product (**3**) on osmylation followed by methylation gives the α-*vic*-glycol almost exclusively. The cyclohexene **1** undoubtedly has the diequatorial conformation, whereas in **3** the tin group can form a pentacoordinated complex with oxygen resulting in a 1,3-diaxial conformation. The conformational difference results in strong steric hindrance by the tin group.

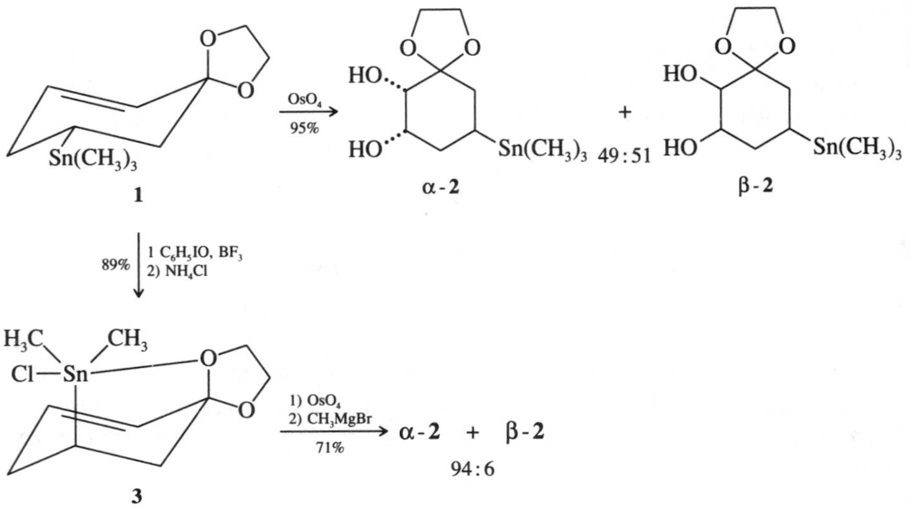

The paper also reports a new method for conversion of $-SnX(CH_3)_2$ into an hydroxyl group with the same configuration by oxidation with alkaline hydrogen peroxide in THF/CH_3OH.

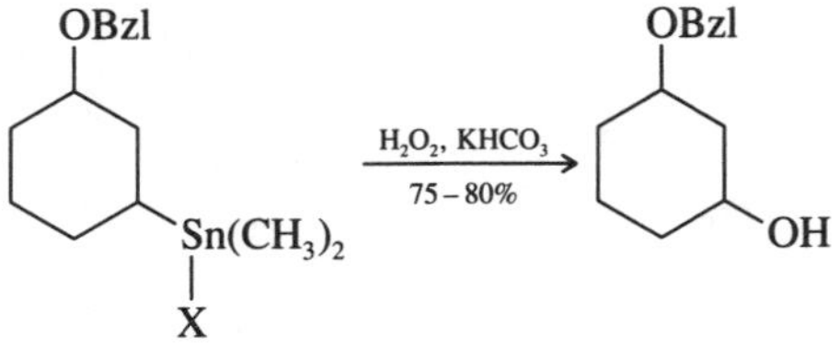

Chiral amino alcohols and diamines.[7] The chiral *vic*-diols available by catalytic asymmetric dihydroxylation of alkenes (**14**, 237–239) can be converted via a derived cyclic sulfite into chiral 1,2-amino alcohols and diamines as shown in equation I. The same transformations are useful in conversion of 1-alkyl- or arylethane-1,2-diols into the corresponding amino alcohols and diamines.

Dihydroxylation of allylic silanes.[8] Osmylation $[OsO_4,(CH_3)_3NO]$ of allylic

(I) C_6H_5 —(OH)—C_6H_5—(OH) $\xrightarrow{SOCl_2, CCl_4}$ $\left[C_6H_5 \begin{array}{c} O \\ SO \\ O \end{array} C_6H_5 \right]$ $\xrightarrow[81\%]{\substack{LiN_3, \\ DMF, 120°}}$

C_6H_5—(OH)—C_6H_5—(N_3)
($\geq$ 96% ee) $\xrightarrow[94\%]{\substack{H_2 \\ Pd/C}}$ C_6H_5—(OH)—C_6H_5—(NH_2)
($\geq$ 96% ee)

1) MsCl
2) LiN_3, DMF, 120°

C_6H_5—(N_3)—C_6H_5—(N_3) $\xrightarrow[80\%]{\substack{H_2, \\ Pd/C}}$ C_6H_5—(NH_2)—C_6H_5—(NH_2)
($\geq$ 96% ee)

silanes bearing an oxygen function at C_1 show *anti*-diastereoselectivity. A free hydroxyl is more effective than an ester or ether group. An increase in the size of the silyl group also favors *anti*-selectivity, but to a less extent (equation I). Similar effects obtain with C_1-oxygenated crotylsilanes, particularly in the case of the (E)-isomers.

(I) $(CH_3)_3Si$—(OR)—=CH_2, CH_3 $\xrightarrow{OsO_4, (CH_3)_3NO}$ $(CH_3)_3Si$—(OR)(OH)—(OH)CH_3 + *syn*-isomer

anti

R = H	65%	>97:3
= Ac	57%	6.5:1

OsO$_4$-catalyzed dihydroxylation of mono allylic silyl ethers shows slight diastereoselectivity, but dihydroxylation of a bis-allylic silyl ether such as **1** provides essentially only one product (**2**), and a second dihydroxylation also provides a single isomer (**3**).[9]

Selective osmylation of trienes.[10] The (tricarbonyl)iron-complexed triene **2**, prepared from the butadiene-tricarbonyliron **1** (**11**, 222), undergoes osmylation to give a single racemic, *cis*-diol **3** in 96% yield. Reaction of **3** with N,N'-carbonyldiimidazole provides the single carbonate **4**.[11] Related carbonates, prepared from D-

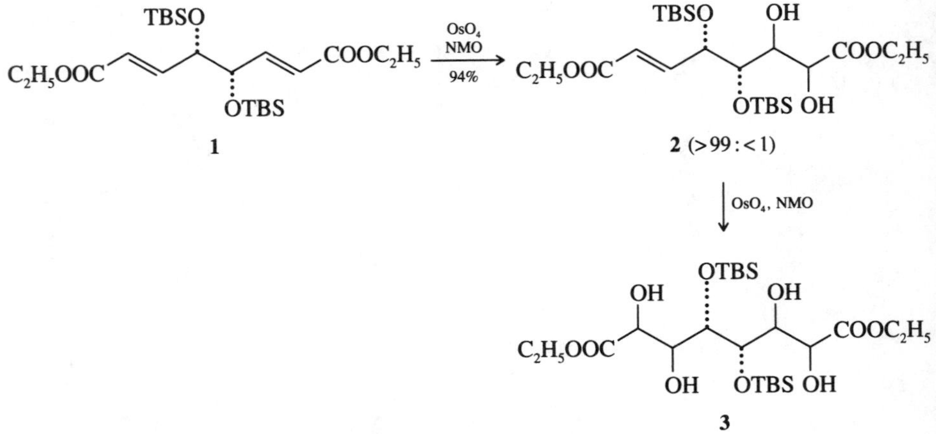

1   **2** (>99 : <1)

3

ribose and xylose, have been used to prepare (5S,6R)-**5** and (5S,6S)-**5**, respectively, which are metabolites of arachidonic acid known as 5,6-DiHETE.

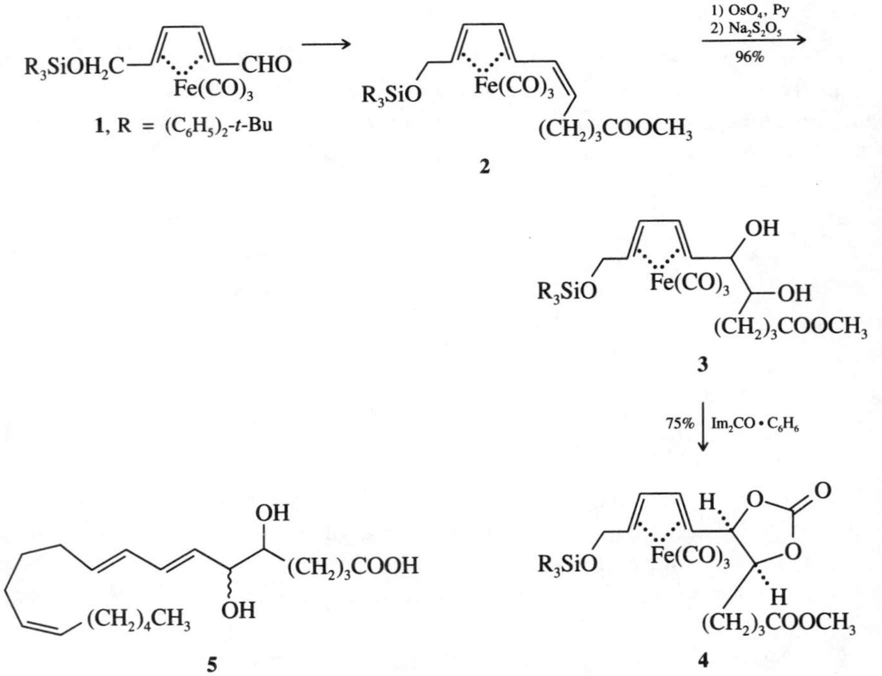

1, R = $(C_6H_5)_2$-*t*-Bu

2

3

5   **4**

The configuration of **4** has been established by X-ray crystallography, and shown to correspond to *anti*-addition to the free bond vicinal to the organometallic group.

Similar stereoselectivity applies to osmylation of the *trans*-isomer of **2**. Since **2** can be resolved,[12] an asymmetric synthesis of **5** should be possible.

[1] B. H. McKee, D. G. Gilheany, and K. B. Sharpless, *Org. Syn.*, submitted (1990).
[2] M. Minato, K. Yamamoto, and J. Tsuji, *J. Org.*, **55**, 766 (1990).
[3] H.-L. Kwong, C. Sorato, Y. Ogino, H. Chen, and K. B. Sharpless, *Tetrahedron Letters*, **31**, 2999 (1990).
[4] B. M. Kim and K. B. Sharpless, *ibid.*, **31**, 3003 (1990).
[5] K. Tomioka, M. Nakajima, and K. Koga, *J.C.S. Chem. Comm.*, 1921 (1989).
[6] M. Ochiai, S. Iwaki, T. Ukita, Y. Matsuura, M. Shiro, and Y. Nagao, *Am. Soc.*, **110**, 4606 (1988).
[7] B. B. Lohray and J. R. Ahuja, *J.C.S. Chem. Comm.*, 95 (1991).
[8] J. S. Panek and P. F. Cirillo, *Am. Soc.*, **112**, 4873 (1990).
[9] S. Saito, Y. Morikawa, and T. Moriwake, *J. Org.*, **55**, 5424 (1990).
[10] A. Gigou, J.-P. Lellouche, J.-P. Beaucourt, L. Toupet, and R. Gree, *Angew. Chem. Int. Ed.*, **28**, 755 (1989).
[11] J. Adams, B. T. Fitzsimmons, Y. Girard, Y. Leblanc, J. F. Evans, J. Rokach, *Am. Soc.*, **107**, 464 (1985).
[12] P. Mangeney, A. Alexakis, and J. F. Normant, *Tetrahedron Letters*, **29**, 2677 (1988).

Oxazaborolidines, 14, 110–111, 156, 239–242.

α,α-*Diaryl-2-pyrrolinemethanols.*[1] These products are obtained in only moderate yield by reaction of an arylmagnesium chlorides with the methyl or ethyl ester of (R)- or (S)-proline. A newer and more dependable route involves conversion of the proline into the N-carboxyanhydride, which is then added to the Grignard reagent to obtain (R)-**1**. Conversion of **1** to the corresponding oxazaborolidine **2** is effected by reaction of **1** with trimethylboroxine in refluxing toluene.

Enantioselective catecholborane reduction of ketones $\diagdown C = O.$[2]   The oxaza-

borolidine **1** can also "function as a chiral catalyst for catecholborane reduction of ketones, which is useful for substrates sensitive to BH_3 and which can proceed at temperatures as low as $-78°$. It is prepared by reaction of (S)-(−)-(diphenylhydroxymethyl)pyrrolidine with butylboronic acid. Reduction of aromatic ketones with catecholborane catalyzed by **1** proceeds at $-78°$ to provide (R)-alcohols in 90–94% ee. This protocol is particularly useful for reduction of α,β-enones to (R)-allylic alcohols in 81–93% ee.

(S)-**1**

Reduction of ketones.   Merck chemists[3] have used oxazaborolidine-catalyzed reduction of a ketone for introduction of chirality in a synthesis of MK-927 (**4**), a carbonic anhydrase inhibitor. They found that even traces of water decreases the enantioselectivity in reductions of **2**. Highest enantioselectivity (98 : 2) is obtained by

careful drying of solutions of **2** in THF with 4-Å molecular sieves and use of neat borane complexed with dimethyl sulfide at temperatures below $-10°$. Under these conditions a variety of ketones are reduced with reproducible enantioselectivity on a large scale. Alkyl aryl ketones are reduced with higher enantioselectivity than dialkyl ketones. Thus cyclohexyl methyl ketone is reduced in the ratio 82 : 18, while methyl phenyl ketone is reduced in the ratio 99 : 1. Substitution on the phenyl group of the oxazaborolidine exerts little effect on the enantioselectivity. Change of the B-methyl group to a phenyl group generally lowers the enantioselectivity.

[1] L. C. Xavier and J. J. Mohan, *Org. Syn.*, submitted (1990).
[2] E. J. Corey and R. K. Bakshi, *Tetrahedron Letters*, **31**, 611 (1990).

[3] T. K. Jones, J. J. Mohan, L. C. Xavier, T. J. Blacklock, D. J. Mathre, P. Sohar, E. T. T. Jones, R. A. Reamer, F. E. Roberts, and E. J. J. Grabowski, *J. Org.*, **56**, 763 (1991).

Oxazolidinones, chiral.

Chiral primary amines.[1] Alkylation of the lithium anion of the N-benzyloxazolidinone 1 (derived from valinol) proceeds with high 1,3-stereoselectivity to provide 2 in 75–96% de, which can be improved by crystallization or chromatography. The products can be degraded to (R)-primary amines (5) by hydrolysis (3) and oxidation to an imine (4), which is then hydrolyzed to an amine. Slight racemization is observed in these last steps.

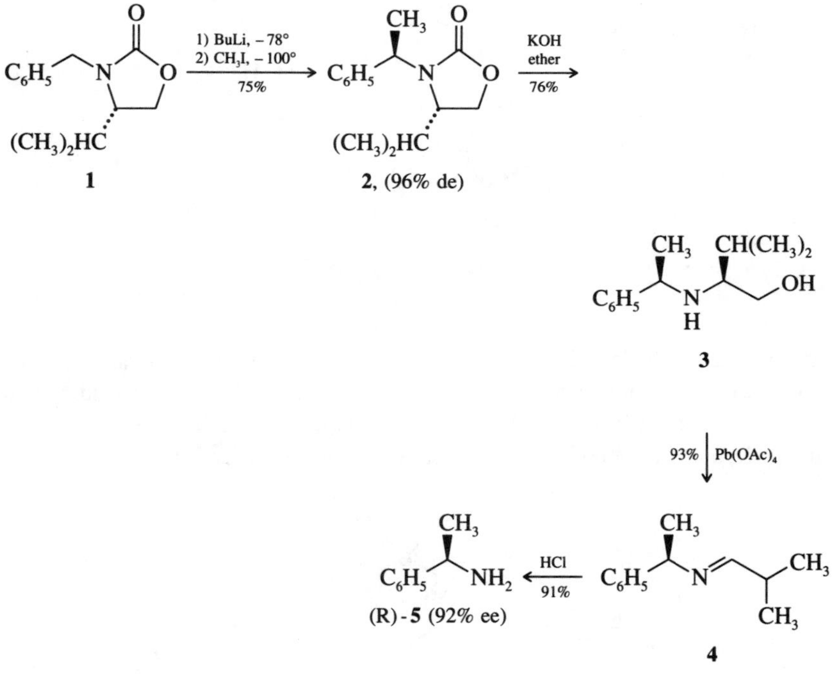

α-Alkyl α-amino acids.[2] A new route to optically active α-alkyl α-amino acids involves alkylation of tricyclic oxazolidinones such as 1, prepared by reaction of salicylaldehyde, L-phenylalanine, and phosgene.[3] Alkylation of the anion of 1 proceeds mainly with retention of configuration to give 2, which is hydrolyzed by LiOH in aqueous dioxane to an α-alkyl α-amino acid (3) in high overall yield and with high optical purity. Similar alkylation of the oxazolidinones formed from L-alanine and L-leucine proceeds with lower enantioselectivity.

α-Azido acids.[4] Full details are now available for asymmetric synthesis of either (R)- or (S)-α-azido carboxylic acids as precursors to α-amino acids. One route employs enolate bromination/azide displacement with tetramethylguanidinium azide (**14**, 242); the complementary route involves electrophilic enolate azidation with trisyl azide (**14**, 327). Removal of the chiral auxiliary can be effected by saponification (LiOH) or transesterification, either before or after reduction of the azide group.

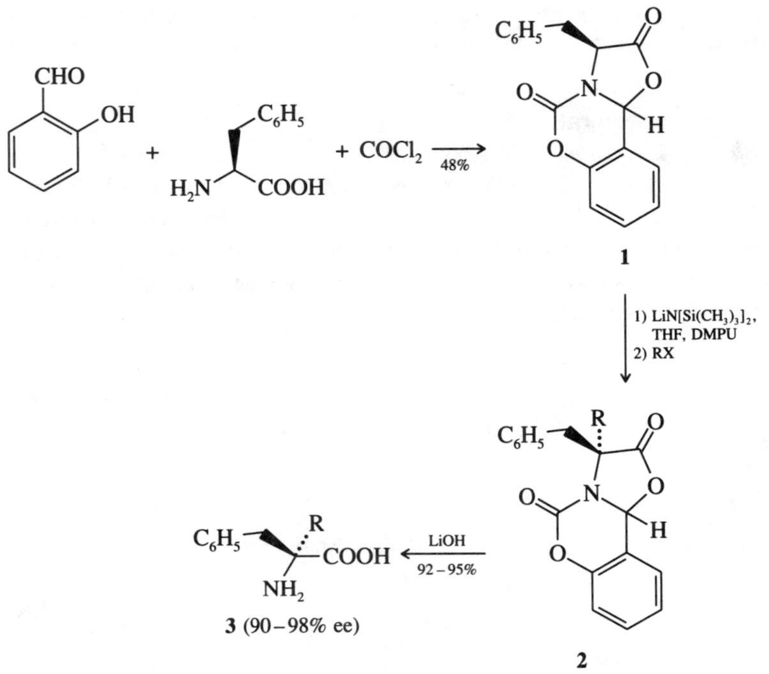

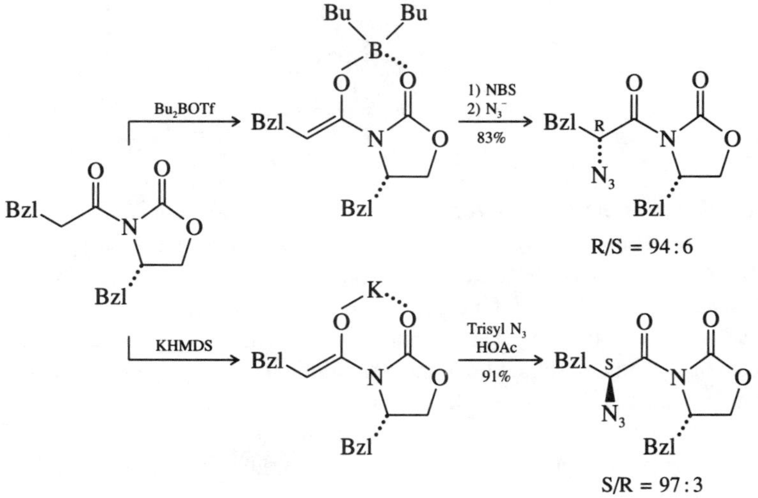

Asymmetric aldol reactions[5] (**11**, 379–380).   The lithium enolate of the N-propionyloxazolidinone (**1**) derived from L-valine reacts with aldehydes with low *syn* vs. *anti*-selectivity, but with fair diastereofacial selectivity attributable to chelation. Transmetallation of the lithium enolate with ClTi(O-*i*-Pr)$_3$ (excess) provides a titanium enolate, which reacts with aldehydes to form mainly the *syn*-aldol resulting from chelation, the diastereomer of the aldol obtained from reactions of the boron enolate (**11**, 379–380). The reversal of stereocontrol is a result of chelation in the titanium reaction, which is not possible with boron enolates. This difference is of practical value, since it can result in products of different configuration from the same chiral auxiliary.

(4S)-Phenyl-2-oxazolidinone.[6]   A one-pot route to this chiral auxiliary involves reduction of L-phenylglycine and BF$_3$ etherate in DME with borane·dimethyl sulfide complex at a temperature maintained at ~82°. The resulting phenylglycinol is then treated with trichloromethyl chloroformate (or the more expensive triphosgene).

[1] R. E. Gawley, K. Rein, and S. Chemburkar, *J. Org.*, **54**, 3002 (1989).
[2] T. M. Zydowsky, E. de Lara, and S. G. Spanton, *ibid.*, **55**, 5437 (1990).
[3] H. Block and P. J. Faulkner, *J. Chem. Soc. C*, 329 (1971).
[4] D. A. Evans, T. C. Britton, J. A. Ellman, and R. L. Dorow, *Am. Soc.*, **112**, 4011 (1990).
[5] M. Nerz-Stormes and E. R. Thornton, *J. Org.*, **56**, 2489 (1991).
[6] L. N. Pridgen, *Org. Syn.*, submitted (1989).

Oxygen, singlet.

Aromatic hydroxylation.[1]   The reaction of 1-isopropylidene-2-indanone (**1**) with singlet oxygen (Rose bengal) in CH$_3$CN at −35° results in an unstable product **2**, which on deoxygenation with P(OC$_2$H$_5$)$_3$ affords the phenol **3** in 55% yield. The initial reaction is presumed to form an endoperoxide (**a**) by a [4+2]cycloaddition, which rearranges to **2**.

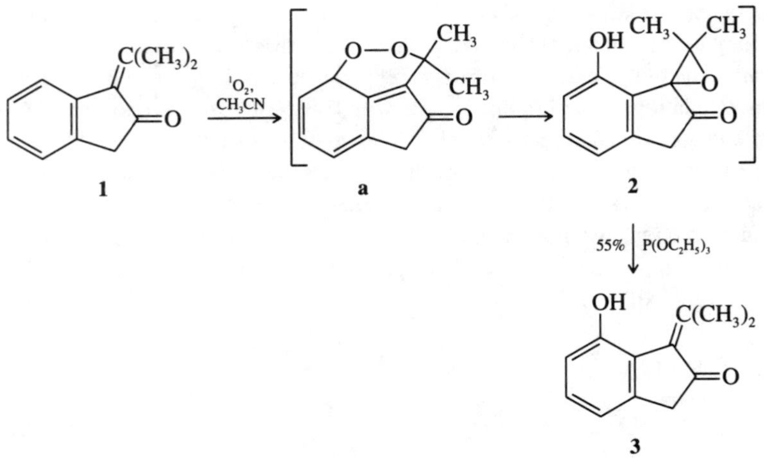

[1] H. E. Ensley, P. Balakrishnan, and C. Hogan, *Tetrahedron Letters*, **30**, 1625 (1989).

P

Palladium(II) acetate.

Vinylation of cycloalkenes.[1] This reaction is possible when catalyzed by 2.5 mole % Pd(OAc)$_2$ in the presence of KOAc and Bu$_4$NCl in DMF. In some cases Pd(OAc)$_2$, P(C$_6$H$_5$)$_3$, and Ag$_2$CO$_3$ in CH$_3$CN shows greater stereoselectivity. 1,4-Dienes are obtained in 50–100% yield.

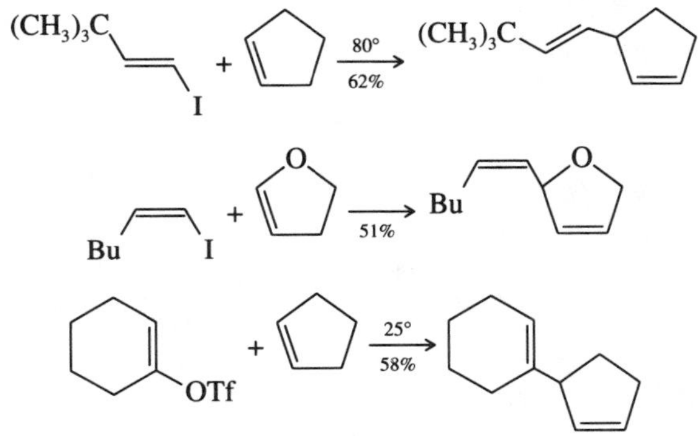

Arylannelation of 1,3-dienes.[2] In the presence of catalytic amounts of Pd(OAc)$_2$, Bu$_4$NCl (1 equiv.), and a base such as Na$_2$CO$_3$ or NaOAc (5 equiv.), an aryl iodide reacts with a 1,3-diene (excess) to afford substituted carbocycles. The overall process is believed to involve an intramolecular carbanion displacement of a π-allylpalladium intermediate.

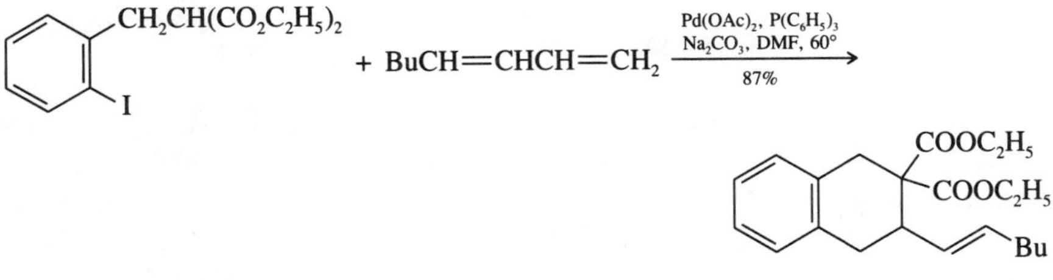

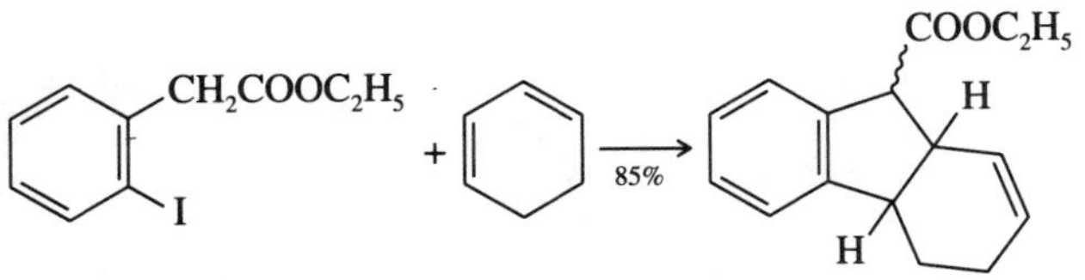

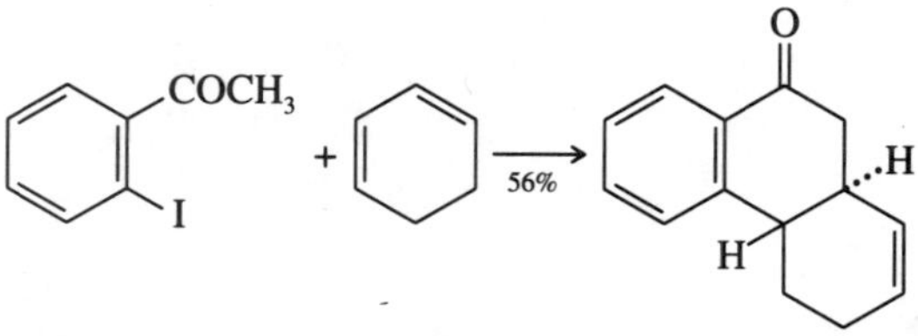

Heteroannelation of 1,3-dienes with a variety of oxygen- or nitrogen-containing aryl iodides affords dihydrobenzofurans or nitrogen heterocycles.[3]

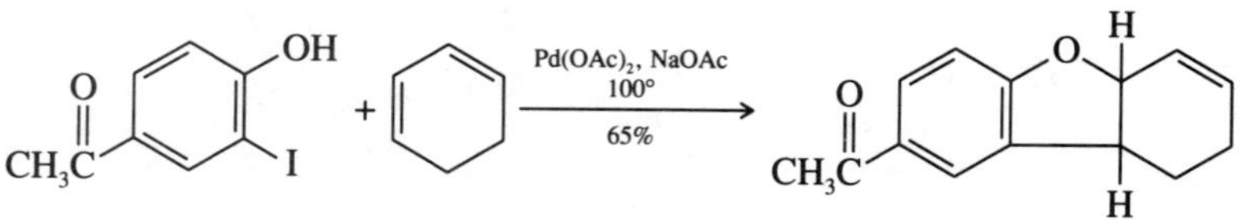

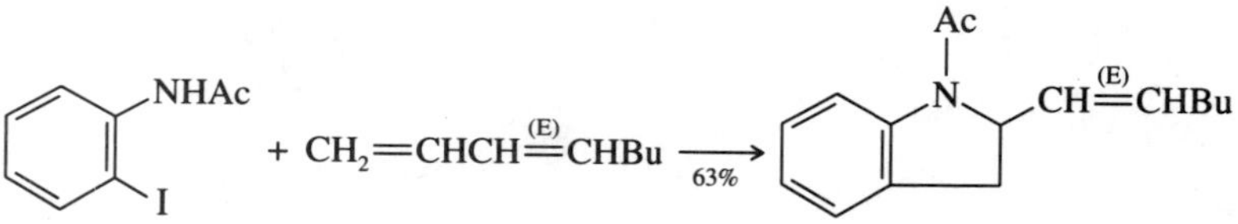

Coupling of ArI and unsaturated epoxides.[4] Palladium-catalyzed coupling of C_6H_5I with unsaturated epoxides in which the two groups are separated by one or more carbon atoms can provide arylated allylic alcohols as the major product. Highest yields obtain with $Pd(OAc)_2$ as catalyst, 1–3 equiv. of an alkali metal formate as

$$(I)\ C_6H_5I + CH_2{=}CH(CH_2)_n CH{-}CH_2 \xrightarrow{Pd(OAc)_2} C_6H_5CH_2(CH_2)_n CH{=}CHCH_2OH$$

n = 1	78%	(E/Z~3:1)
4	62%	
10	44%	

reducing agent, various bases, and alkali metal halides. The yield decreases as the number of carbon atoms separating the double bond and the epoxide increases. Evidently the organopalladium intermediate can migrate along the carbon chain. Substitution on the double bond also lowers yield, as shown in the arylation of 4,5-epoxycyclohexene (equation II).

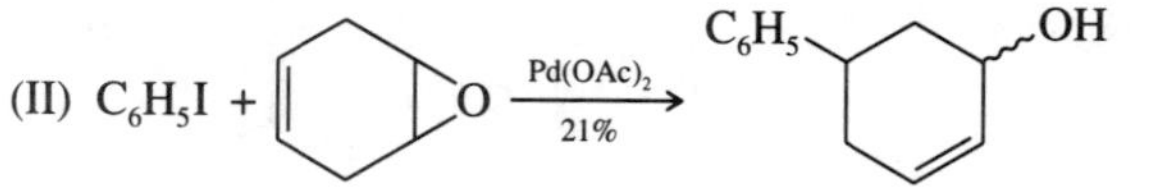

(II) C_6H_5I +

(cis/trans = 67:33)

Bicyclic acetals.[5]   Cyclic allylic alcohols couple with ethyl vinyl ether when treated with Pd(OAc)$_2$. Only a catalytic amount of Pd(II) is required if Cu(OAc)$_2$ is present as a reoxidant. The absence of double-bond isomerization is a useful feature of this coupling.

$$\underset{\underset{\displaystyle CH(OCH_3)_2}{\big|}}{\bigcirc\!\!-OH} + C_2H_5OCH{=}CH_2 \;\xrightarrow[\;71\%\;]{\underset{CH_3CN,\,0°}{Pd(OAc)_2,\,Cu(OAc)_2}}\; \text{bicyclic acetal product}$$

1,4-*Additions to* **1,3-***dienes*[6] (**12**, 367–368; **14**, 249–250; **15**, 245).   This reaction can be used to effect intramolecular cyclization of cyclic 1,3-dienes substituted by a suitable nitrogen nucleophile. Thus reaction of the amido diene **1** with lithium acetate catalyzed by Pd(OAc)$_2$ (with benzoquinone as reoxidant) provides the *cis*-fused heterocycle *cis*-**2**, in which the acetoxy group is *cis* to the ring fusion, formed by an overall *trans*-1,4-oxyamidation of the diene system. Addition of a trace of LiCl improves the yield and results in an overall *cis*-1,4-oxyamidation (equation I). Acetamides and carbamates can also be used in place of amides. 1,4-Chloroamidation can also be effected by use of 2 equiv. of LiCl.

$$(\text{I})\quad \underset{\mathbf{1}}{\text{diene-NHTs}} \;\xrightarrow[\;65\%\;]{\underset{Pd(II)}{LiOAc,\,LiCl}}\; \underset{trans\text{-}\mathbf{2}}{\text{AcO-product}} \;+\; \underset{cis\text{-}\mathbf{2}}{\text{AcO-product}}\quad 93:7$$

Didehydroamino acids, R^{1}CH$=$C(NH$_2$)COOR^2.[7]   Review.[8]   Aryl didehydroamino acids (R^1 = Ar) can be prepared by a modified Heck coupling of N-protected 2-amidoacrylates with aryl iodides catalyzed by Pd(OAc)$_2$ under phase-transfer conditions, which results in the (Z)-didehydroamino acid in moderate to good yield (32–80%). This reaction is particularly useful for preparation of O-benzyl, N-Boc-protected aryl didehydroamino acids.

$$\underset{\underset{\displaystyle NHBoc}{\big|}}{CH_2{=}CCOOBzl} + C_6H_5I \;\xrightarrow[\;80\%\;]{\underset{NaHCO_3,\,DMF,\,80°}{Pd(OAc)_2,\,Bu_4NCl}}\; \underset{\underset{\displaystyle NHBoc}{\big|}}{C_6H_5{-}CH{=}C{-}COOBzl}$$

Cyclopentenones from carbohydrates.[9]   This transformation involves glyco-sides (**2**) obtained by reaction of 6-hydroxy-2,3-dihydro-6H-pyranones (**1**) with an alcohol catalyzed by $ZnCl_2$·etherate (superior to $SnCl_4$). The resulting cyclic acetals (2) rearrange to substituted cyclopentenones (3) when treated with a catalytic amount of $Pd(OAc)_2$ and $NaHCO_3$ (5 equiv.) in DMF. In every case the alkoxy group in the product (3) is *trans* to the free hydroxyl group.

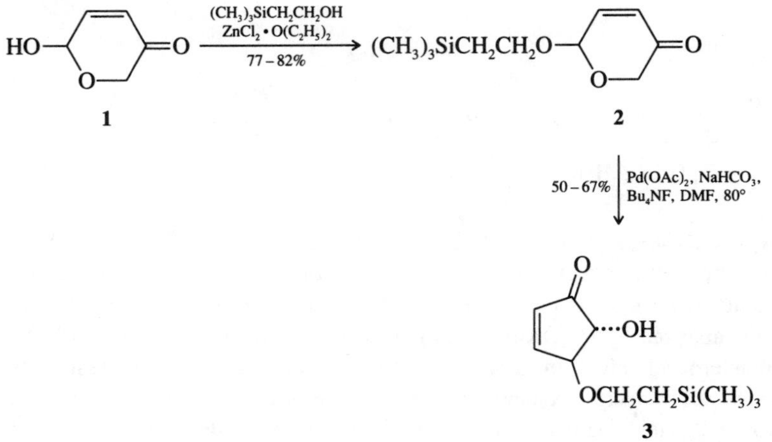

3-Arylcycloalkenes.[10] These products can be obtained by Pd-catalyzed coupling of aryl iodides and cycloalkenes (5 equiv.) in DMF containing tetrabutylammonium chloride and KOAc (2 equiv.) Other acetate bases are less effective. A large excess of the cycloalkene is required to effect monoarylation.

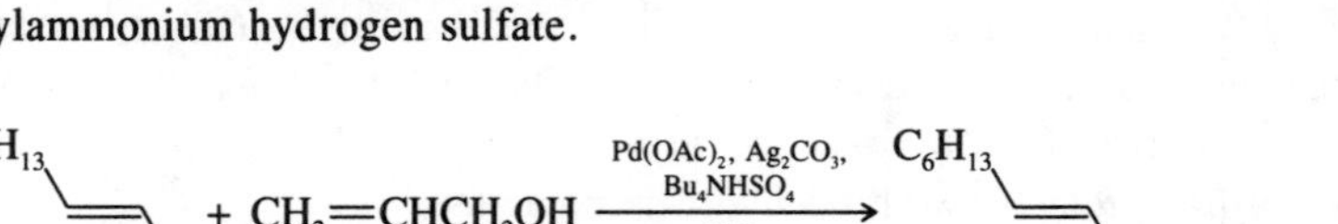

Coupling of vinyl halides with allylic alcohols.[11] $Pd(OAc)_2$ promotes coupling of vinyl halides with primary allylic alcohols in the presence of silver carbonate and tetrabutylammonium hydrogen sulfate.

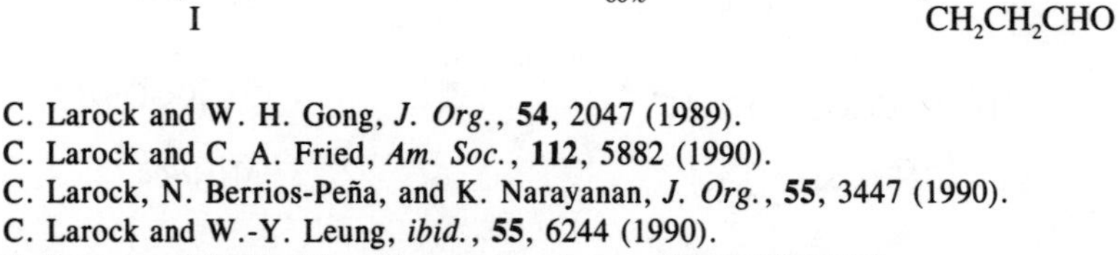

[1] R. C. Larock and W. H. Gong, *J. Org.*, **54**, 2047 (1989).
[2] R. C. Larock and C. A. Fried, *Am. Soc.*, **112**, 5882 (1990).
[3] R. C. Larock, N. Berrios-Peña, and K. Narayanan, *J. Org.*, **55**, 3447 (1990).
[4] R. C. Larock and W.-Y. Leung, *ibid.*, **55**, 6244 (1990).
[5] R. C. Larock and D. E. Stinn, *Tetrahedron Letters*, **30**, 2767 (1989).
[6] J.-E. Bäckvall and P. G. Andersson, *Am. Soc.*, **112**, 3683 (1990).

[7] A.-S. Carlström and T. Frejd, *Synthesis*, 414 (1989).
[8] Review: U. Schmidt, A. Lieberknecht, and J. Wild, *ibid.*, 159 (1988).
[9] B. Mucha and H. M. R. Hoffmann, *Tetrahedron Letters*, **30**, 4489 (1989).
[10] R. C. Larock, H. Song, B. E. Baker, and W. H. Gong, *ibid.*, **30**, 2919 (1988); R. C. Larock and W. H. Gong, *Org. Syn.*, submitted (1989).
[11] T. Jeffery, *Tetrahedron Letters*, **31**, 6641 (1990).

Palladium(II) acetate–N,N′-Bis(benzylidene)ethylenediamine,

$$Pd(OAc)_2 - \left[\begin{array}{l} N{=}CHC_6H_5 \\ N{=}CHC_6H_5 \end{array} \right. \qquad (1)$$

Enyne cyclization.[1] Cyclization of an enyne such as **2** to the diene **3** with Pd(OAc)$_2$ in combination with Ar$_3$P provides mixtures and in low yield. In contrast, cyclization with Pd(OAc)$_2$ and N,N′-bis(benzylidene)ethylenediamine (1) provides the desired diene **3** in 81% yield. The product was used for a synthesis of (−)-sterepolide (4) in 11 steps and 34% overall yield.

2, R = CH$_2$C$_6$H$_4$OCH$_3$-*p*

3

(−)-4

[1] B. M. Trost, P. A. Hipskind, J. Y. L. Chung, and C. Chan, *Angew. Chem. Int. Ed.*, **28**, 1502 (1989).

Palladium(II) acetate/Potassium formate.

Conjugate reduction of α,β-unsaturated carbonyl compounds.[1] The HCOOK/Pd(OAc)$_2$ system is convenient for transfer conjugate reduction of α,β-unsaturated ketones and esters and of 2-buten-4-olides to the corresponding saturated compounds. The reaction is carried out in DMF at 60° using an excess of HCOOK.

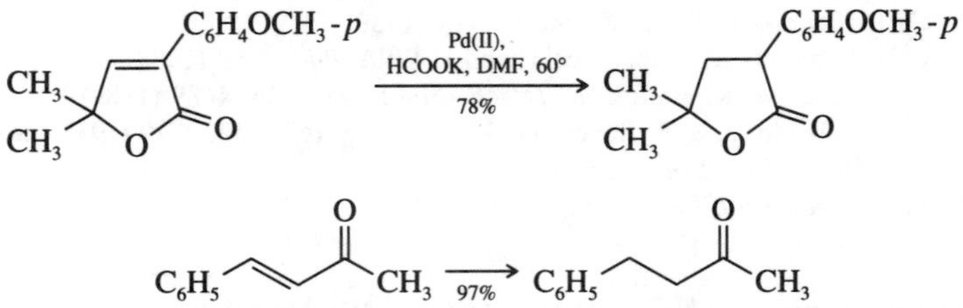

[1] A. Arcadi, E. Bernocchi, S. Cacchi, and F. Marinelli, *Synlett*, 27, (1991).

Palladium(II) acetate–Triethoxysilane.

Hydrogenation.[1] Hydrogenation of water-soluble unsaturated acids is best carried out on the sodium salt in an aqueous solution with a catalytic amount of the catalyst and triethoxysilane as the source of hydrogen. If only 1 equiv. of triethoxysilane is added, triple bonds are selectively reduced to (Z)-alkenes in 70–85% yield.

[1] J. M. Tour and S. L. Pendalwar, *Tetrahedron Letters*, **31**, 4719 (1990).

Palladium(II) acetate–Triphenylphosphine (1).

Heck intramolecular cyclization. Silver carbonate or nitrate was added originally to tandem Heck arylation reactions to depress alkene isomerization, but they also improve selectivity in the β-elimination step. Grigg et al.[1] have used a number of useful additives such as triethylammonium chloride, sodium formate (**15**, 248), phenylzinc chloride, as well as silver(I) and thallium(I) salts. In fact, thallium(I) salts

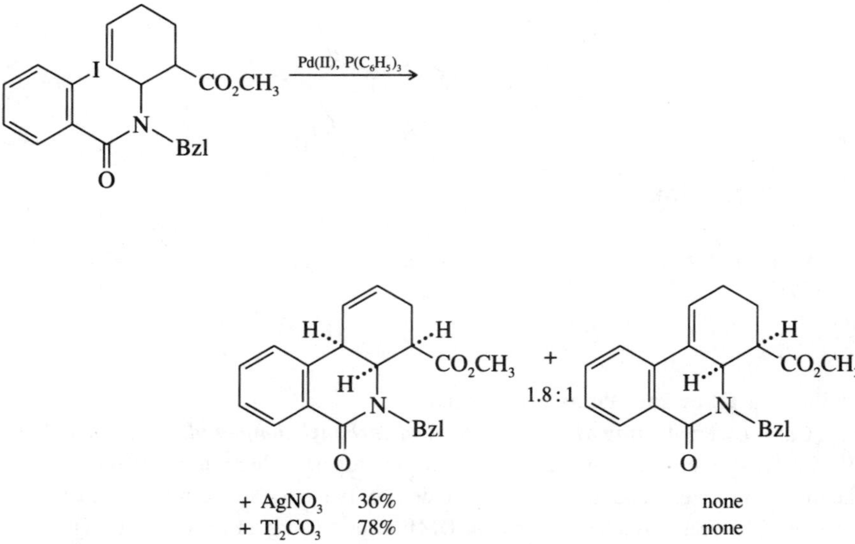

	+ AgNO$_3$	36%	none
	+ Tl$_2$CO$_3$	78%	none

can be as efficacious as silver salts and can increase the rates of cyclization. The effect is attributed to an anion exchange: RPdI + TlX → RPdX + TlI.

Aryl–vinyl coupling; β-aryl-β-amino acids.[2]   A novel route to β-amino acids is based on the formal conjugate addition of an aryl iodide to an enantiomerically pure dihydropyrimidinone such as **1**, prepared by pivaldehyde acetalization of (R)-asparagine (*cf.*, **14**, 69–70). Thus 4-iodoanisole couples with **1** under Heck conditions, Pd(OAc)$_2$, Ar$_3$P, triethylamine, and DMF to give **2** in 78% yield. Reduction of **2**

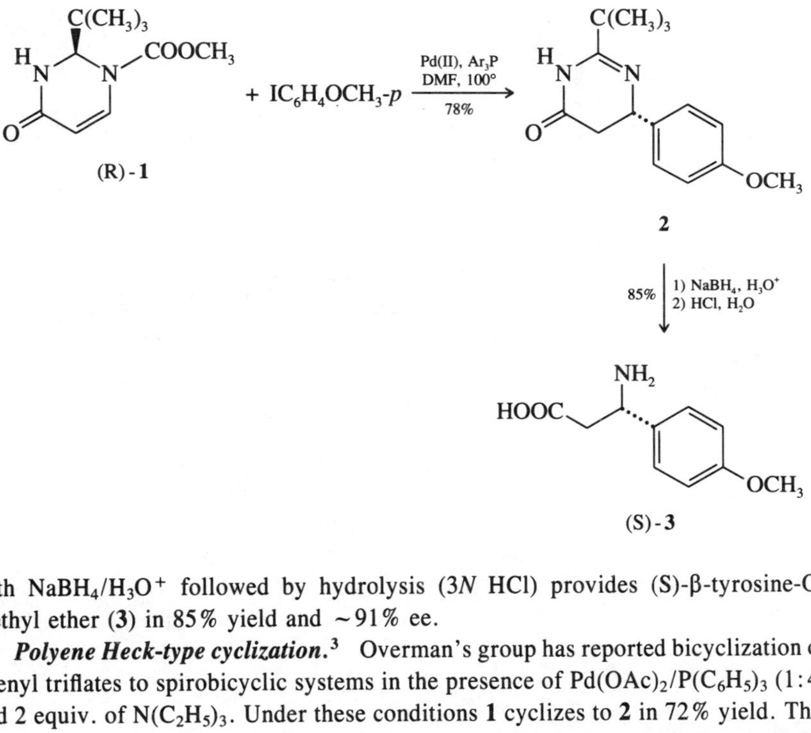

with NaBH$_4$/H$_3$O$^+$ followed by hydrolysis (3*N* HCl) provides (S)-β-tyrosine-O-methyl ether (**3**) in 85% yield and ~91% ee.

Polyene Heck-type cyclization.[3]   Overman's group has reported bicyclization of trienyl triflates to spirobicyclic systems in the presence of Pd(OAc)$_2$/P(C$_6$H$_5$)$_3$ (1:4) and 2 equiv. of N(C$_2$H$_5$)$_3$. Under these conditions **1** cyclizes to **2** in 72% yield. This cyclization is particularly facile when catalyzed by Pd(OAc)$_2$ and (R,R)- or (S,S)-DIOP (**4**, 273) in a 1:1 ratio. In this case, the tricyclic dienone (**2**) is obtained in

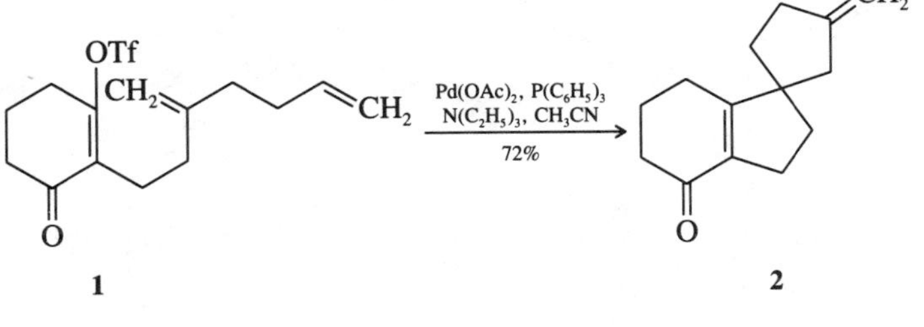

>90% yield and in >45% de. The report includes several other bicyclizations involving formation of spirocyclic centers.

Spirocyclization.[4] Enamides of 2-iodobenzoic acid such as **1** undergo spirocyclization at 45–80° in CH_3CN containing K_2CO_3, $(C_2H_5)_4NCl$ (1 equiv.), and catalytic amounts of $Pd(OAc)_2$ in a 1:2 ratio. The tetraethylammonium chloride permits milder conditions and retards isomerization of the double bond. The reaction can be extended to enamides of type **2**.

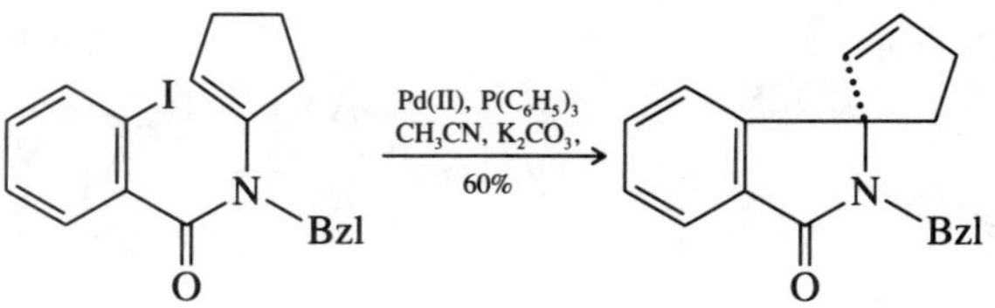

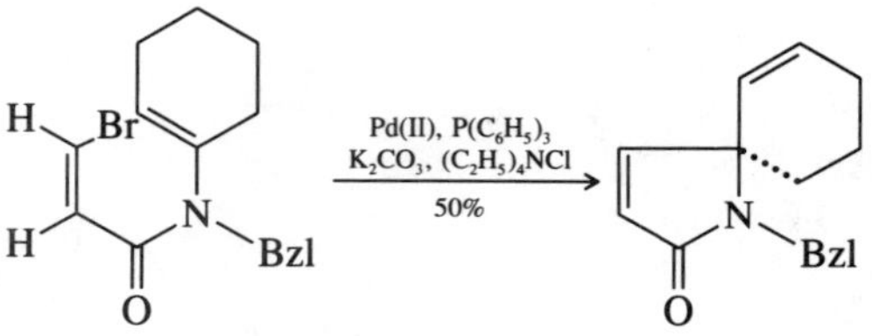

Cyclization of tetraenes.[5] This Pd catalyst effects cyclization of tetraenes containing two 1,3-diene groups in an HX medium to disubstituted cyclopentanes in which the X function is incorporated at one terminus. A number of HX trapping reagents can be used including HOAc, $HN(C_2H_5)_2$, HSO_2Ar. The *trans/cis* ratio is sensitive to the tetraene substrate and to the trapping reagent, and can vary from 1:1 to >20:1.

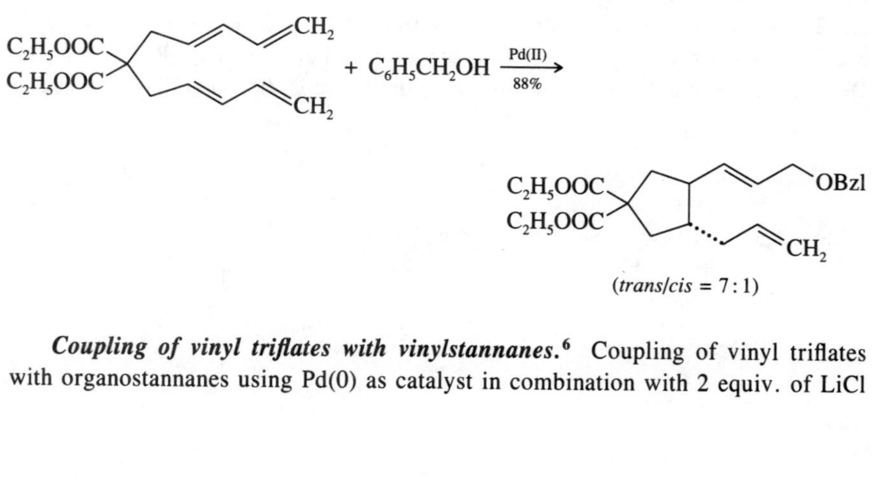

(*trans/cis* = 7:1)

Coupling of vinyl triflates with vinylstannanes.[6] Coupling of vinyl triflates with organostannanes using Pd(0) as catalyst in combination with 2 equiv. of LiCl was first reported by Stille (**14**, 469–470). This coupling has been useful particularly for coupling of a vinyl triflate with a vinyltrialkyltin to give a 1,3-diene with retention of configuration of both partners. However, coupling of the vinyl triflate **1** with

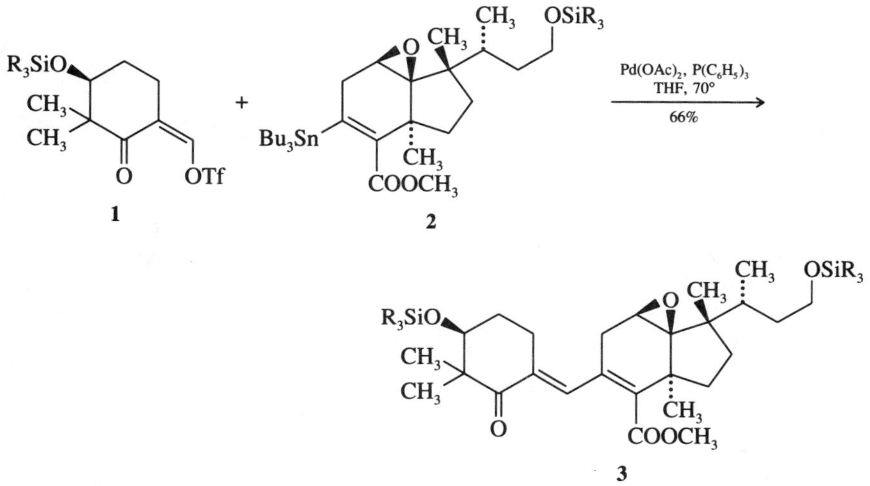

the vinylstannane (**2**) under Stille conditions fails, but succeeds readily under Heck conditions with Pd(OAc)$_2$/P(C$_6$H$_5$)$_3$ (1:2).

Cross-coupling of allenes with 1-alkynes.[7] Pd(OAc)$_2$ in combination with a triarylphosphine (cat. A) can effect cross-coupling of 1-alkynes with 1,1-di- and 1,1,3-trisubstituted allenes to provide as the major product conjugated enynes (equations I, II). The regioselectivity in the case of coupling with methyl 2,3-alkadienoates

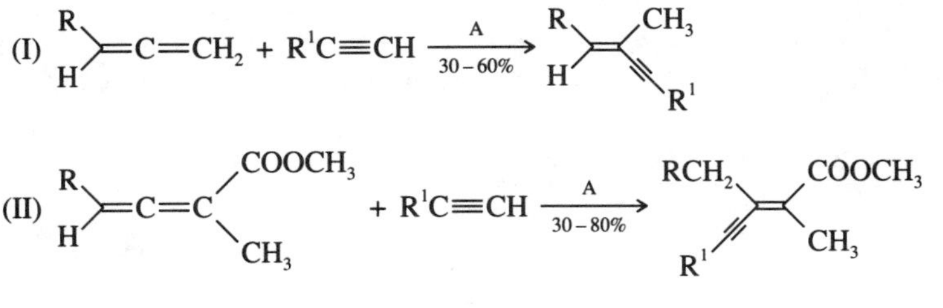

to the fully conjugated system is highly dependent on the choice of catalyst, and can be directed with other catalysts to the nonconjugated enoates. This new reaction can be used to prepare highly unsaturated systems such as 1,5-dien-3-ynes (equation III).

Hydrovinylation of RC≡CR.[8] Reaction of disubstituted alkynes with a vinyl iodide(bromide) catalyzed by Pd(II) and in combination with formic acid and a base

(III) CH$_2$=C=C$\underset{CH_3}{\overset{COOCH_3}{\diagdown}}$ $\xrightarrow{\begin{array}{c} \text{1) (CH}_3\text{)}_3\text{SiC}\equiv\text{CH (83\%),} \\ \text{2) Bu}_4\text{NF (56\%)} \end{array}}$ CH$_3\diagdown\overset{COOCH_3}{}$

1

77% $\downarrow$ 1, A

CH$_3$ COOCH$_3$

CH$_3$ CH$_3$

CH$_3$OOC CH$_3$

(formate reducing system) results in 1,2,4-trisubstituted 1,3-dienes. Configuration of the vinyl partner is retained and *syn*-addition to the alkyne is observed.

C$_6$H$_5$C≡CC$_6$H$_5$ + BuCH=CHX $\xrightarrow[64\%]{\text{Pd(II), HCOOH, N(C}_2\text{H}_5\text{)}_3 \atop 80°}$ $\underset{\substack{\\ C_6H_5 \quad C_6H_5}}{Bu \diagup (E)}$

(E)

[1] R. Grigg, V. Loganathan, V. Santhakumar, V. Sridharan, and A. Teasdale, *Tetrahedron Letters*, **32**, 687 (1991).
[2] J. P. Konopelski, K. S. Chu, and G. R. Negrete, *J. Org.*, **56**, 1355 (1991).
[3] N. E. Carpenter, D. J. Kucera, and L. E. Overman, *ibid.*, **54**, 5846 (1989).
[4] R. Grigg, V. Sridharan, P. Stevenson, and S. Sukirthalingam, *Tetrahedron*, **45**, 3557 (1989).
[5] J. M. Takacs and J. Zhu, *J. Org.*, **54**, 5193 (1989).
[6] E. J. Corey and I. N. Houpis, *Am. Soc.*, **112**, 8997 (1990).
[7] B. M. Trost and G. Kottirsch, *ibid.*, **112**, 2816 (1990).
[8] A. Arcadi, E. Bernocchi, A. Burini, S. Cacchi, F. Marinelli, and B. Pietroni, *Tetrahedron Letters*, **30**, 3465 (1989).

Palladium(II) chloride.

2,5-Disubstituted tetrahydrofurans. Semmelhack[1] has extended his synthesis of tetrahydrofurans by alkoxycarbonylation of 5-hydroxy-1-pentenes (**12**, 372)[2] to

CH$_3$ $\xrightarrow{\begin{array}{c}\text{PdCl}_2, \text{ CuCl}_2 \\ \text{CO, CH}_3\text{OH}\end{array}}$ CH$_3$

CH$_3$ OH =CH$_2$ CH$_3$ O CH$_2$COOCH$_3$

(90% *cis*)   (all *cis*, 90%)

C$_6$H$_5$ $\longrightarrow$ C$_6$H$_5$

CH$_3$ OH =CH$_2$ CH$_3$ O CH$_2$COOCH$_3$

(76% *cis*)   (all *cis*, 100%)

selective synthesis of *cis*- or *trans*-2,5-disubstituted tetrahydrofurans determined by the choice of a substituent at C_3. Thus the configuration of a methyl or a phenyl group at C_3 can determine the *cis*- or *trans*-relationship at C_2 and C_5. A methyl or a phenyl group at C_4 has slight effect on the selectivity. Similar results (equation I) have been reported by another laboratory.[3]

(I)

87%

[1] M. F. Semmelhack and N. Zhang, *J. Org.*, **54**, 4483 (1989).
[2] Review: T. L. B. Boivin, *Tetrahedron*, **43**, 3309 (1987).
[3] M. McCormick, R. Monahan, III, J. Soria, D. Goldsmith, and D. Liotta, *J. Org.*, **54**, 4485 (1989).

Palladium hydroxide, Pd(OH)$_2$.

Hydrogenolysis of allylic acetates.[1] This reaction can be effected in high yield by hydrogenolysis catalyzed by Pd(OH)$_2$ on carbon and with cyclohexene as the

Pd(OH)$_2$/C
C$_6$H$_{10}$, C$_2$H$_5$OH, reflux
65%

hydrogen donor. The readiness of hydrogenolysis decreases in the order primary > secondary > tertiary > acetates, but is increased by steric hindrance.

100%

[1] A. Bianco, P. Passacantilli, and G. Righi, *Tetrahedron Letters*, **30**, 1405 (1989).

(R)-Pantolactone.

Diastereoselective protonation of arylmethylketenes. Merck chemists[1] have described a conversion of (R,S)-2-arylpropionic acids (**1**) into the optically active forms. Thus **1** is converted into the corresponding arylmethylketene (**2**). Addition of a chiral alcohol can give optically active α-hydroxy esters **3** (and the acids). Of a number of chiral alcohols, (S)-ethyl lactate and (R)-isobutyl lactate provide (S)- and (R)-2-arylpropionate esters (**3**) respectively in ~94% de, but (R)-pantolactone (**4**) is significantly more effective and provides (R)-**3** in about 99% de. The essential requirement for diastereoselective protonation of the ketene is apparently a chiral

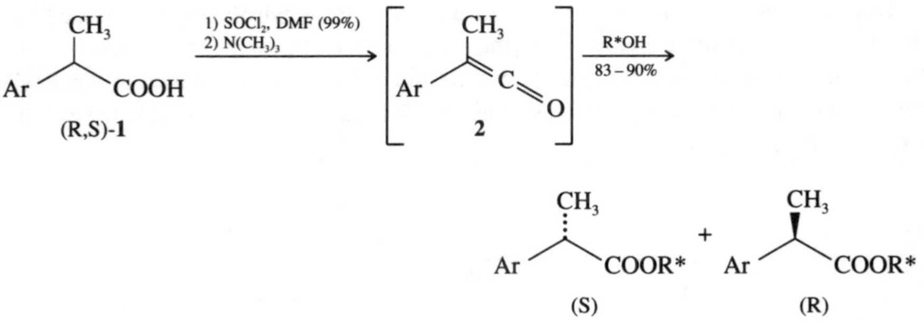

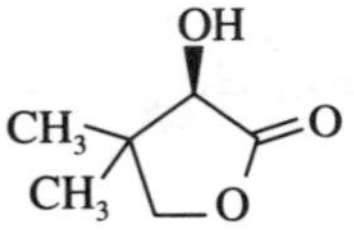

hydroxy group α to a carbonyl and β to a *tert*-alkyl group. The choice of the amine can also affect the diastereoselectivity, with trimethylamine, dimethylethylamine, or N-methylpyrrolidine being most effective.

[1] R. D. Larsen, E. G. Corley, P. Davis, P. J. Reider, and E. J. J. Grabowski, *Am. Soc.*, **111**, 7650 (1989).

(Z)-Pentenylboronates, (**1**).

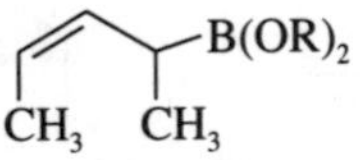

Simple (Z)-pentenylboronates can be prepared by reaction of the (Z)-pentenyl Grignard reagent with different borates. They are of interest because they add to aldehydes to form *syn*-homoallyl alcohols (equation I).[1]

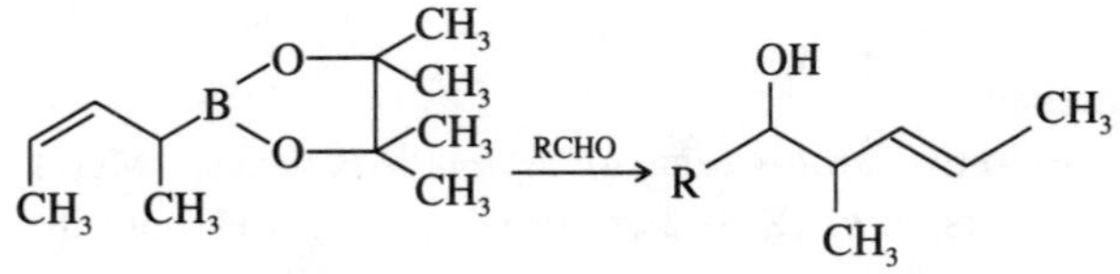

$R = C_2H_5$	92%	96% de
$R = C_6H_5$	95%	94% de

By use of 1,2-dicyclohexyl-1,2-ethanediol as a chiral auxiliary, the chiral (Z)-pentenylboronate **1** has been prepared. This reagent reacts with benzaldehyde with

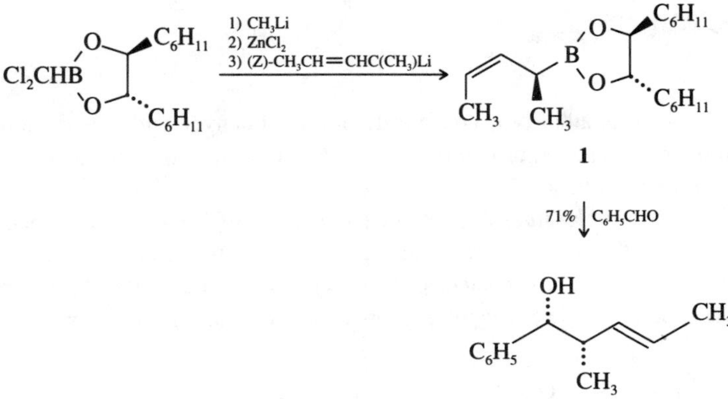

complete transfer of chirality to form *syn*-**2**. It has also been used for a short synthesis of (S,S,S,S)-invictolide (**3**).

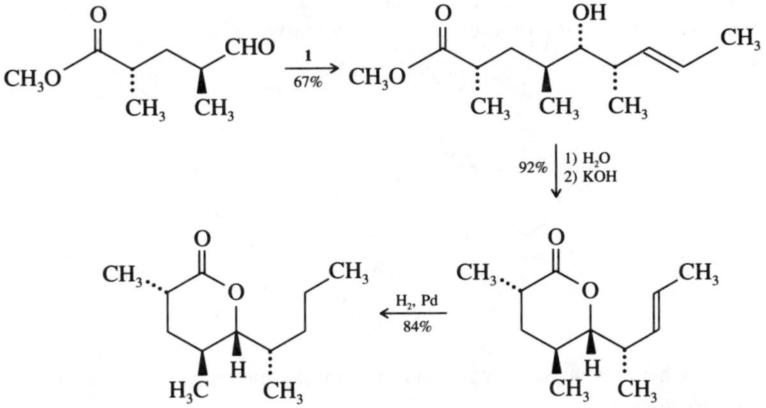

[1] M. W. Andersen, B. Hildebrandt, G. Köster, and R. W. Hoffmann, *Ber.*, **122**, 1777 (1989); R. W. Hoffmann, K. Ditrich, G. Köster, and R. Stürmer, *ibid.*, **122**, 1783 (1989).

Periodinane of Dess-Martin (DMP, 1), 12, 378–379; 15, 252–253.

vic-*Tricarbonyl compounds*. Wasserman[1] has noted that several natural products contain this group or a functionality derivable from it. A short new route to this system[2] involves an aldol-type condensation of an aldehyde with the anion of an α-phenylthio amide. The mixture of products is oxidized to an α,β-diketo amide by the Dess–Martin periodinane.

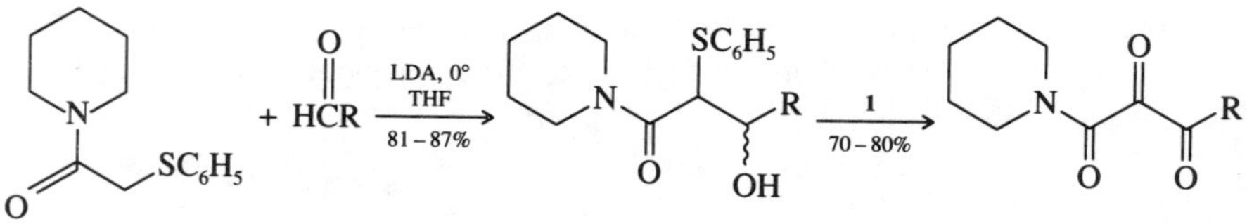

2′- or 3′-Oxonucleosides.[3]   The periodinane (1) of Dess–Martin effects oxidation of 3′,5′- or 2′,5′-disilyl derivatives of nucleosides to the corresponding 5′-silyl derivatives of 2′- or 3′-oxonucleosides in yields as high as 96%. In general, this reagent is superior to Swern-type reagents or a chromium(VI) oxidant.

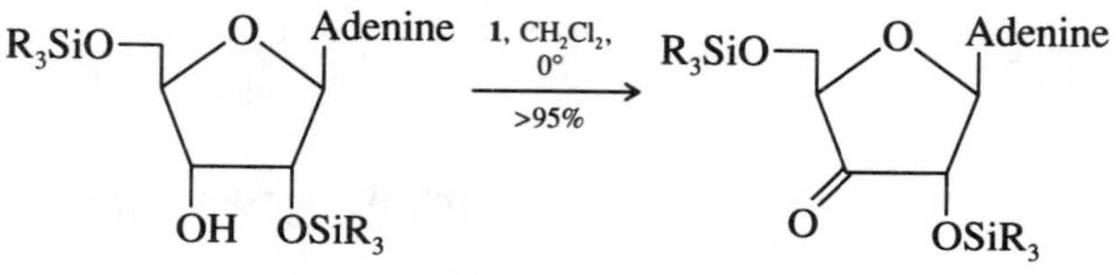

Oxidation of allylic alchols.   Dupuy and Luche[4] recommend this oxidant (1) over MnO_2 especially for oxidation of secondary allylic alcohols to α,β-enones (30–92% yield, 5 examples).

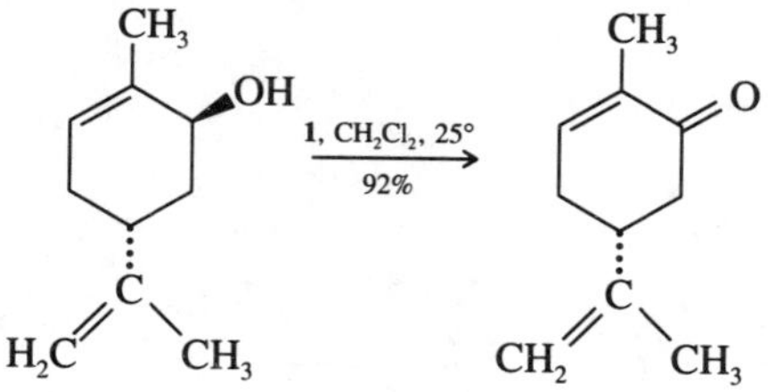

Caution: Chemists at ICE[5] report that the Dess–Martin periodinane can exhibit hazardous explosive properties.

[1] H. H. Wasserman, *Aldrichim. Acta*, **20**, 63 (1987).
[2] R. G. Linde II, L. O. Jeroncic, and S. J. Danishefsky, *J. Org.*, **56**, 2534 (1991).
[3] V. Samano and M. J. Robins, *ibid.*, **55**, 5186 (1990).
[4] C. Dupuy and J. L. Luche, *Tetrahedron*, **45**, 3437 (1989).
[5] J. B. Plumb and D. J. Harper, *C and EN*, July 16, 3 (1990).

Periodic acid–Sodium bisulfite, H_5IO_6–$NaHSO_3$.

Iodohydrins.[1]   This combination (1:2) presumably generates hypoiodous acid, IOH, since it converts alkenes into iodohydrins in moderate to high yield. Bromohydrins can be prepared by use of $NaBrO_3$. Internal alkenes give a mixture of products, but terminal alkenes give Markovnikoff products in which the iodine is bound to the terminal carbon.

[1] H. Ohta, Y. Sakata, T. Takeuchi, and Y. Ishii, *Chem. Letters*, 733 (1990).

Peroxybenzimidic acid (Payne's reagent), **1**.

Stereoselective epoxidation.   Epoxidation of the diene **2** with *m*-chloroperbenzoic acid, monoperphthalic acid, or *t*-butyl hydroperoxide–vanadylacetylacetonate gives a mixture of two epoxides in 89% yield.[1]

In contrast, epoxidation with Payne's reagent results only in **4**. The epoxide **3** has been used to prepare methyl 6α-fluoroshikimate (**5**) by treatment with HF in pyridine and removal of the acetonide group.

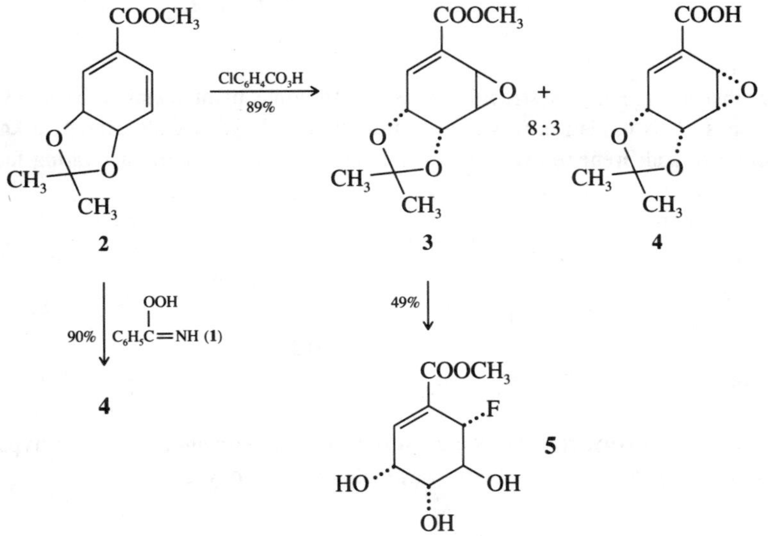

[1] S. Bowles, M. M. Campbell, M. Sainsbury, and G. M. Davies, *Tetrahedron Letters*, **30**, 3711 (1989).

Phenyldiazomethane, $C_6H_5CHN_2$. Preparation.[1]

O- and N-Benzylation.[2]   This reaction can be effected with phenyldiazomethane at $-40°$ in CH_2Cl_2 containing a trace of tetrafluoroboric acid. Alcohols are benzylated somewhat more rapidly than amines; hence selective benzylation of alcohols is possible. Yields are generally satisfactory.

[1] X. Creary, *Org. Syn.*, **64**, 207 (1985).
[2] L. J. Liotta and B. Ganem, *Tetrahedron Letters*, **30**, 4759 (1989).

Phenyliodine(III) bis(trifluoroacetate).

Dihydrobenzofuranes.[1]   Oxidation of a mixture of a *para*-methoxyphenol (**1**) and an electron-rich styrene (**2**) with this oxidant results in a *trans*-disubstituted

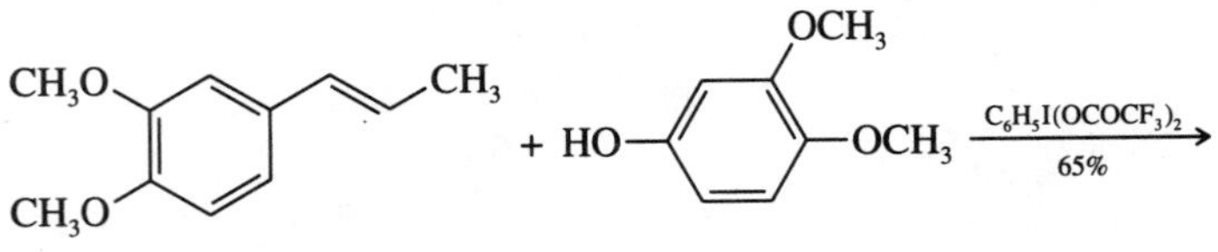

dihydrobenzofuran (**3**), a structure common to some plant metabolites known as neolignans. This oxidation was used to provide the dihydrobenzofuran **4**, a known precursor to kadsurenone (**5**), which is an antagonist to a platelet-activating factor.

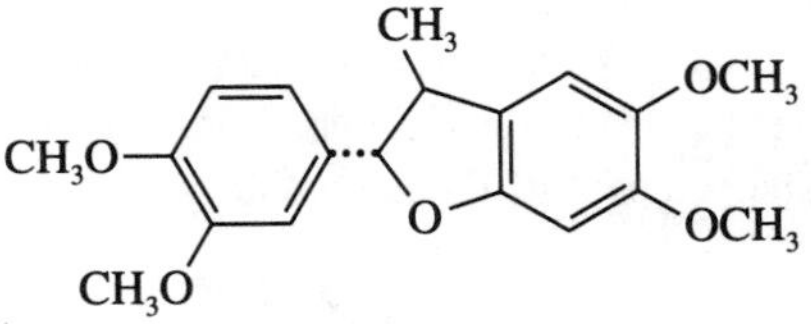

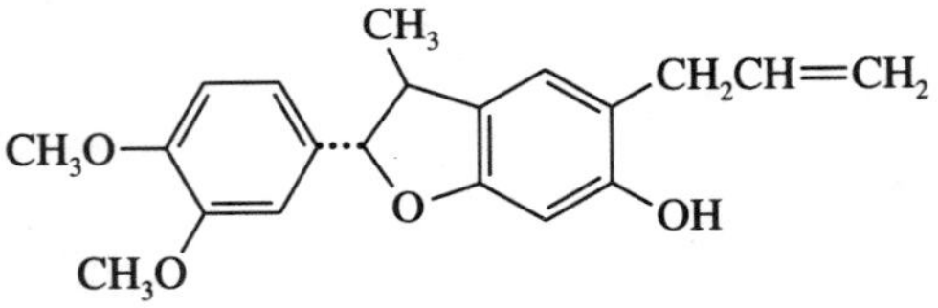

Oxidation of ethynylcarbinols; dihydroxyacetonyl compounds.[2]   This hypervalent iodine reagent oxidizes ethynyl carbinols to dihydroxyacetones.

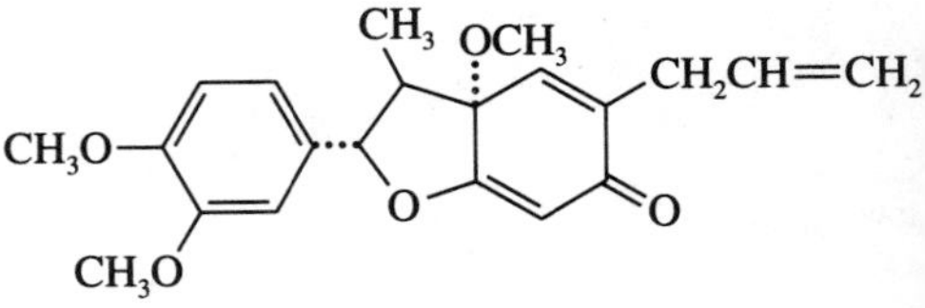

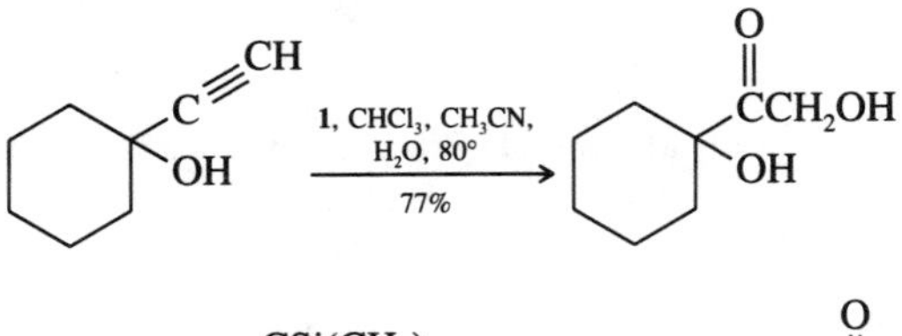

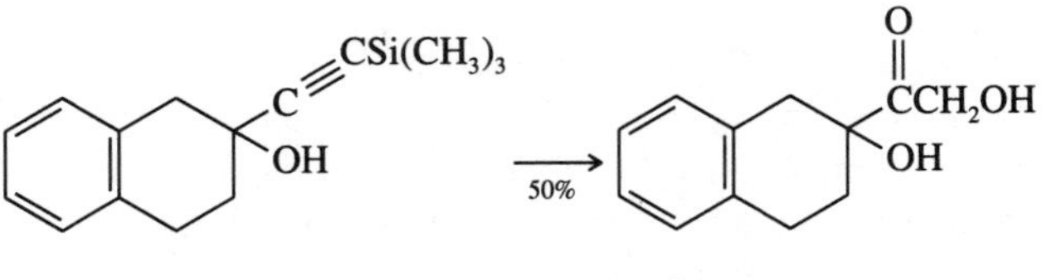

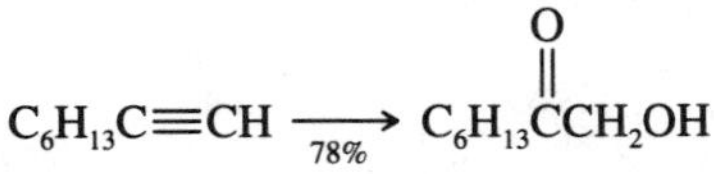

[1] S. Wang, B. D. Gates, and J. S. Swenton, *J. Org.*, **56**, 1979 (1991).
[2] Y. Kita, T. Yakura, H. Terashi, J. Haruta, and Y. Tamura, *Chem. Pharm. Bull.*, **37**, 891 (1989).

Phenyliodine(III) diacetate.

Oxidative rearrangement of aryl methyl ketones.[1]   These ketones have been converted into methyl α-methoxyarylacetates on reaction with $Tl(NO_3)_2$ in trimethyl orthoformate (**7**, 362). This oxidative rearrangement can also be effected with $C_6H_5I(OAc)_2$ (2 equiv.) in trimethyl orthoformate in equally good yield.

$$RC_6H_4COCH_3 \xrightarrow[\text{80-86\%}]{\substack{C_6H_5I(OAc)_2, \\ HC(OCH_3)_3,\ H_3O^+}} RC_6H_4\underset{\underset{OCH_3}{|}}{C}HCOOCH_3$$

[1] O. V. Singh, *Tetrahedron Letters*, **31**, 3055 (1990).

Phenyliodine(III) diacetate–Iodine, 13, 258–259; 14, 242–243.

Angular methylation.[1]   Irradiation of the hemiacetal **1** (formed from an allylic alcohol) with $C_6H_5I(OAc)_2$ and I_2 (1 equiv. of both) results in an iodoformate (**2**), which is reduced to the alcohol **3**.

1   **2**   **3** (*cis/trans* = 4:1)

Transannular cyclization of lactams.[2]   Photolysis of the 8-membered lactam **1** in cyclohexane containing C_6H_5IO and iodine results in the *gem*-diiodide **2** (m. p. 111°, 40%). But the same reaction with 9- or 10-membered lactams results in cyclization to oxoindolizidines or 1-azabicyclo[5.3.0]decanones, respectively, in high

1   **2**

3   4 (82%)   (14%)

yields. Lactams containing eleven or more members show no reaction even at higher temperatures.

[1] G. Stork and R. Mah, *Tetrahedron Letters*, **30**, 3609 (1989).
[2] R. L. Dorta, C. G. Francisco, and E. Suarez, *J.C.S. Chem. Comm.*, 1168 (1989).

Phenyliodine(III) difluoride, $C_6H_5IF_2$ **(1).** The simplest route to this hypervalent reagent is the reaction of iodobenzene dichlorides with aqueous hydrofluoric acid in the presence of HgO (yellow). The *p-t*-butyl derivative of **1** is a stable, crystalline solid.

Fluorination.[1] This reagent effects regio- and stereospecific fluorination of the electron-rich steroid dienamine (**2**) to give the (axial) 6β-fluoro-4-cholestenone-3 (**3**). In the added presence of Cu(acac)$_2$, reaction of **2** with **1** results in 6α-fluoro-4-cholestenone-3 (10%) and 6,6-difluoro-4-cholestenone-3 (15%) as the only products.

2   3

A more significant change obtains in the presence of an electron-transfer reagent such as N-methylviologen. In this case **3** is formed in 21% yield together with 6α-fluoro-4-cholestenone-3 (8%) and 6,6-difluoro-4-cholestenone-3 (13%). Use of a four-fold excess of reagent **1** increases the yield of **3** to 33%. Apparently two radical pathways are possible, both involving ArIF·.

[1] J. J. Edmunds and W. B. Motherwell, *J.C.S. Chem. Comm.*, 881 (1989).

8-Phenylmenthol.

Diastereoselective alkylation of phenols.[1] In the presence of TiCl$_4$, phenols react with glyoxylates at −30 to 20° exclusively at the *ortho*-position to form 2-hydroxymandelic esters. A substituent has no effect on the site of substitution. Use of a chiral glyoxylate, in particular (−)-8-phenylmenthyl glyoxylate, results in high

$$(97-98:3-2)$$

(S)-diastereoselectivity (94% de). Highest yields and diastereoselectivity are obtained with $TiCl_4$ as promoter; use of BCl_3 and $Ti(O-i-Pr)_4$ lowers the diastereoselectivity. A substituent on the phenol can lower the yield, but has little effect on the stereoselectivity. Lower temperatures decrease the yield, but have a negligible effect on the diastereoselection.

[1] F. Bigi, G. Casnati, G. Sartori, C. Dalprato, and R. Bortolini, *Tetrahedron: Asymmetry*, **1**, 861 (1990).

(1S,2R)-(+)-[1-Phenyl-2-(1-piperidinyl)propanol-1, **(1).**

This chiral β-amino alcohol is obtained by reaction of (1S,2R)-norephedrine with 1,5-dibromopentane.

Enantioselective addition $(C_2H_5)_2Zn$ with enones.[1]   In the presence of (1S,2R)-**1** (1 equiv.), diethylzinc undergoes conjugate addition to enones to form (R)-β-ethyl ketones in 81–94% ee. Use of catalytic amounts of **1** decreases the enantioselectivity to 60–80% ee.

(81% ee)

[1] K. Soai, M. Okudo, and M. Okamoto, *Tetrahedron Letters*, **32**, 95 (1991).

(Phenylthiomethylene)triphenylarsorane, $(C_6H_5)_3As=CHSC_6H_5$ **(1).** The ylide precursor is generated from $(C_6H_5)_3As$, NaI, and $ClCH_2SC_6H_5$.

The ylide reacts with aldehydes in THF to form epoxides (**2**) or in THF/HMPA to form phenylthioenol ethers (**3**).[1] The former products rearrange readily to α-phenylthio ketones, the latter to homologated aldehydes.

$$RCHO \xrightarrow[70-90\%]{1, \text{ THF}} \underset{H \quad SC_6H_5}{\overset{R \quad O \quad H}{\diagup\diagdown}} \xrightarrow{SiO_2} \underset{O}{\overset{}{RCCH_2SC_6H_5}}$$

2

$$RCHO \xrightarrow[55-75\%]{1, \text{ THF, HMPA}} \underset{SC_6H_5}{\overset{R}{\diagdown\diagup}} \xrightarrow{H^+} RCH_2CHO$$

3

[1] B. Boubia, C. Mioskowski, S. Manna, and J. R. Falck, *Tetrahedron Letters*, **30**, 6023 (1989).

Phenylthiotrimethylsilane, $C_6H_5SSi(CH_3)_3$ (1). Preparation.[1]

O,S-Acetals.[2] Acyclic O,S-acetals are not readily available. Thus the reaction of methylthiotrimethylsilane with carbonyl compounds catalyzed by a Lewis acid forms the expected O,S-acetal as an intermediate, but the only isolable product is the dithioketal (**8**, 352).

A new, direct route to O,S-acetals is based in part on the ability of trimethylsilyl triflate to mediate synthesis of O,O-acetals from carbonyl compounds and silyl ethers (**10**, 439). Thus reaction of 1:1 mixtures of a silyl ether and phenylthiotrimethylsilane with an aldehyde in the presence of catalytic to stoichiometric amounts of trimethylsilyl triflate can give O,S-acetals in 37–93% yield. Acetone is amenable to this O,S-ketalization, but reactions with cyclohexanone result mainly in O,O-ketals.

$$CH_3CHO + (CH_3)_3SiOCH_2C_6H_4CH_3 + 1 \xrightarrow[CH_2Cl_2, -78°]{(CH_3)_3SiOTf}$$

$$\underset{CH_3 \quad SC_6H_5}{\overset{OCH_2C_6H_4CH_3}{\diagup\diagdown}} + \underset{CH_3 \quad OCH_2C_6H_4CH_3}{\overset{OCH_2C_6H_4CH_3}{\diagup\diagdown}}$$

(81%)   (9%)

[1] R. S. Glass, *J. Organomet. Chem.*, **61**, 83 (1973).
[2] A. Kusche, R. Hoffmann, I. Münster, P. Keiner, and R. Brückner, *Tetrahedron Letters*, **32**, 467 (1991).

(1-Phenylthio-1-trimethylsilyl)allyl-9-borabicyclo[3.3.1]nonane,

$$\overset{Si(CH_3)_3}{\underset{|}{}}$$

$C_6H_5SC{=}CHCH_2$-9-BBN (**1**). The reagent is prepared *in situ* from 9-BBN and

$$\underset{(CH_3)_3Si}{\overset{C_6H_5S}{>}}C=C=CH_2.$$

2-(*Phenylthio*)-1,3-*butadienes* (2).[1] This boron compound (**1**) is an alternative reagent to a similar one based on titanium (**11**, 377) for preparation of either (E)- or (Z)-2-(phenylthio)-1,3-butadienes (**2**). Thus **1** reacts with aldehydes to form adducts (**a**) that undergo either *syn-* or *anti*-elimination of trimethylsilane oxide.

$$n\text{-}C_5H_{11}CHO + 1 \longrightarrow$$

OBBN

$$\underset{C_6H_5S \quad Si(CH_3)_3}{C_5H_{11}\overset{|}{\diagup}\diagdown\diagdown CH_2}$$

(a)

H₂SO₄		NaOH
80%		92%

$$n\text{-}C_5H_{11}\diagdown\diagup\overset{\diagup\diagdown CH_2}{\underset{SC_6H_5}{|}}$$

2 (Z/E = 26:1)

$$n\text{-}C_5H_{11}\diagdown\diagup\overset{SC_6H_5}{\underset{\diagdown CH_2}{}}$$

2 (E/Z = 50:1)

[1] W. H. Pearson, K.-C. Lin, and Y.-F. Poon, *J. Org.*, **54**, 5814 (1989).

3-Phenylthio-2-(trimethylsilylmethyl)propene (1). This conjunctive reagent is prepared from that of Trost (**11**, 258) for methylenecyclopentane annelation.

$$\underset{(CH_3)_3SiCH_2CCH_2OH}{\overset{CH_2}{\overset{\|}{}}} \xrightarrow[\text{2) BuLi, C}_6\text{H}_5\text{SH}]{\text{1) MsCl}} \underset{(CH_3)_3SiCH_2CCH_2SC_6H_5}{\overset{CH_2}{\overset{\|}{}}} \quad \textbf{(1)}$$

[3+3]*Annelation*.[1] This reagent is used for conversion of 3-substituted aldehydes (or acetals) into methylenecyclohexanes. The conversion involves addition of the allylsilane group to the aldehyde or acetal to provide an adduct that undergoes radical cyclization via the allylic sulfide group. The phenylthio group is used to enhance 6-*endo* cyclization over the usual 5-*exo* cyclization. In addition it allows use of bis(tributyltin) as the initiator (**14**, 173–174). Unfortunately the radical cyclization shows only slight stereoselectivity.

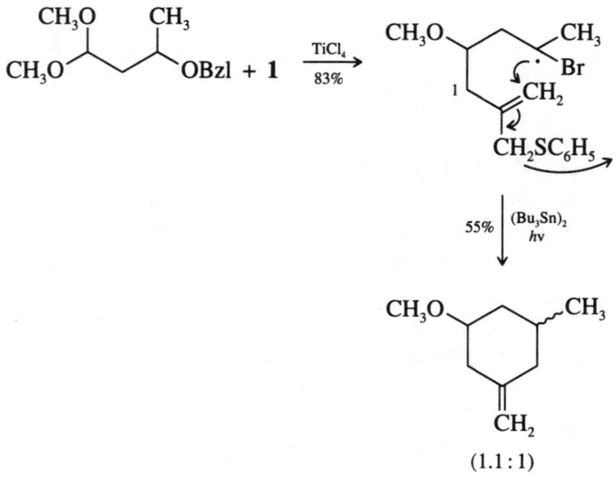

$$CH_3O \quad CH_3$$
$$CH_3O \quad OBzl + 1 \xrightarrow[83\%]{TiCl_4}$$

CH₃O structure with CH₃, Br, CH₂, CH₂SC₆H₅

$$55\% \left| \begin{array}{c} (Bu_3Sn)_2 \\ h\nu \end{array} \right.$$

CH₃O, CH₃, CH₂ cyclohexane structure

(1.1 : 1)

[1] D. E. Ward and B. F. Kaller, *Tetrahedron Letters*, **32**, 843 (1991).

B-3-Pinanyl-9-borabicyclo[3.3.1]nonane (1), 10, 320–321.

Conversion of propargyl esters to dihydrofurans[1] (**11**, 469–470).   The reaction can be used for an enantioselective synthesis of dihydrofurans. Thus reduction of 3-hydroxy-1-alkynyl ketones (**2**) with Midland's reagent, (+)-**1**, provides (4S)-2-butyne-1,4-diols (**3**) in 84–91% ee. Monoacylation (**4**) followed by treatment with

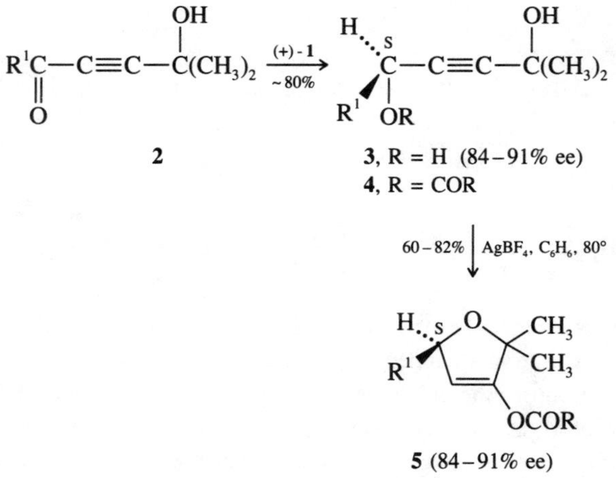

$$R^1C-C\equiv C-C(CH_3)_2 \xrightarrow[\sim 80\%]{(+)-1}$$
$$\underset{O}{\parallel}$$

2

$$\overset{H}{\underset{R^1}{\cdots}}\overset{S}{\underset{OR}{C}}-C\equiv C-C(CH_3)_2$$
$$\overset{OH}{|}$$

3, R = H (84–91% ee)
4, R = COR

$$60-82\% \left| \begin{array}{c} AgBF_4, C_6H_6, 80° \end{array} \right.$$

H, S, O, CH₃, CH₃, R¹, OCOR dihydrofuran structure

5 (84–91% ee)

AgBF₄ in benzene at 80° provides the dihydrofurans **5**, with almost quantitative transfer of stereogenicity.

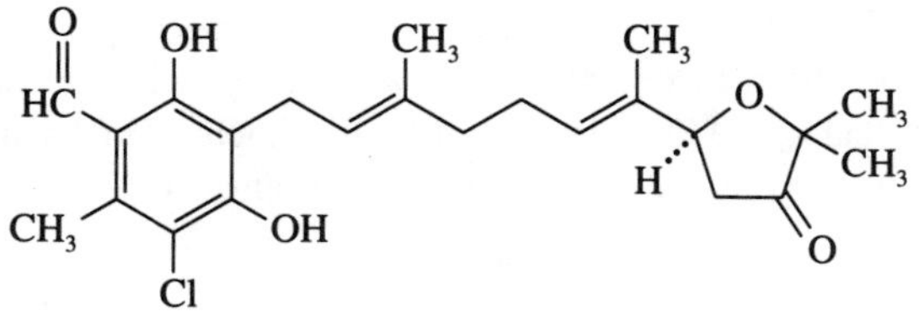

6

This sequence was used for a synthesis of the antibiotic (S)-ascofuranone (**6**).

[1] Y. Shigemasa, M. Yasui, S. Ohrai, M. Sasaki, H. Sashiwa, and H. Saimoto, *J. Org.*, **56**, 910 (1991).

1-Piperidinocyclopropanol (**1**, m. p. 81–82°).
Preparation:

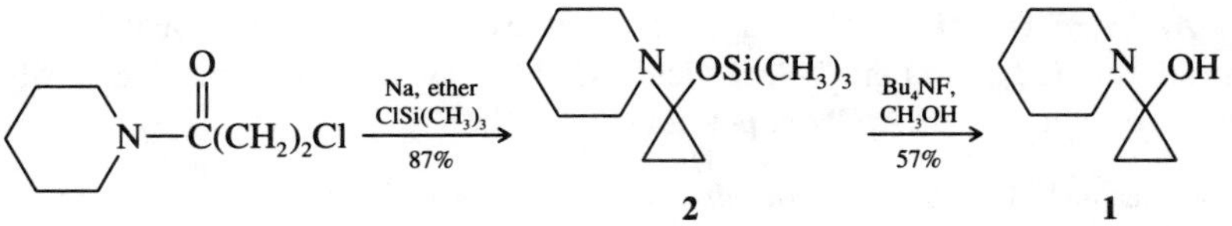

Cyclopropanone equivalent.[1] Unlike cyclopropanone, which is difficult to iso-late, **1** is stable, easily obtained from 3-chloropropionic acid (Aldrich), and is a useful substitute for the ketone. Thus reaction of the silyl ether **2** with RMgBr provides adducts in which the OH group of **1** is replaced by R. The carbinol **1** is also a useful precursor to pyrroles, pyrrolines, and pyrrolizidines (equation I).

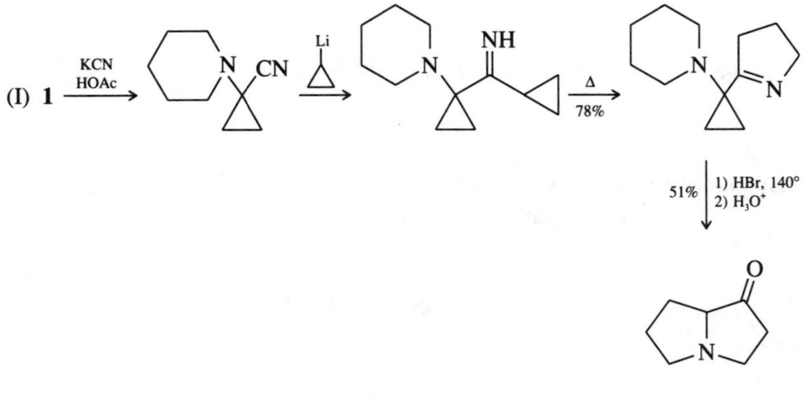

[1] H. H. Wasserman, R. P. Dion, and J. Fukuyama, *Tetrahedron*, **45**, 3203 (1989).

Poly(4-vinylpyridine), 1.

Azulenes.[1] In the presence of this polymer as an acid scavenger, tropylium tetrafluoroborate undergoes [3 + 2]cyclization with simple allenylsilanes to form azulenes. Trimethoxymethylsilane is the preferred scavenger in the case of higher substituted allenes.

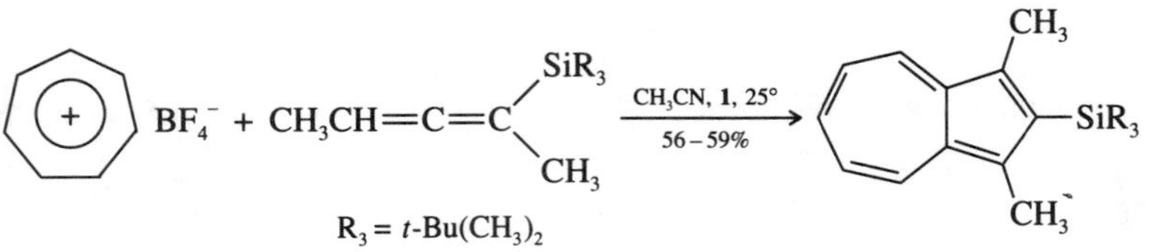

$$R_3 = t\text{-}Bu(CH_3)_2$$

[1] D. A. Becker and R. L. Danheiser, *Am. Soc.*, **111**, 389 (1989).

Potassium 9-*sec*-amyl-9-boratabicyclo[3.3.1]nonane (1), 14, 264.

ArCN → ArCHO.[1] The reagent reduces aromatic nitriles to aldehydes at room temperature (12 hours) in 75–98% yield. Aliphatic nitriles are reduced so slowly that selective reduction of ArCN is possible.

[1] J. S. Cha and M. S. Yoon, *Tetrahedron Letters*, **30**, 3677 (1989).

Potassium fluoride.

α-Diazo ketones or esters.[1] KF supported in Al_2O_3 is an efficient base for diazo transfer reactions of tosyl azide with active methylene groups.

$$CH_3COCH_2COOC_2H_5 + CH_3C_6H_4SO_2N_3 \xrightarrow[94\%]{\substack{KF/Al_2O_3, \\ THF, 20°}} CH_3CO\overset{\overset{N_2}{\|}}{C}COOC_2H_5$$

[1] A. B. Alloum and D. Villemin, *Syn. Comm.*, **19**, 2567 (1989).

Potassium hexamethyldisilazide.

Eliminative cyclization.[1] Treatment of the allylic phosphate ester **1** with this strong base results in eliminative cyclization to the *trans*-cyclopropane **2**, the marine

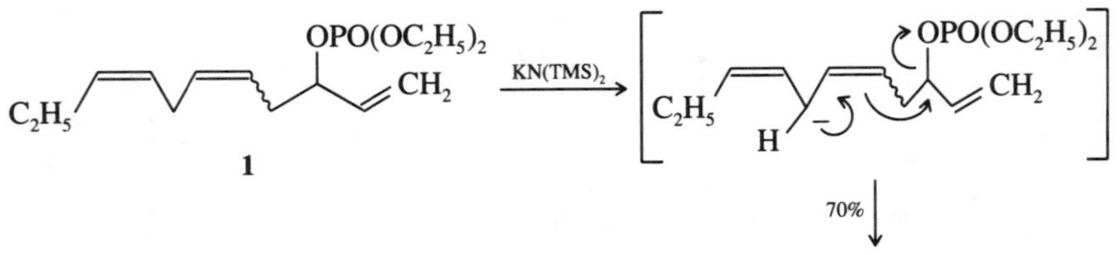

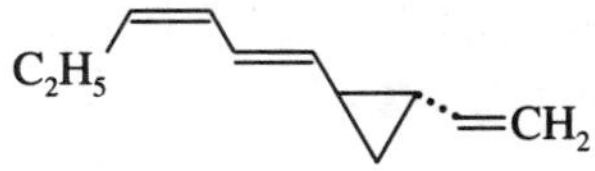

2

gamete attractant dictyopterene B, in 70% yield. The reaction apparently involves removal of the diallylic proton followed by an intramolecular S_N2 reaction.

[1] W. D. Abraham and T. Cohen, *Am. Soc.*, **113**, 2313 (1991).

Potassium hydrogen fluoride, KHF_2.

α-Fluoro ketones.[1] Reaction of KHF_2 and BF_3 etherate (2 equiv. of each) with α,β-epoxy sulfoxides in chloroform results in α-fluoroketones with elimination of the sulfinyl group in ~35–80% yield. Use of $MgCl_2$ in this ring-opening reaction provides α-chloro ketones. Olah's reagent (HF·pyridine) is not useful because of low yields.

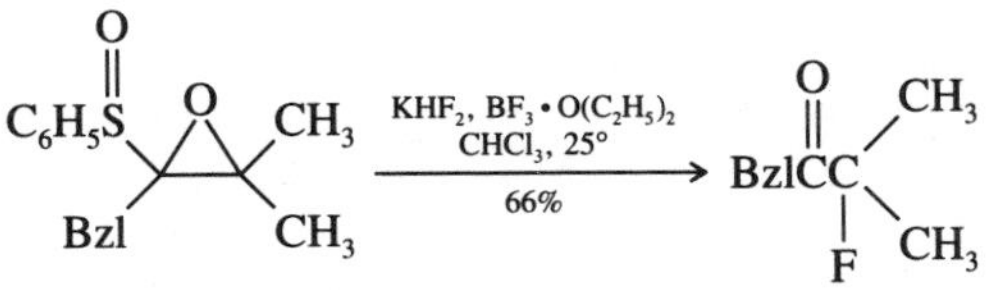

[1] T. Satoh, J. Shishikura, and K. Yamakawa, *Chem. Pharm. Bull.*, **38**, 1798 (1990).

Potassium permanganate–Copper(II) sulfate.

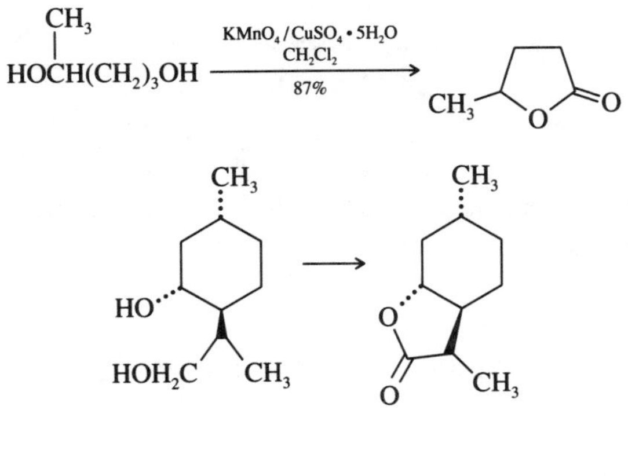

Heterogeneous oxidation of diols to lactones.[1] A mixture of $KMnO_4$ and $CuSO_4 \cdot 5H_2O$ is recommended for heterogeneous oxidation of 1,4- and 1,5-diols to lactones. This oxidation can be highly selective since primary alcohols are oxidized more rapidly than secondary ones. Oxidation of isolated primary alcohols to acids with this system is not useful because the reaction is very slow unless a base is also added.

Oxidation of alkenes.[2] A well-ground mixture of $KMnO_4/CuSO_4 \cdot 5H_2O$ (2:1) suspended in CH_2Cl_2 containing a trace of t-BuOH/H_2O effects oxidation of alkenes at 25° to α-diketones or α-hydroxy ketones in modest to high yield. The t-BuOH/H_2O solvent is usually crucial for successful oxidation. In some cases, epoxides are the major products.

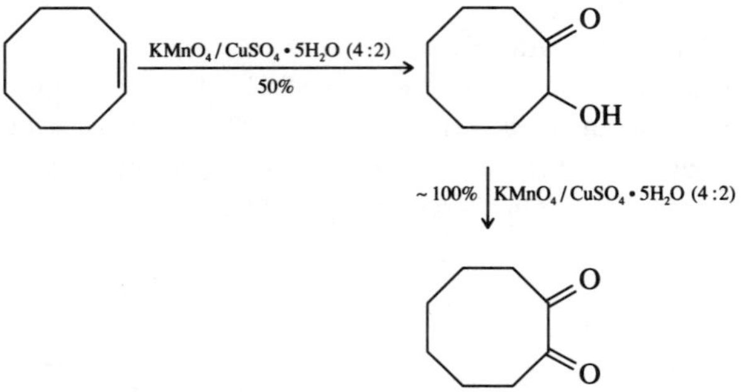

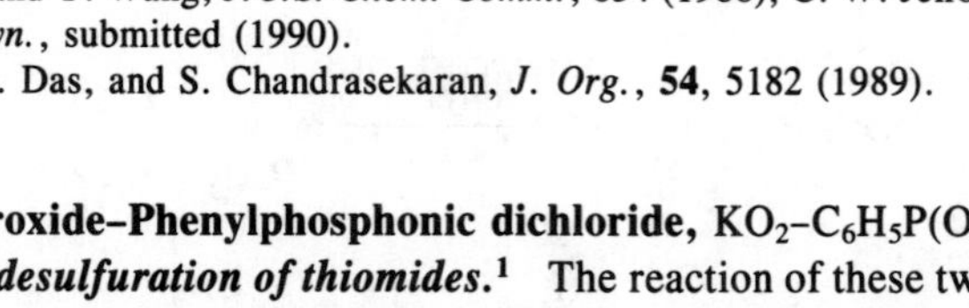

[1] C. W. Jefford and Y. Wang, *J.C.S. Chem. Comm.*, 634 (1988); C. W. Jefford, Y. Li, and Y. Wang, *Org. Syn.*, submitted (1990).
[2] S. Baskaran, J. Das, and S. Chandrasekaran, *J. Org.*, **54**, 5182 (1989).

Potassium peroxide–Phenylphosphonic dichloride, KO_2–$C_6H_5P(O)Cl_2$ **(1).**

Oxidative desulfuration of thiomides.[1] The reaction of these two reagents generates a peroxyphosphonic intermediate that converts thioamides into the corresponding amides in 90–95% yield via a sulfine intermediate.

[1] Y. H. Kim, S. C. Lim, and H. S. Chang, *J.C.S. Chem. Comm.*, 36 (1990).

Potassium peroxymonosulfate (Oxone®).

vic-*Tricarbonyl systems.*[1]   A new route to this system depends on the ability of Oxone® to cleave ylide double bonds more readily than C–C double bonds. Singlet oxygen is less selective than Oxone®. The ylide precursors are available by reaction of acid chlorides with *t*-butyl (triphenylphosphoranylidine)acetate in the presence of bis(trimethylsilyl)acetamide (BSA).[2] Even a trisubstituted double bond is not oxidized by Oxone® in reactions carried out in a two-phase system of benzene/water.

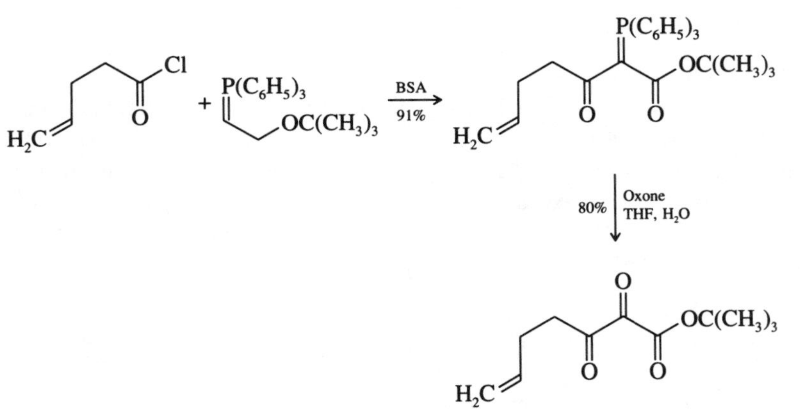

***Oxidation of alicyclic ketones to lactones.*[3]**   Cycloalkanones in CH_2Cl_2 are oxidized to lactones by Oxone® (**1**) mixed with slightly wet alumina as a heterogeneous mixture. Efficient stirring is necessary to obtain reproducible results. Yields are about 80% for 5- and 6-membered ketones, but are low with higher-membered ketones.

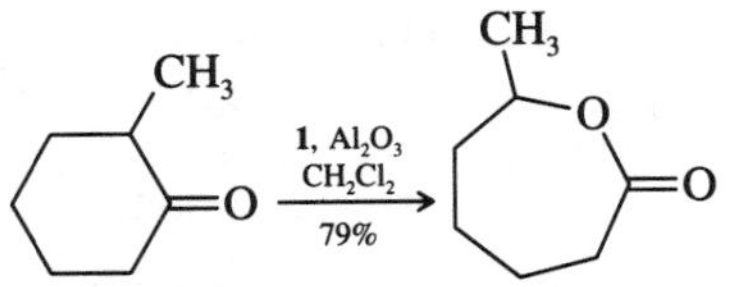

[1] H. H. Wasserman and C. B. Vu, *Tetrahedron Letters*, **31**, 5205 (1990).
[2] H. H. Wasserman, V. M. Rotello, D. R. Williams, and J. W. Benbow, *J. Org.*, **54**, 2785 (1989).
[3] M. Hirano, M. Oose, and T. Morimoto, *Chem. Letters*, 331 (1991).

(η^6-Pyridine)tricarbonylchromium, (1).

Preparation:

2

Regioselective lithiation.[1] This complex undergoes selective lithiation at the *ortho*-position, which can be trapped by methylation to give (2-methylpyridine)tricarbonylchromium. The disilyl complex **2** undergoes selective lithiation at C_4 because of steric effects. Reduction (DIBAH) of **1** provides (1,2-dihydro-pyridine)tricarbonylchromium (**3**) after quenching (MeOH). Reaction of **1** with RLi followed by alkylation with CH_3I provides the complex **4**.

3   4

[1] S. G. Davies and M. R. Shipton, *J.C.S. Chem. Comm.*, 995 (1989).

Pyridinium poly(hydrogen fluoride).

Oxidative fluorination.[1] Reaction of phenol with PbO_2 (2 equiv.) and HF/pyridine (70/30, w/w) in CH_2Cl_2 at 25° provides the dienone **1** in 30% yield. It can undergo Michael addition and aromatization to give substituted *p*-fluorophenols.

Arenes are not oxidized by Pb(IV) reagents, but can undergo anodic oxidative fluorination. In this case $HF/N(C_2H_5)_3$ is the preferred reagent.

Olah and Li[2] have prepared a solid form of this reagent by reaction of anhydrous HF with cross-linked poly-4-vinylpyridine. It is comparable to pyridinium poly(hydrogen fluoride) for hydrofluorination of alkenes and alkynes and fluorination of alcohols, but is easier to handle. Work-up of reactions requires only a simple filtration, and the spent reagent can be regenerated with HF for further use.

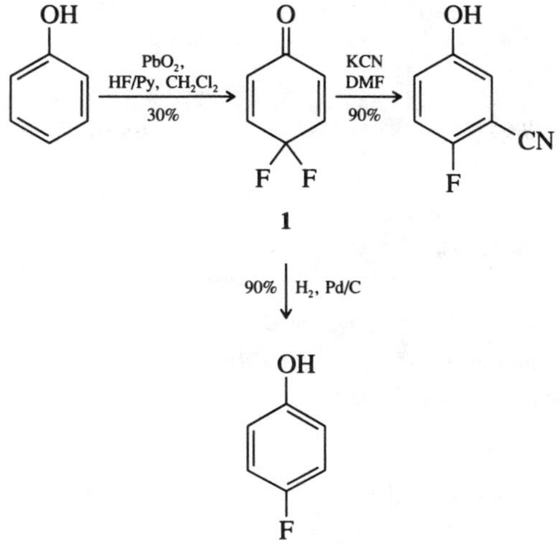

1

¹ J. H. H. Meurs, D. W. Sopher, and W. Eilenberg, *Angew. Chem. Int. Ed.*, **28**, 927 (1989).
² G. A. Olah and X.-Y. Li, *Synlett*, 267 (1990).

Pyridinium *p*-toluenesulfonate (PyH·Ts).

α-*Alkoxy carbonyl compounds.*[1]   The α-*p*-toluenesulfinyl derivatives (**1**) of dimethylhydrazones of aldehydes or ketones react with a variety of alcohols in the presence of this reagent to form the corresponding α-alkoxy hydrazones (**2**). PyH·Ts is more effective than alkylaluminum halide catalysts.

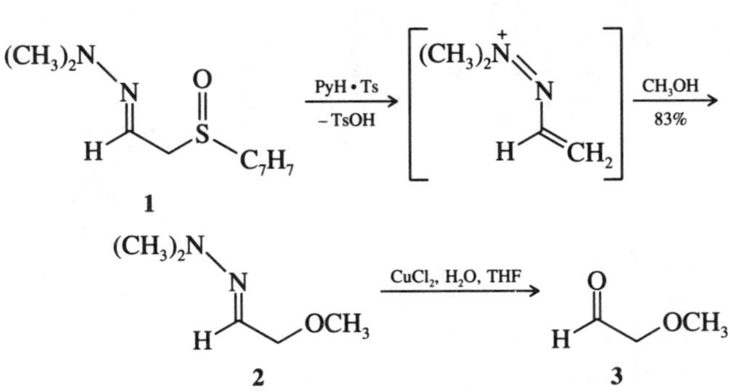

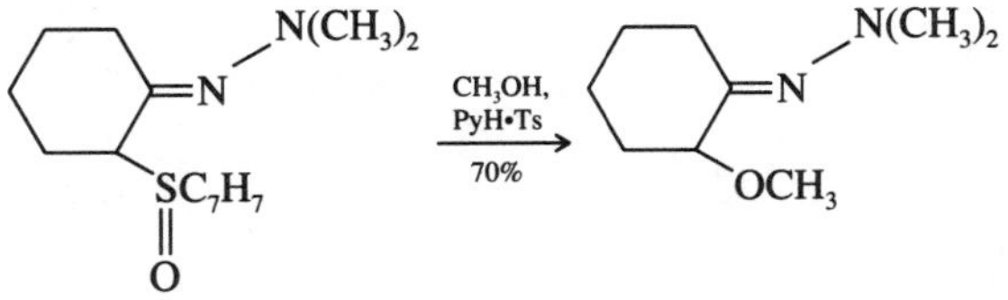

The product can be deprotected by $CuCl_2$ (**7**, 128). The alcohol can be primary, secondary, or tertiary, as well as allylic or benzylic.

[1] P. Pflieger, C. Mioskowski, J. P. Salaun, D. Weissbart, and F. Durst, *Tetrahedron Letters*, **30**, 2791 (1989).

Pyrylium perchlorate, 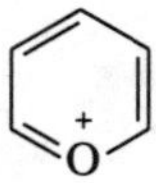 ClO_4^- (**1**), potentially explosive.

(2Z,4E)-Dienals.[1]   This salt reacts with organometallic reagents, particularly Grignard reagents and alkyllithiums by addition at C_2 to form an adduct that undergoes ring opening to a (2Z,4E)-dienal (equation I).

(~97:3)

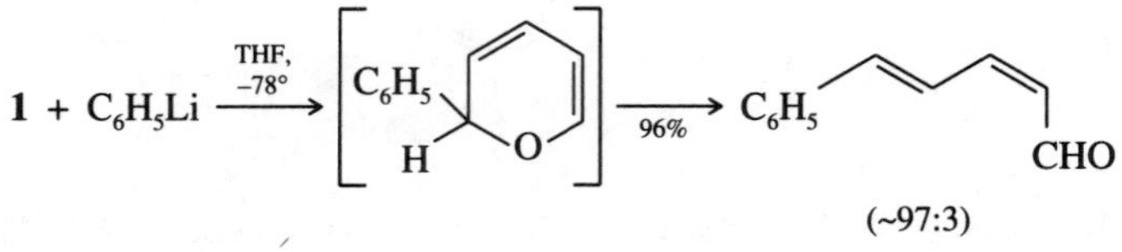

[1] M. Furber, J. M. Herbert, and R. J. K. Taylor, *J.C.S. Perkin I*, 683 (1989).

R

Rhodium(II) carboxylates.

Cyclization/cycloaddition route to oxapolycycles (13, 266). The reaction of 1-diazo-2,5-hexanedione (1)[1] (or of 1-diazo-5-phenyl-2,5-pentanedione[2]) with $Rh_2(OAc)_4$ provides an adduct **a** that reacts with an aldehyde to form the ring system **2**, which can be used as a precursor to brevicomin (**3**).

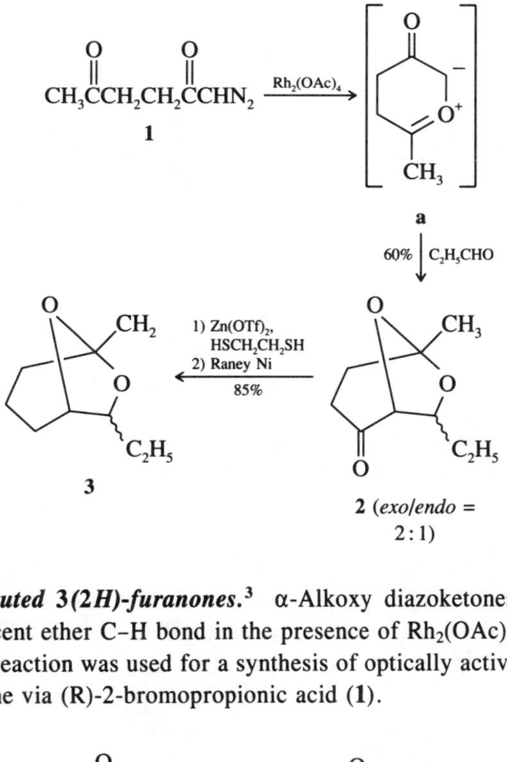

2,5-Disubstituted 3(2H)-furanones.[3] α-Alkoxy diazoketones undergo insertion into an adjacent ether C–H bond in the presence of $Rh_2(OAc)_4$ to form 3(2H)-furanones. This reaction was used for a synthesis of optically active (+)-muscarine (**2**) from D-alanine via (R)-2-bromopropionic acid (**1**).

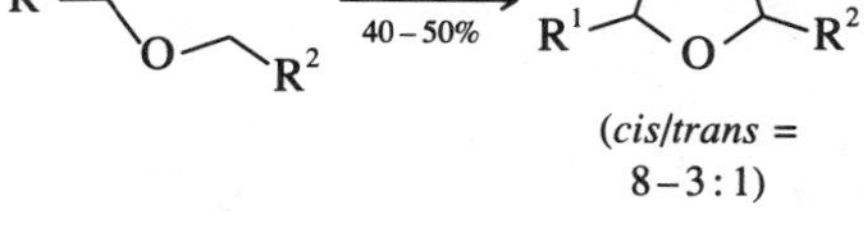

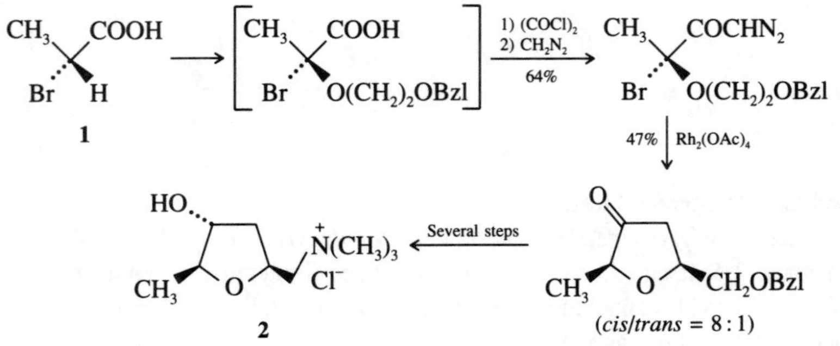

Furans.[4]   Highly substituted furans can be obtained by Rh$_2$(OAc)$_4$-catalyzed reaction of 2-diazo-1,3-dicarbonyls with arylacetylenes. A typical example[5] is formulated.

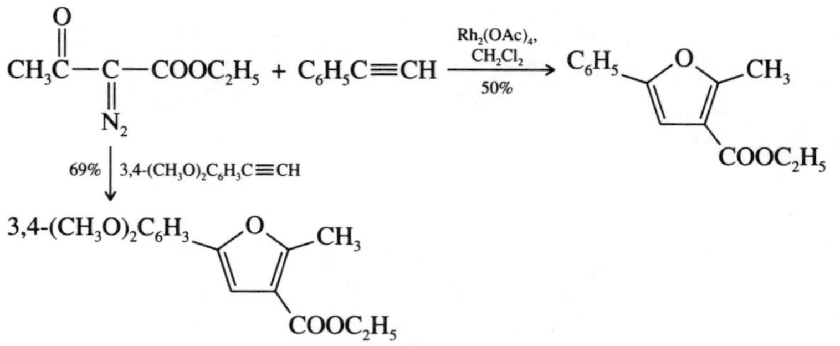

The report stresses the advantages of *p*-acetamidobenzenesulfonyl azide, N$_3$SO$_2$C$_6$H$_4$NHCOCH$_3$-*p*,[6] for preparation of diazo compounds by diazo-transfer. It is safe, inexpensive, and the sulfonamide by-product is removed by simple trituration with H$_2$O.

Carbenoid–monothiophthalimide coupling.[7]   The key step in a synthesis of an isoindolobenzazepine alkaloid **3** involves cyclization of an N-aziridinohydrazone and a monothiophthalimide group catalyzed by Rh$_2$(OAc)$_4$. Thus treatment of **1** with a

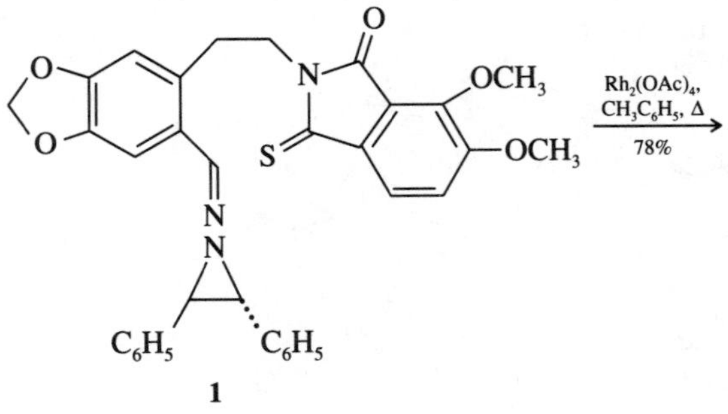

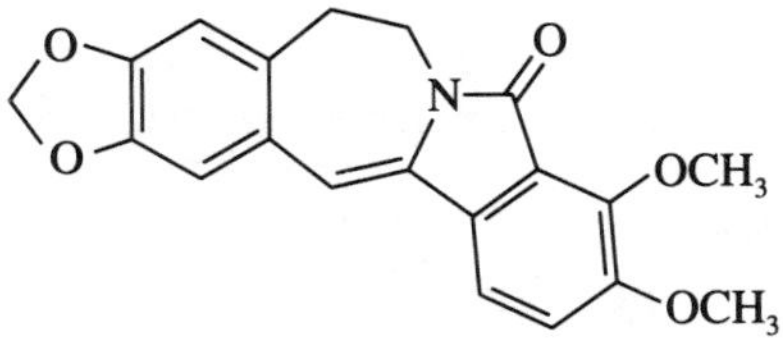

2 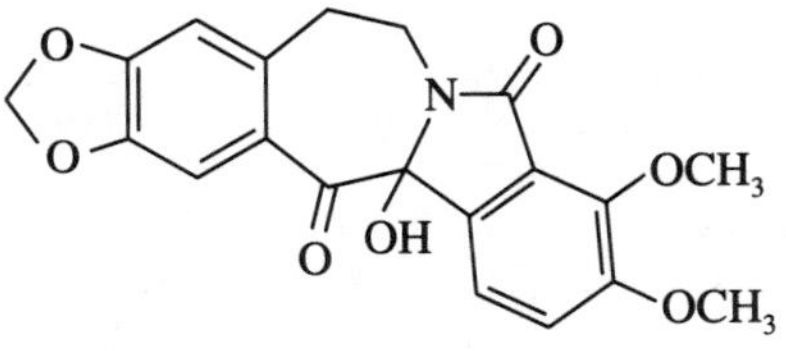   **3**

catalytic amount of $Rh_2(OAc)_4$ in refluxing toluene induces a carbenoid coupling to provide **2**, which is oxidized to chilenine (**3**) by dimethyldioxirane followed by addition of $NaHCO_3$ (38% yield).

Rearrangement of N-nitrosoamides.[8]   N-Nitrosamides (**1**), prepared by acetylation of primary amines followed by nitrosation, are known to decompose in nonpolar solvents at 80–100° to form alkyl acetates with elimination of nitrogen.[9] The presumed diazoalkane intermediate (**a**) can be trapped as a rhodium carbene (**b**), which undergoes rearrangement to an alkene (equation I). The overall result is a mild, nonbasic version of the classical Hofmann degradation of amines.

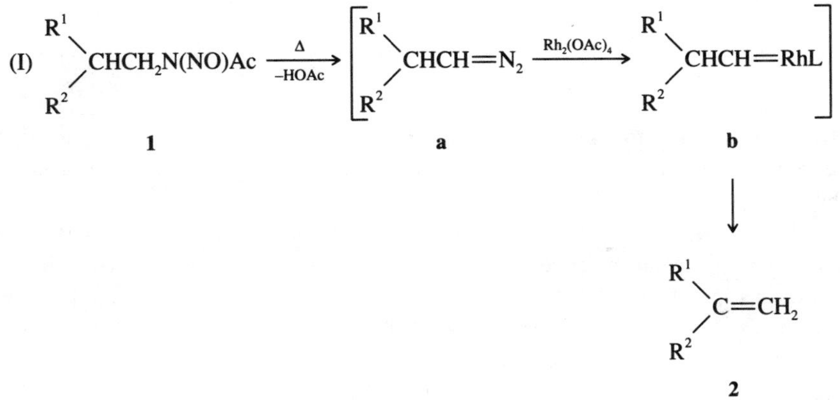

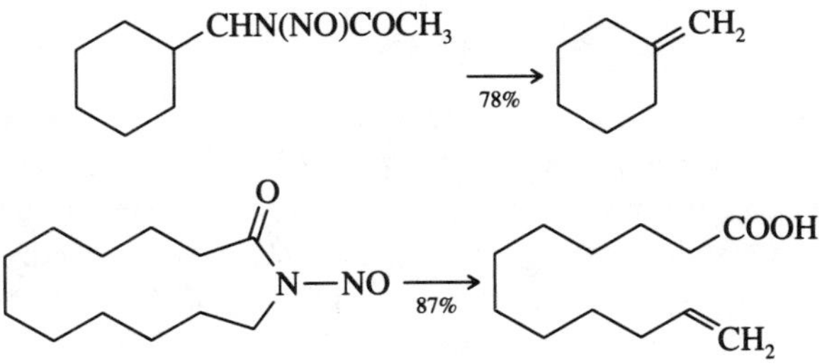

This reaction is applicable to nitrosolactams to provide ω-unsaturated fatty acids (second example).

[1] A. Padwa, R. L. Chinn, and L. Zhi, *Tetrahedron Letters*, **30**, 1491 (1989).

[2] A. Padwa, G. E. Fryxell, and L. Zhi, *J. Org.*, **53**, 2875 (1988).

[3] J. Adams, M.-A. Poupart, L. Grenier, C. Schaller, N. Ouimet, and R. Frenette, *Tetrahedron Letters*, **30**, 1749, 1753 (1989).

[4] H. M. L. Davies and K. R. Romines, *Tetrahedron*, **44**, 3343 (1988).

[5] H. M. L. Davies, W. R. Cantrell, Jr., and J. S. Baum, *Org. Syn.*, submitted (1989).

[6] J. B. Hendrickson, and W. A. Wolf, *J. Org.*, **33**, 3610 (1968).

[7] F. C. Fang and S. J. Danishefsky, *Tetrahedron Letters*, **30**, 2747 (1989).

[8] A. G. Godfrey and B. Ganem, *Am. Soc.*, **112**, 3717 (1990).

[9] N. Nikolaides and B. Ganem, *J. Org.*, **54**, 5996 (1989).

Rhodium(III) chloride.

Dimethylketene trimethylsilyl acetals.[1] Hydrosilylation of methacrylates with trimethylsilane catalyzed by rhodium(III) chloride results in rearrangement to dimethylketene trimethylsilyl acetals in 65–85% yield.

$$CH_2=C\begin{array}{c}CH_3\\\\COOR\end{array} + (CH_3)_3SiH \xrightarrow[65-85\%]{\substack{RhCl_3 \cdot 6H_2O \\ THF}} (CH_3)_2C=C\begin{array}{c}OSi(CH_3)_3\\\\OR\end{array}$$

[1] A. Revis and T. K. Hilty, *J. Org.*, **55**, 2972 (1990).

Rhodium(II) perfluorobutyrate, $Rh_2(pfb)_4$.

This reagent is obtained as a bright yellow-green solid by transesterification of $Rh_2(OAc)_4$ with perfluorobutyric acid and the anhydride.

Alcoholysis of R_3SiH.[1] Rhodium(II) perfluorobutyrate is more effective than $Rh_2(OAc)_4$ as a catalyst for reaction of primary or secondary alcohols with trialkylsilanes at 25° to form silyl ethers. Tertiary alcohols are inert under these conditions. Selective reactions with only primary alcohols can be realized with *t*-butyldimethylsilane but not with dimethylphenylsilane.

[1] M. P. Doyle, K. G. High, V. Bagheri, R. J. Pieters, P. J. Lewis, and M. M. Pearson, *J. Org.*, **55**, 6082 (1990).

Ruthenium(VIII) oxide (ruthenium tetroxide).

Catalytic oxidation of alkenes.[1] RuO_4 is useful for this reaction, but since it is expensive and toxic, it is usually generated *in situ* from a catalytic amount of $RuCl_3$ with $NaIO_4$ as the stoichiometric oxidant (**11**, 462–463). In another version of a catalytic reaction, O_2 is used as the primary oxidant for autoxidation of acetaldehyde to peracetic acid, which then oxidizes RuO_2 *in situ* to RuO_4. This RuO_2–CH_3CHO–O_2 system in acetone at 40° is particularly useful for oxidative cleavage of terminal alkenes and α,β-unsaturated carbonyl compounds.

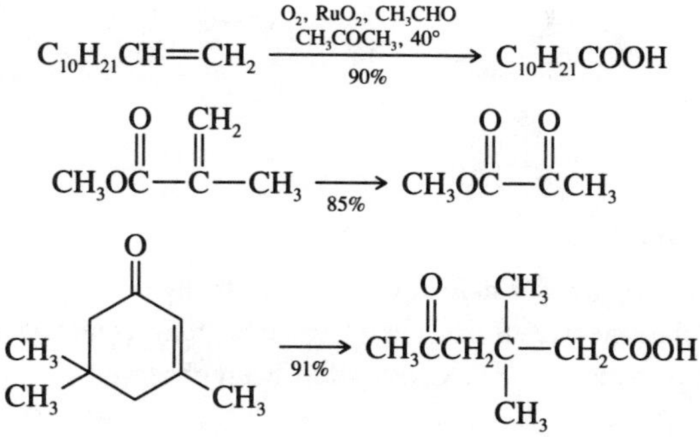

$$C_{10}H_{21}CH{=}CH_2 \xrightarrow[\substack{90\%}]{\substack{O_2,\ RuO_2,\ CH_3CHO \\ CH_3COCH_3,\ 40°}} C_{10}H_{21}COOH$$

$$\underset{\displaystyle CH_3O\overset{\displaystyle O}{\overset{\|}{C}}{-}\overset{\displaystyle CH_2}{\overset{\|}{C}}{-}CH_3}{} \xrightarrow[85\%]{} \underset{\displaystyle CH_3O\overset{\displaystyle O}{\overset{\|}{C}}{-}\overset{\displaystyle O}{\overset{\|}{C}}CH_3}{}$$

$$\xrightarrow[91\%]{} CH_3\overset{\displaystyle O}{\overset{\|}{C}}CH_2\underset{\displaystyle CH_3}{\overset{\displaystyle CH_3}{C}}{-}CH_2COOH$$

[1] K. Kaneda, S. Haruna, T. Imanaka, and K. Kawamoto, *J.C.S. Chem. Comm.*, 1467 (1990).

S

Samarium(II) iodide.

Pinacol coupling of dialdehydes.[1] 1,6-Dialdehydes, obtained by periodate cleavage of cyclohexane-1,2-diols, when treated with SmI_2 undergo coupling to *cis*-diols in good yield. Similar *cis*-selectivity has been observed in coupling with $TiCl_3$

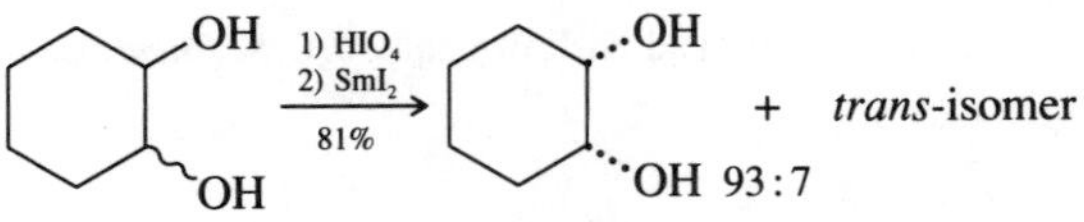

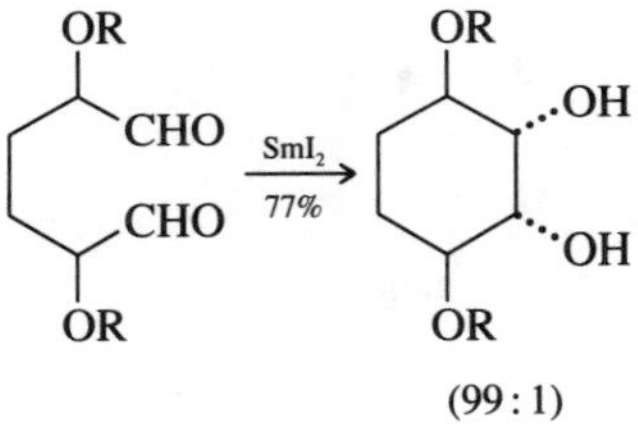

(99:1)

and Zn/Cu (**15**, 317). The presence of α-alkoxy groups can effect the orientation of the *cis*-diol in favor of the diol with an orientation opposite to that of the substituents. This pinacol coupling can be used to effect a diol inversion of carbohydrate derivatives as shown in equation (I).

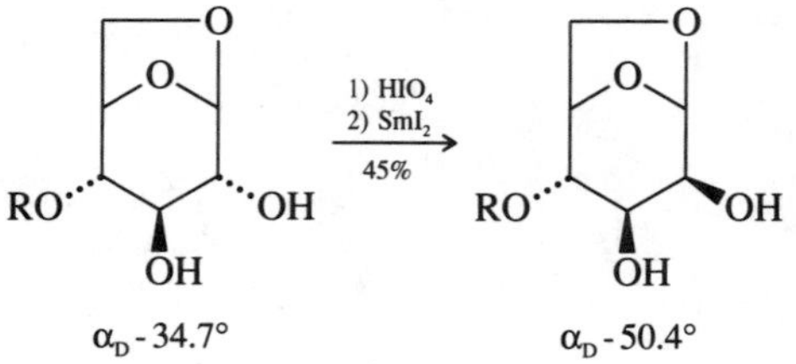

Reductive coupling of imines.[2] Aromatic ketimines are reduced to the corresponding secondary amines by SmI_2, but aromatic aldimines couple to 1,2-diamines (equation II).

anti-1,3-*Diol monoesters.*[3] SmI_2 can effect a Tishchenko-type reaction between a β-hydroxy ketone and an aldehyde to afford a monoester of an *anti*-diol in high

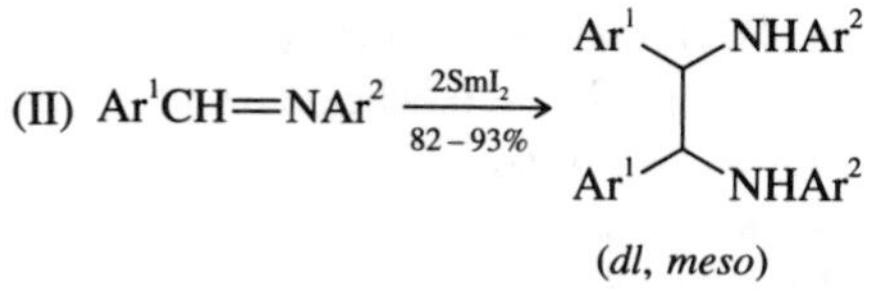

$$\text{(II) } Ar^1CH{=}NAr^2 \xrightarrow[82-93\%]{2SmI_2}$$

(dl, meso)

yield and diastereoselectivity (anti/syn, $>99:1$). The reaction probably involves intramolecular delivery of hydride from the aldehyde to the ketone via a complex of the hydroxy ketone and the aldehyde with the samarium catalyst. SmI_3 and $SmCl_3$ are less effective than SmI_2, which should be freshly prepared. Reduction of syn- and anti-α-methyl β-hydroxy ketones shows equally high asymmetric reduction.

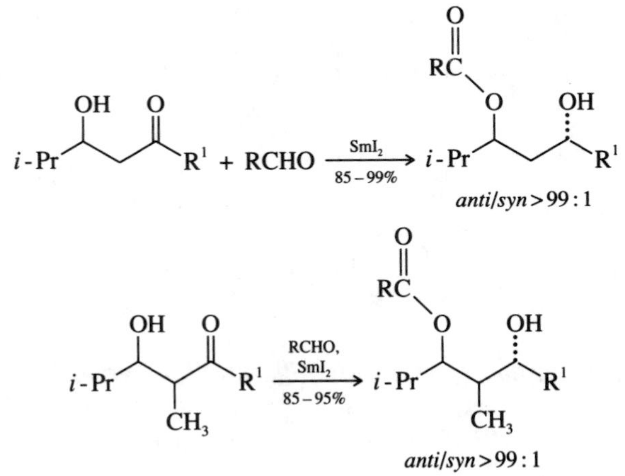

anti/syn $>99:1$

anti/syn $>99:1$

Pyranose → cyclopentanes.[4] One method for transformation of a carbohydrate to a carbocycle involves reaction of a lactol (1) with a Wittig reagent followed by

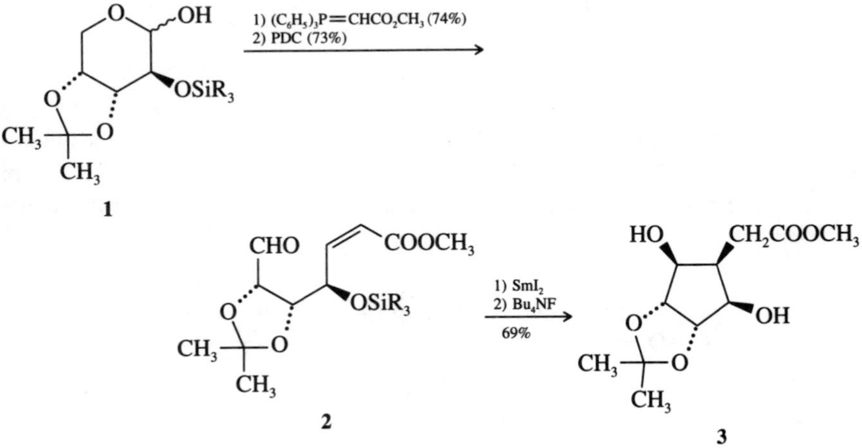

1) $(C_6H_5)_3P{=}CHCO_2CH_3$ (74%)
2) PDC (73%)

1

1) SmI_2
2) Bu_4NF

69%

2

3

oxidation to an aldehyde (2). Intramolecular cyclization of 2 induced with SmI_2 gives a single cyclopentane 3 (99:1). The stereochemistry of cyclization depends on the geometry of double bond. A *cis*-alkene favors a *syn*-product.

Coupling of ketones with alkenes.[5] This reaction can be effected with SmI_2 catalyzed by HMPA (14, 280–281). The alkene can be an activated terminal one (CH_2=CHOAc), a conjugated diene, or a silyl dienol ether.

$$C_6H_5(CH_2)_2\overset{\overset{\displaystyle O}{\|}}{C}CH_3 + CH_2{=}CHOAc \xrightarrow[\substack{HMPA, 25° \\ 62\%}]{SmI_2, THF,} C_6H_5(CH_2)_2\overset{\overset{\displaystyle CH_3}{\diagup}\,\overset{\displaystyle OH}{\diagdown}}{C}CH_2CH_2OAc$$

$$1$$

$$1 + CH_2{=}\overset{\overset{\displaystyle CH_3}{|}}{\underset{\underset{\displaystyle CH_3}{|}}{C}}CH{=}CH_2 \xrightarrow{99\%} C_6H_5(CH_2)_2\overset{CH_3\,OH}{C}CH_2\overset{CH_3}{\underset{\underset{\displaystyle CH_3}{|}}{CH}}CH{=}CH_2 + C_6H_5(CH_2)_2\overset{CH_3\,OH\,CH_3}{C}CH_2C{=}C(CH_3)_2$$

$$2:1$$

$$1 + \text{[anthracene]} \xrightarrow{95\%} \text{[product]}$$

Reductive cyclization of carbonyls with an alkyne group.[6] SmI_2 can effect coupling of aldehydes or ketones with an alkyne substituted by a phenyl or ethoxycarbonyl group. Addition of HMPA and *t*-butanol is required for satisfactory yields.

$$\underset{O}{\underset{\|}{(CH_2)_n}}{\diagdown}_R\,\text{—}C{\equiv}C\text{—}COOC_2H_5 \xrightarrow[60-72\%]{SmI_2} (CH_2)_n\overset{COOC_2H_5}{\diagdown}\underset{OH}{\overset{R}{\diagup}}$$

$$n = 1, 2, R = H, CH_3$$

$R^1R^2C{=}O \rightarrow R^1CH(CN)R^2$.[7] The cyanophosphates 1, readily obtained by reaction of ketones with diethyl phosphorocyanidate and LiCN (14, 187–188), are reduced by SmI_2 (excess) and *t*-BuOH (1 equiv.) in THF at 25° to nitriles in 80–

$$\underset{R^2}{\overset{R^1}{\diagdown}}C{=}O \xrightarrow[LiCN]{(C_2H_5O)_2P(O)CN} \underset{R^1\quad R^2}{\overset{NC\diagup\overset{\overset{\displaystyle O}{\|}}{OP(OC_2H_5)_2}}{\diagdown}} \xrightarrow[\substack{t\text{-BuOH} \\ 80-100\%}]{SmI_2} \underset{R^2}{\overset{R^1}{\diagdown}}CHCN$$

$$1$$

100% overall yield. The same sequence provides α,β-unsaturated nitriles from α,β-enones in 75–97% yield.

Deoxygenation of α-alkoxy(hydroxy) esters.[8] This deoxygenation can be effected by SmI_2-HMPA in THF with an alcohol as the proton source. The reaction is

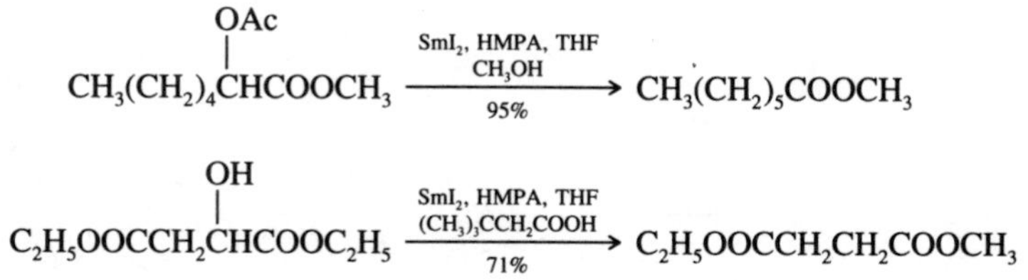

$$\underset{\overset{|}{OAc}}{CH_3(CH_2)_4CHCOOCH_3} \xrightarrow[\overset{CH_3OH}{95\%}]{SmI_2,\ HMPA,\ THF} CH_3(CH_2)_5COOCH_3$$

$$\underset{\overset{|}{OH}}{C_2H_5OOCCH_2CHCOOC_2H_5} \xrightarrow[\overset{(CH_3)_3CCH_2COOH}{71\%}]{SmI_2,\ HMPA,\ THF} C_2H_5OOCCH_2CH_2COOCH_3$$

particularly efficient with α-acetoxy and α-methoxy esters. Dehydroxylation of an α-hydroxy ester requires a carboxylic acid (pivalic acid) as the proton source.

Dehydroxylation of (R,R)-tartrates by this method gives only (R)-malates, but in this case ethylene glycol is the preferred proton source.

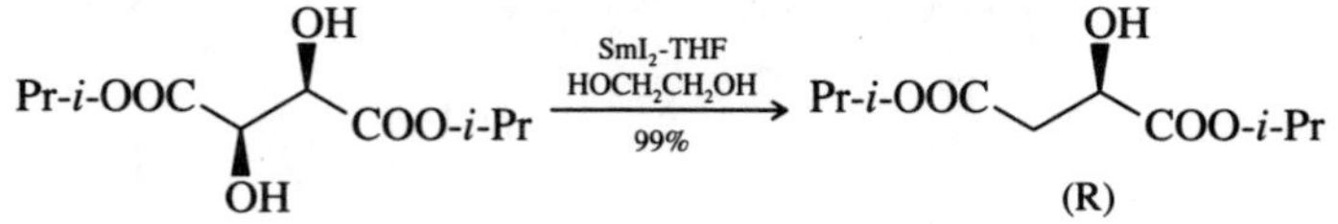

Alkene synthesis.[9] The key step in the Julia synthesis of alkenes (**11**, 473–475) involves reductive elimination of a β-hydroxy sulfone with sodium amalgam. A recent modification involves elimination of a β-hydroxy imidazolyl sulfone with SmI_2 (equation I).[1] Both syntheses are particularly useful for preparation of disubstituted alkenes and conjugated dienes and trienes. Both methods of elimination favor formation of (E)-alkenes. In a direct comparison, a higher yield was obtained with SmI_2 than with Na(Hg).

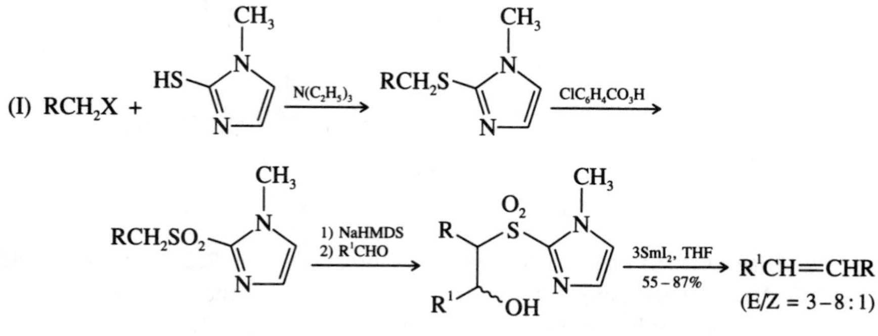

Aryl radical cyclization.[10] SmI$_2$ in HMPA/THF at 25° can effect cyclization of 1-allyloxy-2-iodobenzene to a Sm(III) intermediate (**a**) that can be trapped by electrophiles, including aldehydes or ketones. The report suggests that a similar mechanism operates in the Barbier-type coupling: generation of an alkyl radical followed by formation of RSmI$_2$, which adds to a carbonyl compound to form an adduct that is hydrolyzed to an alcohol.

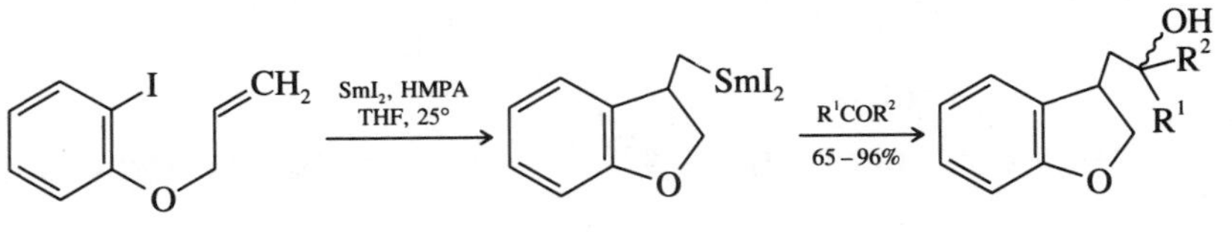

α,α′-Dihydroxy ketones.[11] Reaction of benzyl chloromethyl ether, 2,6-xylyl isocyanide, and SmI$_2$ in THF/HMPA at −15° results in an intermediate (**a**) that reacts with aldehydes or ketones to form derivatives of α,α′-dihydroxy imines such as **1**. These can be converted to α,α′-dihydroxy ketones by hydrolysis. This method was used to convert a protected D-glyceraldehyde (**2**) to D-ribulose (**3**), page 299.

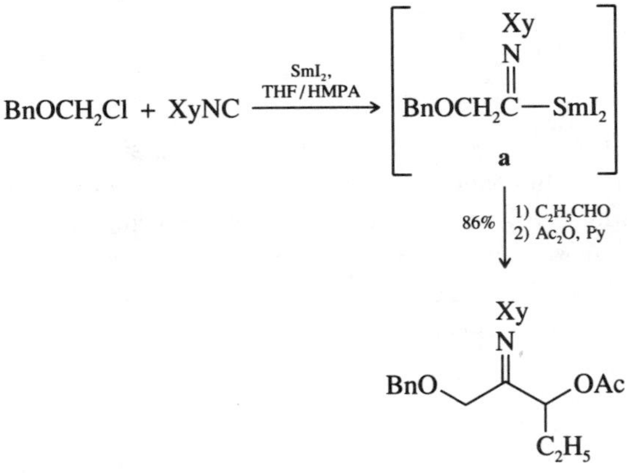

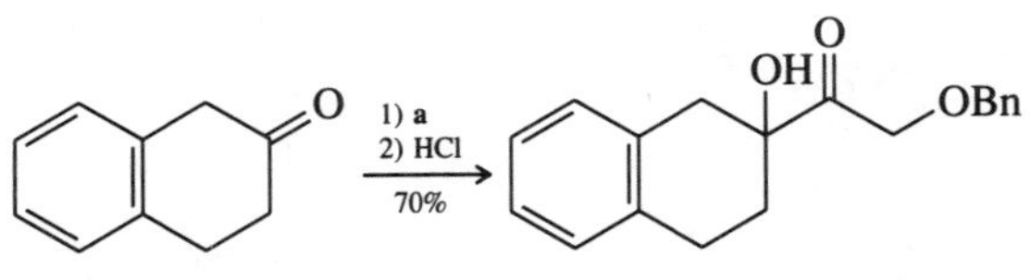

vic-Diketones.[12] The reaction of ethyl bromide with 2,6-xylyl isocyanide (XyNC, 2 equiv.) and SmI$_2$ (2.5 equiv.) in THF/HMPA results in an intermediate, formulated as **a**, that is hydrolyzed by H$_2$O at 0° to a diimine (**2**). The intermediate also reacts with aldehydes or ketones to provide a hydroxy diimine such as **3**. The intermediate (**a**) even reacts with esters to provide acylated diimines, which can be hydrolyzed to 1,2,3-tricarbonyl compounds (probably as the hydrate).

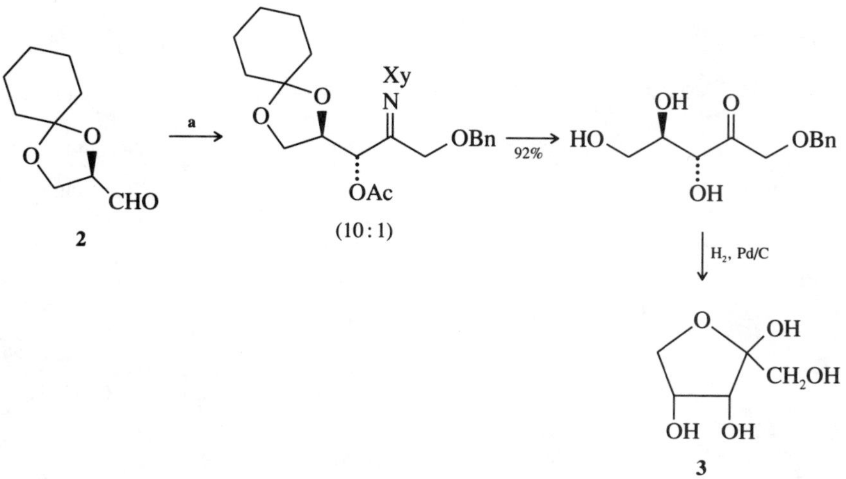

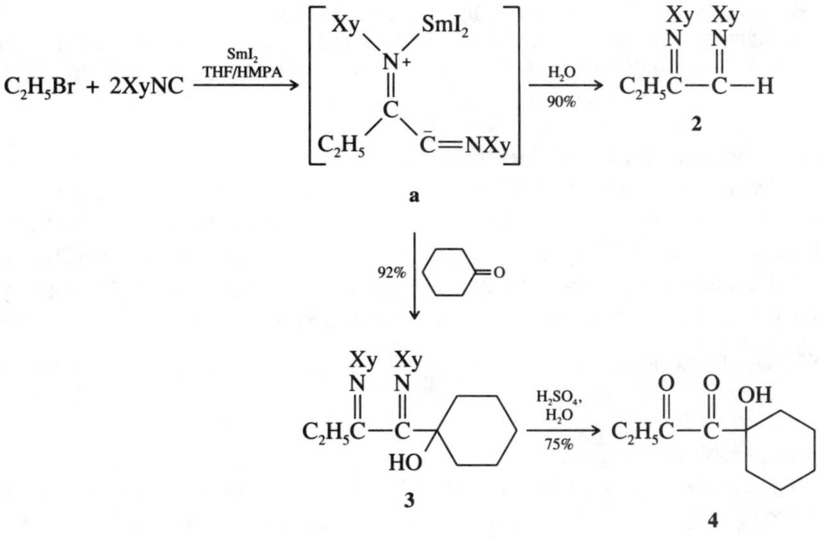

Review.[13] Soderquist has reviewed the uses of SmI_2 for synthesis (51 references).

$$\text{a} + CH_3COCOOC_2H_5 \xrightarrow[63\%]{} \overset{\overset{Xy}{\underset{\|}{N}}}{C_2H_5C}-\overset{\overset{Xy}{\underset{\|}{N}}}{C}-\overset{\overset{O}{\underset{\|}{}}}{COCH_3} \xrightarrow[H_2O]{H_2SO_4,} \overset{\overset{O}{\underset{\|}{}}}{C_2H_5C}-\overset{\overset{O}{\underset{\|}{}}}{C}-\overset{\overset{O}{\underset{\|}{}}}{COCH_3}$$

[1] J. L. Chiara, W. Cabri, and S. Hanessian, *Tetrahedron Letters*, 1125 (1991).
[2] T. Imamoto and S. Nishimura, *Chem. Letters*, 1141 (1990).
[3] D. A. Evans and A. H. Hoveyda, *Am. Soc.*, **112**, 6447 (1990).
[4] E. J. Enholm and A. Trivellas, *ibid.*, **111**, 6463 (1989).
[5] O. Ujikawa, J. Inanaga, and M. Yamaguchi, *Tetrahedron Letters*, **30**, 2837 (1989).
[6] S. C. Shim, J.-T. Hwang, H.-Y. Kang, and M. H. Chang, *ibid.*, **31**, 4765 (1990).
[7] R. Yoneda, S. Harusawa, and T. Kurihara, *ibid.*, **30**, 3681 (1989).
[8] K. Kusuda, J. Inanaga, and M. Yamaguchi, *ibid.*, **30**, 2945 (1989).
[9] A. S. Kende and J. S. Mendoza, *ibid.*, **31**, 7105 (1990).
[10] D. P. Curran, T. L. Fevig, and M. J. Totleben, *Synlett*, 773 (1990).
[11] M. Murakami, T. Kawano, and Y. Ito, *Am. Soc.*, **112**, 2437 (1990).
[12] M. Murakami, H. Masuda, T. Kawano, H. Nakamura, and Y. Ito, *J. Org.*, **56**, 1 (1991).
[13] J. A. Soderquist, *Aldrichim. Acta*, **24**, 15 (1991).

Samarium(III) chloride–Chlorotrimethylsilane

Cleavage of acetals.[1]   A variety of acetals, 1,3-dioxolanes, and even a 1,3-dioxane, can be cleaved by reaction with $SmCl_3$ and $ClSi(CH_3)_3$ ($\sim 1:1$) in CH_2Cl_2, THF, or CH_3CN. The effective reagent is considered to be $(CH_3)_3Si^+SmCl_4^-$. Dimethylacetals can be cleaved by use of only a catalytic amount of $SmCl_3$ combined with 1.5–2 equiv. of the silane. Yields are generally above 80%.

[1] Y. Ukaji, N. Koumoto, and T. Fujisawa, *Chem. Letters*, 1623 (1989).

Silver perchlorate, $AgClO_4$.

Cyclization of hydroxy dithioketals.   Nicolaou *et al.*[1] have used this reaction for construction of oxocenes. The most satisfactory method involves activation of the sulfur of the starting material (1) with $AgClO_4$ to produce an oxocene derivative (2), followed by C–S cleavage. This second step can be effected with $(C_6H_5)_3SnH$ (AIBN) or by oxidation to the corresponding sulfoxide or sulfone, which then is cleaved with

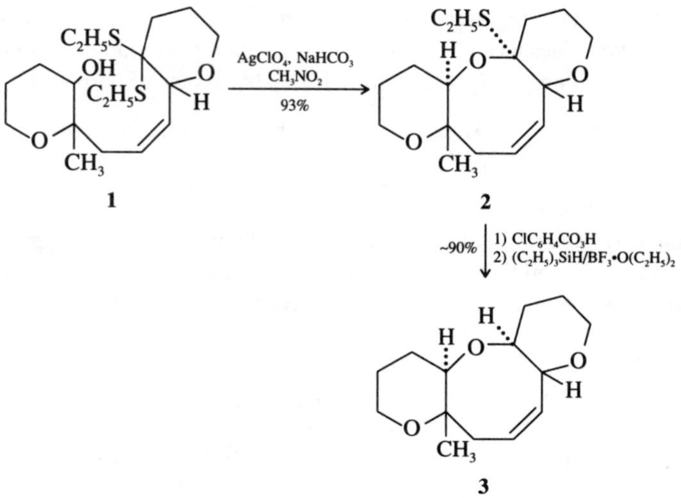

1

2

~90% | 1) ClC$_6$H$_4$CO$_3$H
       | 2) (C$_2$H$_5$)$_3$SiH/BF$_3$·O(C$_2$H$_5$)$_2$

3

(C$_2$H$_5$)$_3$SiH/BF$_3$·O(C$_2$H$_5$)$_2$. The presence of a *cis*-double bond is an essential for this cyclization. Another limitation is that yields are low when extended to synthesis of a nine-membered oxocene.

[1] K. C. Nicolaou, C. V. C. Prasad, C.-K. Hwang, M. E. Duggan, and C. A. Veale, *Am. Soc.*, **111**, 5321 (1989).

Silver trifluoroacetate, AgOCOCF$_3$ (1).

Regiocontrolled alkylation.[1] The usual Lewis acids are ineffective for alkylation of 2-(trimethylsilyloxy)furan (2), but use of several silver salts, of which 1 is the

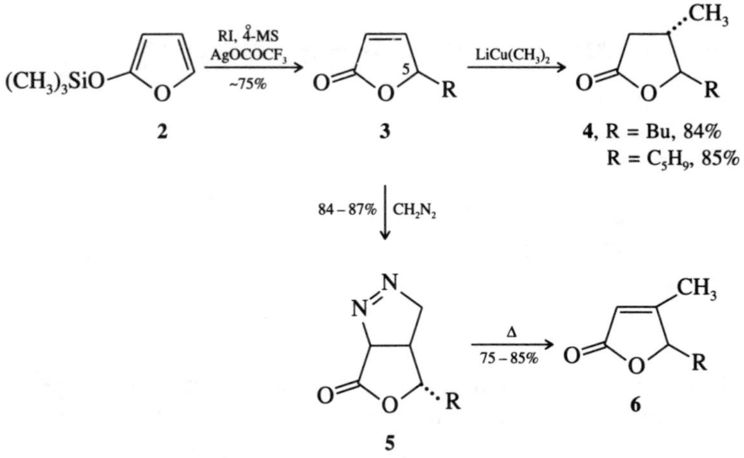

2

3

4, R = Bu, 84%
   R = C$_5$H$_9$, 85%

84–87% | CH$_2$N$_2$

5

6

most useful, results in regioselective alkylation of C_5 to give the alkylbutenolides **3**. The dialkylbutenolides **6** can be obtained by 1,3-dipolar addition of diazomethane to **3** to give **5**, which on pyrolysis provides 4,5-dialkylfuran-2(5H)-ones **6**.

[1] C. W. Jefford, A. W. Sledeski, and J. Boukouvalas. *Helv.*, **72**, 1362 (1989).

Silver trifluoromethanesulfonate.

Amino cyclization.[1]   In the presence of AgOTf or AgBF$_4$ allenic amines cyclize to 2-substituted pyrrolidines. A stereogenic center adjacent to N can induce asymmetric induction, which is dependent on the concentration of Ag(I).

(80% de)

[1] D. N. A. Fox, D. Lathbury, M. F. Mahon, K. C. Molloy, and T. Gallagher, *J.C.S. Chem. Comm.*, 1073 (1989).

O-Silyl ketene N,O-acetals.

Aldol condensation.[1]   These O-silyl enol derivatives of amides are available by hydrosilylation of α,β-unsaturated amides catalyzed by Wilkinson's catalyst. A typical reagent of this type, **1**, reacts with aldehydes in the absence of a catalyst to form aldol adducts (**2**) with unusual *anti*-selectivity. This silyl aldol reaction can be ex-

tended to a synthesis of an optically active *anti*-aldol. Thus the (S)-prolinol propionamide (**10**, 332) on metallation (LDA) and reaction with (CH$_3$)$_2$SiCl$_2$ affords the bicyclic siloxane **3** a single product. This cyclic N,O-acetal reacts with aldehydes to form essentially only one enantiomerically pure *anti*-aldol **4**. The high stereoselectivity is attributed to interaction between the Si and the amide carbonyl group resulting in a high-energy transition state with hypervalent silicon.

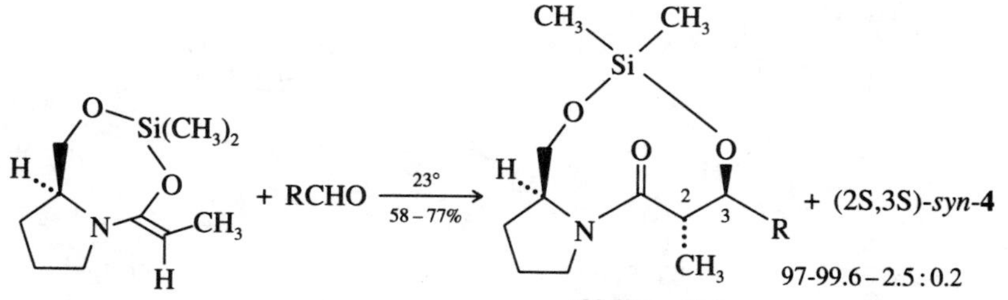

[1] A. G. Myers and K. L. Widdowson, *Am. Soc.*, **112**, 9672 (1990).

Sodium–Ammonia.

Reductive decyanation.[1] This reaction is a key step in a route to *syn*-1,3-diol acetonides from β-trimethylsilyloxy aldehydes (**1**). Reaction of **1** with trimethylsilyl cyanide followed by acetonation gives a 1:1 mixture of a protected cyanohydrin (**2**). This mixture is converted into a single isomer (**3**) on alkylation of the anion of the cyanohydrin acetonide. Reductive decyanation with Na–NH₃ at −78° produces a *syn*-diol acetonide (**4**). The apparent retention of configuration in the reduction results from preferential formation of an intermediate axial anion.

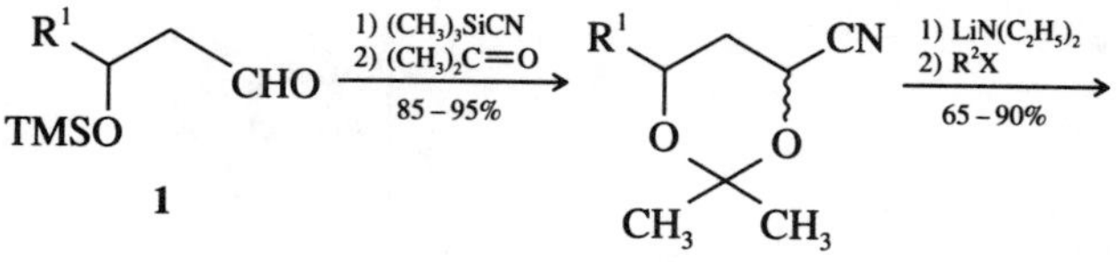

This reaction can be extended to synthesis of alternating polyol chains found in polyene macrolide antibiotics. Thus reaction of the dibromide **5** with 2 equiv. of the cyanohydrin anion of **6** provides, after reductive cyanation, the protected polyol **7** in good yield.

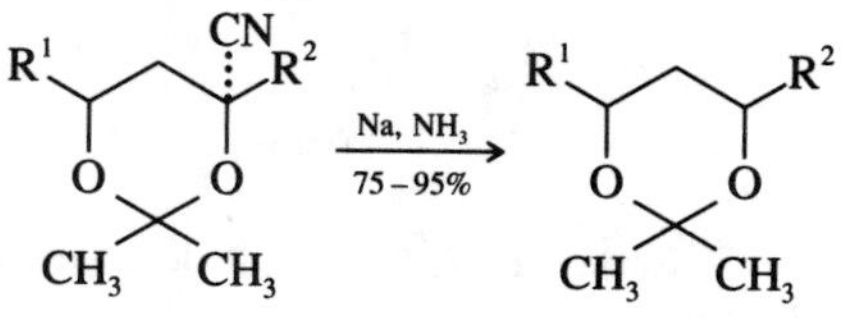

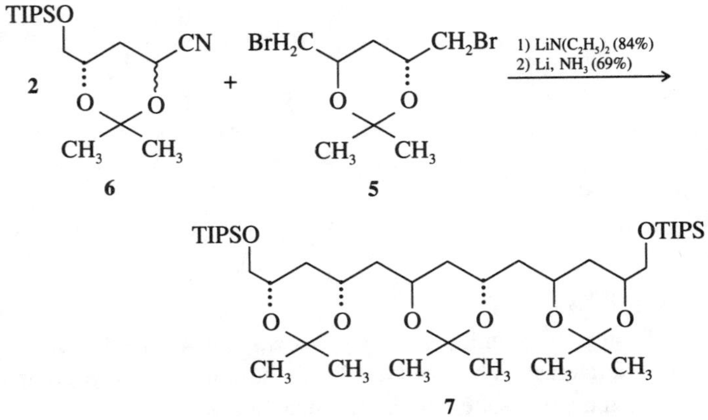

[1] S. D. Rychnovsky, S. Zeller, D. J. Skalitzky, and G. Griesgraber, *J. Org.*, **55**, 5550 (1990).

Sodium borohydride.

Stereoselective reduction. A new synthesis of (3S,4S)-statine (**4**) from N,N-dibenzyl-D-valine (**1**) depends on reduction of a β-keto ester (**2**) with sodium borohydride with nonchelation control owing to the adjacent N,N-dibenzylamino group.

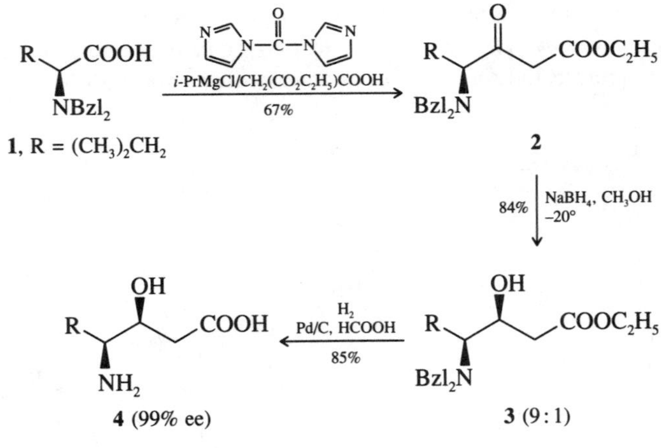

[1] M. T. Reetz, M. W. Drewes, B. R. Matthews, and K. Lennick, *J.C.S. Chem. Comm.*, 1474 (1989).

Sodium borohydride–(L)-Tartaric acid.

Enantioselective reduction of ketones.[1] Sodium borohydride aged with L-tartaric acid can effect enantioselective reduction of ketones bearing an α-substituent capable of chelation. An α-methoxy group is most effective (84% ee). Similar enantioselective reduction of α- and β-keto esters (81–85% ee) is possible.

[1] M. Yatagai and T. Ohnuki, *J.C.S. Perkin I*, 1826 (1990).

Sodium cyanoborohydride.

Reductive amination.   Conversion of ketones or aldehydes to amines is usually accomplished by reduction of the carbonyl compound with sodium cyanoborohydride in the presence of an amine (Borch reduction, **4**, 448–449). However, yields are generally poor in reactions of hindered or acid-sensitive ketones, aromatic amines, or trifluoromethyl ketones. Yields can be improved markedly by treatment of the ketone and amine first with $TiCl_4$[1] or $Ti(O\text{-}i\text{-}Pr)_4$[2] in CH_2Cl_2 or benzene to form the imine or enamine and then with $NaCNBH_3$ in CH_3OH to effect reduction. Note that primary amines can be obtained by use of hexamethyldisilazane as a substitute for ammonia (last example).

Piperidine synthesis.[3]   The important glycohydrolase inhibitor 1-deoxynojirimycin (**3**) can be prepared in two steps from 5-keto-D-glucose (**1**). Thus **1** on treatment with benzhydrylamine and sodium cyanoborohydride in CH_3OH at 0° undergoes double reductive amination to form essentially only one product (**2**), which on deprotection provides the piperidine **3**. The high stereoselectivity is attributed to hydroxyl-directed hydride delivery.

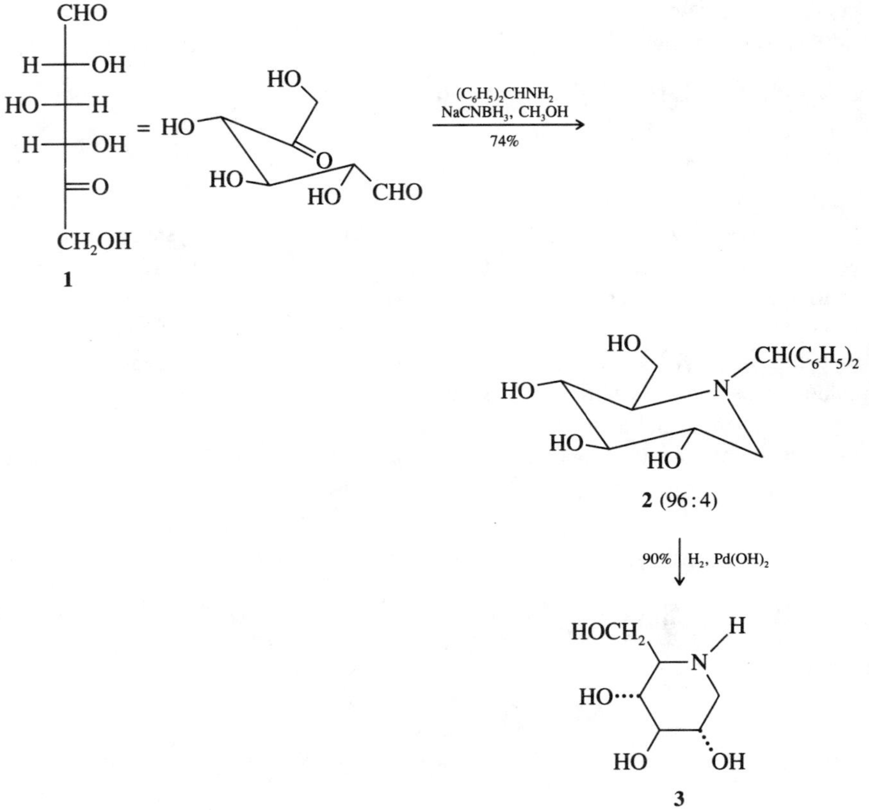

A similar double-reductive amination of 5-keto-D-fructose provides a route to pyrrolidines (equation I). This reductive amination also shows some stereoselectivity for the glucitol isomer.

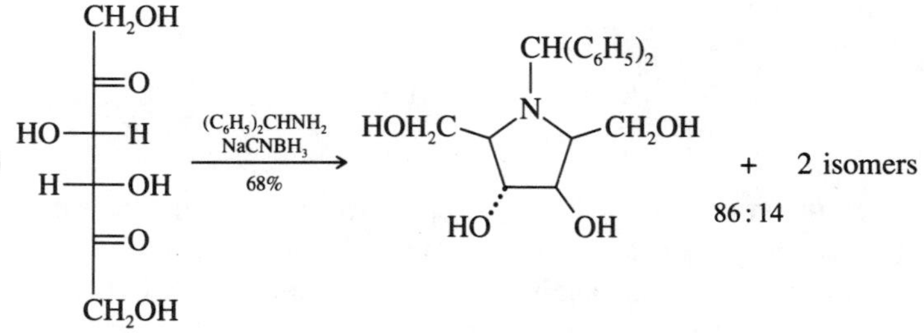

[1] C. L. Barney, E. W. Huber, and J. R. McCarthy, *Tetrahedron Letters*, **31**, 5547 (1990).
[2] R. J. Mattson, K. M. Pham, D. J. Leuck, and K. A. Cowen, *J. Org.*, **55**, 2552 (1990).
[3] A. B. Reitz and E. W. Baxter, *Tetrahedron Letters*, **31**, 6777 (1990).

Sodium hexamethyldisilazide, $NaN[Si(CH_3)_3]_2$.

Deprotonation of ketones.[1]   This amide is more useful than the corresponding lithium amide or LDA for preparation of (Z)-enolates under thermodynamic control.

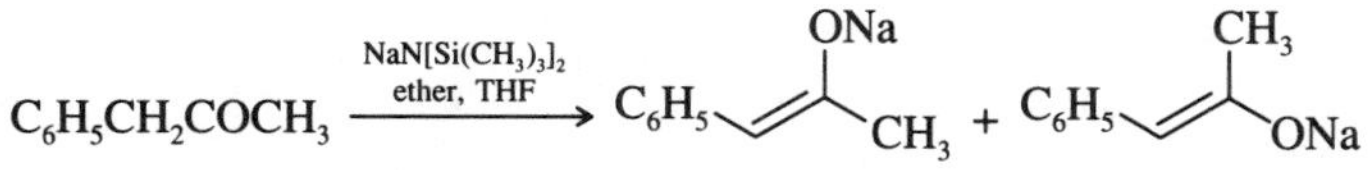

	(Z)	(E)
$-100°$	15%	85%
$22°$	100%	0

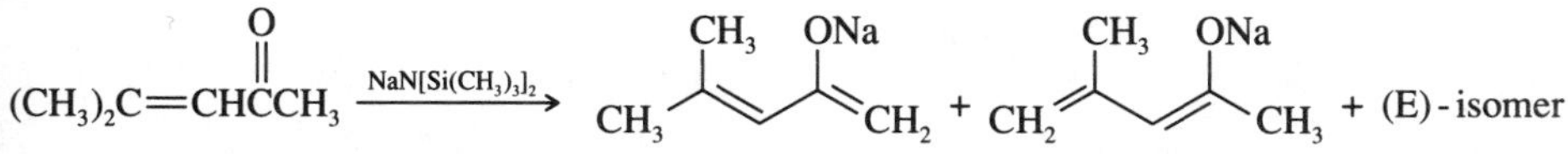

	(Z)	
$-70°$	65%	35%	0
$20°$	0	95%	5

[1] M. Gaudemar and M. Bellassoued, *Tetrahedron Letters*, **30**, 2779 (1989).

Sodium hydride.

Coupling with homophthalic anhydrides (**12**, 448).[1]   This reaction has been used for the first asymmetric synthesis of $(-)$-γ-rhodomycinone (**4**) by coupling of a chiral AB unit (**1**) with 4-acetoxy-5-methoxyhomophthalic anhydride (**2**), a precursor to the CD-unit. The product (**3**) is converted into **4** by deacetylation (CF_3COOH, 93% yield) and rearrangement of the quinone group ($AlCl_3$, 66% yield). The high regioselectivity of the base-induced coupling is ascribed to chelation between secondary hydroxyl group of **1** and the carbonyl group at C_1 of **2**.

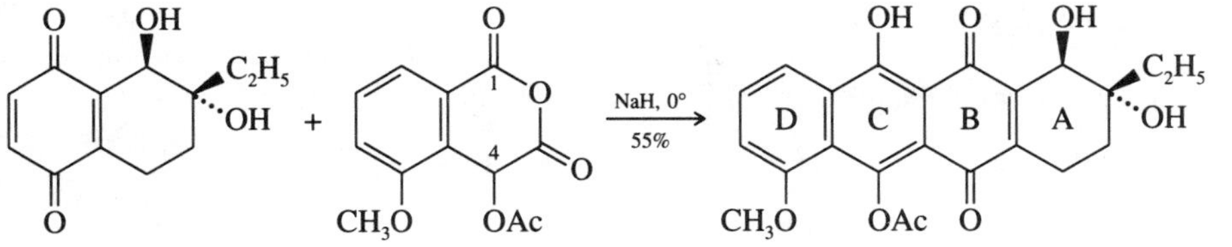

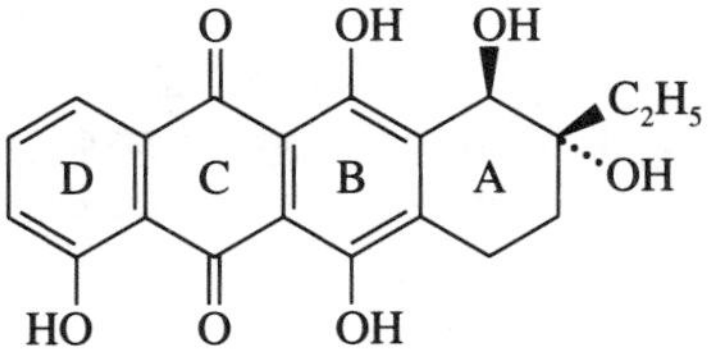

Dehydrohalogenation of 6-halohexopyranosides.[2] This reaction has been effected with DBU or AgF, but NaH in DMF at 0–50° is as efficient. Hydroxyl, azide, benzyl, and ester groups are stable under these conditions. Yields of 5,6-hexenopyranosides are 60–83%.

[1] H. Fujioka, H. Yamamoto, H. Kondo, H. Annoura, and Y. Kita, *J.C.S. Chem. Comm.*, 1509 (1989).
[2] F. Chrétien, *Syn. Comm.*, **19**, 1015 (1989).

Sodium hydride–Sodium *t*-amyl oxide–Nickel acetate (NiCRA, **10**, 365; **14**, 288).

NiCRA-bpy.[1] A NiCRA complex containing 2,2'-bipyridine (bpy) effects homocoupling of aryl halides (Ullmann coupling), often in high yield, which can be improved in some cases by addition of KI or NaI. When used in a catalytic amount, reduction to an alkane is the main side reduction.

[1] M. Lourak, R. Vanderesse, Y. Fort, and P. Caubère, *J. Org.*, **54**, 4840 (1989); M. Lourak, Y. Fort, R. Vanderesse, and P. Caubère, *ibid.*, **54**, 4848 (1989).

Sodium hypochlorite, NaOCl.

RCH(NH₂)COOH → RCHO.[1] Tryptophane is oxidized to indole-3-acetaldehyde by Chlorox® in C_6H_6 at 50–55°. The crude aldehyde thus obtained (33–60% yield) is purified via a bisulfite adduct. The overall yield of pure product is 60%. Careful control of the pH to >7.7 is essential since the substrate is unstable to base.

[1] R. A. Gray and W. M. Welch, *Org. Syn.*, submitted (1989).

Sodium nitrite, $NaNO_2$.

RCH₂NH₂ → RCH₂OH. One method for this conversion is the thermal rearrangement of N-nitrosoamides derived from RCH_2NH_2. An improved procedure is based on the observation that the N-nitrosoamides derived from a trihaloacetamide of RCH_2NH_2 can rearrange at 0° in the presence of HOAc.[1]

$$RCH_2NH_2 \xrightarrow{CCl_3COCl} RCH_2NHCOCl_3 \xrightarrow[70-90\%]{\substack{1)\ NaNO_2,\ HOAc,\ Ac_2O,\ 0° \\ 2)\ OH^-}} RCH_2OH$$

[1] N. Nikolaides and B. Ganem, *J. Org.*, **54**, 5996 (1989).

Sodium phenylselenotriethoxyborate, $Na^+[C_6H_5SeB(OCH_2CH_3)_3]^-$ (1); prepared by reduction of diphenyl diselenide with $NaBH_4$ in ethanol.

Reduction.[1] α,β-Epoxy ketones are selectively reduced to β-hydroxy ketones, even when the substrate (2) also contains an enone group. Reduction of 2 with Zn/Cu also results in the same product (3), but in low yield as well as a number of products including the fully saturated ketone 4.

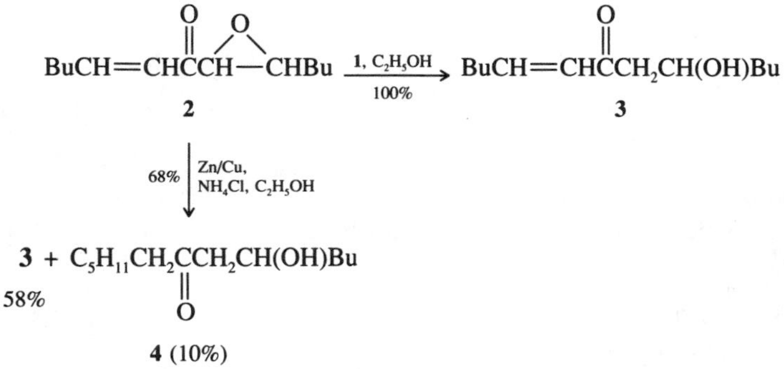

[1] M. Miyashita, T. Suzuki, and A. Yoshikoshi, *Tetrahedron Letters*, **30**, 1819 (1989).

Sodium tetrachloropalladate(II), Na_2PdCl_4.

Ethoxycarbonylation.[1]   Allylic bromides (chlorides) undergo carbonylation in ethanolic sodium ethoxide when catalyzed by Na_2PdCl_4 in combination with bis-(diphenylphosphine)ethane, dppe, to provide β,γ-unsaturated esters in 70–95% yield.

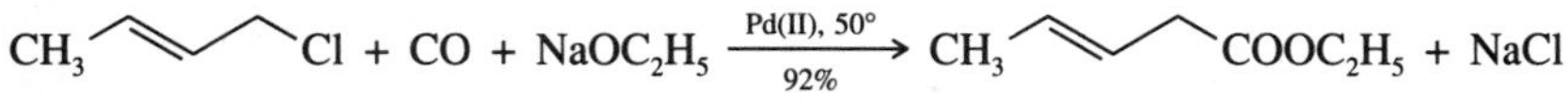

[1] J. Kiji, T. Okano, H. Konishi, and W. Nishiumi, *Chem. Letters*, 1873 (1989).

Sodium triacetoxyborohydride, $NaBH(OAc)_3$.

Reductive amination of carbonyls.[1]   This borohydride is generally superior to sodium cyanoborohydride for reductive aminations with weakly basic amines.

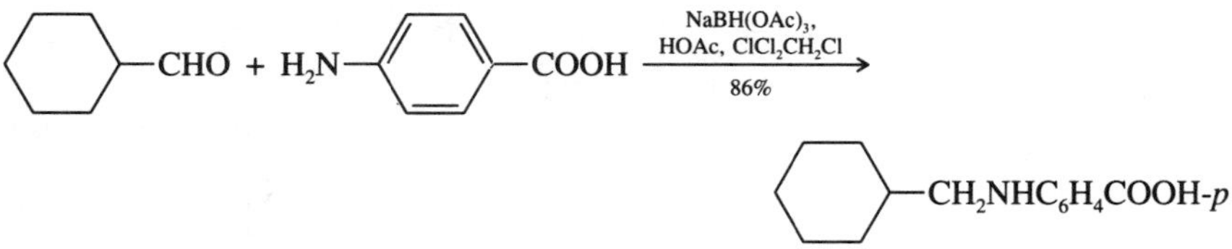

[1] A. F. Abdel-Magid and C. A. Maryanoff, *Syn Lett*, 537 (1990); A. F. Abdel-Magid, C. A. Maryanoff, and K. G. Carson, *Tetrahedron Letters*, **31**, 5595 (1990).

Sodium perborate, $NaBO_3 \cdot 4H_2O$.

Oxidation of organoboranes.[1] This oxidation is generally conducted with H_2O_2 (30%) and 3 *N* NaOH at 50°. Comparable yields can be obtained by use of this milder oxidant, which is less expensive, more stable, and far safer than H_2O_2. The yield of the reaction shown in equation (I) is 35% when H_2O_2 is used in place of sodium borate.

$$
\text{(I)} \quad ClCH_2\overset{\overset{\displaystyle CH_3}{|}}{C}=CH_2 \xrightarrow[\underset{83\%}{}]{\begin{array}{l}1)\ BH_3\cdot S(CH_3)_2\\ 2)\ NaBO_3\cdot 4H_2O,\ RT,\ 2\ hr\end{array}} ClCH_2\overset{\overset{\displaystyle CH_3}{|}}{C}HCH_2OH
$$

Hydroxylation of arenes.[2] Sodium perborate in combination with trifluoromethanesulfonic acid is an attractive reagent for electrophilic oxidation of arenes to phenols.

[1] G. W. Kabalka, T. M. Shoup, and N. M. Goudgaon, *Tetrahedron Letters*, **30**, 1483 (1989); *J. Org.*, **54**, 5930 (1989).
[2] G. K. S. Prakash, N. Krass, Q. Wang, and G. A. Olah, *Synlett*, 39 (1991).

N-Sulfonyl-1-aza-1,3-butadienes.

Diels–Alder reactions with electron-rich alkenes.[1] Simple α,β-unsaturated imines (1-aza-1,3-butadienes) do not undergo Diels–Alder reactions with dienophiles. In contrast, the N-phenylsulfonyl imines derived from an aldehyde or ketone undergo Diels–Alder reactions under forcing conditions with electron-rich dienophiles to

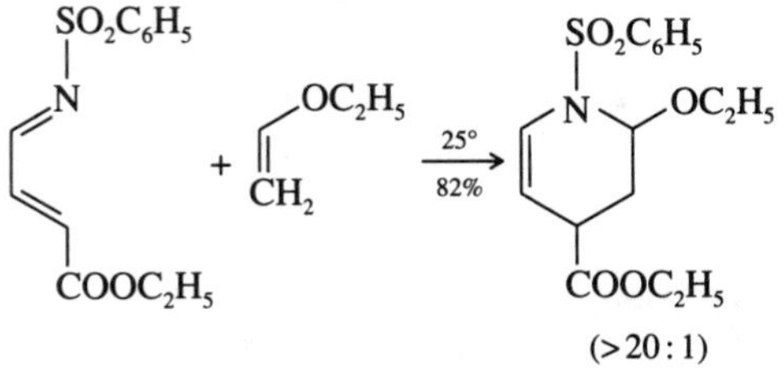

(>20:1)

provide usually a single product (*endo*) with preservation of the alkene geometry. Thus this reaction can provide a diastereoselective route to 1,2,3,4-tetrahydropyridines. The addition of an electron-withdrawing group ($COOC_2H_5$) at the 2- or 4-position of the diene accelerates the rate of [4 + 2]cycloaddition without diminishing the *endo*-selectivity of the parent azadienes.

[1] D. L. Boger, W. L. Corbett, T. T. Curran, and A. M. Kasper, *Am. Soc.*, **113**, 1713 (1991).

Sulfuryl chloride.

α-Keto-β-lactams.[1] 3-(Phenylthio)-2-azetidinones (**1**) undergo a Pummerer-type reaction with SO_2Cl_2 to provide 3-chloro-3-(phenylthio)-β-lactams (**2**) in yields of 78–95%. These are hydrolyzed to the 2,3-diones (**3**) in 85–90% yield by moist silica gel and $ZnCl_2$ (catalyst).

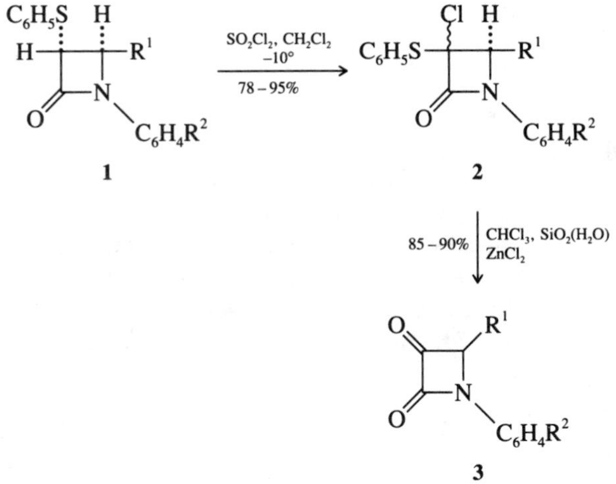

[1] J. M. van der Veen, S. S. Bari, L. Krishnan, M. S. Manhas, and A. K. Bose, *J. Org.*, **54**, 5758 (1989).

T

Tantalum(V) chloride–Zinc.

Trisubstituted allylic alcohols. A low-valent tantalum prepared by reduction of $TaCl_5$ with Zn in DME/benzene adds to alkynes to form a complex that reacts with aldehydes to form (E)-allylic alcohols. The regioselectivity is determined by the bulkiness of the groups on the alkyne and of the R group of the aldehyde.

$$n\text{-}C_5H_{11}\text{—}C\equiv C\text{—}C_5H_{11}\text{-}n \xrightarrow{\text{TaCl}_5,\ \text{Zn}} \left[\begin{array}{c} n\text{-}C_5H_{11} \quad\quad C_5H_{11}\text{-}n \\ C{=}C \\ TaL_n \end{array} \right]$$

96% ↓ 1) $C_6H_5(CH_2)_2CHO$
　　　2) NaOH, H_2O

$$n\text{-}C_5H_{11} \quad\quad C_5H_{11}\text{-}n$$
$$H \quad\quad (CH_2)_2C_6H_5$$
$$HO$$

[1] K. Takai, Y. Kataoka, and K. Utimoto, *J. Org.*, **55**, 1707 (1990).

(R,R)-Tartaric acid.

Addition of $(C_2H_5)_2Zn$ to RCHO.[1] The diol **1**, prepared by Barbier addition of C_6H_5MgBr (2 equiv.) to the acetonide of dimethyl (R,R)-tartrate,[2] converts $Ti(OC_2H_5)_4$ into the optically active spirotitanate **2**. In the presence of 0.05–2.0 equiv. of **2**, diethylzinc reacts with anisaldehyde in toluene at 0° to form the (R)-alcohol **3** (equation I). The enantioselectivity and the chemical yield increases with an increase in **2**. Surprisingly, the enantioselectivity is reversed in reactions of the

1, m.p. 192°, $\alpha_D = 68.5°$

$+ \ Ti(OC_2H_5)_4 \longrightarrow$

2

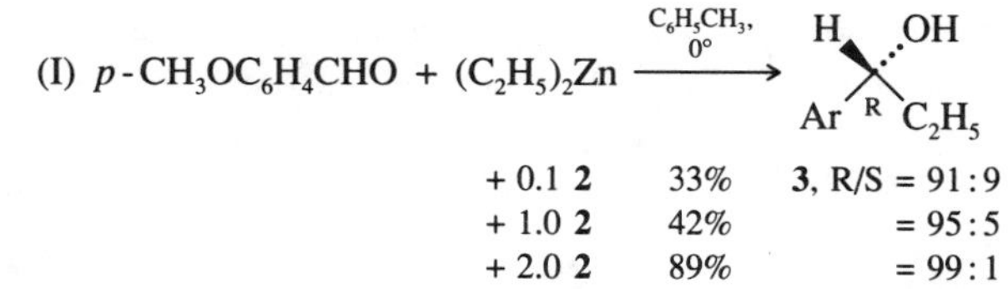

(I) p-CH3OC6H4CHO + (C2H5)2Zn
C6H5CH3, 0°
H OH
Ar R C2H5
+ 0.1 2 33% 3, R/S = 91:9
+ 1.0 2 42% = 95:5
+ 2.0 2 89% = 99:1

aldehyde with diethylzinc in the presence of 0.1 equiv. of **2** by addition of 1.2 equiv. of titanium(IV) isoproxide. Thus the selectivity induced by **2** can be reversed rather than diluted by addition of an achiral titanate (equation II). Furthermore, in these reactions even ether or THF can be used in place of toluene.

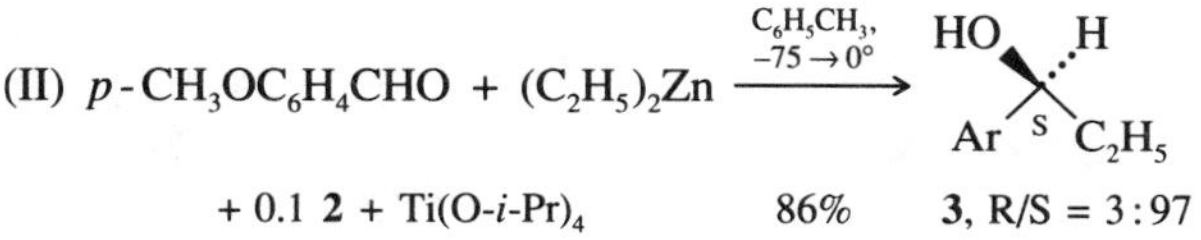

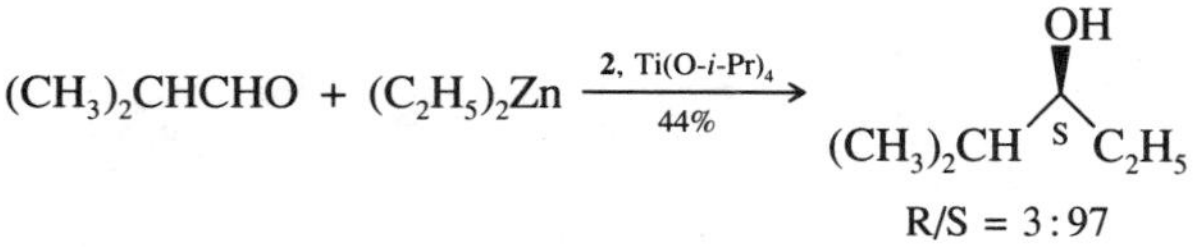

(II) p-CH3OC6H4CHO + (C2H5)2Zn
C6H5CH3, −75 → 0°
HO H
Ar S C2H5
+ 0.1 2 + Ti(O-i-Pr)4 86% 3, R/S = 3:97

(CH3)2CHCHO + (C2H5)2Zn
2, Ti(O-i-Pr)4
44%
OH
(CH3)2CH S C2H5
R/S = 3:97

In addition, the chiral titanate **2** can effect enantioselective addition of CH_3Li or CH_3MgBr to aryl aldehydes to provide (R)-alcohols (equation III).

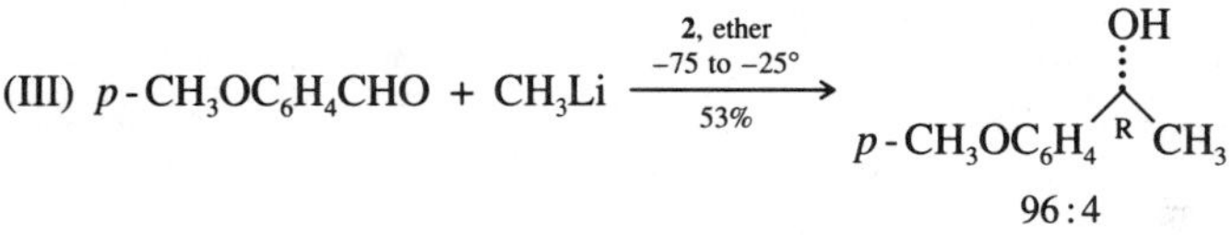

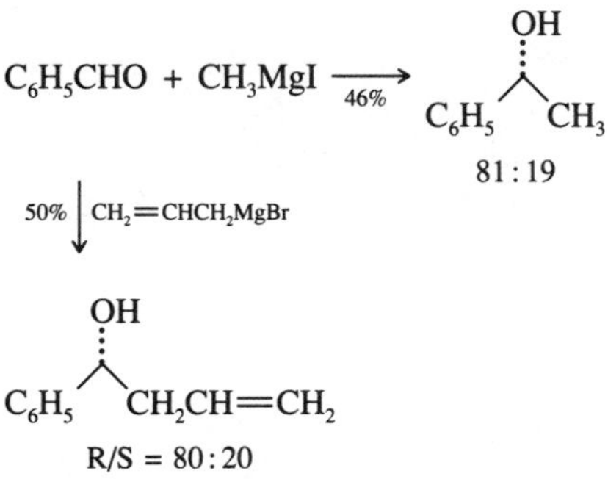

(III) p-CH3OC6H4CHO + CH3Li
2, ether
−75 to −25°
53%
OH
p-CH3OC6H4 R CH3
96:4

C6H5CHO + CH3MgI
46%
OH
C6H5 CH3
81:19
50% CH2=CHCH2MgBr
OH
C6H5 CH2CH=CH2
R/S = 80:20

Seebach attributes the high efficiency of **2** for enantioselective reactions to the hydroxy(diphenyl)methyl group, which is present in several chiral auxiliaries.

Asymmetric Diels–Alder reactions.[3] The observation that simple acyloxyboranes such as $H_2BOCOCH=CH_2$, prepared by reaction of BH_3 with acrylic acid, can serve as Lewis acid catalysts for reactions of the α,β-unsaturated acids with cyclopentadiene (15, 2) has been extended to the preparation of chiral acyloxyboranes derived from tartaric acid. The complex formulated as 3, prepared by reaction of BH_3 with the monoacylated tartaric acid 2, catalyzes asymmetric Diels–Alder reactions of α,β-enals with cyclopentadiene with high enantioselectivity. The process is applicable to various dienes and aldehydes with enantioselectivities generally of 80–97% ee.

(2R, 3R) - 2   3

96% ee
exo/endo = 9 : 1

Asymmetric aldol reactions.[4]   The borane complex 3 can also serve as the Lewis acid catalyst for the aldol reaction of enol silyl ethers with aldehydes (Mukaiyama reactions).[5] Asymmetric induction is modest (80–85% ee) in reactions of enol ethers of methyl ketones, but can be as high as 96% ee in reactions of enol ethers of ethyl ketones. Moreover, the reaction is *syn*-selective, regardless of the geometry of the enol. However, the asymmetric induction is solvent-dependent, being higher in nitroethane than in dichloromethane.

Asymmetric [2 + 2]cycloaddition.[6] The chiral titanium reagent 4, prepared from $TiCl_2(O-i-Pr)_2$ and the chiral diol 1, derived from L-tartaric acid and known to effect asymmetric Diels–Alder reactions (14, 232–233), can also promote asymmetric [2 + 2]cycloaddition of 1,1-bis(methylthio)ethylene to α,β-unsaturated N-acyloxazolidinones.

1

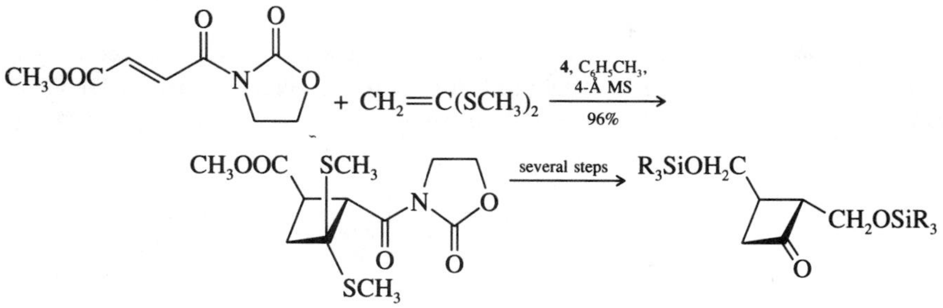

(98% ee)

No reaction occurs with vinyl ethers, silyl enol ethers, or ketene silyl acetals, usually used in thermal [2 + 2]cycloaddition reactions, but the present case is the first example of the preparation of a chiral cyclobutanone by a cycloaddition route.

Asymmetric intramolecular Diels–Alder reaction.[7] This chiral Ti catalyst system **4** is also effective for enantioselective intramolecular Diels–Alder reactions (equation I). In this particular case, a dithiane group accelerates the rate and enhances the *endo*-selectivity, and is comparable to the *gem*-dialkyl effect.

(I)

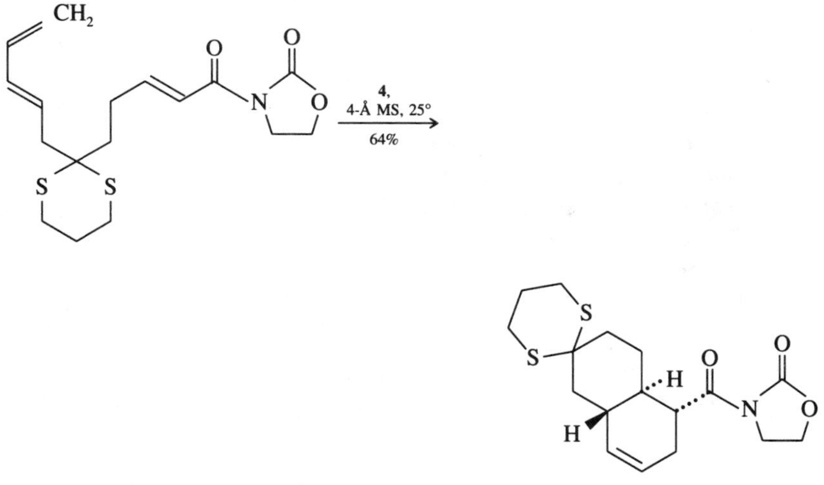

(86% ee)

[1] B. Schmidt and D. Seebach, *Angew. Chem. Int. Ed.*, **30**, 99 (1991).
[2] D. Seebach, A. K. Beck, R. Imwinkelried, S. Roggo, and A. Wonnacott, *Helv.*, **70**, 954 (1987).
[3] K. Furuta, S. Shimizu, Y. Miwa, and H. Yamamoto, *J. Org.*, 1481 (1989).
[4] K. Furuta, T. Maruyama, and H. Yamamoto, *Am. Soc.*, **113**, 1041 (1991).
[5] T. Mukaiyama, *Org. React.*, **28**, 203 (1982).
[6] Y. Hayashi and K. Narasaka, *Chem. Letters*, 793 (1989).
[7] N. Iwasawa, J. Sugimori, Y. Kawase, and K. Narasaka, *ibid.*, 1947 (1989).

Tellurium(IV) chloride.

α,β-*Enones*.[1] TeCl$_4$ cleaves 1-alkyl-1-silyloxycyclopropanes (**1**) selectively in CH$_2$Cl$_2$ at 0° to form β-trichlorotelluro ketones (**2**) in essentially quantitative yield. Addition of DMSO (or a variety of bases) induces dehydrotelluration to give α-

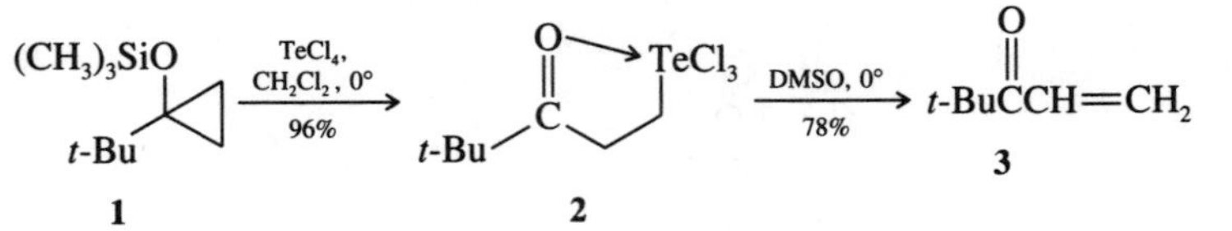

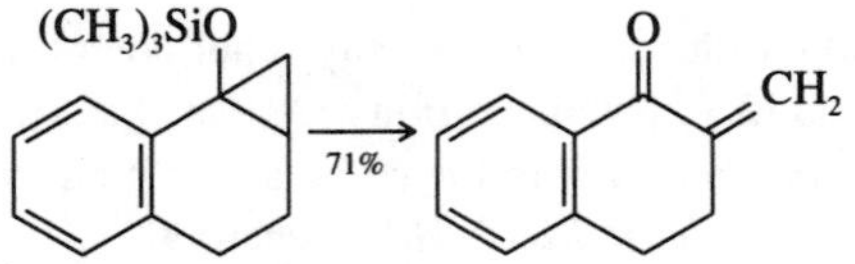

methylene ketones. Similar reactions have been observed when TeCl$_4$ is replaced SnCl$_4$ (**13**, **124**), but dehydrotelluration proceeds much more readily than dehydrostannation.

[1] H. Nakahira, I. Ryu, L. Han, N. Kambe, and N. Sonoda, *Tetrahedron Letters*, **32**, 229 (1991).

2,4,4,6-Tetrabromo-2,5-cyclohexadienone (1).

Brominative cyclization.[1] Even though brominative cyclization of γ,δ-unsaturated alcohols usually results in tetrahydrofurans (**12**, 457–458), reaction of the

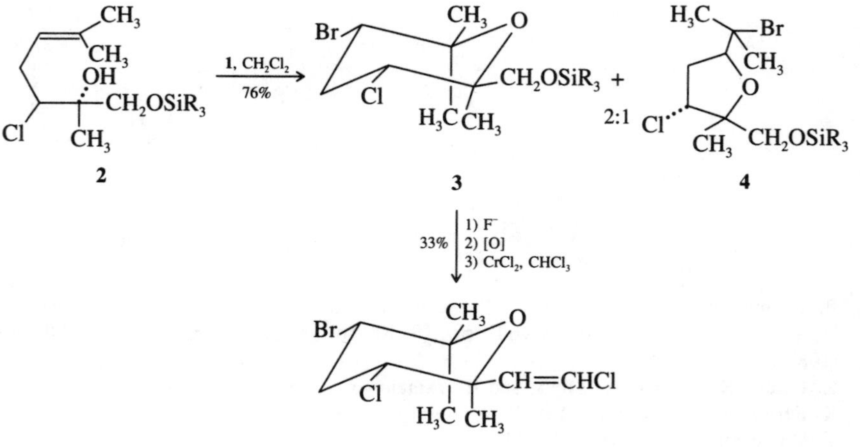

5

unsaturated chlorohydrin **2** with **1** results mainly in the tetrahydropyran **3**, probably because of the preference for an equatorial chlorine atom. Indeed the des-chloro analog of **2** is cyclized by **1** to a 1:1 mixture of a bromotetrahydropyran and a bromotetrahydrofuran. The major product (**3**) was converted into the cytotoxic monoterpene aplysiapyranoid D (**5**) by desilylation, Swern oxidation, and chlorovinylation.

[1] M. E. Jung and W. Lew, *J. Org.*, **56**, 1347 (1991).

Tetrachlorosilane, $SiCl_4$.

Dithioacetalization.[1]   This weak Lewis acid is a useful catalyst for this reaction. Moreover, it is useful for complete selectivity in dithioacetalization of aromatic aldehydes in the presence of aromatic ketones and for preferential dithioacetalization of aliphatic aldehydes in the presence of aliphatic ketones.

[1] B. Ku and D. Y. Oh, *Syn. Comm.*, **19**, 433 (1989).

Tetrakis(triphenylphosphine)nickel(0).

ArC≡N.[1]   Aryl triflates react with KCN at 60° in CH_3CN in the presence of Ni(0) (5 mole %), generated *in situ* from $Br_2Ni[P(C_6H_5)_3]_2$ and zinc, to form aryl nitriles in 69–78% yield. This coupling is not useful in the case of haloaryl triflates because of coupling of KCN with the halo groups. Pd catalysts are not useful.

[1] M. R. I. Chambers and D. A. Widdowson, *J.C.S. Perkin I*, 1365 (1989).

Tetrakis(triphenylphosphine)palladium(0).

Biphenyls.[1]   Pd(0)-catalyzed coupling of aryltributyltins and aryl triflates provides a useful route to substituted biphenyls. Lithium chloride is required for reasonable yields; it inhibits decomposition of the catalyst. A wide range of functional groups is tolerated on both components. Aryl halides undergo this coupling, but triflates have some advantages because of the many available substituted phenols.

Under the same conditions, aryl triflates couple with arylboronic acids to form biphenyls.[2]

Coupling of alkylsilanes with aryl triflates.[3]   Coupling of organosilanes with triflates is possible with fluoride ion (2 equiv.) and a Pd(0) catalyst. The striking feature of this reaction is that even alkyltrifluorosilanes can participate, as well as alkenyl- and aryltrifluorosilanes. Presumably a pentacoordinated siliconate is the effective participant. The stereochemistry has been examined in the case of the optically active silane **1**. The reaction proceeds in THF in 31–51% yield. At temperatures of 50°, the reaction proceeds mainly with retention. The optical yields decrease as the temperature is increased, and at temperatures > 75°, inversion predominates. The effect of solvents is even more striking. Addition of HMPA results in inversion, whereas DMF or DMSO favor retention.

(Z,E)-2-Bromo-1,3-dienes.[4]   1,1-Dibromoalkenes and E-vinylboronic acids couple to (Z,E)-2-bromo-1,3-dienes in the presence of this Pd(0) catalyst and TlOH as base, conditions developed earlier for coupling of vinyl bromides with vinyl-

boranes to give 1,3-dienes with retention of configuration of both partners. Base-sensitive groups are not affected.[1] Most efficient and stereoselective couplings obtain when the 1,1-dibromoalkene is substituted by an allyl alkoxy group.

These (Z,E)-2-bromo-1,3-dienes are of interest because they can show high stereoselectivity in intramolecular Diels–Alder reactions resulting in substituted bicyclic systems.[5]

Alkenyl sulfides.[6]   An attractive route to vinyl sulfides involves cross coupling of a primary 9-alkyl-9-BBN with 1-bromo-1-phenylthioethene catalyzed by Pd(0) in the presence of a base, NaOH or K_2HPO_4. This coupling can be extended to (E)- and

(Z)-2-bromo-1-phenylthio-1-alkenes to afford (E)- and (Z)-vinyl sulfides, respectively.

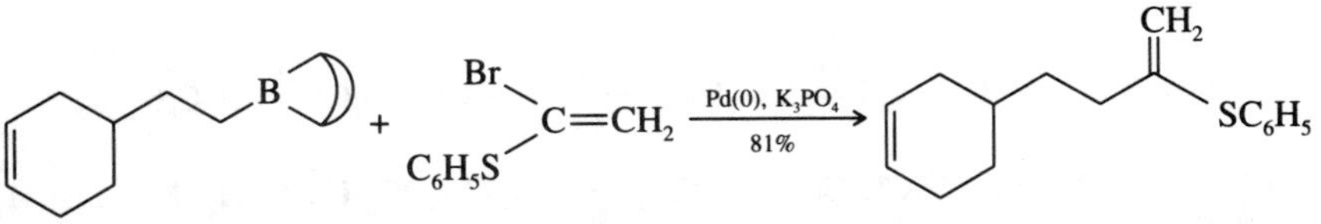

ArOTf + AlR$_3$ → ArR. [7]   This coupling proceeds in generally high yield when catalyzed by a Pd(0) catalyst. Thus the reaction of 1-naphthyl triflate in THF with tri-*i*-butylaluminum (2 equiv.) catalyzed by Pd(0) provides 1-isobutylnaphthalene in 67% yield.

1,2-Diene-4-ynes. [8]   Propargylic carbonates in reactions catalyzed by Pd(0) can react as allenyl complexes. Thus carbonylation of 2-alkynyl carbonates catalyzed by Pd(0) results in 2,3-dienylcarboxylates as the major or only product (equation I). Similarly, coupling of 2-alkynyl carbonates with terminal acetylenes in the presence of CuI and LiCl and catalyzed by Pd(0) provides 1,2-diene-4-ynes in 60–83% yield (equation II).

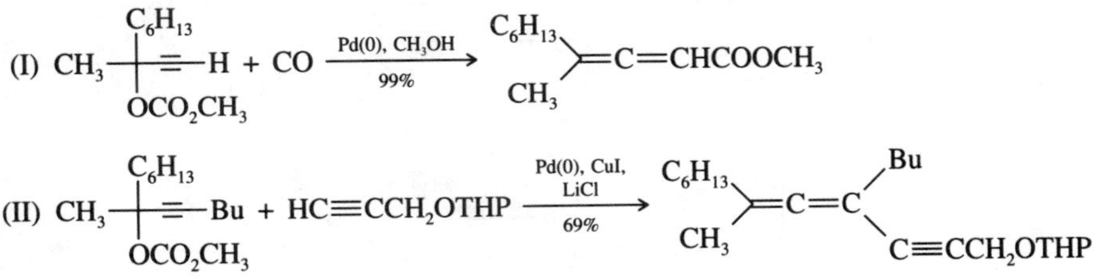

Indole synthesis. [9]   A new route to 2-substituted indoles is based on the Pd(0) coupling of alkynylstannanes with 2-bromoanilines or 2-trifloxyanilines to give 2-alkynylanilines. Intramolecular cyclization of these products with a variety of Pd(II) catalysts provides substituted indoles.

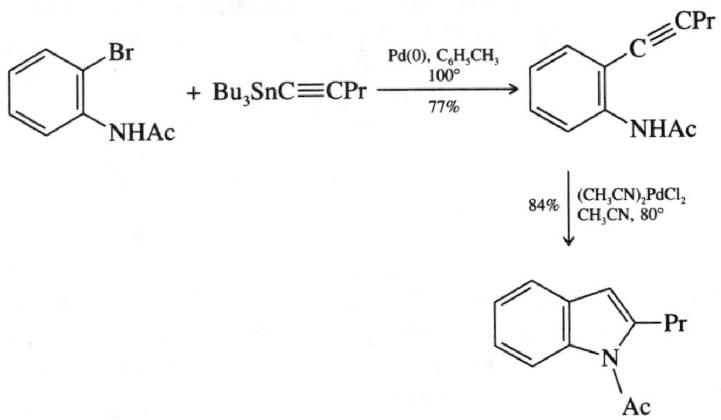

Sulfones.[10] Pd-catalyzed coupling of organostannanes with sulfonyl chlorides provides a direct synthesis of sulfones, particularly vinyl and allyl sulfones.

$$CH_3SO_2Cl + Bu_3SnCH\overset{(E)}{=\!=}CHC_6H_5 \xrightarrow[90\%]{\substack{Pd(0), \\ THF, 60-70°}} CH_3SO_2CH\overset{(E)}{=\!=}CHC_6H_5$$

$$p\text{-}CH_3OC_6H_4SO_2Cl + Bu_3SnCH_2CH\overset{(E)}{=\!=}CHCH_3 \xrightarrow[60\%]{} p\text{-}CH_3OC_6H_4SO_2CH_2CH\overset{(E)}{=\!=}CHCH_3$$

Carbonylation of aryl and vinyl iodides.[11] Pd-catalyzed carbonylation in a basic medium of aryl or vinyl iodides containing enolizable carbonyl groups can result in reaction with the *in situ* generated enolate to form enol esters or enol lactones.

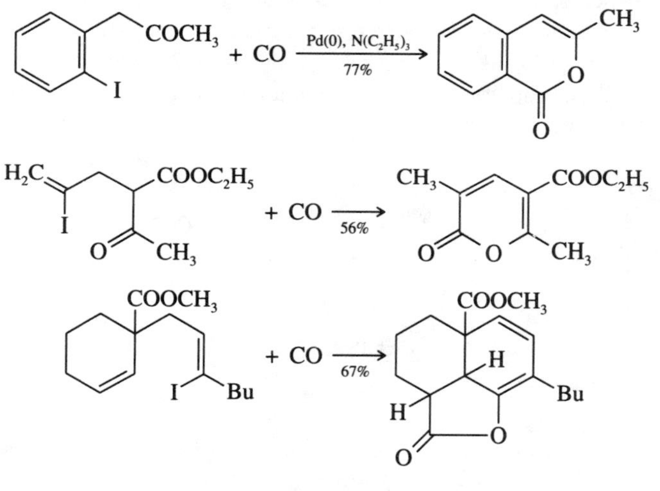

Benzylpalladation.[12] Benzyl halides do not undergo usual radical cyclizations effected with organic halides, but do undergo Pd-catalyzed cyclization with alkenes and alkynes. A typical reaction is formulated in equation (I). The leaving group can also be Br, I, or OMs, but not $OCOOCH_3$ or $OCOCH_3$. Attempts to cyclize **1** with

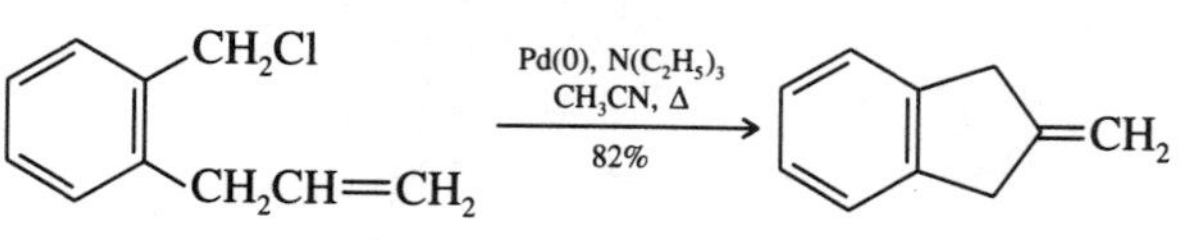

Bu_3SnH (AIBN) fail, but the corresponding bromide or iodide does undergo hydrostannation in moderate yield. Consequently, the cyclization of **1** to **2** is probably not a radical reaction but involves benzylpalladation. The paper includes several other examples of this cyclic carbopalladation such as the conversion of **3** to **4**.

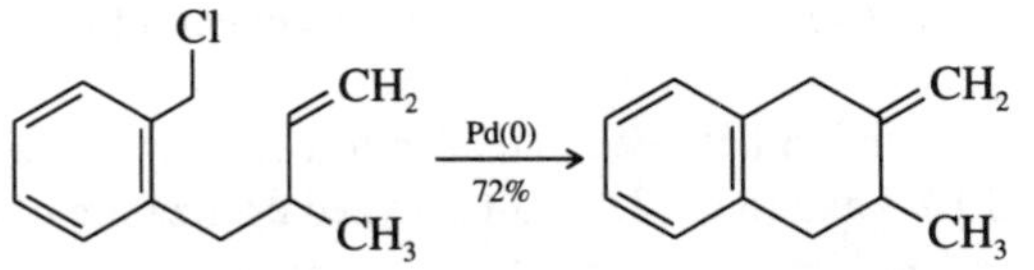

3   **4**

Intramolecular Heck cyclizations.[13] Pd(0)-catalyzed cyclization of substrates
containing one alkenyl and one vinyl iodide group is a well-established route to
bicyclic products (**14**, 297–298; **15**, 301–302). Tandem cyclization to tricyclic products is possible when the substrate contains a triple bond. Thus, in the presence of a
Pd(0) catalyst and 2 equiv. of triethylamine, **1** undergoes tricyclization to **2** in high

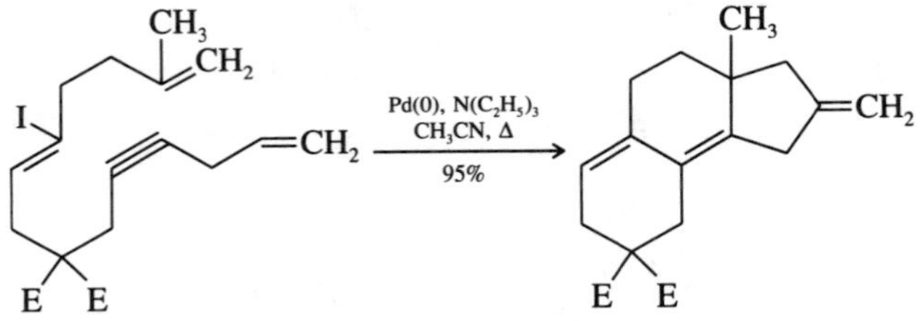

1 (E = COOC$_2$H$_5$)   **2**

yield. This "zipper" type reaction has been extended to tetracyclization of a substrate containing two triple bonds. Thus **3** under the same conditions cyclizes to **4** as
the major product.

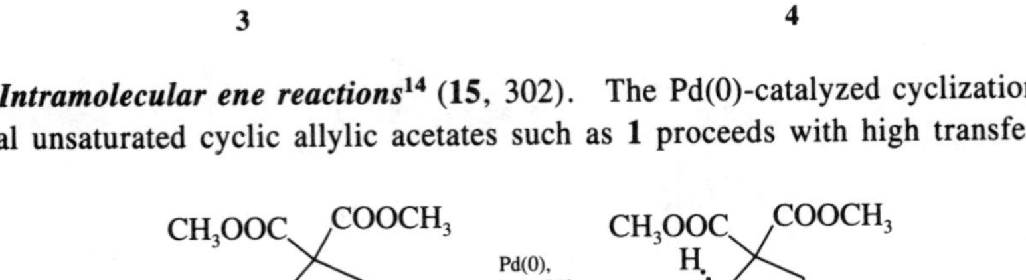

3   **4**

Intramolecular ene reactions[14] (**15**, 302).   The Pd(0)-catalyzed cyclization of
chiral unsaturated cyclic allylic acetates such as **1** proceeds with high transfer of

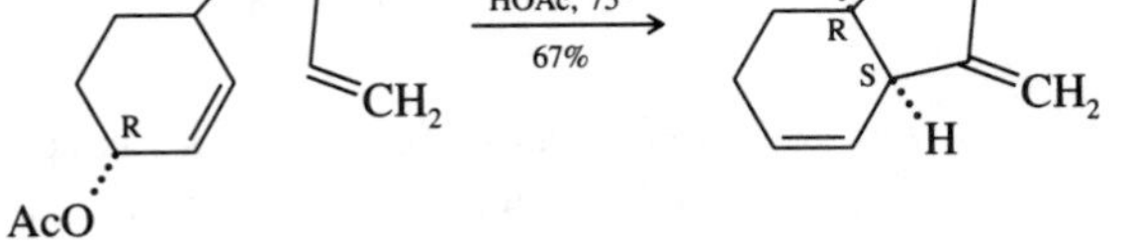

1 (100% ee)   **2** (>92% ee)

chirality to give **2**. The overall result implies displacement of the acetate group with inversion.

This ene-type cyclization has been extended to chiral unsaturated acyclic allylic acetates such as (S,E)-**3** and (S,Z)-**3**.[15] Pd-catalyzed cyclization of (S,E)-**3** results in the (S)-pyrrolidine **4** in high optical purity by apparent displacement of the acetate

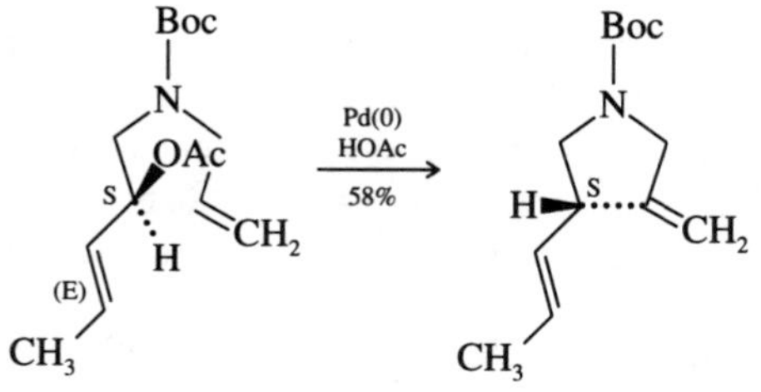

(S,E)-**3** (98.6% ee)   (S,E)-**4** (>96% ee)

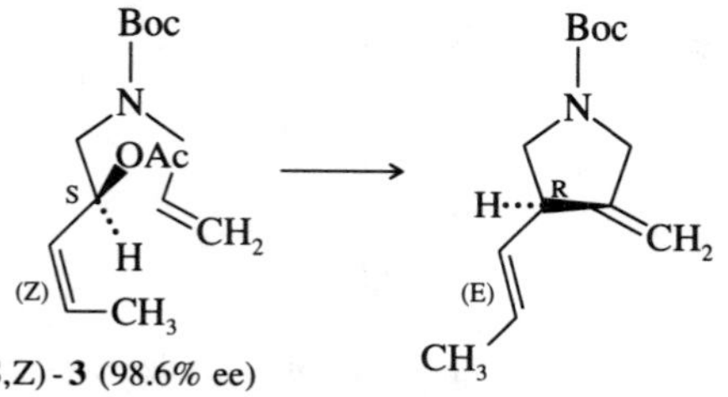

(S,Z)-**3** (98.6% ee)   (R,E)-**4** (>96 ee)

group with inversion. Cyclization of (S,Z)-**3** under the same conditions occurs with net retention at the acetoxy center and Z → E isomerization to give (R)-**4** also in high optical purity. Evidently, in this case the π-allyl complex intermediate can undergo isomerization. Actually the acetoxydiene (S)-**5** lacking the terminal methyl group of **3** cyclizes to a racemic product (**6**).

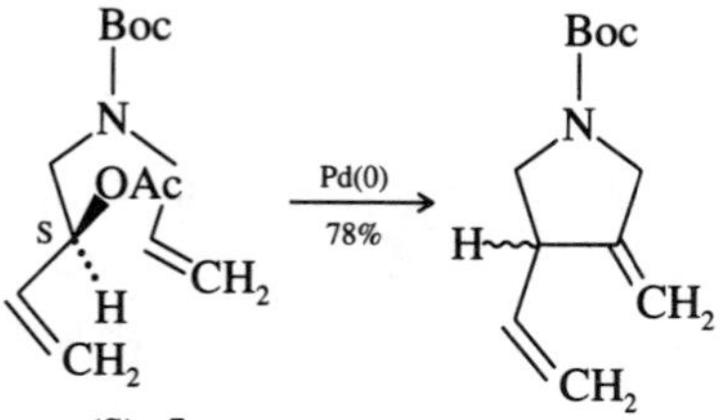

(S)-**5**   (R,S)-**6**

***Coupling of 1-alkynes and vinyl halides*[16]** (**6**, 61).   This coupling reaction has been used extensively for synthesis of isomers of lipoxin A$_4$ and B$_4$, linear trihydroxy eicosanoids with a conjugated tetraene system. One valuable feature is the retention of geometry of the vinyl halide, as well as the high yield. Thus coupling of **1** and **2** provides **3**, a convenient precursor to optically active 7-*cis*,11-*trans*-LXA$_4$ methyl ester (**4**).

HO─CH═CH─Br **1**   +   HC≡C─CH(OSiR$_3$)─CH(OSiR$_3$)─(CH$_2$)$_3$COOCH$_3$ **2** $\xrightarrow[89\%]{\text{Pd(0), CuI,} \atop (C_2H_5)_2NH}$

HOCH$_2$─CH═CH─C≡C─CH(OSiR$_3$)─CH(OSiR$_3$)─(CH$_2$)$_3$COOCH$_3$ **3**

4

[1] A. M. Echavarren and J. K. Stille, *Am. Soc.*, **109**, 5478 (1987); J. K. Stille, A. M. Echavarren, R. M. Williams, and J. A. Hendrix, *Org. Syn.*, [submitted (1990).]

[2] J. Fu and V. Snieckus, *Tetrahedron Letters*, **31**, 1665 (1990).

[3] Y. Hatanaka and T. Hiyama, *ibid.*, **31**, 2719 (1990); *Am. Soc.*, **112**, 7793 (1990).

[4] W. R. Roush, K. J. Moriarty, and B. B. Brown, *Tetrahedron Letters*, **31**, 6509 (1990).

[5] W. R. Roush and R. J. Riva, *J. Org.*, **53**, 710 (1988).

[6] Y. Hoshino, T. Ishiyama, N. Miyaura, and A. Suzuki, *Tetrahedron Letters*, **29**, 3983 (1988); T. Ishiyama, N. Miyaura, and A. Suzuki, *Org. Syn.*, submitted (1990).

[7] K. Hirota, Y. Isobe, and Y. Maki, *J.C.S. Perkin I*, 2513 (1989).

[8] J. Tsuji, T. Sugiura, and I. Minami, *Tetrahedron Letters*, **27**, 731 (1986); T. Mandai, T. Nakata, H. Murayama, H. Yamaoki, M. Ogawa, M. Kawada, and J. Tsuji, *ibid.*, **31**, 7179 (1990).

[9] D. E. Rudisill and J. K. Stille, *J. Org.*, **54**, 5856 (1989).

[10] S. S. Labadie, *ibid.*, **54**, 2496 (1989).

[11] I. Shimoyama, Y. Zhang, G. Wu, and E. Negishi, *Tetrahedron Letters*, **31**, 2841 (1990).

[12] G. Wu, F. Lamaty, and E. Negishi, *J. Org.*, **54**, 2507 (1989).

[13] Y. Zhang, G. Wu, G. Agnel, and E. Negishi, *Am. Soc.*, **112**, 8590 (1990).

[14] W. Oppolzer, J.-M. Gaudin, and T. N. Birkinshaw, *Tetrahedron Letters*, **29**, 4705 (1988); W. Oppolzer, T. H. Keller, M. Bedoya-Zurita and C. Stone, *ibid.*, **30**, 5883 (1989).

[15] W. Oppolzer, T. N. Birkinshaw, and G. Bernardinelli, *ibid.*, **31**, 6995 (1990).

[16] K. C. Nicolaou, B. E. Marron, C. A. Veale, S. E. Webber, and C. N. Serhan, *J. Org.*, **54**, 5527 (1989).

Tetramethylethylenediaminezinc dichloride,

Conjugate addition of RMgX to enones.[1] This complex can serve as a catalyst for this reaction. The related complex in which one chlorine is replaced by methoxy is somewhat more reactive. Yields of adducts are similar to those obtained by use of 1 equiv. of trialkylzincates.

[1] J. F. G. A. Jansen and B. L. Feringa, *J.C.S. Chem. Comm.*, 741 (1989).

Tetramethylammonium triacetoxyborohydride, $(CH_3)_4NHB(OAc)_3$ (1).

Reduction of a chiral imide (2).[1] Highly selective reduction is possible for the chiral protected *cis*-dihydroxytartarimide 2 with this reagent (equation I). Lithium aluminum hydride shows only slight diastereoselectivity (22%), and use of SMEAH provides 3 in about 64% de. However, tetramethylammonium triacetyxyborohydride

does not reduce a silyl-protected 2. In contrast, *t*-butyldimethylsilyl-2 is reduced by lithium tri-*sec*-butylborohydride to 4 in >98% de. Note that the MEM ether of 2 shows only slight selectivity with either $LiAlH_4$ or lithium tri-*sec*-butylborohydride.

[1] S. A. Miller and A. R. Chamberlin, *J. Org.*, **54**, 2502 (1989).

2,2,6,6-Tetramethylpiperidinyl-1-oxyl (TEMPO, 1) and the 4-methoxy derivative (2), 12, 479–480; 13, 183.

Oxidation of diols.[1,2] Both 1 and 2 can serve as effective catalysts for oxidation of diols by NaOCl in aqueous CH_2Cl_2. This method is useful for selective oxidation

$$CH_3CHOH(CH_2)_8CH_2OH \xrightarrow[\text{CH}_2\text{Cl}_2/\text{H}_2\text{O, KBr}]{\text{NaOCl, 2}} CH_3CHOH(CH_2)_8CHO + CH_3CO(CH_2)_8CHO$$
$$\phantom{CH_3CHOH(CH_2)_8CH_2OH \xrightarrow{}} \underset{68\%}{} \qquad\qquad \underset{10\%}{}$$

of primary/secondary diols to hydroxy aldehydes. Oxidation of very hydrophilic diols is best conducted with added LiCl to provide a *pseudo*-solid–liquid system.

[1] P. L. Anelli, S. Banfi, F. Montanari, and S. Quici, *J. Org.*, **54**, 2970 (1989).
[2] R. Siedlecka, J. Skarzewski, and K. Młochowski, *Tetrahedron Letters*, **31**, 2177 (1990).

Tetrapropylammonium perruthenate, 14, 302.

Oxidation of alcohols. Griffith and Ley[1] report various improvements for use of this reagent for catalytic oxidation of alcohols in combination with N-methyl-morpholine N-oxide as reoxidant. In general, the yields with this oxidant are higher than those obtained by the Swern reagent.

[1] W. P. Griffith and S. V. Ley, *Aldrichim. Acta*, **23**, 13 (1990).

Thallium(III) acetate.

Dehydrogenation of enones. Tl(OAc)$_3$ can be as effective as DDQ or benzene-selenenic anhydride for dehydrogenation of α,β-enones to 1,4-dienones. However, in some cases the reaction involves dehydrogenation, acetylation, and aromatization to catechols, as in the oxidation of **1** to **2**.

[1] A. K. Banerjee, M. C. Carrasco, and C. A. Peña-Matheud, *Rec. Trav.*, **108**, 94 (1989).

Thallium(III) nitrate (TTN).

Oxidative phenolic coupling.[1] The vancomycin antibiotics are polypeptides with bridging diphenyl ether groups. Evans *et al.* have shown in model systems such as **1** that cyclization to *o*-halophenolic peptides (**2**) can be accomplished by oxidation with thallium(III) nitrate in THF–methanol or CH_2Cl_2–methanol followed by $CrCl_2$ reduction of a *para*-quinol intermediate (**a**). In three cases the yield of cyclic products was 40–48%.

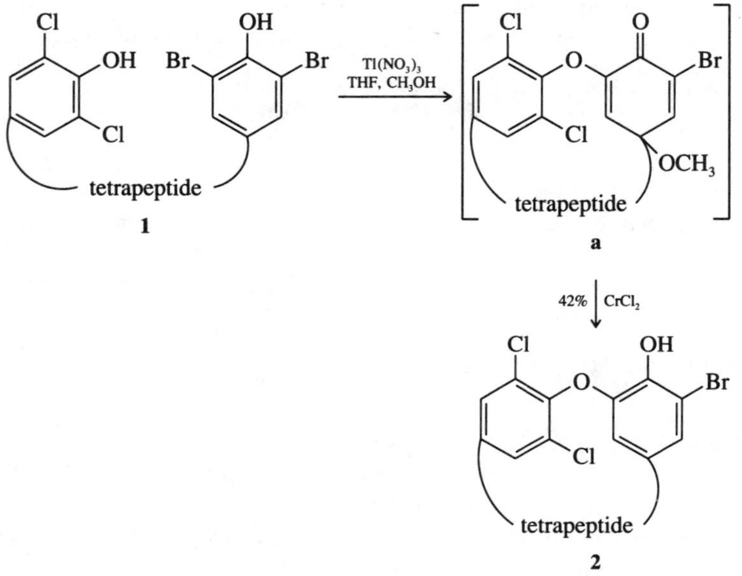

Isoflavones.[2] Treatment of the flavanone **1** with TTN in methanol/chloroform containing perchloric acid (essential) results in an isoflavone (**2**) as the major product, formed via a 2,3-aryl migration. Substitution of a halogen group at the *para*-position of the phenyl group increases the yield of the isoflavone.

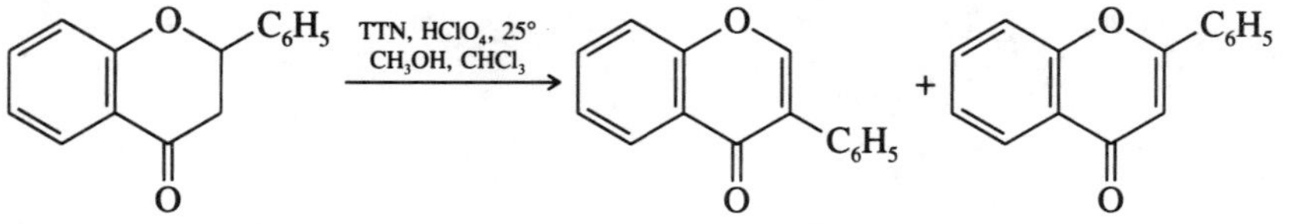

[1] D. A. Evans, J. A. Ellman, and K. M. DeVries, *Am. Soc.*, **111**, 8912 (1989).
[2] T. Kinoshita, K. Ichinose, and U. Sankawa, *Tetrahedron Letters*, **31**, 7355 (1990).

Thexylborane (ThxBH₂).

Intramolecular hydroboration of allyl vinyl ethers.[1] The hydroboration and subsequent oxidation of allyl vinyl ethers **1** with ThxBH₂ (2 equiv.) leads to 1,3-diols with almost exclusive *syn* selectivity. High *syn* selectivity obtains regardless of the bulk of R¹, but is lowered when R¹ is phenyl. Apparently, electronic effects of R¹ are important for stereoselectivity.

1, R¹ = Bu, R² = H	83%	>200:1
R¹ = Bu, R² = CH₃	30%	16:1
R¹ = C₆H₁₁, R² = H	90%	>200:1
R¹ = C₆H₅, R² = H	89%	16:1

Hydroboration–reduction of enones.[2] Hydride reduction of a carbonyl group can be used to induce asymmetric intramolecular hydroboration of a double bond via a cyclic transition state. Thus reaction of the enone **1** with thexylborane (1 equiv.) followed by oxidation provides the 1,5-diol **2** with high 1,4-*syn* selectivity. A similar reaction with the homologous enone provides a 1,6-diol with modest 1,5-*syn* selectivity (*syn/anti* = 6.6:1).

    Reduction of 1,4-diketones with thexylborane (1 equiv.) is also stereoselective, with *anti*-selectivity of 17–47:1. Reduction of 1,5- and 1,6-diketones shows no stereoselectivity, but reduction of 1,3-diketones can show modest *anti*-stereoselectivity.

[1] T. Harada, Y. Matsuda, J. Uchimura, and A. Oku, *J.C.S. Chem. Comm.*, 1429 (1989).
[2] T. Harada, Y. Matsuda, S. Imanaka, and A. Oku, *ibid.*, 1641 (1990).

Thiophenol.

β-Lactones; alkenes.[1] A new one-step preparation of β-lactones is based on the fact that the enolates of phenylthiol esters react with carbonyl compounds to form β-

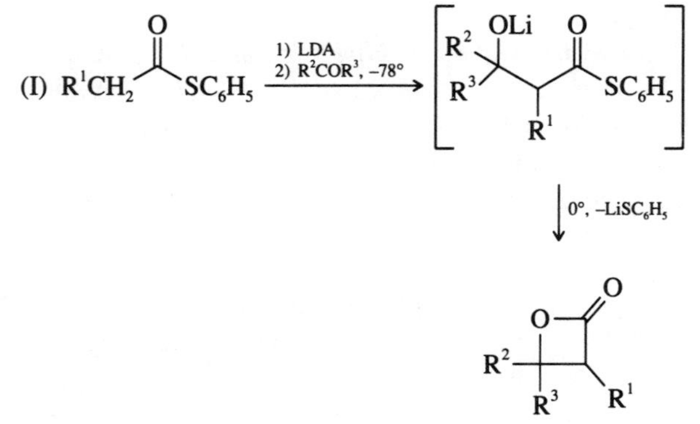

(I) $R^1CH_2\overset{\overset{O}{\|}}{C}SC_6H_5$ $\xrightarrow[\text{2) }R^2COR^3, -78°]{\text{1) LDA}}$ $\left[\begin{array}{c} R^2 \\ R^3 \end{array} \overset{OLi}{\underset{R^1}{\bigg\backslash}} \overset{\overset{O}{\|}}{C} SC_6H_5 \right]$

$\downarrow$ 0°, $-LiSC_6H_5$

lactones directly (equation I). This spontaneous cyclization is not observed with aliphatic thiol esters. This β-lactone synthesis shows moderate to good stereoselectivity for formation of the less sterically hindered diastereomer. The stereoselectivity can be improved by use of esters of 2,6-dimethylbenzenethiol.

β-Lactones are known to lose CO_2 stereospecifically when heated (80–160°) to

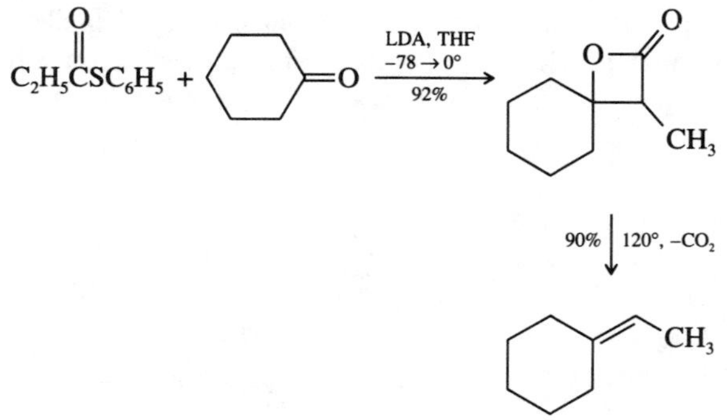

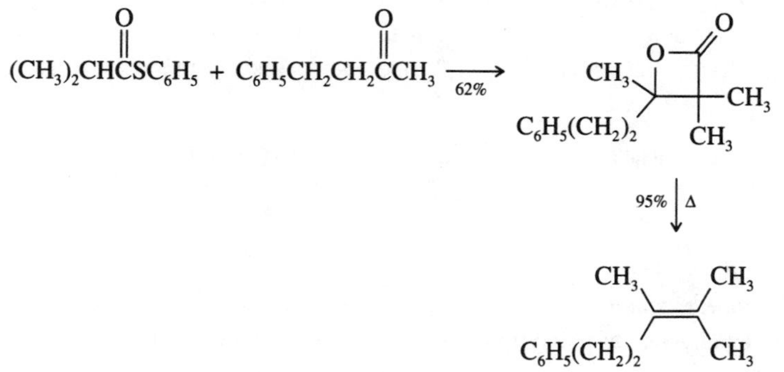

$C_2H_5\overset{\overset{O}{\|}}{C}SC_6H_5$ + (cyclohexanone) $\xrightarrow[\text{92\%}]{\text{LDA, THF} \atop -78 \to 0°}$ (β-lactone spiro product)

90% $\Big\downarrow$ 120°, $-CO_2$

(methylenecyclohexane product with CH_3)

$(CH_3)_2CH\overset{\overset{O}{\|}}{C}SC_6H_5$ + $C_6H_5CH_2CH_2\overset{\overset{O}{\|}}{C}CH_3$ $\xrightarrow{\text{62\%}}$ (β-lactone product)

95% $\Big\downarrow$ Δ

(alkene product)

provide alkenes. This new synthesis thus provides an attractive route to a wide variety of tri- and tetrasubstituted alkenes.

[1] R. L. Danheiser and J. S. Nowick, *J. Org.*, **56**, 1176 (1991).

Tin(II) chloride.

Diastereoselective carbonyl allylation.[1] Pd(II)-catalyzed allylation of aldehydes when carried out with $SnCl_2$ (3 equiv.) can result in α- and/or γ-addition. In addition, γ-addition can be *syn*- and/or *anti*-selective. The regio- and diastereoselectivity can be controlled by the solvent. All three products are obtained from reaction of (E)-2-butenol-1 with C_6H_5CHO in THF. Addition of H_2O permits reactions at $-20°$ and results only in γ-addition with *anti*-selectivity. Reactions of DMSO/H_2O can be controlled by the amount of water to provide either *syn*- or *anti*-γ-addition.

$$CH_3 \diagdown\diagup \diagdown OH + C_6H_5CHO \xrightarrow{\text{Pd(II), SnCl}_2}$$

$$CH_3CH{=}CHCH_2\underset{\underset{C_6H_5}{|}}{\overset{\overset{\displaystyle OH}{|}}{C}} \quad + \quad CH_2{=}CH\diagdown\underset{\underset{CH_3}{|}}{\overset{\overset{\displaystyle CH_3}{|}}{\diagup}}\overset{\overset{\displaystyle OH}{|}}{\diagdown}C_6H_5 \quad + \text{ γ, \textit{anti}-}$$

$$\alpha \qquad\qquad\qquad \text{γ, \textit{syn}-}$$

THF, 25°	72%	53 : 22 : 25
THF + H_2O, –20°	35%	0 : 6 : 94
DMSO + H_2O (1 : 1)	84%	0 : 86 : 14
DMSO + H_2O (1 : 5)	70%	0 : 16 : 84

[1] Y. Masuyama, J. P. Takahara, and Y. Kurusu, *Tetrahedron Letters*, **30**, 3437 (1989).

Tin(II) trifluoromethanesulfonate–Dibutyltin diacetate.

Aldol reactions[1] (**15**, 314–315).   In the presence of a complex (**1**) obtained from a chiral diamine such as (S)-1-ethyl-2-[(piperidinyl)methyl]pyrrolidine (**13**, 302), and Sn(OTf)$_2$ and Bu$_2$Sn(OAc)$_2$, aldehydes react with silyl enol ethers of α-benzyloxy thioesters to form *anti*-α,β-dihydroxy thioesters in high diastereoselectivity.

$$RCHO + \underset{\underset{BzlO}{}\qquad\underset{SC_2H_5}{}}{\diagup\diagdown\diagup\diagdown}^{OSi(CH_3)_3} \xrightarrow[70-83\%]{1, CH_2Cl_2} R\overset{\overset{\displaystyle OH}{|}}{\underset{\underset{OBzl}{\vdots}}{C}}\overset{\overset{\displaystyle O}{\|}}{C}SC_2H_5$$

anti-**2** (96% de)

[1] T. Mukaiyama, H. Uchiro, I. Shiina, and S. Kobayashi, *Chem. Letters*, 1019 (1990).

Titanium(III) chloride.

α,β-Dihydroxy ketones.[1]   Reduction of 1,2-diketones with aqueous TiCl$_3$ (2 equiv.) in HOAc or CH$_3$OH results mainly in an α-hydroxy ketone or the dimer. The same reduction in the presence of an aldehyde results in (*syn, anti*)-α,β-dihydroxy ketones by reaction of a ketyl radical from the diketone with the aldehyde (equation I). The yields decrease with increase in the size of R^3; the *syn/anti* ratio also is dependent on the bulk of R^3.

$$\text{(I)} \quad \underset{\substack{\| \\ O}}{R^1C}\!-\!\underset{\substack{\| \\ O}}{CR^2} + HCOR^3 \xrightarrow[30-100\%]{TiCl_3,\ H^+} \underset{\substack{\| \\ O}}{R^1C}\!-\!\underset{\substack{| \\ OH}}{\overset{R^2}{C}}\!-\!\underset{\substack{| \\ OH}}{\overset{H}{C}}\!-\!R^3$$

$$C_6H_5COCOC_6H_5 + CH_3CHO \xrightarrow[97\%]{TiCl_3,\ H^+} \underset{\substack{\| \\ O}}{C_6H_5C}\!-\!\underset{\substack{| \\ OH}}{\overset{C_6H_5}{C}}\!-\!\overset{CH_3}{CHOH}$$

(*syn/anti* = 60 : 40)

[1] A. Clerici and O. Porta, *J. Org.*, **54**, 3872 (1989).

Titanium(III) chloride–Lithium aluminum hydride (14, 307–308).

Trienols.   Solladié has extended his reductive elimination of allylic diols with TiCl$_3$–LiAlH$_4$ (2:1) to a synthesis of optically active trienols from chiral allylic diols.

$$53\% \left\downarrow \begin{array}{l} 1\ \text{Ti(0)}, -78 \rightarrow 25° \\ 2\ \text{Bu}_4\text{NF} \end{array}\right.$$

[1] G. Solladié and C. Hamdouchi, *Synlett*, 66 (1989).

Titanium(III) chloride–Potassium/Graphite, 14, 308–309.

Enone–aldehyde coupling.   A recent synthesis of the ring system **2** of mevinolin (**3**),[1] a fungal metabolite used for treatment of hypercholesterolemia, depends on successful coupling of enone aldehydes such as **1**. The McMurry coupling is highly sensitive to the $C_8K/TiCl_3$ ratio and to the relative amount of substrates and titanium reagent, but consistently high yields can be obtained by use of a large excess of the low-valent titanium obtained from $TiCl_3$ and C_8K in the ratio $2:1$.

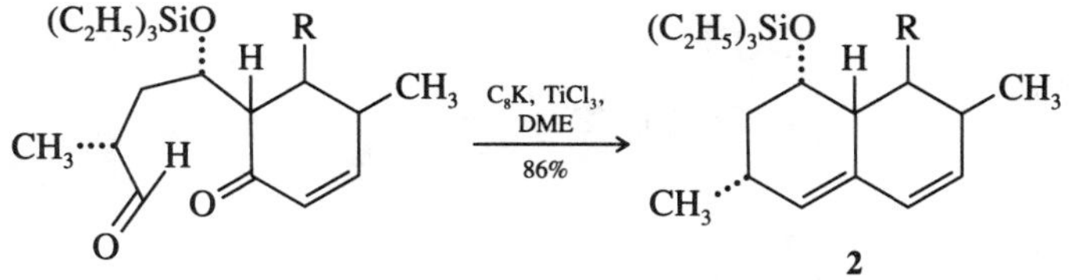

[1] D. L. J. Clive, K. S. K. Murthy, A. G. H. Wee, J. S. Prasad, G. V. J. da Silva, M. Majewski, P. C. Anderson, C. F. Evans, R. D. Haugen, L. D. Heerze, and J. R. Barrie, *Am. Soc.*, **112**, 3018 (1990).

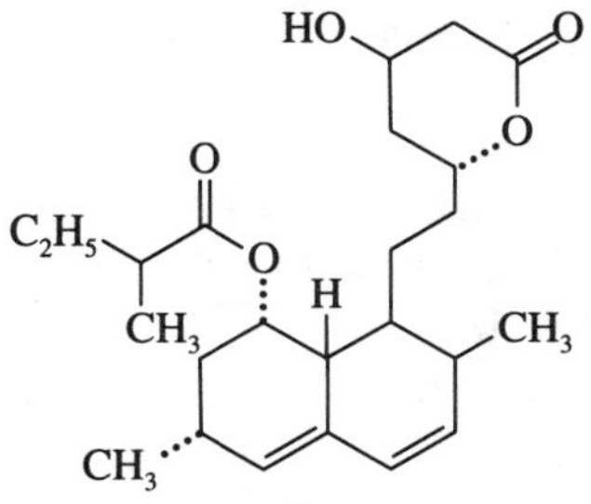

Titanium(III) chloride–Sodium borohydride.

Reduction of α-oximino esters.[1]   This combination (3.15:2.8) reduces α-oximino esters (1 equiv.) to α-amino esters in a buffered solution (pH 7). L-Tartaric acid is a useful buffer; it does not induce any enantiomeric selectivity, but affords clean reactions.

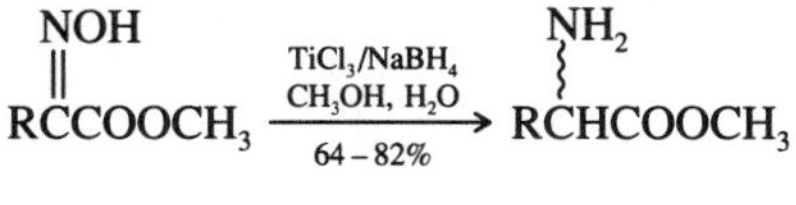

[1] C. Hoffman, R. S. Tanke, and M. J. Miller, *J. Org.*, **54**, 3750 (1989).

Titanium(III) chloride–Zinc/copper.

McMurry coupling.[1]   This reaction has been carried out by McMurry *et al.* with a low-valent Ti reagent from $TiCl_3/LiAlH_4$ in THF, $TiCl_3/K$, $TiCl_3/Li$, and finally from $TiCl_3/ZnCu$ (**8**, 483; **9**, 466). In addition, $TiCl_4/Zn/Py$ and $TiCl_3/K/graphite$

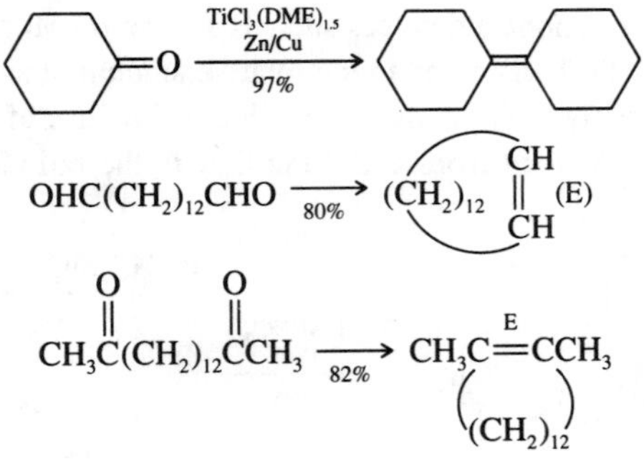

have also been used in other laboratories. The now-preferred Ti reagent is the $TiCl_3(DME)_{1.5}$ solvate, prepared by refluxing $TiCl_3$ in refluxing $CH_3OCH_2CH_2OCH_3$ for 2 days. The coupling reagent from this solvate of $TiCl_3$ and Zn/Cu gives consistently high yields.

Betschart and Seebach[2] have reviewed the reagents obtained by reduction of a number of Ti(III) and Ti(IV) salts with a variety of reductants. All are probably forms of metallic, soluble Ti(0), and have similar potential for organic synthesis (195 references from 1971 to 1989).

[1] J. E. McMurry, T. Lectka, and J. G. Rico, *J. Org.*, **54**, 3748 (1989).
[2] C. Betschart and D. Seebach, *Chimia*, **43**, 39 (1989).

Titanium(IV) chloride.

Titanium enolates. These enolates have generally been prepared by transmetalation of alkali–metal enolates or silyl enolate ethers. Surprisingly, Evans *et al.*[1] find that a titanium enolate can be prepared directly from the oxazolidinone **1** by reaction with $TiCl_4$ (1 equiv.) in CH_2Cl_2 and shortly thereafter with ethyldiisopropylamine (or triethylamine) at $0°$. The enolate may actually be an ate complex (**2a**)

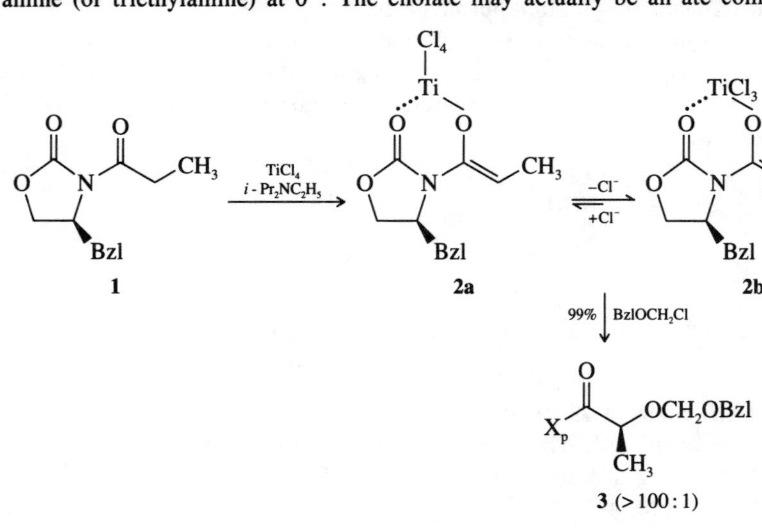

rather than the expected trichlorotitanium enolate (**2b**), but in any case it reacts with an electrophile in the same way as shown for **2b** generated from alkali–metal enolates. TiCl$_4$ can be replaced by *i*-PrOTiCl$_3$, and in this case triethylamine can be used as base and can be added at the same time.

In addition to the simplified operation, this direct generation of titanium enolates has some advantages such as improved yields and enantioselectivities, and its compatibility with a wider range of electrophiles. One limitation is that it is less useful for unhindered ketones, because of self-condensation.

The tetrachloroenolates of chiral ethyl ketones undergo highly diastereoselective reactions with aldehydes.[2] In fact, the *syn/anti* diastereoselectivity is comparable to that obtained with the analogous boron enolate, but the isolated yield of the *syn*-aldol adduct is often higher. The Ti enolates are generally prepared by addition of TiCl$_4$ to the ketone in CH$_2$Cl$_2$ at $-17°$, followed by dropwise addition of N(C$_2$H$_5$)$_3$ or C$_2$H$_5$N(*i*-Pr)$_2$.

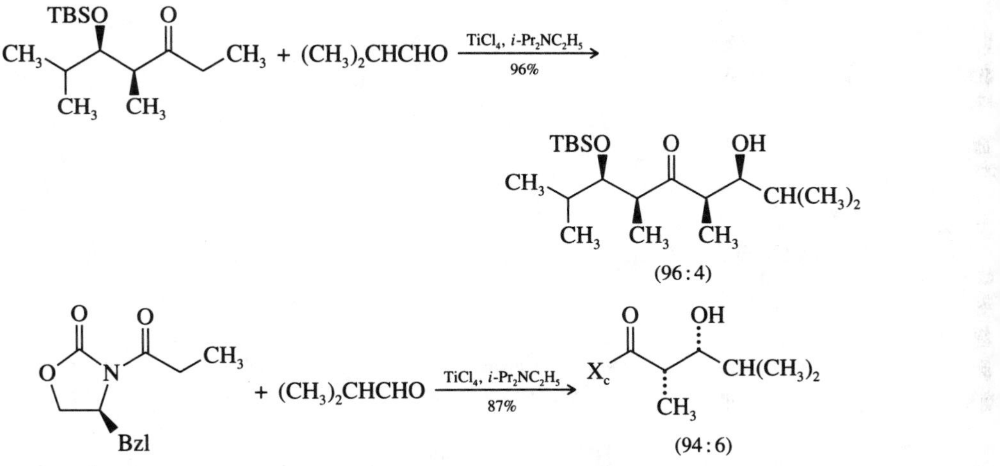

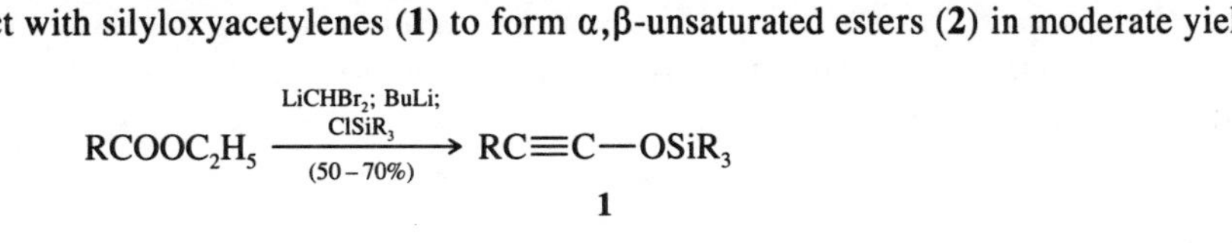

Disubstituted α,β-*unsaturated esters.*[3]   In the presence of TiCl$_4$, aldehydes react with silyloxyacetylenes (**1**) to form α,β-unsaturated esters (**2**) in moderate yield.

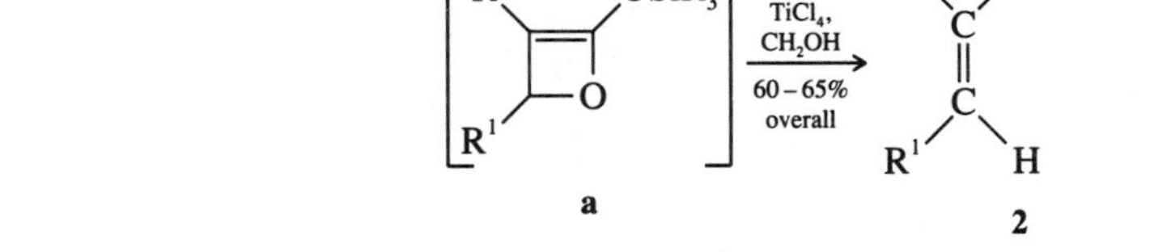

The silyloxy acetylenes can be prepared in one step in moderate yield from esters[4] by sequential reaction with dibromomethyllithium, BuLi, and a chlorotrialkylsilane, usually chloro(triisopropyl)silane. A valuable feature of this reaction is the high (E)-selectivity, which could result from an oxetene intermediate (a).

Diastereoselective reaction of allylsilanes with α-amino aldehydes.[5] The diastereoselectivity of the reaction of allylsilane with the protected α-amino aldehyde **1** depends on the quantity of TiCl$_4$ used. Thus *anti*-selectivity predominates when 1 equiv. of TiCl$_4$ is used, *but* syn-selectivity obtains with less than 1 equiv. The α-amino aldehyde **3** shows *syn*-selectivity with stoichiometric or catalytic amounts of TiCl$_4$. These results suggest that aldehydes such as **1** can form both 1:1 and 2:1 complexes with TiCl$_4$.

+ TiCl$_4$ (1 equiv.)	84%	20:1
+ TiCl$_4$ (0.5 equiv.)	45%	1:8
+ TiCl$_4$ (2 equiv.)	69%	1:1

+ TiCl$_4$ (1 equiv.)	82%	1:14
+ TiCl$_4$ (0.6 equiv.)	46%	1:1.5

Dihydropyrans.[6] The silyl ethers of homopropargyl alcohols in the presence of TiCl$_4$ condense with aldehydes to form dihydropyrans. This reaction is highly dependent on the substituents of the silyl group, with best results obtained with the bis-silyl ether (**1**) of 4-pentynol-2. Use of AlCl$_3$ as catalyst or of the free alcohol results in at least three acyclic products.

Titanoxycyclopropanes; tetrahydrofurans.[7]   Methyl 2-siloxycyclopropanecar-
boxylate (1) provides the pale-yellow titanoxycyclopropane 2, identified by ^{13}C-
NMR. Reactions of 2 with benzaldehyde at low temperatures provides a single
homoaldol (3). Reduction of 3 with a silane and BF_3 etherate provides a single

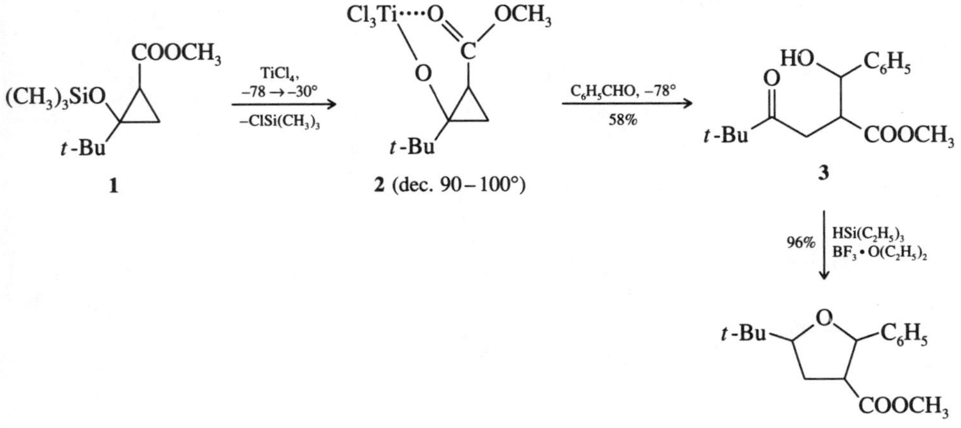

1
2 (dec. 90–100°)
3
4

tetrahydrofuran (4). Reaction of 2 with isobutyric aldehyde provides 5 as a mixture
of two isomers in 79% yield. The titanoxycyclopropane also reacts with acetophe-
none to give essentially one product.

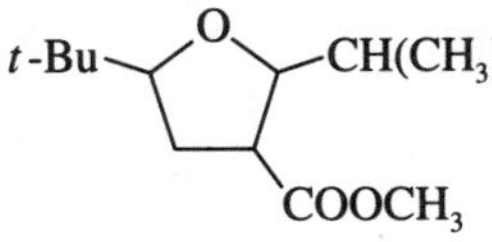

5 (79%, 90:10)

Oxepanes.[8]   Addition of a 2,3-epoxypropyl 2-[(tributylstannyl)methyl]-2-pro-
penyl ether such as 1 to a dilute solution of $TiCl_4$ in CH_2Cl_2 results in attack of the
allylic metal on the epoxide to provide the oxepane 2 in 95% yield. Similar cycliza-

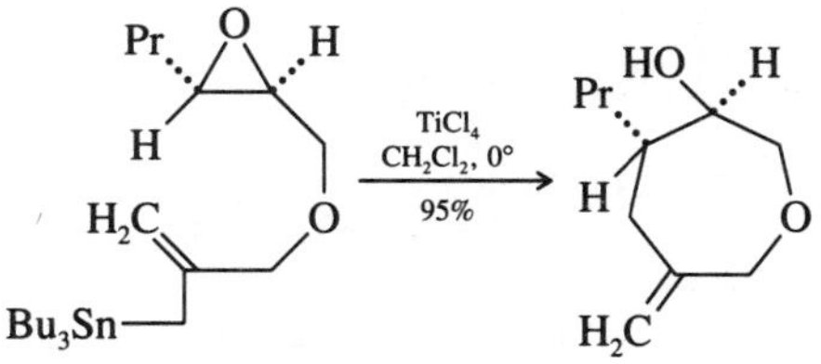

1
2

tion obtains when Sn is replaced by Si.   Of a variety of Lewis acids including Ti(O-i-Pr)$_4$, SnCl$_4$, AlCl$_3$, only TiCl$_4$ is generally useful for this cyclization. Of further interest, the process is diastereospecific if the substrate is added slowly to the Lewis acid (1 equiv.).

The precursors are available by alkylation of substituted 2,3-epoxy alkoxides with methyl 2-[(tributylstannyl)methyl]-2-propenesulfonate (**3**)[9] followed by chromatography.

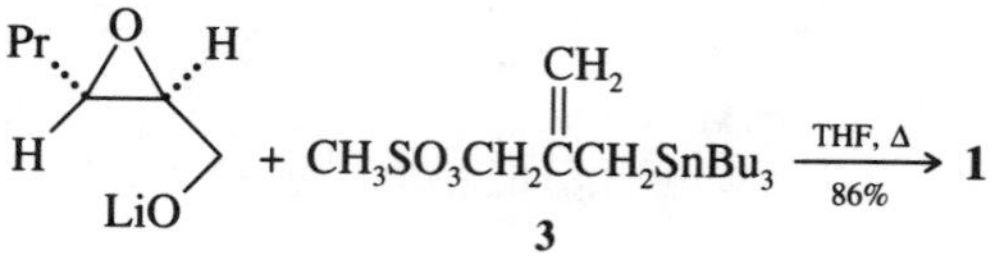

Conjugate propynylation of enones.[10]   In the presence of TiCl$_4$ (1 equiv.) allenyltriphenylstannanes add 1,4 to cyclic and acyclic α,β-enones. Other Lewis acids are ineffective. Use of ZnI$_2$ can result in 1,2-addition.

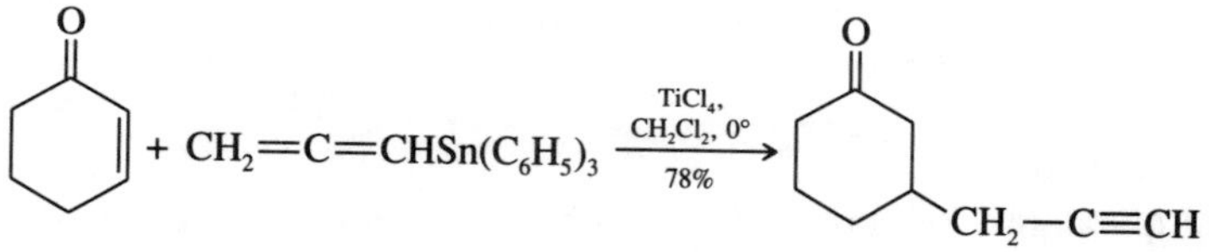

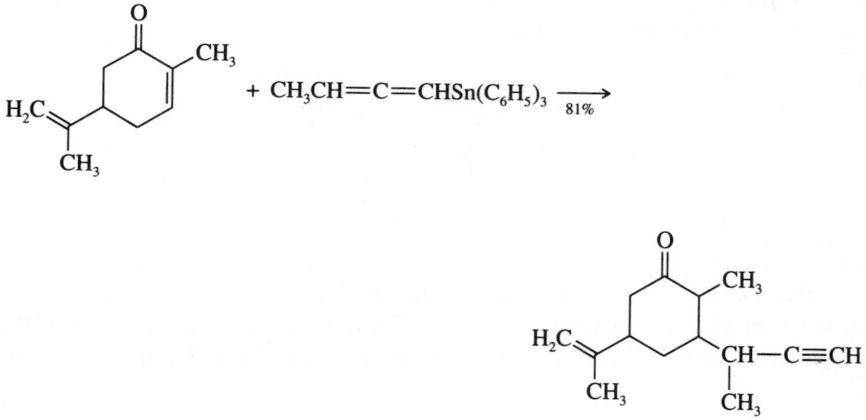

[1] D. A. Evans, F. Urpí, T. C. Somers, J. S. Clark, and M. T. Bilodeau, *Am. Soc.*, **112**, 8215 (1990).

[2] D. A. Evans, D. L. Rieger, M. T. Bilodeau, and F. Urpí, *ibid.*, **113**, 1047 (1991).

[3] C. J. Kowalski and S. Sakdarat, *J. Org.*, **55**, 1977 (1990).

[4] C. J. Kowalski, G. S. Lal, and M. S. Haque, *Am. Soc.*, **108**, 7127 (1986).

[5] S. Kiyooka, M. Nakano, F. Shiota, and R. Fujiyama, *J. Org.*, **54**, 5409 (1989).

[6] T.-H. Chan and P. Arya, *Tetrahedron Letters*, **30**, 4065 (1989).

[7] H.-U. Reissig, H. Holzinger, and G. Glomsda, *Tetrahedron*, **45**, 3139 (1989).

[8] G. A. Molander and D. C. Shubert, *Am. Soc.*, **109**, 576 (1987); G. A. Molander and S. W. Andrews, *J. Org.*, **54**, 3114 (1989).

[9] B. M. Trost, D. M. T. Chan, and T. N. Manninga, *Org. Syn.*, **62**, 58 (1984).

[10] J. Haruta, K. Nishi, S. Matsuda, Y. Tamura, and Y. Kita, *J.C.S. Chem. Comm.*, 1065 (1989).

Titanium(IV) chloride–Aluminum.

Allylation of imines.[1] A low-valent Ti(0) species generated by reduction of $TiCl_4$ with aluminum foil in THF can effect allylation of imines with allyl bromide, even when used in catalytic amounts (0.05 equiv.). This combination of a catalytic amount of $TiCl_4$ with 1 equiv. of aluminum presumably generates Al(III) and Ti(0), which reacts with the allyl halide to form an allyltitanium, the reactive species.

Use of an optically active imine from L-valine results in an optically active homo-allylamine.

[1] H. Tanaka, K. Inoue, U. Pokorski, M. Taniguchi, and S. Torii, *Tetrahedron Letters*, **31**, 3023 (1990).

Titanium(IV) chloride–Titanium(IV) isopropoxide (1, 1:1).

[2 + 2]Cycloadditions.[1] Acyclic enones do not usually undergo photochemical [2 + 2]cycloaddition with alkenes, but this Ti(IV) Lewis acid does promote this cycloaddition in the case of methoxymethyl vinyl ketone and, to a less extent, of methyl vinyl ketone. Usually 2 equiv. of **1** or of $TiCl_2(O\text{-}i\text{-}Pr)_2$ is required.

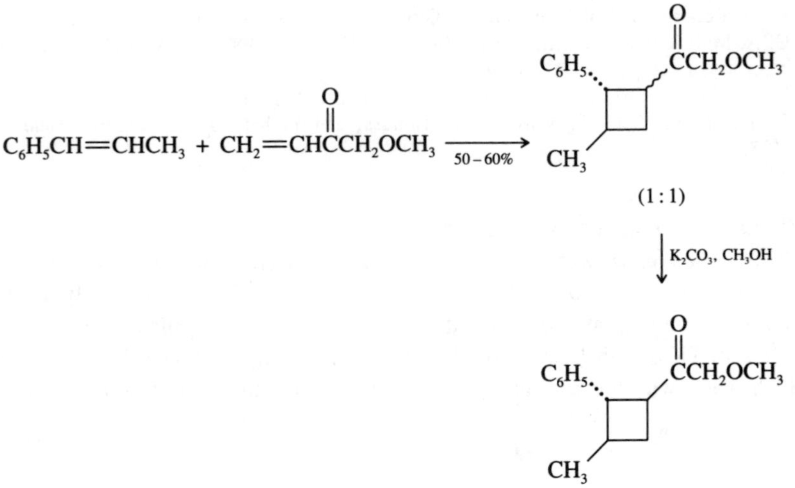

C6H5CH=CHCH3 + CH2=CHCCH2OCH3
50–60%
C6H5
CCH2OCH3
CH3
(1 : 1)
K2CO3, CH3OH

Pterocarpans.[2] This Lewis acid (**1**) effects [2 + 2]cycloaddition of 2*H*-chromens with 2-alkoxy-1,4-benzoquinones to form cyclobutanes, which rearrange *in situ* to the pterocarpan skeleton.

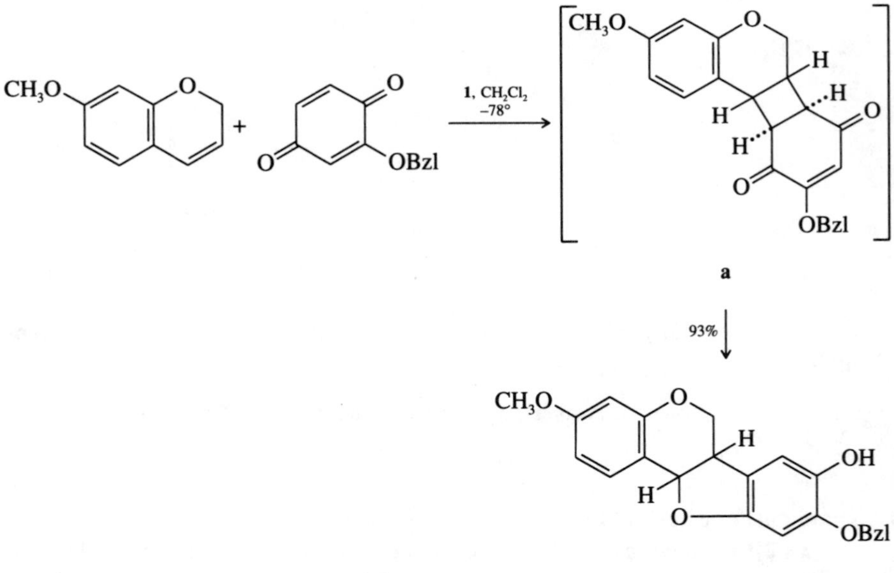

CH3O
OBzl
1, CH2Cl2
−78°
a
93%
OH

[1] T. A. Engler, M. H. Ali, and D. V. Velde, *Tetrahedron Letters*, **30**, 1761 (1989).
[2] T. A. Engler, K. D. Combrink, and J. P. Reddy, *J.C.S. Chem. Comm.*, 454 (1989).

Titanium(IV) chloride–Zinc/copper couple.

α,β-Dihydroxy esters.[1]   The low-valent Ti compound formed from $TiCl_4$–Zn/Cu converts alkyl glyoxylates into a Ti enediolate that reacts with carbonyl compounds to form α,β-dihydroxy esters. The reaction is generally performed at -23 to $-45°$ in CH_2Cl_2/DME, since DME alone promotes self-coupling of the carbonyl compound.

$$BzlO_2CCHO + ArCHO \xrightarrow[80-95\%]{\substack{TiCl_4 - Zn/Cu, \\ CH_2Cl_2, \ DME}} BzlO_2C\overset{\overset{OH}{|}}{C}H-\overset{\overset{OH}{|}}{C}HAr$$

$$(syn,anti = 72-82:28-18)$$

$$BzlO_2CCHO + C_6H_5COCH_3 \xrightarrow[63\%]{} BzlO_2C\overset{\overset{OH}{|}}{C}H-\underset{\underset{CH_3}{|}}{\overset{\overset{OH}{|}}{C}}C_6H_5$$

$$(syn,anti = 30:70)$$

[1] T. Mukaiyama, H. Sugimura, T. Ohno, and S. Kobayashi, *Chem. Letters*, 1401 (1989).

Titanium(IV) isopropoxide.

Halohydrins.   In the presence of 1 equiv. of Ti(O-*i*-Pr)$_4$, bromine or iodine converts epoxides of *trans*-allylic or homoallylic alcohols to halohydrins at $0-25°$ in high yield with marked regioselectivity. The presence of the free hydroxyl group is essential for regioselectivity.

[1] E. Alvarez, M. T. Nuñez, and V. S. Martín, *J. Org.*, **55**, 3429 (1990).

Titanium(IV) isopropoxide–L-(+)-Diisopropyl tartrate (DIPT).

Enantioselective trimethylsilylcyanation.[1]   The addition of $(CH_3)_3SiCN$ to benzaldehydes in the presence of an equimolar or catalytic amount of a Sharpless-like

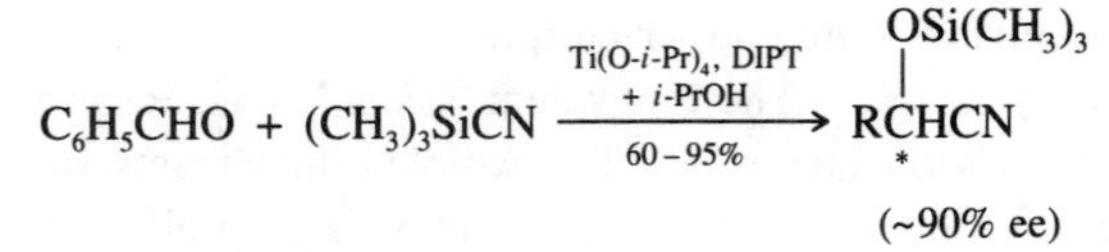

$$C_6H_5CHO + (CH_3)_3SiCN \xrightarrow[\substack{60-95\%}]{\substack{Ti(O\text{-}i\text{-}Pr)_4,\ DIPT \\ +\ i\text{-}PrOH}} \overset{OSi(CH_3)_3}{\underset{*}{RCHCN}}$$

(~90% ee)

catalyst (A) prepared from Ti(O-*i*-Pr)$_4$ and L-DIPT (1:1) proceeds with only moderate enantioselectivity, which is not improved by addition of molecular sieves or by removal of isopropanol from catalyst A by freeze-drying. However, addition of 1–2 equiv. of isopropanol per Ti to this freeze-dried catalyst provides a catalyst (D) which is highly effective, both in equimolar and catalytic amounts.

[1] M. Hayashi, T. Matsuda, and N. Oguni, *J.C.S. Chem. Comm.*, 1364 (1990).

p-Toluenesulfonylhydrazine.

Alkene synthesis. A modified version of an earlier synthesis of alkenes from the tosylhydrazones of aldehydes and certain alkyllithium reagents (9, 472–473) employs the N-*t*-butyldimethylsilyl derivatives (1) of tosylhydrazones of α,β-enals, which undergo only 1,2-addition of an alkyllithium at −78°. The adducts (2) decom-

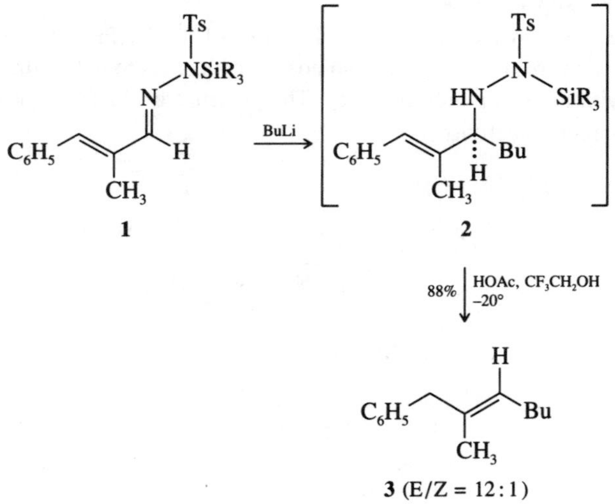

3 (E/Z = 12:1)

pose at −20° when treated with acetic acid and CF$_3$CH$_2$OH with elimination of *p*-toluenesulfinic acid, the silyl group, and nitrogen with a 1,5-sigmatropic rearrangement to provide a mixture of an (E)- and (Z)-alkene (3) in the ratio 12:1, in high overall yield (88%).

Alkenes can also be obtained by reaction of the silylated *p*-tosylhydrazone of saturated aldehydes with a vinyllithium. Of further interest, no epimerization occurs at centers adjacent to the aldehyde group. The (E)-configuration is favored in the case of both di- and trisubstituted alkenes.

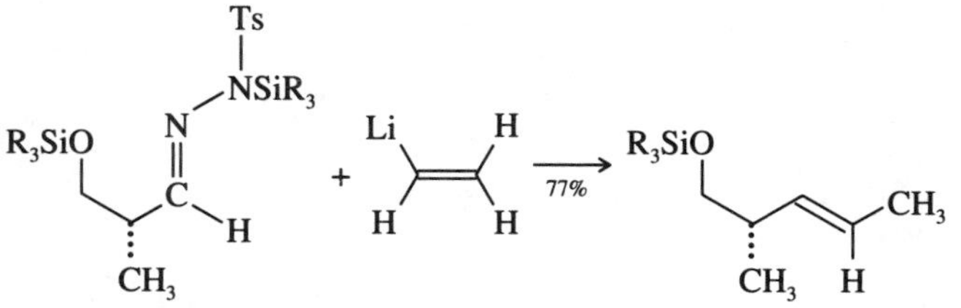

$[R_3 = (C_6H_5)_2C(CH_3)_3]$ (E/Z = >20:1)

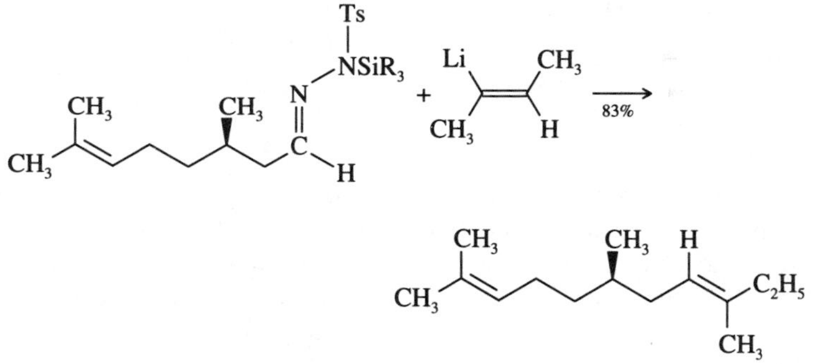

(E/Z = >20:1)

[1] A. G. Myers and P. J. Kukkola, *Am. Soc.*, **112**, 8208 (1990).

Tributyl[(methoxymethoxy)methyl]tin, $Bu_3SnCH_2OCH_2OCH_3$ (**1**). Preparation:

$$Bu_3SnH \xrightarrow[2)\ (HCHO)_n]{1)\ LDA,\ THF} [Bu_3SnCH_2OH] \xrightarrow[77\%]{CH_3OCH_2OCH_3,\ BF_3\cdot O(C_2H_5)_2,\ 4\text{Å MS}} \mathbf{1}$$

Hydroxymethyl anion equivalent (*cf.*, **8**, 495).[2] This tin reagent undergoes transmetallation on treatment with LDA or BuLi at −78° to provide the corresponding lithium reagent (**2**) which adds to ketones to form a monoprotected derivative of a *gem*-diol. The methoxymethyl group of the adduct is cleaved on treatment with dilute acid.

$$1 \xrightarrow{\text{BuLi}} \underset{\textbf{2}}{\text{LiCH}_2\text{OCH}_2\text{OCH}_3} \xrightarrow[92\%]{}$$

HO   CH$_2$OCH$_2$OCH$_3$

$$\Big\downarrow{\scriptstyle 85\%} \quad \begin{array}{l} \text{HCl, CH}_3\text{OH,} \\ 55° \end{array}$$

HO   CH$_2$OH

[1] R. L. Danheiser, K. R. Romines, H. Koyama, S. K. Gee, C. R. Johnson, and J. R. Medich, *Org. Syn.*, submitted (1990).

[2] C. R. Johnson, J. R. Medich, R. L. Danheiser, K. R. Romines, H. Koyama, and S. K. Gee, *ibid.*, submitted (1990).

Tributylarsine, Bu$_3$As.

Catalytic Wittig-type reaction.[1] Arsonium ylides, Ar$_3$As$^+$CHX$^-$, are useful Wittig-type reagents (**13**, 137; **14**, 170), and can be more reactive than the corresponding phosphonium ylides. One disadvantage is the toxicity of arsenic salts. However, arsonium ylides can be generated *in situ* and can be used in catalytic amounts in the presence of triphenyl phosphite, which regenerates Bu$_3$As from tributylarsine oxide. Thus the reaction of various aldehydes with methyl bromoacetate in the presence of triphenyl phosphite (1 equiv.) and K$_2$CO$_3$ (1 equiv.) and Bu$_3$As (0.2 equiv.) provides α,β-unsaturated esters in 60–87% yield and in E/Z ratios of 97–99:3–1.

$$\text{RCHO} + \text{BrCH}_2\text{COOCH}_3 + (\text{C}_6\text{H}_5\text{O})_3\text{P} + \text{K}_2\text{CO}_3 \xrightarrow[60-87\%]{\overset{\text{Bu}_3\text{As}}{\text{THF}-\text{CH}_3\text{CN, 25°}}}$$

$$\overset{\text{(E)}}{\text{RCH}}=\text{CHCOOCH}_3 + (\text{C}_6\text{H}_5\text{O})_3\text{P}=\text{O}$$

[1] L. Shi, W. Wang, Y. Wang, and Y.-Z. Huang, *J. Org.*, **54**, 2027 (1989).

Tributyl(isoprenyl)tin, Bu$_3$SnCH$_2$CH=C(CH$_3$)$_2$ (**1**).

This allyl trialkyltin can be prepared[1] by addition of hydrochloric acid (36%) to isoprene to form 4-chloro-2-methyl-2-butene, which undergoes an ultrasound Barbier-type cross coupling with Bu$_3$SnCl. This method is useful for preparation of other allylic tributyltins.

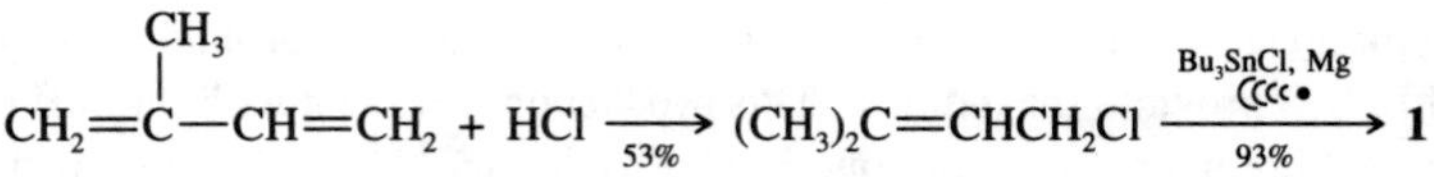

$$\underset{CH_2=C-CH=CH_2}{\overset{CH_3}{|}} + HCl \xrightarrow[53\%]{} (CH_3)_2C=CHCH_2Cl \xrightarrow[93\%]{\overset{Bu_3SnCl, Mg}{\text{(Ccc \bullet)}}} 1$$

Allylation of quinones. [2] In the presence of BF_3 etherate various allyl tributyltins add to benzo- and naphthoquinones to form a mixture of the corresponding quinones or hydroquinones. An example is the synthesis of ubiquinone (**2**).

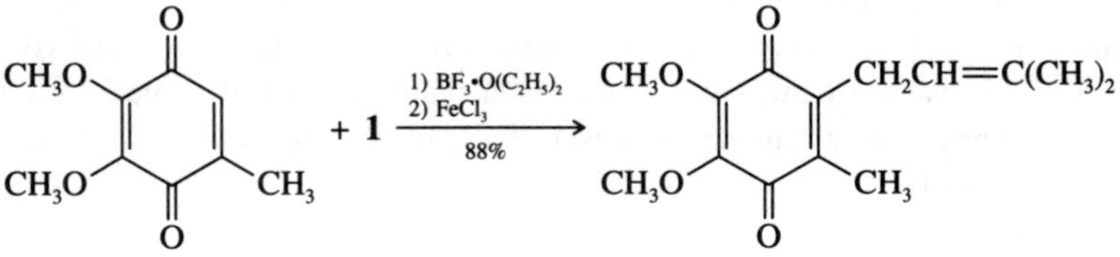

[1] Y. Naruta, Y. Nishigaichi, and K. Maruyama, *Org. Syn.*, submitted (1990).
[2] Y. Naruta and K. Maruyama, *Org. Syn.*, submitted (1990).

Tributyltin hydride.
Radical ring expansion [1] (**14**, 317–318). This reaction has been used for synthesis of muscone (**3**) from cyclododecanone. Alkylation of the ketone with 3-chloro-

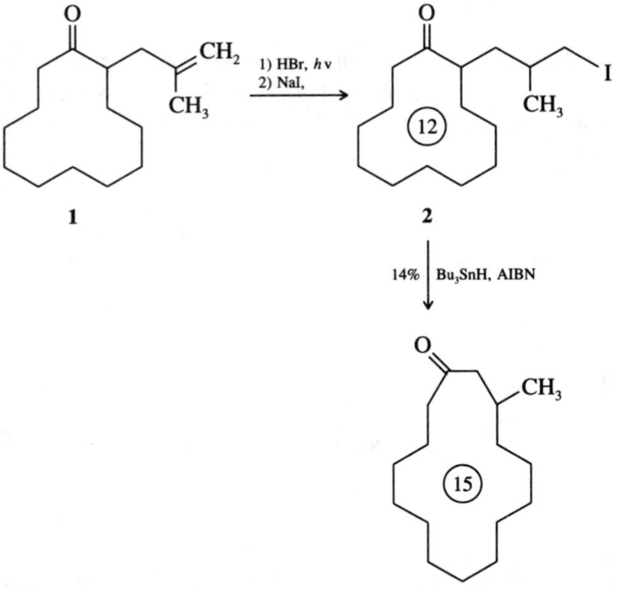

2-methylpropene provides **1**, which on addition of hydrogen bromide and replacement of Br by I provides the iodide **2**. This product on treatment with Bu$_3$SnH (AIBN) under high dilution provides muscone **3** (14% yield) and the product of direct reduction of **2** (63% yield). This sequence was used for a synthesis of (R)-muscone by alkylation of cyclododecanone with (S)-3-bromo-2-methylpropanol (available from Aldrich) to give a product that can be converted into an optically active form of **2**. Conversion of this chiral iodide provided (R)-muscone, also in low yield (15%).

Stereoselective radical reduction or allylation.[2]   Radical reduction of β-alkoxy-α-halo esters such as **1** shows marked *anti*-stereoselectivity with Bu$_3$SnH (AIBN) at 50°, which is markedly improved when conducted under photochemical irradiation at −78°. Similar but even higher stereoselectivity obtains in reduction of the tetrahydrofuran derivative **2**.

Allylation of these β-alkoxy-α-halo esters with allyltributyltin initiated with AIBN at 60° also shows good to impressive stereoselectivity, which can be improved by use of triethylborane as initiator at −78°. However, these reactions are generally slower than radical reductions.

Stereoselective reduction of acyclic α-bromo esters.[3]   Highly diastereoselective radical reduction of α-bromo esters obtains when an electronegative group (F or OCH$_3$) is present at the β-position, particularly at low temperatures.

(I) C$_6$H$_5$—C(H, R ; H$_3$C, Br)—COOC$_2$H$_5$ $\xrightarrow[\text{AIBN, } h\nu]{\text{Bu}_3\text{SnH, } -78°}$ C$_6$H$_5$—C(H, R ; H, CH$_3$)—COOC$_2$H$_5$ + C$_6$H$_5$—C(H, R ; H$_3$C, H)—COOC$_2$H$_5$

dl1, a, R = CH$_3$	85%	2:1
b, R = OCH$_3$	90%	32:1
c, R = F	88%	20:1

syn-*Hydrostannylation of alkynes*.[4]   This reaction can be effected with Bu$_3$SnH in the presence of Pd[P(C$_6$H$_5$)$_3$]$_4$ at 25° to provide vinylstannanes. In the case of a 1-alkyne, both possible adducts are usually formed, but some regioselectivity is generally observed in hydrostannylation of internal alkynes.

AcOCH$_2$C≡CCH$_2$OAc + Bu$_3$SnH $\xrightarrow[93\%]{\text{Pd(0),} \ \text{C}_6\text{H}_6, 25°}$ (AcOCH$_2$)(H)C=C(CH$_2$OAc)(SnBu$_3$)

HOCH$_2$C≡CCH$_3$ + Bu$_3$SnH ⟶ (HOCH$_2$)(Bu$_3$Sn)C=C(CH$_3$)(H) + (HOCH$_2$)(H)C=C(CH$_3$)(SnBu$_3$)

(74%)   (14%)

syn-1,4-*Diols*.[5]   The radical formed from (bromomethyl)dimethylsilyl ethers of homoallylic alcohols with mono- or dimethyl-substituted double bonds undergoes exclusive cyclization to 6-membered *cis*-siloxanes. These are converted to *cis*-1,4-diols on oxidation with H$_2$O$_2$ (30%) in the presence of KHCO$_3$ (equation I). However, the same reaction when applied to a substrate lacking a substituent at the olefin terminus results in cyclization to a seven-membered ring, which is oxidized to a 1,5-

(I) c-C$_6$H$_{13}$ ... (H$_3$C)(CH$_3$)Si—O—CH$_2$Br ... olefin with H, CH$_3$ $\xrightarrow[90\%]{\text{Bu}_3\text{SnH} \ \text{AIBN}}$ c-C$_6$H$_{13}$ ... (CH$_3$)(CH$_3$)Si—O ring—CH$_3$

83–85% $\bigg\downarrow$ $\dfrac{\text{H}_2\text{O}_2, \text{KHCO}_3}{\text{THF/CH}_3\text{OH, } \Delta}$

c-C$_6$H$_{13}$—CH(OH)—CH$_2$—CH(CH$_2$OH)—CH$_2$—CH$_3$

(*cis*)

(II)

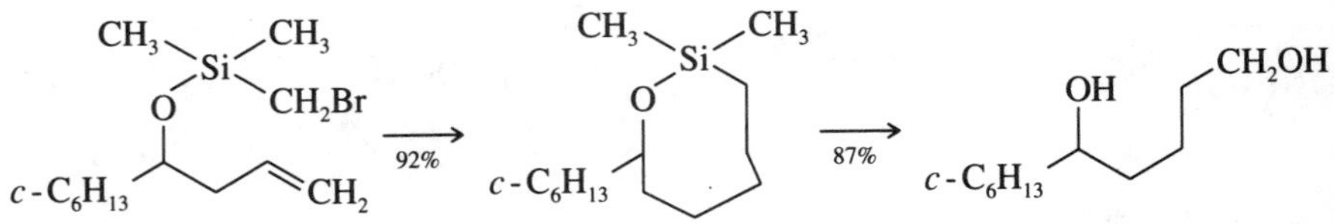

diol. This cyclization provides a route to steroids with dihydroxylated side chains (equation III).

(III)

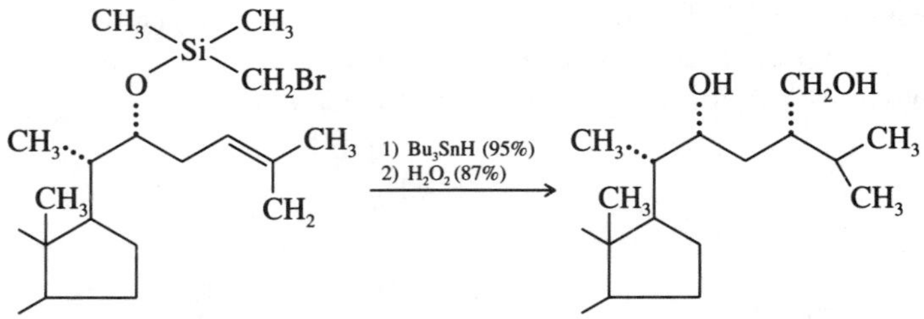

Cyclization via epoxide fragmentation.[6] Treatment of the epoxythionoimidaz-olide **1** with Bu_3SnH and AIBN generates an alkoxy radical by fragmentation of the epoxy group with cyclization to a *cis*-fused bicyclic system (**2**).

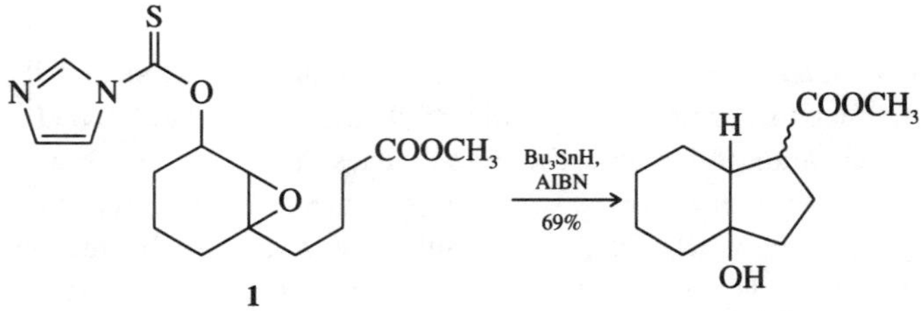

Deoxygenation of R_2CHOH[7] (**11**, 550).   Radical deoxygenation of secondary alcohols via thiocarbonyl esters was introduced by Barton and McCombie. More recently, phenoxythiocarbonyl esters have been recommended for the radical deoxy-genation (**10**, 306–307). Particularly rapid and quantitative reduction can be obtained using 2,4,6-trichlorophenoxy- or pentafluorophenoxythiocarbonyl esters.

RX → RCHO.[8] Carbonylation of alkyl or aryl radicals generated with Bu_3SnH/AIBN from halides is usually not useful because reduction to the alkane is more facile. However, this radical conversion of RX to RCHO can be effected in yields of 40–70% if Bu_3SnH is used in low concentrations (~ 0.05 M) and with 15–90 atm. of CO.

Acyl radicals from $RCOSeC_6H_5$.[9] Acyl radicals generated from phenylseleno-esters can participate in inter- and intramolecular alkene addition. The latter tandem reaction can be used for macrocyclization under standard high-dilution conditions of phenylselenoesters containing a terminal unsubstituted alkene group.

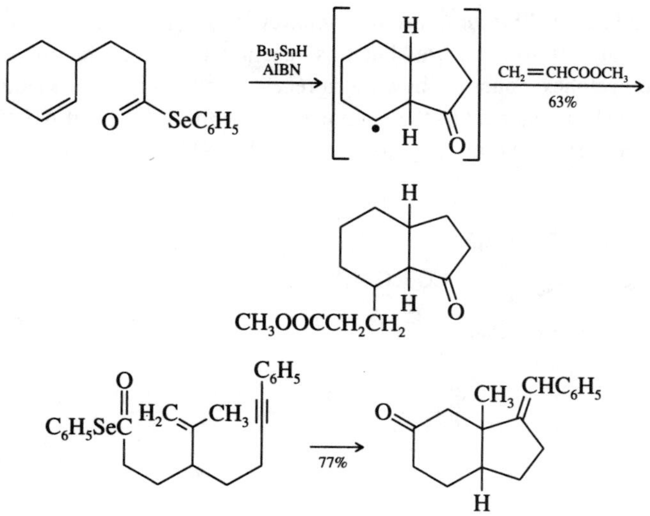

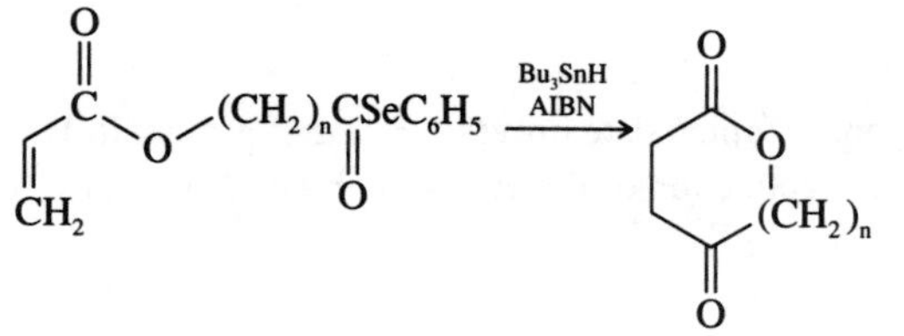

n = 11	68%
n = 7	46%

Intermolecular free radical reactions.[10] Giese notes the diastereoselectivity of reactions of acrylonitrile with cyclic 5- and 6-membered ring radicals can be controlled by adjacent substituents. Thus an axial β-substituent can favor axial attack, whereas an equatorial β-substituent favors equatorial attack in the case of 6-membered cyclic radicals. Glucosyl radicals, regardless of the precursor, yield α-substituted products (88:12).

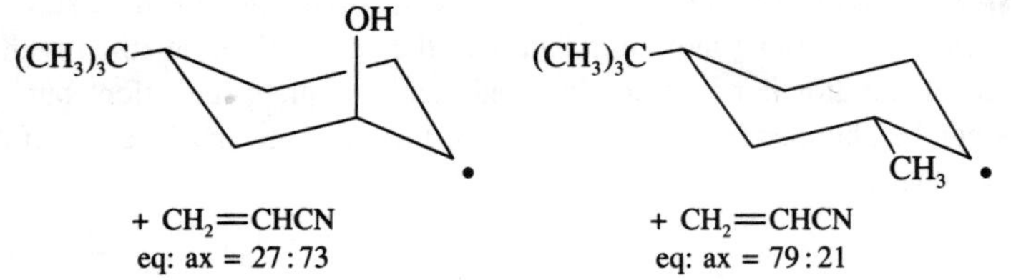

+ CH$_2$=CHCN
eq: ax = 27:73

+ CH$_2$=CHCN
eq: ax = 79:21

Acyl radical cyclization to cyclohexanones.[11]   Acyl radical cyclization to five-membered rings is well known, but this reaction is also useful for synthesis of substituted cyclohexanones as shown by a recent synthesis of the α-methylenecyclo-hexanone **3**. Thus treatment of the selenol ester **1** with Bu₃SnH and AIBN in C₆H₆ at 80° provides a 1:1 mixture of cyclohexanones **2** in 91% yield. Oxidation of **2** with

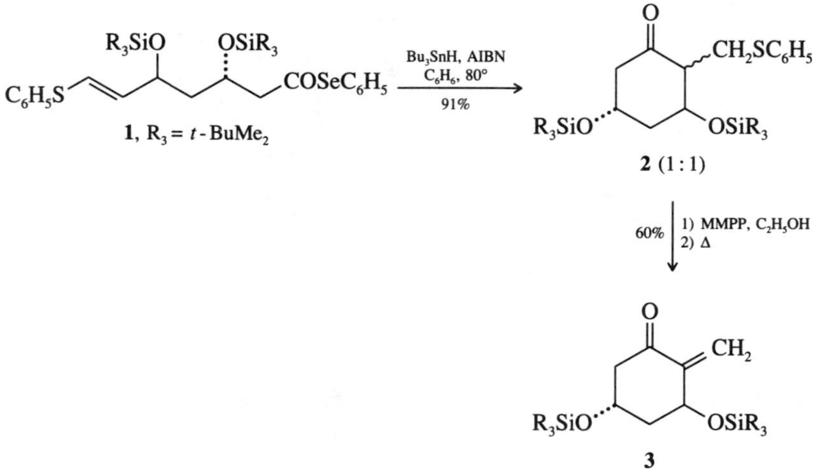

magnesium monoperoxyphthalate provides **3**, a unit present in 1α,25-dihydroxy vita-min D₃. The same cyclization of the *syn*-isomer **4** of **1** provides cyclohexanones **5** in 81% yield.

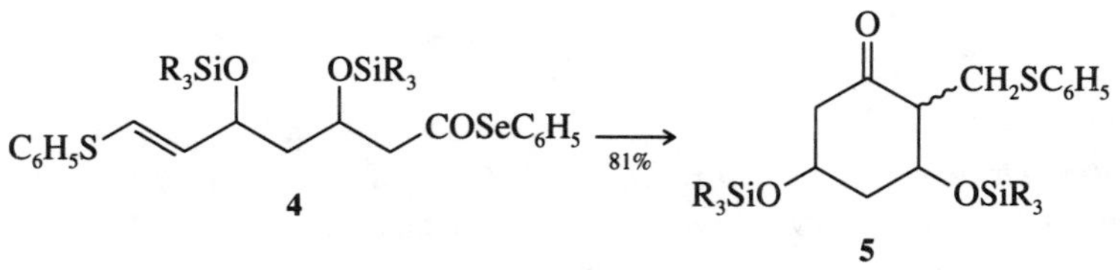

β-*Glycosides*.[12]   Hemithioortho esters such as **1** are reduced by Bu₃SnH (AIBN) mainly to β-glucosides (**2**). The method can also produce the maltose disaccharide **4** as a 10:1 mixture of anomers. The effect of anomeric C_2-substituents is not clear.

Desulfonylation of β-keto sulfones.[13]   Radical reduction of these sulfones with Bu₃SnH and AIBN can be more efficient than the conventional method with Al/Hg. An excess of initiator is necessary for ready and complete reduction, but yields of 80–95% can be obtained by use of 4 equiv. of the stannane and 2 equiv. of AIBN in refluxing toluene.

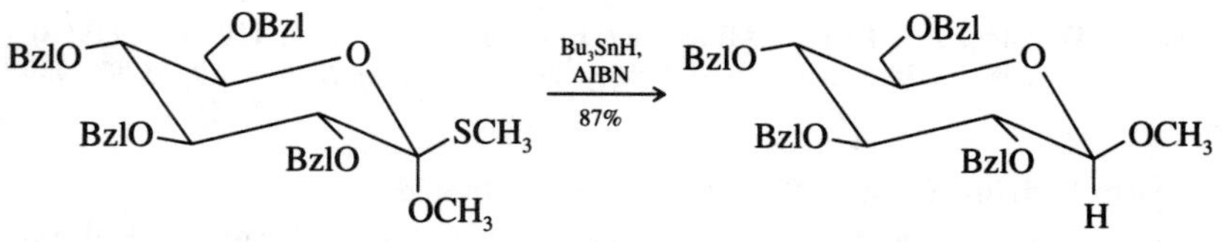

BzlO—, OBzl, O — Bu₃SnH, AIBN, 87% →

(Structures)

1

2 (β/α = 12 : 1)

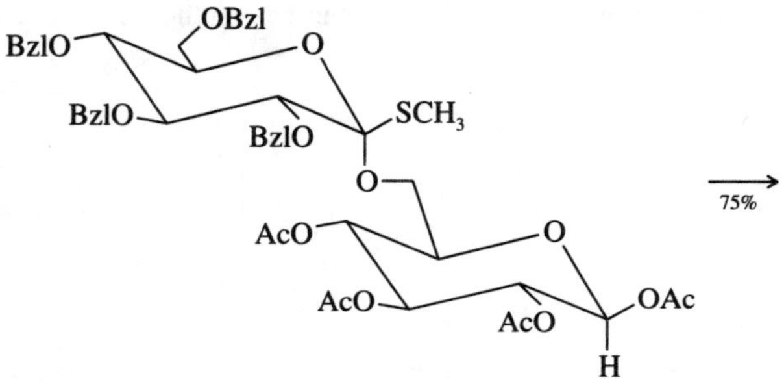

3

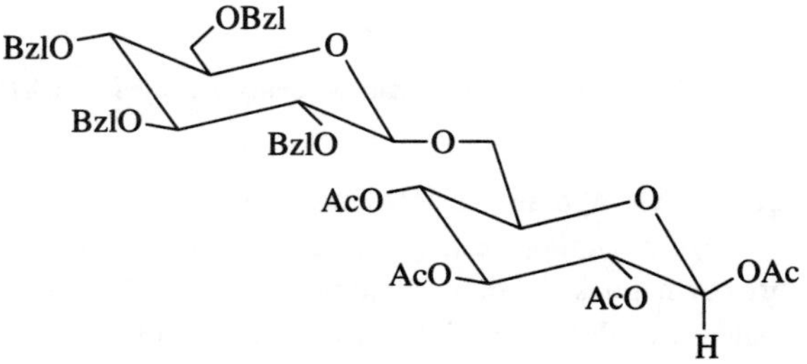

75% →

4 (10 : 1)

[1] P. Dowd and S.-C. Choi, *Tetrahedron Letters*, **32**, 565 (1991).

[2] Y. Guindon, J.-F. Lavallée, L. Boisvert, C. Chabot, D. Delorme, C. Yoakim, D. Hall, R. Lemieux, and B. Simoneau, *Tetrahedron Letters*, **32**, 27 (1991).

[3] Y. Guindon, C. Yoakim, R. Lemieux, L. Boisvert, D. Delorme, and J.-F. Lavallée, *Tetrahedron Letters*, **31**, 2845 (1990).

[4] H. Miyake and K. Yamamura, *Chem. Letters*, 981 (1989).

[5] M. Koreeda and L. G. Hamann, *Am. Soc.*, **112**, 8175 (1990).

[6] V. H. Rawal, R. C. Newton, and V. Krishnamurthy, *J. Org.*, **55**, 5181 (1990).

[7] D. H. R. Barton and J. C. Jaszberenyi, *Tetrahedron Letters*, **30**, 2619 (1989).

[8] I. Ryu, K. Kusano, A. Ogawa, N. Kambe, and N. Sonoda, *Am. Soc.*, **112**, 1295 (1990); I. Ryu, K. Kusano, N. Masumi, H. Yamazaki, A. Ogawa, N. Sonoda, *Tetrahedron Letters*, **31**, 6887 (1990).

[9] D. L. Boger and R. J. Mathvink, *Am. Soc.*, **112**, 4003, 4008 (1990).

[10] B. Giese, *Angew. Chem. Int. Ed.*, **28**, 969 (1989).

[11] D. Batty, D. Crich, and S. M. Fortt, *J.C.S. Chem. Comm.*, 1366, 1696 (1989).

[12] D. Kahne, D. Yang, J. J. Lim, R. Miller, and E. Paguaga, *Am. Soc.*, **110**, 8716 (1988).
[13] A. B. Smith, III, K. J. Hale, and J. P. McCauley, Jr. *Tetrahedron Letters*, **30**, 5579 (1989).

Tributyltin hydride–Copper(I) iodide–Lithium chloride.

Conjugate reduction of α,β-enones.   The combination of Bu$_3$SnH and CuI solubilized with LiCl (2.5 equiv.) results in a hydridocuprate (**1**) that reduces α,β-enones to the ketones in 67–98% yield. Addition of ClSi(CH$_3$)$_3$ can be beneficial.

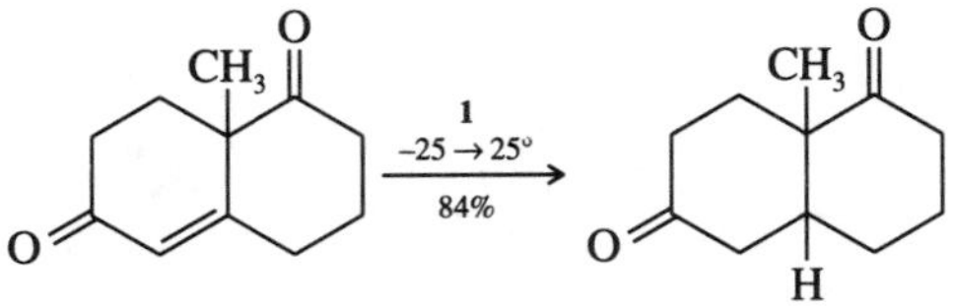

(*cis/trans* = 12 : 1)

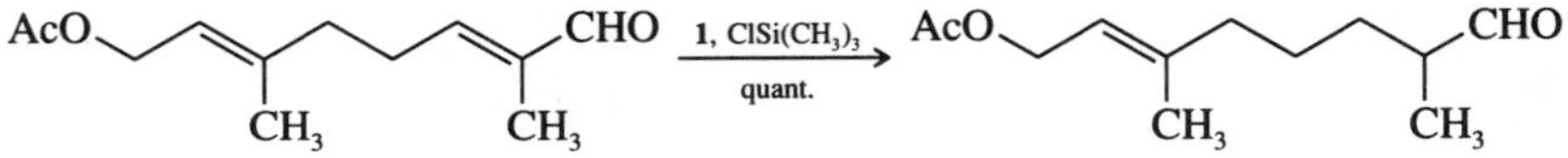

[1] B. H. Lipshutz, C. S. Ung, and S. Sengupta, *Synlett*, 64 (1989).

Tributyltin hydride–Triethylborane.

Hydrodehalogenation.   Triethylborane (10 mole %) can initiate reduction of organic halides, particularly iodides or bromides, by Bu$_3$SnH even at −78°. Alkenyl iodides are also reduced under these conditions. Aryl halides require room temperature or higher. Yields are comparable to those obtained by radical initiators.

[1] K. Miura, Y. Ichinose, K. Nozaki, K. Fugami, K. Oshima, and K. Utimoto, *Bull. Chem. Soc. Japan*, **62**, 143 (1989).

Tributyltinlithium, Bu$_3$SnLi.

Cyclic ethers from lactones.[1]   The ready addition of RLi to the C=S group of thionolactones (**14**, 9) has been used to prepare functionalized cyclic ethers from lactones of the same ring size. The first step involves Lawesson thionation of lactone

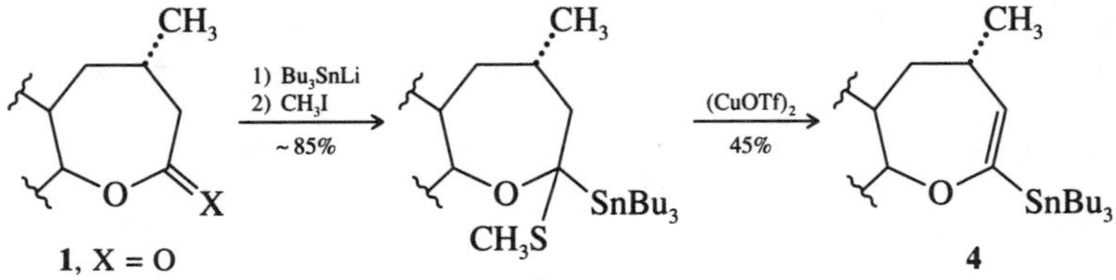

(1) to provide a thionolactone (2). Addition of Bu_3SnLi followed by CH_3I provides a methylthio ether (3), which eliminates methyl mercaptan on treatment with copper(I) triflate to provide a vinyltin 4. This product can be used as is or can undergo transmetallation to a vinyllithium, which undergoes ready reaction with various electrophiles. The sequence from thionolactone to vinyltins is general.

Acylstannanes.[2] These tin reagents can be prepared by reaction of carboxylic esters with Bu_3SnLi (1) and BF_3 etherate (2 equiv. of each) in THF at $-78°$. The Lewis acid is not necessary in the reaction of 1 with thiol esters, R^1COSR^2. Acid chlorides are inferior to esters in this synthesis.

$$C_6H_5COOC_2H_5 + Bu_3SnLi \xrightarrow[60\%]{BF_3 \cdot O(C_2H_5)_2} C_6H_5COSnBu_3$$

[1] K. C. Nicolaou, D. G. McGarry, and P. K. Sommers, *Am. Soc.*, **112**, 3696 (1990).
[2] A. Capperucci, A. Degl'Innocenti, C. Faggi, G. Reginato, A. Ricci, P. Dembech, and G. Seconi, *J. Org.*, **54**, 2966 (1989).

Tributyltin methoxide, Bu_3SnOCH_3. Supplier: Aldrich.

Stannylquinones.[1] The quinone synthesis based on addition of alkynyllithiums to substituted cyclobutenediones (13, 209–210, 284) can provide stannylquinones. Thus thermolysis of the alkynylcyclobutenol 1 with Bu_3SnOCH_3 results in rearrangement to the stannylquinone 2. As expected, these stannylquinones undergo palladium-catalyzed cross-coupling with organic halides (Stille reaction, **14**, 35), particularly with allylic halides.

[1] L. S. Liebeskind and B. S. Foster, *Am. Soc.*, **112**, 8612 (1990).

Tri-μ-carbonylhexacarbonyldiiron, $Fe_2(CO)_9$.

Alkyne–alkene carbonylative coupling. Intramolecular carbonylative coupling of dialkynes catalyzed by $Fe(CO)_5$ provides a route to cyclopentadienones (equation I). The more difficult carbonylative alkyne–alkene coupling to provide cyclopentenones (Pauson–Khand reaction) can also be effected with $Fe(CO)_5$, but in modest yield. In an improved coupling, acetone is treated with $Fe_2(CO)_9$ to form Fe-

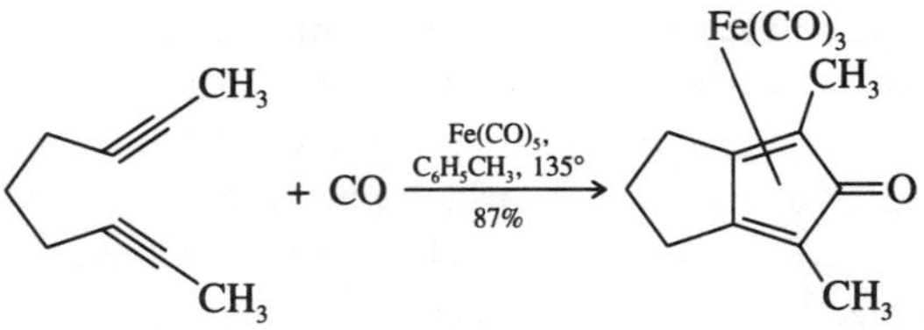

(CO)$_4$(acetone). This complex reacts with an enyne such as **1** in toluene at 145° under 55 psi of CO to form a cyclopentenone (**2**) in high yield.

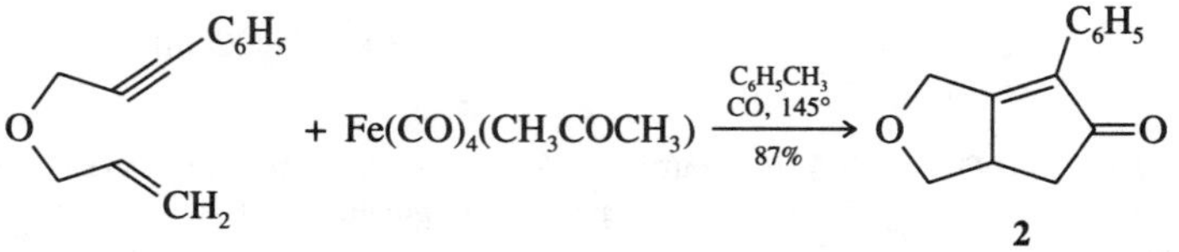

[1] A. J. Pearson and R. A. Dubbert, *J.C.S. Chem. Comm.*, 202 (1991).

Tri-μ-carbonylnonacarbonyltetrarhodium, Rh$_4$(CO)$_{12}$.

Three-component aldol synthesis.[1] This rhodium carbonyl can promote aldol coupling of enol silyl ethers with aldehydes or ketones. It can also effect coupling of an enone, an aldehyde, and a trialkylsilane to provide a silyl aldol. In the case of an enolizable aldehyde, yields are improved by addition of a phosphine ligand such as

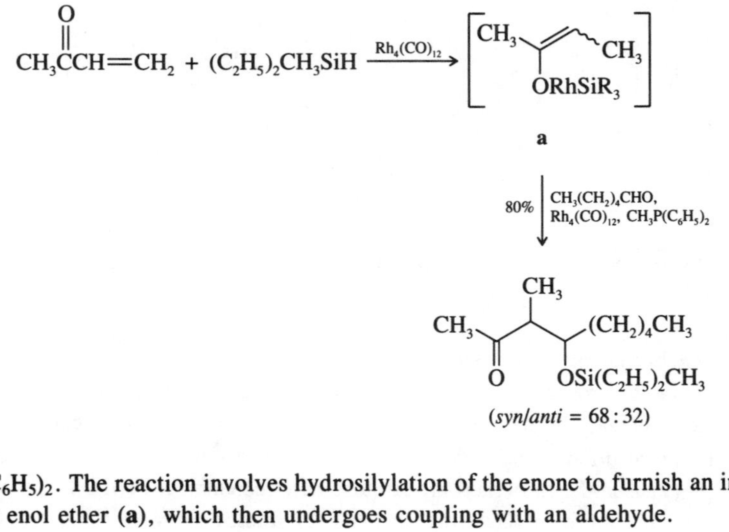

(*syn/anti* = 68 : 32)

CH$_3$P(C$_6$H$_5$)$_2$. The reaction involves hydrosilylation of the enone to furnish an intermediate enol ether (**a**), which then undergoes coupling with an aldehyde.

[1] I. Matsuda, K. Takahashi, and S. Sato, *Tetrahedron Letters*, **31**, 5331 (1990).

2,4,6-Trichlorobenzoyl chloride, 9, 478–479; 11, 552.

Macrolactonization.[1]   High dilution is usually essential for macrolactonization, but the 14-membered lactone, 9-dihydroerythronolide A (**3**), can be obtained in almost quantitative yield by lactonization of the seco-acid **2** via the mixed anhydride formed with 2,4,6-trichlorobenzoyl chloride (**1**) with triethylamine and 4-dimethyl-

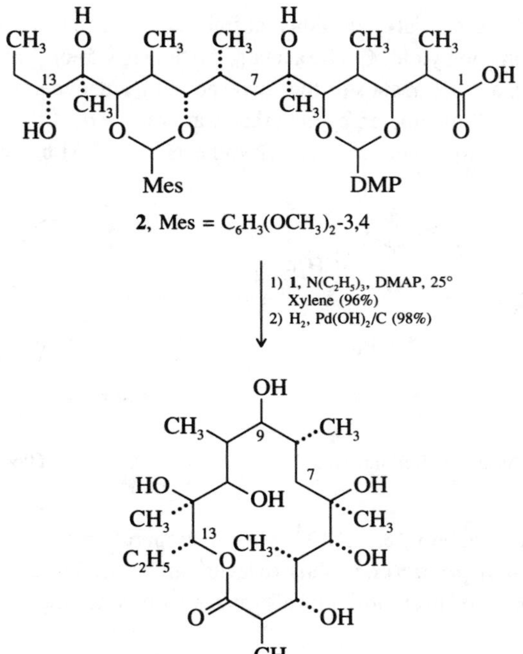

2, Mes = $C_6H_3(OCH_3)_2$-3,4

1) **1**, $N(C_2H_5)_3$, DMAP, 25°
   Xylene (96%)
2) H_2, Pd(OH)$_2$/C (98%)

3

aminopyridine. Lactonization of **2** is also possible with mesitylenesulfonyl chloride, but in somewhat lower yield.

[1] M. Hikota, H. Tone, K. Horita, and O. Yonemitsu, *J. Org.*, **55**, 7 (1990); M. Hikota, Y. Sakurai, K. Horita, and O. Yonemitsu, *Tetrahedron Letters*, **31**, 6367 (1990).

Trichloroisopropoxytitanium, Cl₃TiO-*i*-Pr (1).

β-Alkoxy cyclic ethers.[1]   This Lewis acid is particularly effective for an intramolecular reaction of a γ-alkoxy allyltin group with an acetal group (*cf.* **14**, 16). This reaction has been developed as a route to β-alkoxy cyclic ethers. Thus in the presence

2

3 (*trans/cis* = 1.3 : 1)

of **1** (2 equiv.), the ω-trialkyltin ether acetal (**2**) cyclizes to an α-vinyl, β-alkoxy cyclic ether (**3**) in good yield. Cyclic acetals, particularly 5- or 7-membered ones, are more useful than acyclic acetals or the corresponding aldehyde. This cyclization can result in 6-, 7-, or 8-membered cyclic ethers, as well as α,α′,β-trisubstituted cyclic ethers. It was used for a stereoselective synthesis of a [6,8] bicyclic ether (**4**).

4 (α,β-*trans*/α,β-*cis* = 11 : 1)

[1] J. Yamada, T. Asano, I. Kadota, and Y. Yamamoto, *J. Org.*, **55**, 6066 (1990).

Trichloromethyl carbonate, $(Cl_3CO)_2C=O$. Preparation.[1]

α-Chloro chloroformates.[2]   This reagent can be used as a substitute for phosgene in a synthesis of α-chloro chloroformates from aldehydes.

$$CH_3(CH_2)_8CHO + (Cl_3CO)_2C=O \xrightarrow[94\%]{\substack{Py, CCl_4 \\ -10-40°}} CH_3(CH_2)_8CH(Cl)O\overset{\overset{\displaystyle O}{\|}}{C}Cl$$

[1] H. Eckert and B. Forster, *Angew. Chem. Int. Ed.*, **26**, 894 (1987).
[2] M. J. Coghlan and B. A. Caley, *Tetrahedron Letters*, **30**, 2033 (1989).

Triethoxysilane–Palladium(II) acetate.
    Polymerization of triethoxysilane in water catalyzed by $Pd(OAc)_2$ results in a palladium-deposited siloxane that can be used as a heterogeneous catalyst for reductions with hydrogen.
    Hydrogenation.[1]   Alkenes, including α,β-unsaturated ketones and esters, can be reduced by triethoxysilane and 5 mole % of $Pd(OAc)_2$ in THF/H_2O (5 : 1) at 25°.

Water is essential for this reduction. This system reduces alkynes to *cis*-alkenes with high stereoselectivity. Addition of methyl propynoate (10 mole %) increases the reactivity and can effect reduction of alkynes to alkanes.

[1] J. M. Tour, J. P. Cooper, and S. L. Pendalwar, *J. Org.*, **55**, 3452 (1990).

Triethylgallium, $(C_2H_5)_3Ga$.

Lactams from ω-amino carboxylic acids. Five- to eight-membered lactams can be formed from ω-amino carboxylic acids by treatment with $(C_2H_5)_3Ga$ even in the absence of high dilution. Triethylaluminum is less useful because of thermal decomposition.

$$H_2N(CH_2)_5COOH \xrightarrow[81\%]{\substack{(C_2H_5)_3Ga, \\ C_6H_5CH_3, \Delta}} \text{(azepan-2-one)}$$

[1] Y. Yamamoto and T. Furuta, *Chem. Letters*, 797 (1989).

Triethyl phenylsulfinylorthoacetate, $C_6H_5\overset{\displaystyle O}{\overset{\|}{S}}CH_2C(OC_2H_5)_3$ (1). The reagent is obtained by reaction of 2,2,2-trichloroethyl phenyl sulfoxide or 2,2-dichlorovinyl phenyl sulfoxide with sodium ethoxide.

Claisen rearrangement of allylic alcohols to ethyl dienoates. Claisen rearrangement of allylic alcohols with an orthoacetate is known to provide 2-carbon homologated γ,δ-unsaturated esters (**6**, 607–608). Reaction with the phenylsulfinyl-orthoacetate **1** is accompanied by an *in situ* sulfoxide elimination to provide 2-carbon homologated dienoate esters (equation I). This novel reagent was used to convert the

(I) $CH_2{=}CHCHCH_3$ + 1 $\xrightarrow[79\%]{H^+, 150°}$ $C_2H_5OOC\diagdown\diagup\diagdown\diagup CH_3$

        |
        OH

$(-)$ **2**   **3**

allylic alcohol (−)-**2** into the precursor (**3**) for the A-ring of 1α,25-dihydroxyvitamin D₃.

¹ G. H. Posner and C. M. Kinter, *J. Org.*, **55**, 3967 (1990).

Triethylsilane, b. p. 108°.

Trifluoroacetic acid deprotection. Trialkylsilanes in combination with an acid as a hydride donor are useful for ionic hydrogenation. DuPont chemists recommend triethyl- or triisopropylsilane as scavengers for carbocationic by-products formed in TFA deprotection of amino acids. In fact, triethylsilane can be more effective than ethanedithiol, a commonly used scavenger.

¹ D. A. Pearson, M. Blanchette, M. L. Baker, and C. A. Guindon, *Tetrahedron Letters*, **30**, 2739 (1989).

Triethylsilane–Dodecanethiol, $(C_2H_5)_3SiH–CH_3(CH_2)_{11}SH$.

Reduction of RX to RH.[1] In the presence of di-*t*-butylhyponitrite (initiator) and a thiol,[2] triethylsilane reduces alkyl chlorides, bromides, or iodides to alkanes in >91% yield by a chain reaction in which the thiol effects transfer of H from the silane to an alkyl radical. This reduction is generally effected with a R_3SnH, which is toxic and more difficult to remove from the products.

¹ R. P. Allen, B. P. Roberts, and C. R. Willis, *J.C.S. Chem. Comm.*, 1387 (1989).
² H. Kiefer and T. G. Traylor, *Tetrahedron Letters*, 6163 (1966).

Triethylsilane–Palladium/C.

RCOOH → RCHO.[1] This reaction can be conducted in two steps: conversion of RCOOH to the ethyl thiolester, which is reduced by triethylsilane catalyzed by Pd/C to an aldehyde (equation I). This method is suitable for conversion of optically active amino acids to amino aldehydes without racemization.

$$(I) \quad RCOOH \xrightarrow[\substack{60-85\%}]{\substack{1)\ C_2H_5OCOCl,\ N(C_2H_5)_3 \\ 2)\ C_2H_5SH,\ DMAP}} R\overset{\overset{\textstyle O}{\|}}{C}SC_2H_5$$

$$\Big\downarrow \substack{(C_2H_5)_3SiH,\ Pd/C, \\ acetone} \quad 75-92\%$$

$$RCHO + (C_2H_5)_3SiSC_2H_5$$

¹ T. Fukuyama, S.-C. Lin, and L. Li, *Am. Soc.*, **112**, 7050 (1990).

Triethylsilane–Trifluoromethanesulfonic acid.

Hydrogenation of alkenes.[1] Ionic hydrogenation of alkenes has generally been conducted with a trialkylsilane and an acid no stronger than trifluoroacetic acid.

$$(CH_3)_2C{=}CHC_2H_5 \xrightarrow[\substack{CH_2Cl_2, \ -75° \\ 98\%}]{(C_2H_5)_3SiH, \ CF_3SO_3H}} (CH_3)_2CHCH_2C_2H_5$$

Actually hydrogenation with the silane and CF_3SO_3H proceeds in almost quantitative yield at $-75°$ in 5 minutes. Similar results are obtained by use of CF_3SO_3H and $CpMoH(CO)_3$ as the hydride donor, and in fact even tetrasubstituted alkenes can be reduced to alkanes.

$$(CH_3)_2C{=}C(CH_3)_2 \xrightarrow[\substack{-75°, \ 5 \ min.}]{CpMoH(CO)_3, \ CF_3SO_3H} (CH_3)_2CH{-}CH(CH_3)_2$$

[1] R. M. Bullock and B. J. Rappoli, *J.C.S. Chem. Comm.*, 1447 (1989).

Trifluoroethyl trifluoroacetate, $CF_3CO_2CH_2CF_3$ (1).

α-Diazo ketones.[1] Direct diazo transfer to ketone enolates is generally not productive, and α-diazo ketones are generally prepared by reaction of sulfonyl azides with α-formyl ketones with elimination of a sulfonylformamide. Danheiser's group finds that α-diazo ketones are obtained in significantly higher yield via α-trifluoro-acetates of ketones, particularly if the enolate is generated with lithium hexamethyl-

disilazide in THF at $-78°$. Trifluoroethyl trifluoroacetate is superior to ethyl trifluoroacetic anhydride for trifluoroacetylation. Although tosyl azide is the traditional reagent for diazo transfer, methanesulfonyl azide is a convenient reagent, and can be prepared in $>90\%$ yield by reaction of MsCl with NaN_3 in acetone.

[1] R. L. Danheiser, R. F. Miller, R. G. Brisbois, and S. Z. Park, *J. Org.*, **55**, 1959 (1990).

Trifluoromethanesulfonic anhydride.

α-Glycosylation.[1] Triflic anhydride is highly effective for α-glycosylation of 2,3,4,6-tetrabenzyl-β-D-glucopyranosyl fluoride with a dissacharide. It is superior to TiF_4, $AgOTf/SnCl_2$, or trimethylsilyl triflate, listed in the order of decreasing reactivity.

Glycosylation of phenols and even hindered alcohols is possible by reaction with an O-protected 1-phenylsulfinylglucose (**1**, excess) and triflic anhydride at $-78°$ (equation I).[2] The α/β ratio is influenced by the protective groups and the solvent. Use of CH_2Cl_2 or propionitrile favors β-glycosylation, whereas toluene favors α-glycosylation.

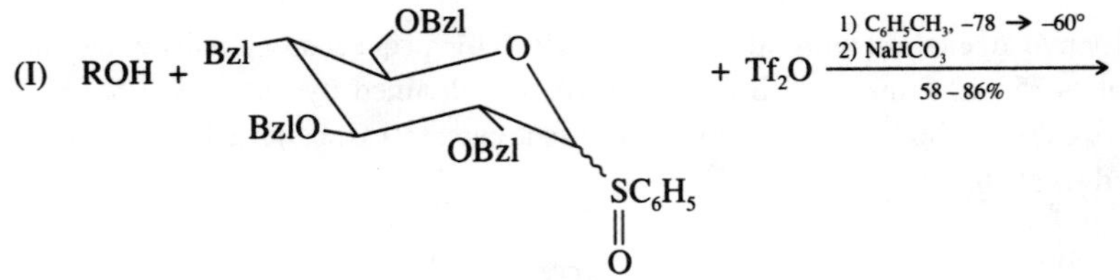

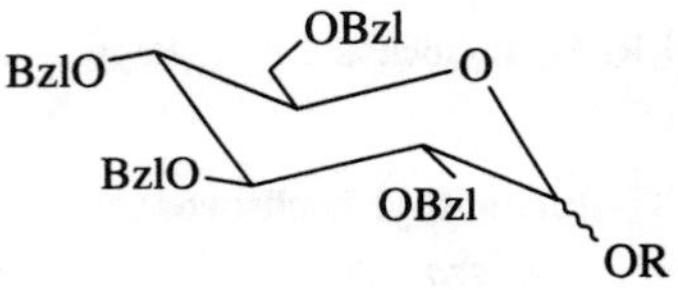

$(\alpha/\beta = 27:1)$

Dehydration of formamides to isonitriles.[3]   This reaction can be effected in high yield at $-78°$ with the combination of triflic anhydride and a tertiary amine. It is effective even in the case of vinyl formamides.

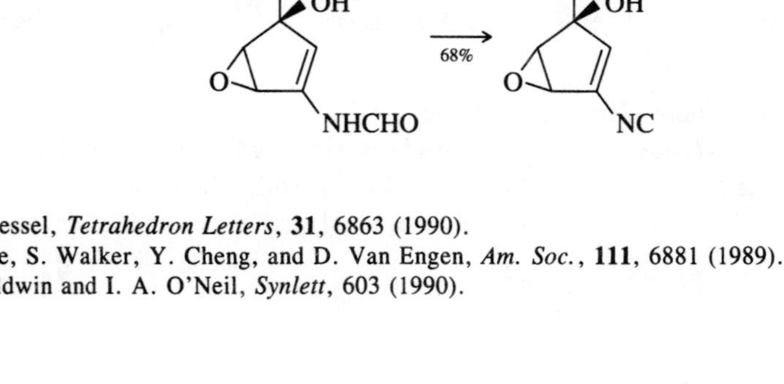

[1] H. P. Wessel, *Tetrahedron Letters*, **31**, 6863 (1990).
[2] D. Kahne, S. Walker, Y. Cheng, and D. Van Engen, *Am. Soc.*, **111**, 6881 (1989).
[3] J. E. Baldwin and I. A. O'Neil, *Synlett*, 603 (1990).

Triisobutylaluminum, $[(CH_3)_2CHCH_2]_3Al$ (**1**).

Annelation of acetals. A few years ago Mori and Yamamoto[1] reported that this trialkylaluminum converts the acetal **2** to the enol ether **3** in high yield. Since then, the reaction has been shown to be applicable to 1,3-dioxanes of acyclic and cyclic

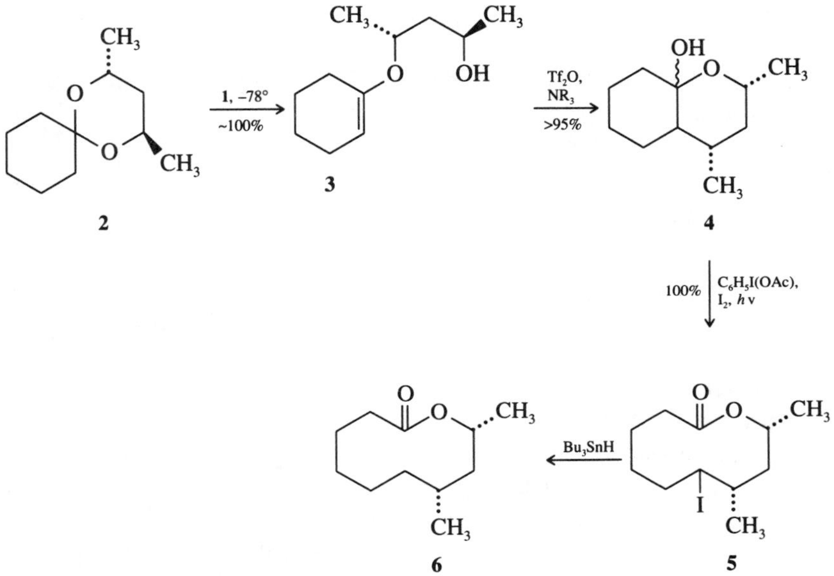

ketones and to provide a general and stereoselective route to lactones.[2] Thus the enol ether **3** cyclizes to a hemiacetal (**4**) on treatment with triflic anhydride and ethyldiiso-propylamine. Hemiacetals of this type are readily converted into iodolactones (**5**) on photolysis with $C_6H_5I(OAc)_2$ and I_2. The cyclization step proceeds with complete inversion at the hydroxy groups as shown by formation of a single lactone (**6**) after reduction with Bu_3SnH.

This novel annelation not only provides a route to chiral lactones, but was applied to a synthesis of a diol (**8**) from the acetal **7** as a precursor to (−)-lardolure (**9**).

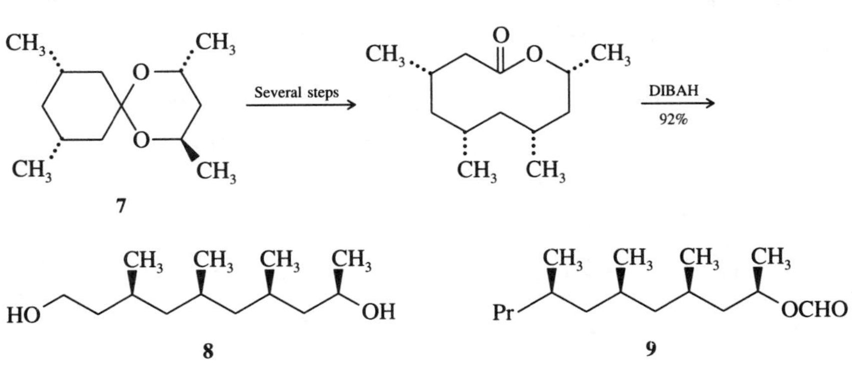

[1] A. Mori and H. Yamamoto, *J. Org.*, **50**, 5444 (1985).
[2] M. Kaino, Y. Naruse, K. Ishihara, and H. Yamamoto, *J. Org.*, **55**, 5814 (1990).

Triisopropyl phosphite, $P(O-i-Pr)_3$.

$>C=O \longrightarrow >CH_2$.[1] Aldehydes and ketones are reduced to hydrocarbons when heated for about 24 hours in refluxing triisopropyl phosphite with continuous removal of acetone. In general, yields are higher from aldehydes than from ketones, and highly hindered ketones are not reduced.

$$R^1R^2C=O + P(O-i-Pr)_3 \xrightarrow[50-86\%]{} R^1R^2CH_2 + 2(CH_3)_2C=O$$

[1] G. A. Olah and A. Wu, *Synlett*, 54 (1990).

Trimethylamine N-oxide.

$RCH_2Br \rightarrow RCHO$.[1] Anhydrous trimethylamine N-oxide[2] in DMSO can oxidize primary bromides (but not the chlorides) to aldehydes at 25° in yields of about 80% (5 hours reaction). Allylic and benzylic chlorides and bromides are oxidized more rapidly at 0° in a mixed solvent of $DMSO/CH_2Cl_2$.

Benzylic hydroxylation.[3] Direct introduction of the C_{10}-hydroxyl group into anthracycline glycosides can be effected with $(CH_3)_3NO$. Thus 1 is converted into oxaunomycin (2) in 82% yield from reactions in DMF. The yield is only 53% when acetone is the solvent, and is also lower (34%) in the absence of the C_{11}-hydroxyl group.

[1] A. G. Godfrey and B. Ganem, *Tetrahedron Letters*, **31**, 4825 (1990).
[2] J. A. Soderquist and C. A. Anderson, *ibid.*, **27**, 3961 (1986).
[3] S. Nakajima, H. Kawai, and M. Sakakibara, *ibid.*, **30**, 4857 (1989).

Trimethylsilyldiazomethane, $(CH_3)_3SiCHN_2$.

(Z)-1-Trimethylsilyl-1-alkenes.[1] The α-trimethylsilyl diazoalkanes (**2**), prepared by reaction of primary halides (**1**) with the lithium anion of trimethylsilyl diazomethane, decompose on treatment with rhodium(II) pivalate [superior in this case to rhodium(II) acetate] to (Z)-1-trimethylsilyl-1-alkenes.

$$RCH_2X \xrightarrow{(CH_3)_3SiC(Li)N_2} RCH_2\overset{\overset{\displaystyle N_2}{\|}}{C}Si(CH_3)_3$$

$$\mathbf{1} \qquad\qquad\qquad \mathbf{2}$$

$$78-96\% \downarrow Rh_2(OCO\text{-}t\text{-Bu})_4$$

$$(Z/E = 80-97:20-3)$$

[1] T. Aoyama and T. Shioiri, *Chem. Pharm. Bull.*, **37**, 2261 (1989).

β-Trimethylsilylethoxymethyl chloride (SEM-Cl), $ClCH_2OCH_2CH_2Si(CH_3)_3$.

Aldol-type reactions.[1] Reaction of lithium enolates with SEM chloride results in α-hydroxymethyl ketones protected as the β-trimethylsilylethyl ethers, which can undergo deprotection with Bu_4NF or TFA. Yields are in the range of 55–80%.

Directed metallation.[2] SEM ethers of phenols are as useful as MOM (methoxymethyl) ethers for directed *ortho*-metallation of aromatic systems, and have the added advantage of removal by fluoride ion. This group is also useful for directed lithiation of hydroxypyridines.

[1] D. Crich and L. B. L. Lim, *Synlett*, 117 (1990).
[2] S. Sengupta and V. Snieckus, *Tetrahedron Letters*, **31**, 4267 (1990).

(2-Trimethylsilyloxy)allyltrimethylsilane, $CH_2=C\overset{\diagup OSi(CH_3)_3}{\diagdown CH_2Si(CH_3)_3}$ (**1**).

Preparation:

$$CH_3COCl \xrightarrow[80\%]{(CH_3)_3SiCH_2MgCl-CuI} CH_3\overset{\overset{\displaystyle O}{\|}}{C}CH_2Si(CH_3)_3 \xrightarrow[87\%]{\substack{1)\ LDA \\ 2)\ (CH_3)_3SiCl}} \mathbf{1}$$

Double alkylation.[1]   When activated by TiCl$_4$ (1 equiv.) **1** reacts with acetals or carbonyl compounds to give products of double alkylation, probably via a 1,3-C → O Si rearrangement. Thus **1** can function as the α,α'-dianion of acetone.

$$C_6H_5CH(OCH_3)_2 + 1 \xrightarrow[69\%]{TiCl_4, -78°} C_6H_5\overset{\displaystyle OCH_3}{\underset{\displaystyle |}{C}}HCH_2\overset{\displaystyle O}{\underset{\displaystyle ||}{C}}CH_2\overset{\displaystyle OCH_3}{\underset{\displaystyle |}{C}}HC_6H_5$$

$$C_6H_5CHO + 1 \xrightarrow[63\%]{TiCl_4, -78°} C_6H_5\overset{\displaystyle OH}{\underset{\displaystyle |}{C}}HCH_2\overset{\displaystyle O}{\underset{\displaystyle ||}{C}}CH_2\overset{\displaystyle OH}{\underset{\displaystyle |}{C}}HC_6H_5$$

[1] A. Hosomi, H. Hayashida, and Y. Tominaga, *J. Org.*, **54**, 3254 (1989).

2-(Trimethylsilyl)thiazole, (**1**), b.p. 73–75°/15 mm.)

The reagent is prepared by treatment of 2-bromothiazole (Aldrich) with BuLi followed by ClSi(CH$_3$)$_3$ (85% yield).

Formyl anion equivalent.[1]   This reagent can be used as the equivalent of the formyl anion. Thus it reacts with an aldehyde such as **2** in the absence of a catalyst to form mainly the *anti*-adduct. After protection of the hydroxy group, the thiazole group can be converted to a formyl group by N-methylation, reduction, and hydrolysis.

2

3 (≥95% de)

96% | C$_6$H$_5$CH$_2$Br

1) CH$_3$I
2) NaCNBH$_3$
3) HgCl$_2$, H$_2$O
62%

5

4

¹ A. Dondoni, G. Fantin, and M. Fogagnolo, *Org. Syn.*, submitted (1990).

Trimethylsilyl trifluoromethanesulfonate, TMSOTf.

*Cyclic α-amino acids.*¹ The unsaturated imine 1, prepared as shown, on treatment with TMS triflate (1 equiv.) cyclizes to a mixture of *trans*- and *cis*-2, with marked preference for the former cyclic amino acid. The selectivity is dependent on the solvent. It is highest (33:1) in toluene, but the highest yield (55%) and cleanest reaction is obtained in *t*-butyl methyl ether even though the diastereoselectivity is lower (18:1). Lewis acids do not initiate this cyclization.

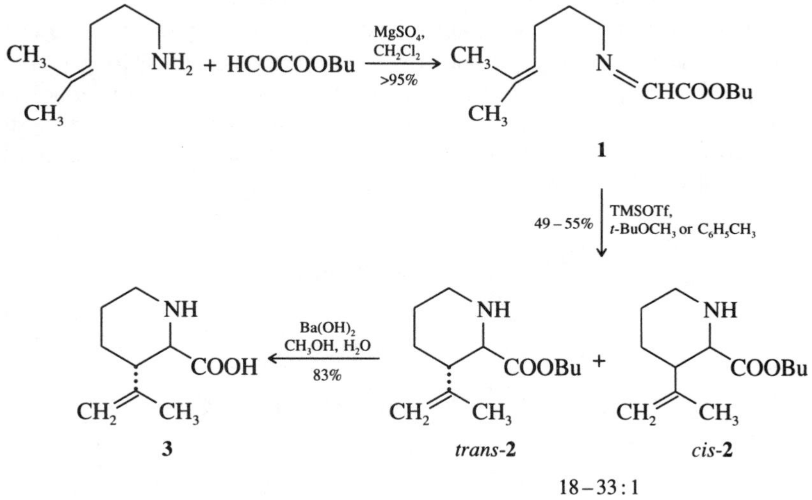

1

3   *trans*-2   *cis*-2

18–33 : 1

The doubly activated imines **4** can be cyclized by Lewis acids or TMS triflate to provide the cyclic amino acid derivative **5** after saponification and decarboxylation.

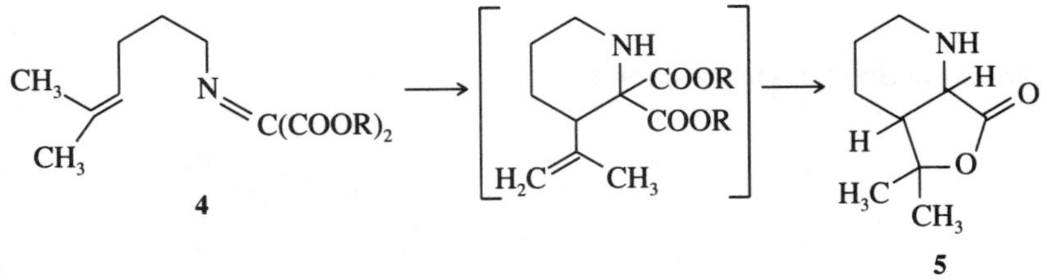

4   **5**

*Conjugate addition of alkynylzinc bromides.*² Alkynylzinc bromides in combination with 1 equiv. of a trialkylsilyl triflate undergo conjugate addition to cyclic and acyclic α,β-enones at −40° in THF or ether. A mixture of 1,2- and 1,4-adducts is formed from β-disubstituted α,β-enones.

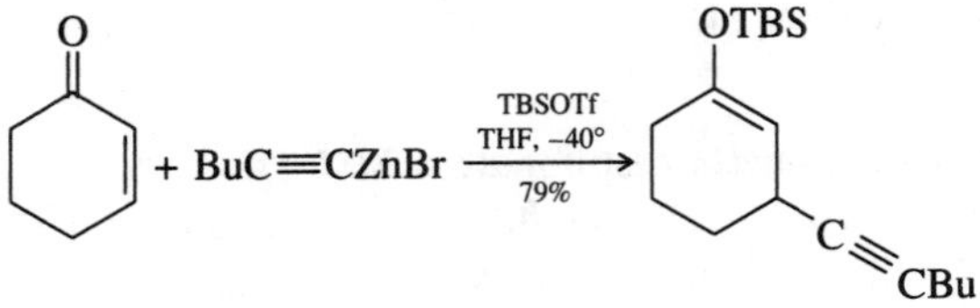

[1] L. F. Tietze and M. Bratz, *Synthesis*, 439 (1989).
[2] S. Kim and J. M. Lee, *Tetrahedron Letters*, **31**, 7627 (1990).

Trimethylsilyl trifluoromethanesulfonate–Titanium(IV) chloride.

Spiroannelation of silyl enol ethers[1] (**15**, 237–238). Spiroannelation can be effected by reaction of a silyl enol ether with an alkyltin reagent bearing a terminal acetal group in the presence of trimethylsilyl triflate and $TiCl_4$. An intermolecular reaction is followed by intramolecular ring closure to provide a spiro ring system, best isolated after oxidation of the secondary alcohol group to a ketone. This method provides access to [4.4], [4.5], [4.6], [5.5], and [5.6] spirocyclic systems.

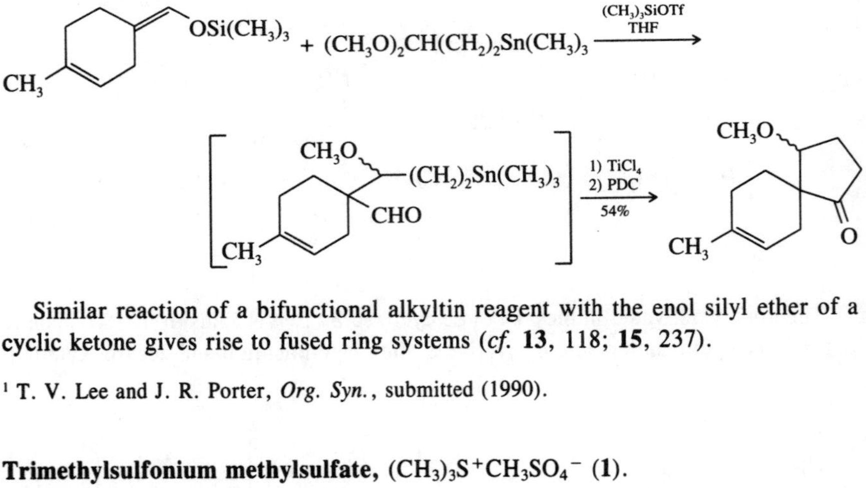

Similar reaction of a bifunctional alkyltin reagent with the enol silyl ether of a cyclic ketone gives rise to fused ring systems (*cf.* **13**, 118; **15**, 237).

[1] T. V. Lee and J. R. Porter, *Org. Syn.*, submitted (1990).

Trimethylsulfonium methylsulfate, $(CH_3)_3S^+CH_3SO_4^-$ (1).

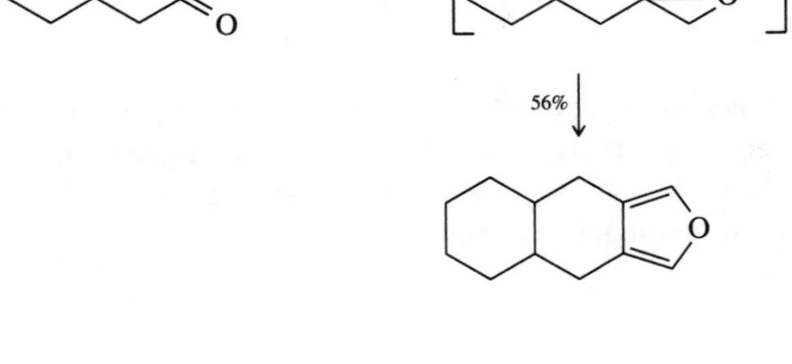

Furan synthesis.[1] The original synthesis of 3,4-disubstituted furans by Garst using dimethylsulfonium methylide (4, 196–197) can be improved by use of **1** under phase-transfer conditions.

[1] M. E. Price and N. E. Schore, *J. Org.*, **54**, 2777 (1989).

Triphenylcarbenium hexachloroantimonate, $TrSbCl_6$.

α-Substituted cyclic ethers.[1] In the presence of this Lewis acid or of a mixed system of $SbCl_5$–$ClSi(CH_3)_3$ and SnI_2 (15, 14–15), the silyl ketene acetal 1-(*t*-butyldi-methylsilyloxy)-1-ethoxyethene (**1**) and a silyl nucleophile react with lactones to form α-substituted cyclic ethers.

$(n = 2-4)$ **1**

75–95% $(C_2H_5)_3SiH$

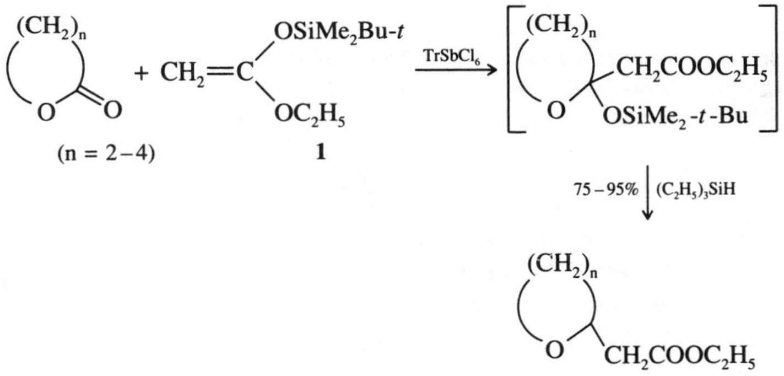

This reaction can also be used to convert 6-alkyl-1,3-dioxan-4-ones, available from β-hydroxy carboxylic acids,[2] to *syn*-1,3-diols.

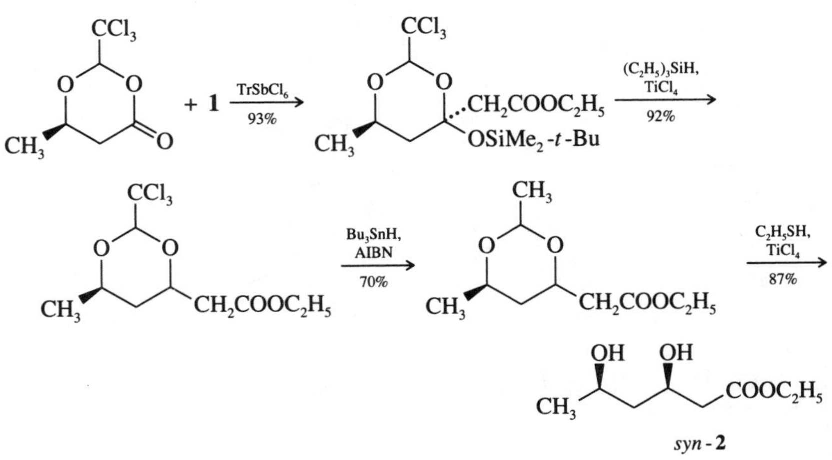

syn - **2**

β-*Amino esters.*[1]   A number of metal salts catalyze the reaction of ketene silyl acetals with imines to form β-amino esters, usually as a mixture of *syn-* and *anti*-isomers. The highest *anti*-selectivity and chemical yield are observed with trityl hexachloroantimonate in the case of N-phenylimines, but iron(II) iodide is equally effective in the case of N-benzyl- or N-diphenylmethylimines.

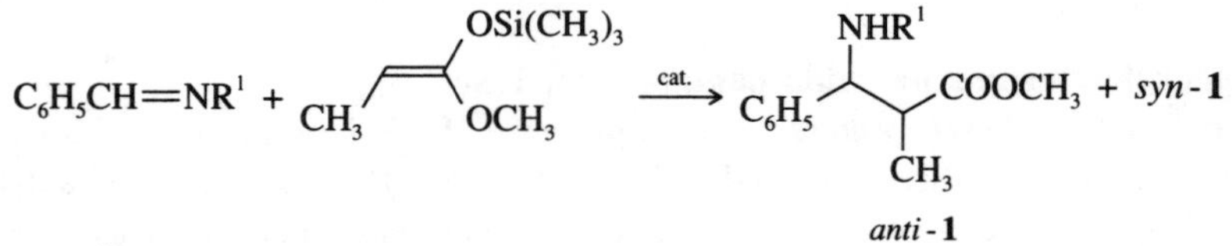

$R^1 = C_6H_5$	FeI_2	92%	82:18
$R^1 = C_6H_5$	$TrSbCl_6$	95%	92:8
$R^1 = Bzl$	FeI_2	77%	91:9
$R^1 = Bzl$	$TrSbCl_6$	71%	93:7

[1] K. Homma, H. Takenoshita, and T. Mukaiyama, *Bull. Chem. Soc.*, **63**, 1898 (1990).
[2] D. Seebach, R. Imwinkelried, and G. Stucky, *Helv.*, **70**, 448 (1987).
[3] T. Mukaiyama, H. Akamatsu, and J. S. Han, *Chem. Letters*, 889 (1990).

Triphenylphosphine/Carbon tetrabromide.

1-Bromo-1-alkynes.[1]   These products are formed in 92–96% yield by reaction of terminal alkynes with CBr_4 (3 equiv.) and $P(C_6H_5)_3$ (6 equiv.) in CH_2Cl_2.

Cyclization of unsaturated ethers.[2]   Reaction of γ- or δ-unsaturated THP or MOM ethers with this phosphine reagent in CH_2Cl_2 at 25° results in cyclization to tetrahydropyrans.

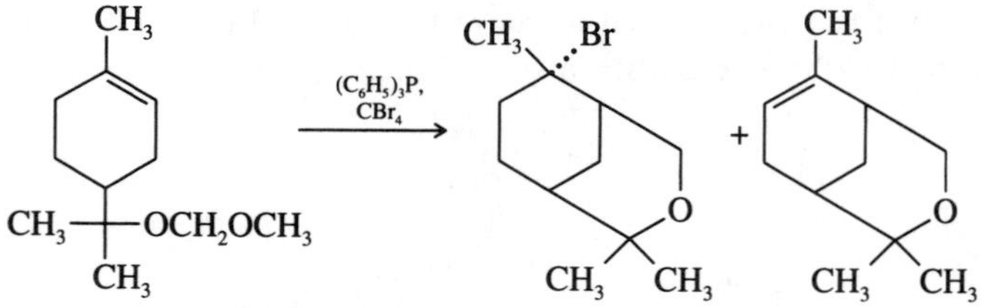

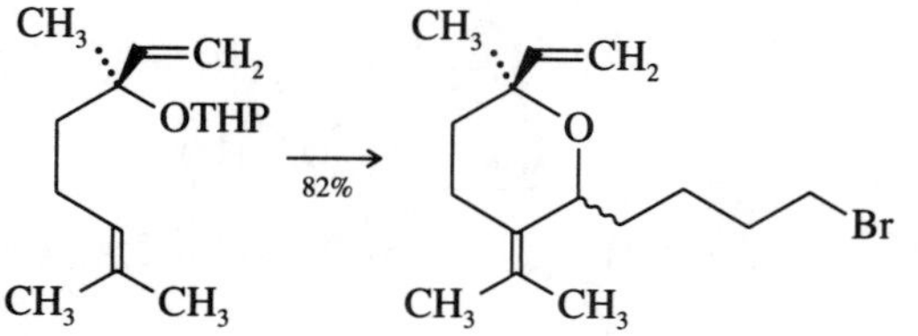

[1] A. Wagner, M. P. Heitz, and C. Mioskowski, *Tetrahedron Letters*, **31**, 3141 (1990).
[2] *Idem., ibid.*, **30**, 1971 (1989).

Triphenylphosphine/Carbon tetrachloride (1).

Bischler–Napieralski cyclodehydration of amides.[1] This reaction is tradition-ally effected with P_2O_5 or $ZnCl_2$. It can also be effected under neutral conditions with $(C_6H_5)_3P$ and CCl_4 (equation I).

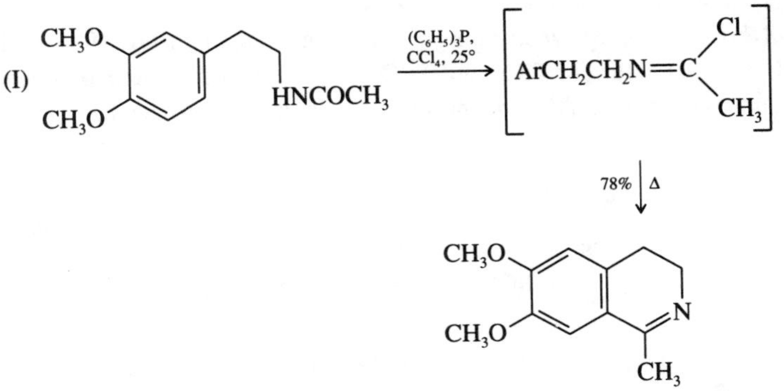

This version also permits a one-pot synthesis of dihydroisoquinolines or β-carbo-lines from 2-arylethylamines and carboxylic acids.

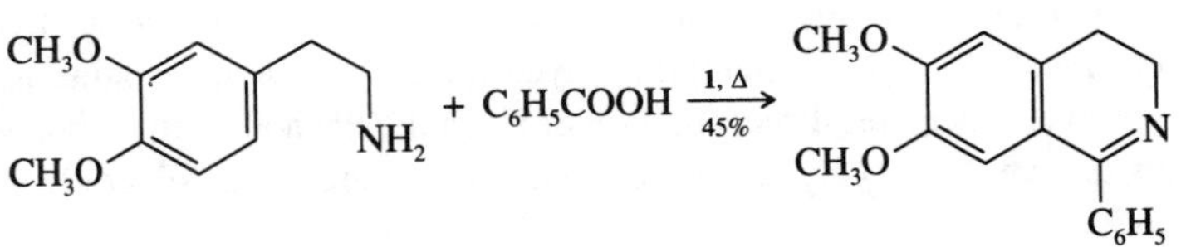

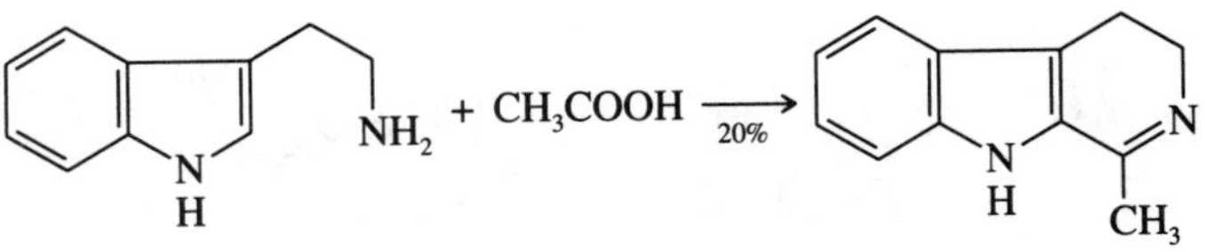

Wittig-type cyclization.[2] 2-Acyloxy- or 2-benzoylbenzoic acids undergo cycli-zation to 3-chloroflavones or 2-chloroindenones, respectively, when treated with excess $(C_6H_5)_3P/CCl_4$ and $N(C_2H_5)_3$ in CH_2Cl_2 in 40–60% yield. This reaction proba-bly involves formation of a chloro ylide (a) via the acid chloride.

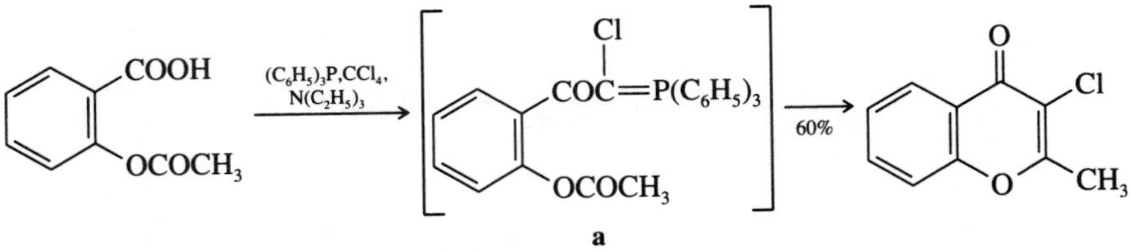

[1] A. Bhattacharjya, P. Chattopadhyay, M. Bhaumik, and S. C. Pakrashi, *J. Chem. Res. (S)*, 228 (1989).
[2] H. Vorbrüggen, B. D. Bohn, and K. Krolikiewiez, *Tetrahedron*, **46**, 3489 (1990).

Triphenylphosphine–Titanium(IV) chloride.

α′-Alkylidenation of cyclic enones.[1]   The combination of $P(C_6H_5)_3$ and 1 equiv. of $TiCl_4$ or of a mixture of $TiCl_4$ and $Ti(O-i-Pr)_4$ activates cyclic α,β-enones for an aldol condensation with aldehydes to form α'-alkylidenocycloenones.

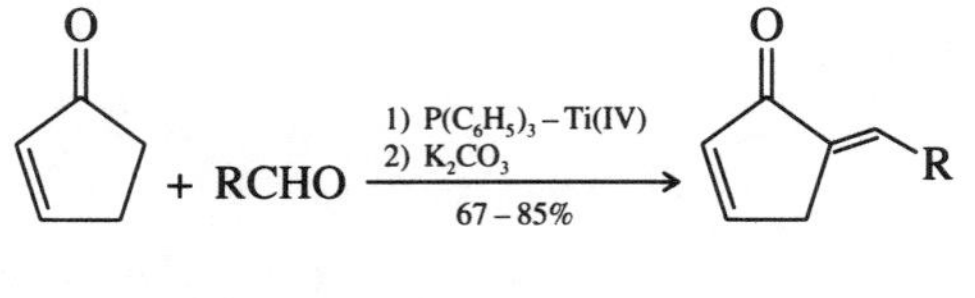

[1] T. Takanami, K. Suda, and H. Ohmori, *Tetrahedron Letters*, **31**, 677 (1990).

Triphenylphosphine hydrobromide, $(C_6H_5)_3P \cdot HBr$.

Glycosylation of glucals.[1]   Preparation of 2-deoxyglycosides by glycosylation of glucals is difficult because the usual Lewis acid catalysts provoke allylic rearrangement resulting in 2,3-unsaturated glycosides. Triphenylphosphine hydrobromides, however, promotes direct addition of alcohols, phenols, and carboxylic acids to glucals to give 2-deoxyglycosides in moderate to good yield, generally with high α-selectivity.

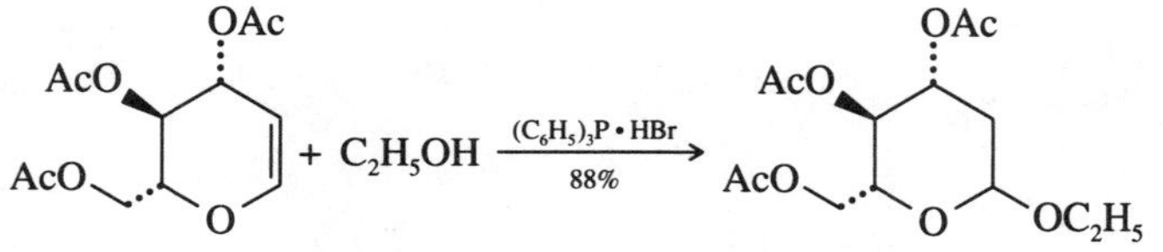

$(\alpha'/\beta = 78:22)$

[1] V. Bolitt, C. Mioskowski, S.-G. Lee, and J. R. Falck, *J. Org.*, **55**, 5812 (1990).

Triphenylphosphine oxide–Trifluoromethanesulfonic anhydride $(C_6H_5)_3\overset{+}{P})_2O$
$2OTf^-$ **(1).** The reagent **(1)** is prepared *in situ* in 1,2-dichloroethane.

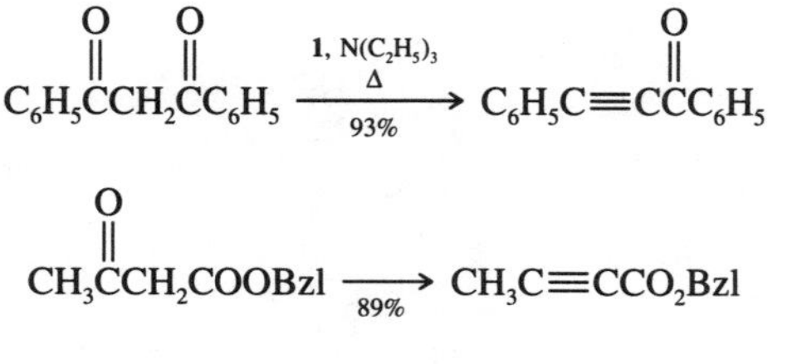

Alkynes.[1] In combination with triethylamine, the reagent effects dehydration of activated ketones to alkynes.

[1] J. B. Hendrickson and M. S. Hussoin, *Synthesis*, 217 (1989).

Triphenyltin hydride–Triethylborane.

Cyclopropylcarbinyl radicals.[1] Generation of a radical adjacent to a cyclopropyl group is attended by cleavage of the cyclopropane ring. The precursors are available by cyclopropanation of a cyclic allylic alcohol followed by replacement of

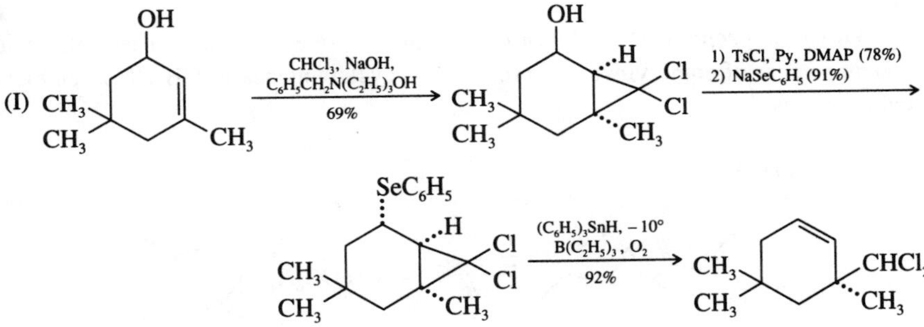

the hydroxyl group by a group that undergoes ready homolysis (such as phenylseleno) with a stannane and an initiator, photolysis, AIBN, or triethylborane/air (**15**, **337**). An example is shown in equation (I).

Another example:

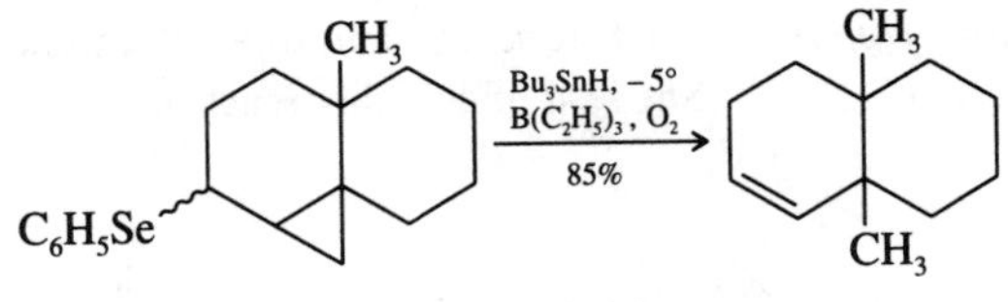

[1] D. L. J. Clive and S. J. Daigneault, *J. Org.*, in press (1991); S. Daigneault, C. J. Nichols, and D. L. J. Clive, *Org. Syn.*, submitted (1990).

Tris(4-bromophenyl)aminium hexahaloantimonate, $(BrC_6H_4)_3\overset{+\cdot}{N}:SbX_6^-$ (1; **14**, 338).

Glycosylation.[1] When treated with **1**, RS- or ArS-glycosides undergo glycosylation with primary and secondary hydroxyl groups of O-glycosides to give β-O-linked disaccharides. Acetonitrile is the only suitable solvent.

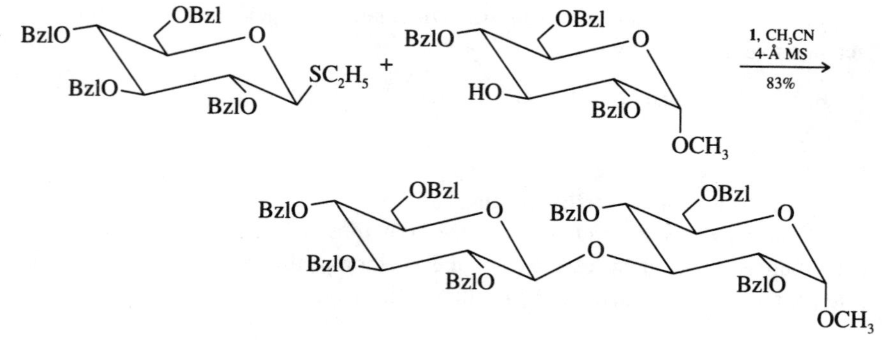

Vinylcyclopropane → ***cyclopentene.***[2] This one-electron oxidant permits rapid rearrangement of some vinylcyclopropanes to cyclopentenes in CH_3CN at ambient temperatures.

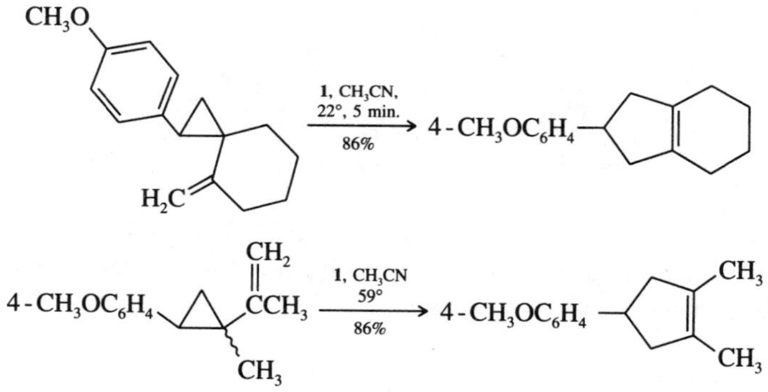

Isomerization of epoxides to ketones.[3] This radical cation [as well as trityl hexachlorostibnate, $(C_6H_5)_3C^+SbCl_6^-$] effects this reaction.

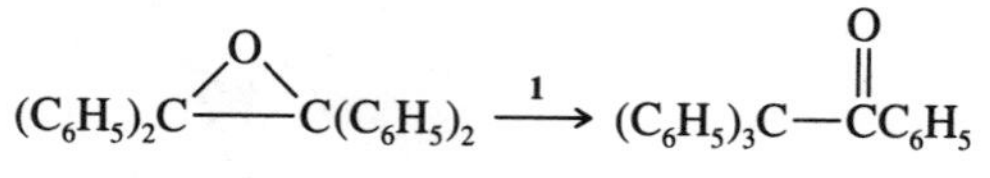

[1] A. Marra, J.-M. Mallet, C. Amatore, and P. Sinäy, *Synlett*, 572 (1990).
[2] J. P. Dinnocenzo and D. A. Conlon, *Am. Soc.*, **110**, 2324 (1988).
[3] L. Lopez and L. Troisi, *Tetrahedron Letters*, **30**, 3097 (1989).

Tris(μ-chloro)hexakis(tretrahydrofuran)divanadium hexachlorodizincate [V$_2$-Cl$_3$(THF$_6$)$_6$][Zn$_2$Cl$_6$] (**1**). This green, dimeric vanadium(II) reagent is obtained in 95% yield by reduction of VCl$_3$(THF)$_3$ with zinc dust.[1]

Cross pinacol coupling of aldehydes.   Pedersen *et al.*[2] have reported the first general method for stereoselective pinacol coupling of aldehydes by use of this vanadium(II) reagent. Aryl aldehydes couple in the presence of **1** to pinacols in >90% yield and with high diastereoselectivity (*dl*/meso = 12–100:1). Aliphatic aldehydes under similar conditions couple only slowly, but depending on the struc-

$R^1 = C_6H_5CH_2CH_2$	$R^2, R^3 = H$	67%	4:1
$= i\text{-}Pr$	$R^2, R^3 = H$	73%	8:1
$= t\text{-}Bu$	$R^2, R^3 = H$	42%	>100:1
$= i\text{-}Pr$	$R^2 = CH_3, R^3 = H$	81%	14:1
$= i\text{-}Bu$	$R^2 = H, R^3 = CH_3$	80%	4:1

ture, can show markedly different rates. Of greater interest, **1** can effect intermolecular coupling of aliphatic aldehydes when one aldehyde is activated by chelation to one vanadium(II) center of the reagent. Thus 3-formylpropanamides (**2a**) couple with simple aliphatic aldehydes to form 1,2-diols with *syn*-selectivity (4:1). The *syn*-selectivity increases with α-branching in either aldehyde. Attempts to improve the diastereoselectivity by use of low temperatures fails because then homocoupling of the chelated aldehyde prevails over cross coupling.

3,3-*Disubstituted allylic alcohols*.[3]   Pinacol coupling of aldehydes with α-(diphenylphosphinoyl)acetaldehydes (**2**) can be used for a stereoselective route to 3,3-disubstituted allylic alcohols. A general route to the phosphinoylacetaldehydes from (α-chloromethyl)diphenylphosphine oxide is formulated in equation (I). These chelating aldehydes undergo vanadium(II)-promoted pinacol cross-coupling with nonchelating aldehydes to provide only *syn*-diols in high yield (equation II). Moreover, this cross coupling is effected with diastereofacial selectivity at the prochiral center even when the substituents are only methyl versus ethyl. This diastereofacial selectivity increases with branching of the R^2 group and is highest when R^1 = isopropyl. These pinacols on Wittig–Horner *syn*-elimination provide (E)-3,3-disubstituted allylic alcohols with high stereospecificity.

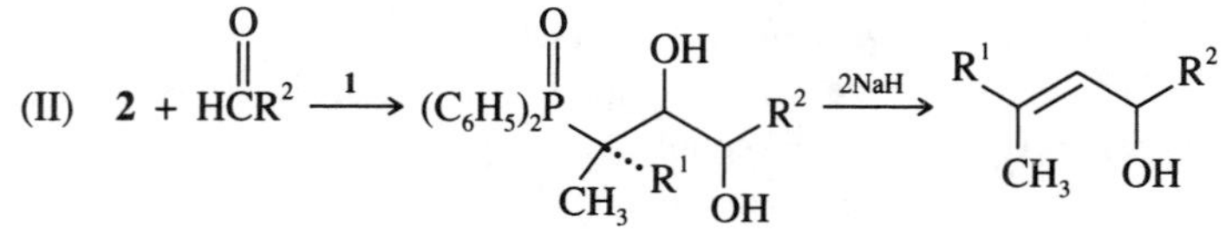

		ds ratio	
$R^1 = C_2H_5$	$R^2 = BzlCH_2$	82%	14:1	92%
$R^1 = Bzl$	$R^2 = CH_3$	94%	99:1	91%
$R^1 = i\text{-}Pr$	$R^2 = i\text{-}Bu$	82%	14:1	92%

3-Amino-1,2-diols.[4] In the presence of this vanadium(II) reagent, N-protected α-amino aldehydes undergo pinacol coupling with aliphatic aldehydes to provide *syn, syn*-3-amino-1,2-diols as the major product (5–50:1).

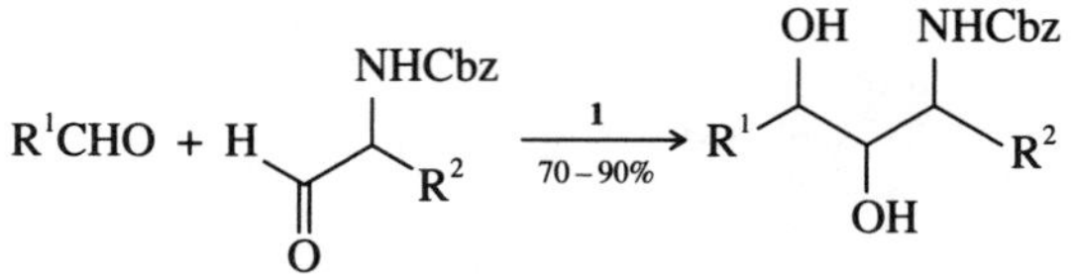

[1] F. A. Cotton, S. A. Duraj, and W. J. Roth, *Inorg. Chem.*, **24**, 913 (1985).
[2] J. H. Freudenberger, A. W. Konradi, and S. F. Pedersen, *Am. Soc.*, **111**, 8014 (1989).
[3] J. Park and S. F. Pedersen, *J. Org.*, **55**, 5924 (1990).
[4] A. W. Konradi and S. F. Pedersen, *ibid.*, **55**, 4506 (1990).

Tris(dibenzylideneacetone)dipalladium (chloroform), $(dba)_3Pd_2 \cdot CHCl_3$ (1).

Polyolefin cyclization.[1] This Pd(0) catalyst in combination with a phosphine ligand and acetic acid effects cyclization of polyunsaturated substrates to polycyclic products in one step (zipper reaction). However, a triple bond is required for initiation. This polyolefin cyclization has been used to prepare tetra- and pentaspirocycles, as a mixture of only two stereoisomers.

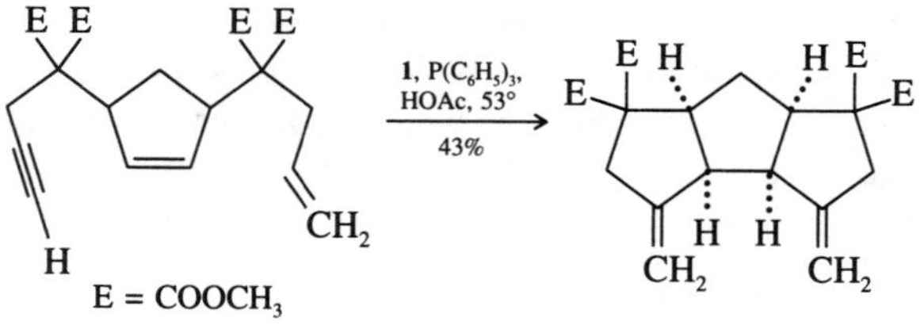

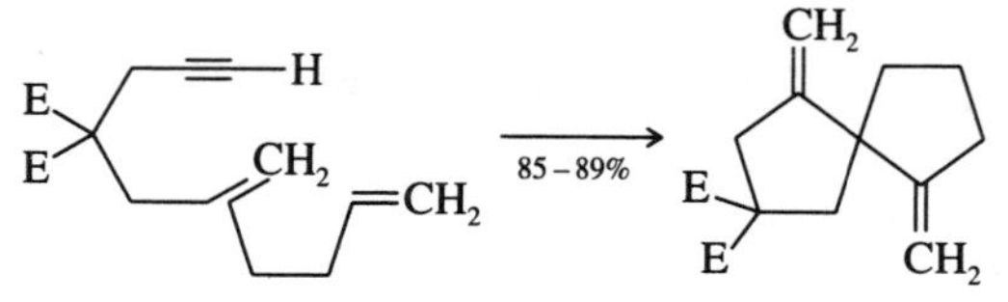

[1] B. M. Trost and Y. Shi, *Am. Soc.*, **113**, 701 (1991).

Tris[2-(2-methoxyethoxy)ethyl]amine (TDA-1), **13**, 336-337; **14**, 341-342.

Hydrogenation catalyst.[1] TDA-1 can serve as a catalytic ligand for hydrogenation of unsaturated substrates with $PdCl_2$ or $RhCl_3 \cdot nH_2O$ at 20°. Yields are generally as high as those obtained with expensive complexes of Pd or Rh.

N-Arylation of amides.[2] Bromo- or chlorobenzenes arylate amides in the presence of TDA-1 and CuCl (1:5).

[1] D. Villemin and M. Letulle, *Syn. Comm.*, **19**, 2833 (1989).
[2] A. Greiner, *Synthesis*, 312 (1989).

Tris(methoxyethoxypropyl)tin hydride, $[CH_3O(CH_2)_2O(CH_2)_3]_3SnH$ (**1**), prepared from methoxyethyl allyl ether, $CH_3OCH_2CH_2OCH_2CH=CH_2$.

Radical reductions of alkyl halides in water.[1] Radical reactions with Bu_3SnH are limited to organic solvents, but this tin hydride (**1**) is sufficiently soluble in water to reduce alkyl bromides or iodides under free-radical conditions in water or in organic solvents.

[1] J. Light and R. Breslow, *Tetrahedron Letters*, **31**, 2957 (1990).

Tris(2-methoxymethoxyphenyl)phosphine,

. Preparation.[1]

cis-Stilbenes.[2] Preparation of stilbenes by reaction of (triphenylphosphino)-arylmethanides with aryl aldehydes under usual Wittig conditions results in stilbenes as a mixture of *cis*- and *trans*-isomers close to 1:1. Use of tris(2-methoxymethoxyphenyl)phosphine in place of triphenylphosphine increases the *cis/trans* ratio markedly, particularly in reactions carried out at −75°.

[1] S. Jeganathan, M. Tsukamoto, and M. Schlosser, *Synthesis*, 109 (1990).
[2] M. Tsukamoto and M. Schlosser, *Synlett*, 605 (1990).

Tris(2-methyl-2-phenylpropyl)tin hydride (trineophyltin hydride), $HSn[CH_2\text{-}C(CH_3)_2C_6H_5]_3$ (1), m. p. 50–51°. The hydride is prepared by borane reduction of bis(trineophyltin) oxide in refluxing THF.

Stereoselective radical reductions of halides.[1]   This hydride when exposed to

4 : 1

AIBN selectively reduces the α-halo group, whether Cl, Br, or I, of pivaloyloxy-methyl 6,6-dihalopenicillanates. The order of reactivity is I > Br > Cl > F.

[1] E. G. Mata, O. A. Mascaretti, A. E. Zuniga, A. B. Chopa, and J. C. Podesta, *Tetrahedron Letters*, **30**, 3905 (1989).

Tris[(trimethylsilyl)methyl]aluminum, $[(CH_3)_3SiCH_2]_3Al$ (1). This reagent is generated *in situ* by reaction of $AlCl_3$ with $[(CH_3)_3SiCH_2]Li$. The neat reagent is pyrophoric.

Allyltrimethylsilanes.[1]   In the presence of $Pd[P(C_6H_5)_3]_4$, this alane reacts with vinyl or aryl triflates to form allyltrimethylsilanes in 60–85% yield with transfer of one $(CH_3)_3SiCH_2$ group.

[1] M. G. Saulnier, J. F. Kadow, M. M. Tun, D. R. Langley, and D. M. Vyas, *Am. Soc.*, **111**, 8320 (1989).

Tris(trimethylsilyl)silane, 15, 358–359.

Radical reductions with $[(CH_3)_3Si]_3SiH$.[1]   In combination with an initiator, this silane (1) reduces not only alkyl halides, but also thionoesters (used for deoxygenation of secondary alcohols) and selenides. However, cleavage of C–S bonds is inefficient. It is more efficient than Bu₃SnH for reduction of isocyanides to the corresponding hydrocarbons.

The silane also can be used for hydrosilylation of dialkyl ketones and acrylonitrile.

(I) $=O$ + **1** $\xrightarrow[90\%]{\text{AIBN}\ C_6H_5CH_3,\ \Delta}$ $-OSi[Si(CH_3)_3]_3$

This silane can also be used to effect addition of alkyl radicals to alkenes if it is added slowly to a solution of the precursor to the radical and the alkene (equation II).

(II) $c\text{-}C_6H_{11}I + CH_2{=}CHCN \xrightarrow[90\%]{\text{1, AIBN}\ \Delta} c\text{-}C_6H_{11}CH_2CH_2CN$

Catalytic reduction of RX.[2] This silane can be used in catalytic amounts in the radical reduction of alkyl halides if NaBH₄ is added to regenerate the silane from the $[(CH_3)_3Si]_3SiX$ formed on reduction of RX by **1**. Yields in the catalytic hydrodehalogenation process are only slightly lower than those obtained by use of an equimolar amount of the silane.

[1] M. Ballestri, C. Chatgilialoglu, K. B. Clark, and D. Griller, B. Giese, and B. Kopping, *J. Org.*, **56**, 678 (1991).
[2] M. Lesage, C. Chatgilialoglu, and D. Griller, *Tetrahedron Letters*, **30**, 2733 (1989).

Trityl bromide (triphenylmethyl bromide).

N-Trityl α-amino acids.[1] TrBr is markedly superior to TrCl for tritylation of α-amino acids. The reaction of the acids with TrBr (2 equiv.) and (excess) triethylamine in CHCl₃/DMF (2:1) provides the N-tritylamino trityl ester, which is then hydrolyzed *in situ* with methanol at 50°. Yields are 80–86%.

[1] M. Mutter and R. Hersperger, *Synthesis*, 198 (1989).

Trityl perchlorate.

Mukaiyama–Michael reaction[1] (**13**, 339–340; **14**, 344–345). The conjugate addition of enol silyl ethers of optically active ketones to α,β-enones catalyzed by trityl perchlorate can proceed with high diastereoselectivity. Thus the (Z)-enol silyl ether (**2**) of the ketone (R)-**1** reacts with enone **3** to give the 1,5-ketone **4** with high anti-selectivity. Removal of the silicon substituent with aqueous HBF$_4$ results in an intramolecular aldol reaction followed by dehydration to give a cyclohexenone (5).

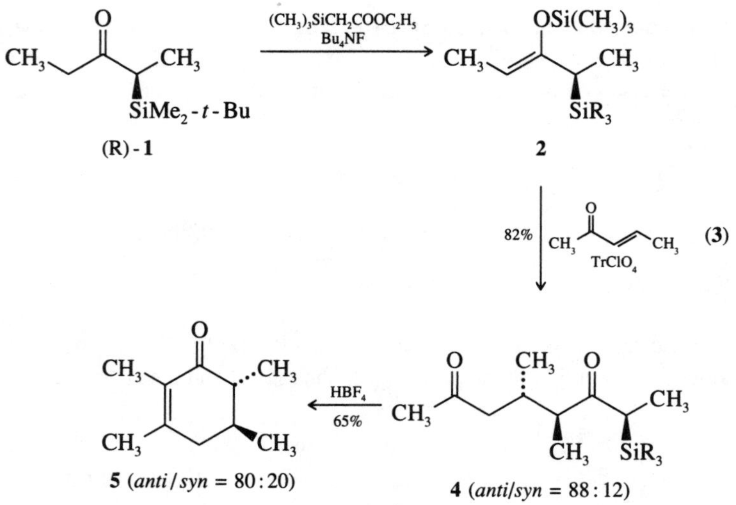

[1] B. B. Lohray and R. Zimbiniski, *Tetrahedron Letters*, **31**, 7273 (1990).

U

Ultrasound,))) .

Review.[1] This review includes a discussion of the three types of ultrasonic generators: whistlers, cleaning baths, and probe disruptors. The last is the most efficient, but the most expensive. Cleaning baths are inexpensive, but are limited in the temperature range to that of the liquid used, generally water. The review concludes that sonication is most useful in heterogeneous reactions, particularly those of organometallics. The references (235) date from 1953 to the present time, with most in the last 10 years.

Luche *et al.*[2] point out that sonication can result in an overall rate increase owing to mechanical effects such as agitation, but that sonication can also influence the mechanism of a reaction. They note that in the latter case sonication leads to acceleration of single electron transfers and can result in a different product than that obtained in the absence of sonication. In general, ionic reactions are not sensitive to ultrasound.

Oxymercuration.[3] Commercial mercury(II) acetate is the usual reagent for this reaction, although higher yields and greater selectivity obtain with the more expensive mercury(II) trifluoroacetate. A variety of mercury(II) salts can be prepared by sonication of a mixture of HgO with a carboxylic acid (2 equiv.) in solvents such as CH_2Cl_2, hexane, and aqueous THF. They can also be prepared *in situ* by oxymercuration, in which case only 1 equiv. of the acid is required since 1 equiv. of the acid is formed on oxymercuration. And this oxymercuration can result in increased selectivity in monooxymercuration of a diunsaturated substrate such as limonene (**1**).

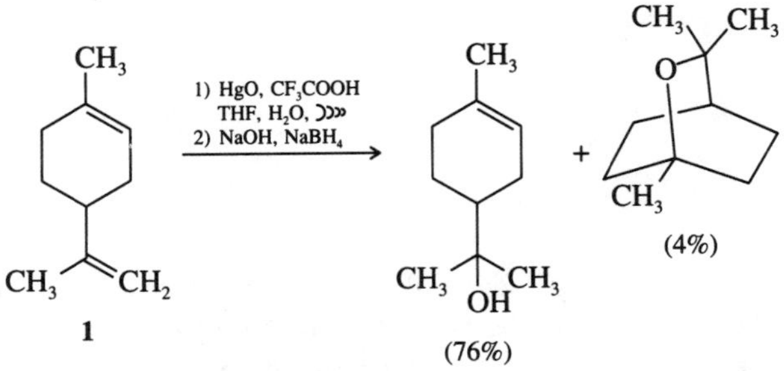

Hydrostannylation.[4] This reaction with a tin hydride normally requires an initiator, usually AIBN but also $B(C_2H_5)_3$ (**14**, 314). It can also be initiated with high intensity ultrasound, and such reactions show large rate acceleration (100–600 times) and take place even at temperatures of $-50°$. Sonication is also effective for radical reductions and cyclizations (last example).

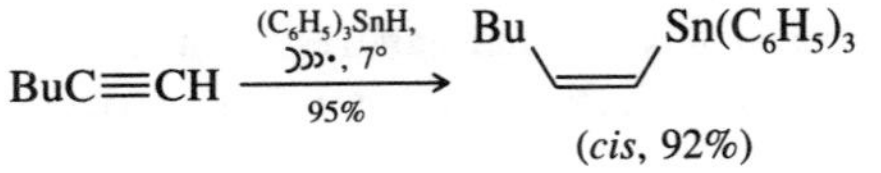

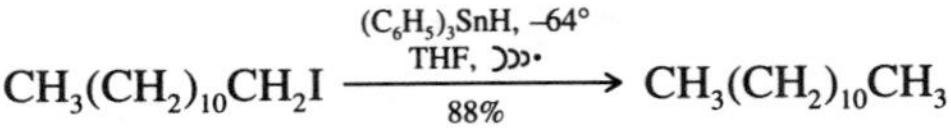

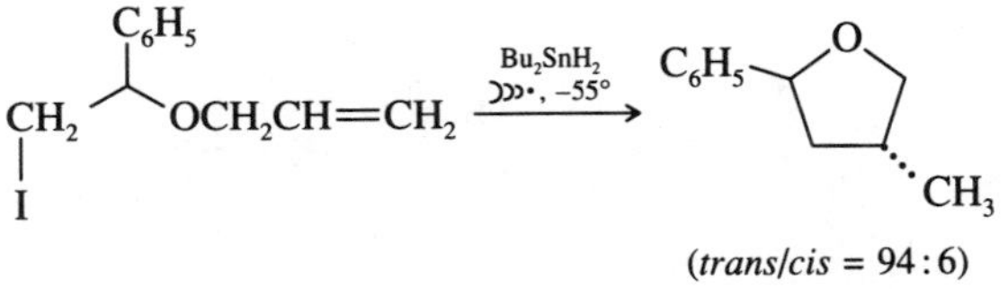

Acyloin condensation.[5] The acyloin condensation of 1,4- to 1,6-diesters according to the Rühlmann version using chlorotrimethylsilane and highly dispersed sodium is markedly simplified by sonochemical activation. Technical grade $ClSi(CH_3)_3$ and small cubic pieces of sodium can be used.

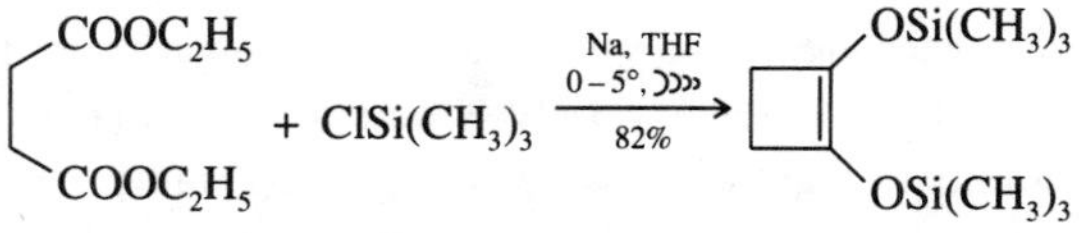

[1] C. Einhorn, J. Einhorn, and J. L. Luche, *Synthesis*, 787 (1989).
[2] J. L. Luche, C. Einhorn, J. Einhorn, and J. V. Sinisterra-Gago, *Tetrahedron Letters*, **31**, 4125 (1990); C. Einhorn, J. Einhorn, M. J. Dickens, and J. L. Luche, *ibid.*, **31**, 4129 (1990).
[3] J. Einhorn, C. Einhorn, and J. L. Luche, *J. Org.*, **54**, 4479 (1989).

[4] E. Nakamura, D. Machii and T. Inubushi, *Am. Soc.*, **111**, 6849 (1989).
[5] A. Fadel, J.-L. Canet, and J. Salaün, *Synlett*, 89 (1990).

Urea–Hydrogen peroxide, H_2NCONH_2–H_2O_2, (**1**, m.p. 90–93°). This crystalline solid is formed on crystallization of urea from aqueous H_2O_2. It is commercially available from Interox Chemicals Ltd. or Aldrich.

Alternative to anhydrous H_2O_2.[1] This addition complex in combination with acetic anhydride or trifluoroacetic anhydride (0.25–0.50 equiv.) can serve as an alternative to anhydrous H_2O_2 for epoxidation, Baeyer–Villiger reactions, and conversion of nitrogen heterocycles to N-oxides. The complex is reasonably stable, but it should be used with care.

The complex converts N,N-bis(trimethylsilyl)urea to bis(trimethylsilyl) peroxide in 86% yield.[2]

[1] M. S. Cooper, H. Heaney, A. J. Newbold, and W. R. Sanderson, *Synlett*, 533 (1990).
[2] W. P. Jackson, *Synlett*, 536 (1990).

V

L-Valinol.

Kinetic resolution of (±)-(o-anisaldehyde)Cr(CO)₃ (1).[1]   This arenechromium carbonyl has been resolved by chromatography of a derivative, but a more convenient method involves reaction with L-valinol to form two imines (**a, b**) with widely different R_f values. The fastest moving fraction (**a**) is a precursor to (−)-**1**, whereas the slower moving fraction provides (+)-**1** on hydrolysis.

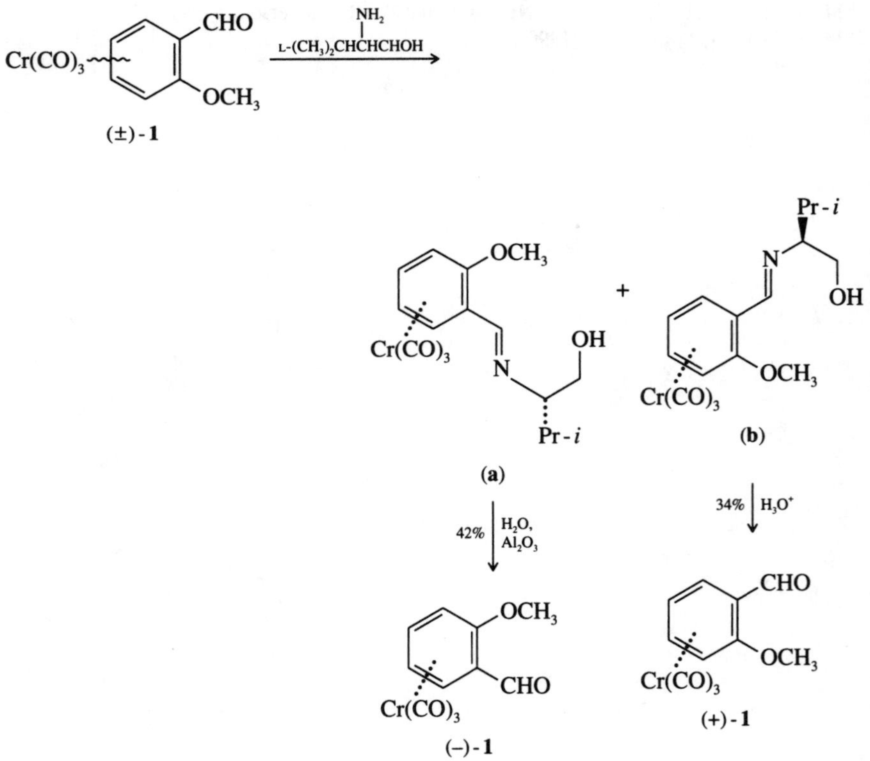

[1] S. G. Davies and C. L. Goodfellow, *Org. Syn.*, submitted (1989).

Vanadium oxytrichloride, VOCl₃.

Oxidative decarboxylation–deoxygenation of β-hydroxy carboxylic acids.[1]
When heated in refluxing chlorobenzene with VOCl₃, β-hydroxy carboxylic acids
undergo decarboxylation and deoxygenation with formation of an alkene. This process is useful for preparation of tri- and tetrasubstituted alkenes, but it has some
limitations, including dehydration, isomerization, migration of the double bond, and
formation of ketonic by-products. In general, yields of the desired alkene are im-

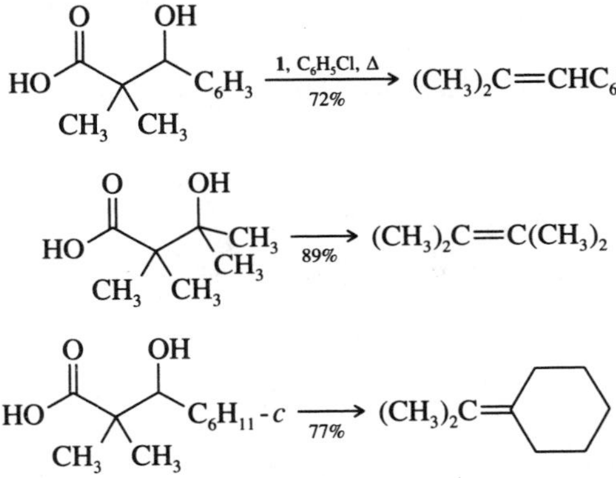

proved by use of trichloro(*p*-tolylimino)vanadium(V), $CH_3C_6H_4N{=}VCl_3$ (1). Proton Sponge (1 equiv.) can be added to prevent isomerization of the double bond.

[1] I. K. Meier and J. Schwartz, *Am. Soc.*, **111**, 3069 (1989); *idem, J. Org.*, **55**, 5619 (1990).

Vanadium oxytrifluoride, VOF₃.

Asymmetric phenolic oxidative coupling.[1] Oxidation of the chiral oxazolidine
1 with VOF₃ results in efficient *para-para* coupling to a spirodienone **2**. The *trans-*

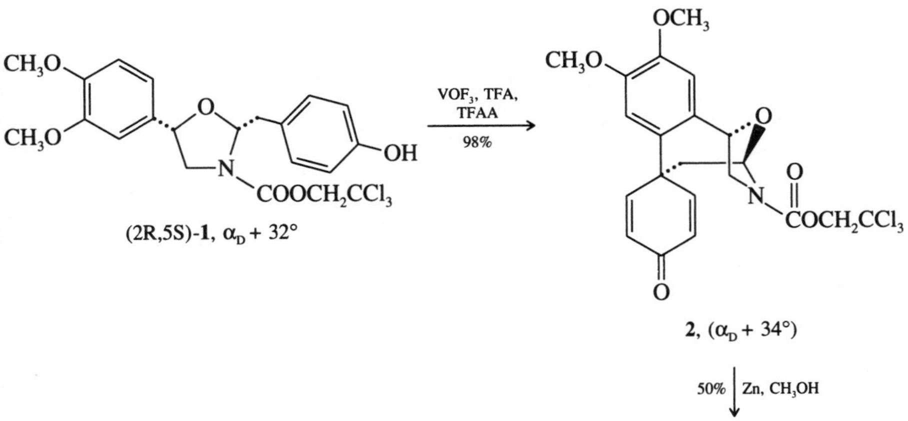

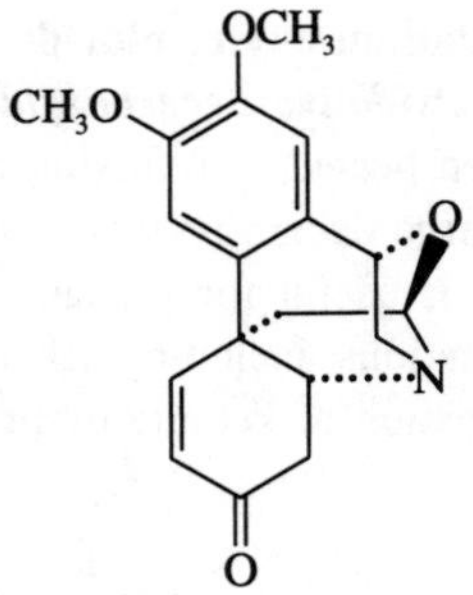

3

isomer of **1** does not undergo coupling. Reductive removal of the trichloroethoxycarbonyl group results in the pentacyclic amine **3**.

[1] J. D. White, R. J. Butlin, H.-G. Hahn, and A. T. Johnson, *Am. Soc.*, **112**, 8595 (1990).

W

Water.

Claisen rearrangement. Grieco *et al.*[1] have found that water as solvent can accelerate the rate of Claisen rearrangement of several substrates, and thereby allow rearrangement at lower temperatures with higher yields. Thus **1** rearranges to **2** in H_2O at 60° in 85% yield. Rearrangement of the corresponding methyl ester in benzene at 60° requires >100 hours and gives the rearranged ester in 64% yield. The paper presents several other examples of the advantages of water as solvent or cosolvent. The rate acceleration may be associated with stabilization of a polar transition state.

NaOOC(CH₂)₆ — O — CH₂ / CH₂ $\xrightarrow[85\%]{H_2O,\ 60°}$ CHO ... (CH₂)₆COONa

1   **2**

[1] P. A. Grieco, E. B. Brandes, S. McCann, and J. D. Clark, *J. Org.*, **54**, 5849 (1989).

Wittig and Wittig–Horner reagents.

Review.[1] This review covers use of phosphonium ylides, phosphoryl carbanions, and phosphine oxide carbanions in synthesis with particular emphasis on (Z)- and (E)-selectivities (558 references).

[1] B. E. Maryanoff and A. B. Reitz, *Chem. Rev.*, **89**, 863 (1989).

Y

Ytterbium(0).

Chiral 1,3-diols. Coupling of diaryl ketones with chiral epoxides mediated by Yb (**15**, 366) provides chiral 1,3-diols in high optical yields. Generally two diols are formed by coupling at both carbon atoms, but coupling with styrene oxide occurs mainly at the more-substituted carbon.

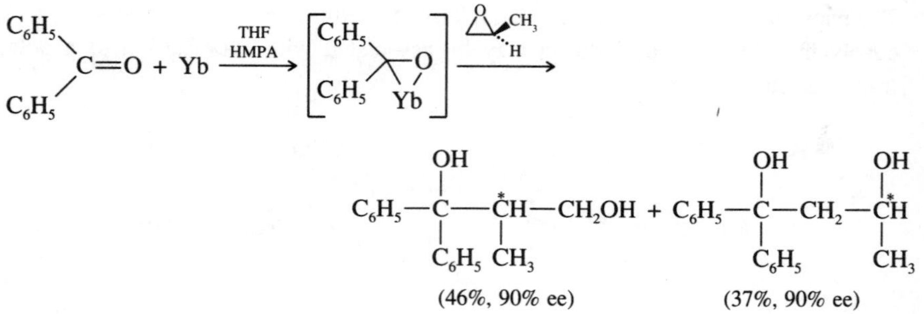

$$C_6H_5-\underset{\underset{C_6H_5}{|}}{\overset{\overset{OH}{|}}{C}}-\overset{*}{\underset{\underset{CH_3}{|}}{C}}H-CH_2OH \; + \; C_6H_5-\underset{\underset{C_6H_5}{|}}{\overset{\overset{OH}{|}}{C}}-CH_2-\overset{*}{\underset{\underset{CH_3}{|}}{C}}H$$

(46%, 90% ee) (37%, 90% ee)

[1] K. Takaki, S. Tanaka, F. Beppu, Y. Tsubaki, and Y. Fujiwara, *Chem. Letters*, 1427 (1990).

Ytterbium(III) chloride, $YbCl_3$.

Reversal of diastereoselectivity of reactions with RMgX and RLi.[1] The chiral 2-acyl-1,3-oxathiane **1** (**12**, 237–239) undergoes diastereoselective addition of

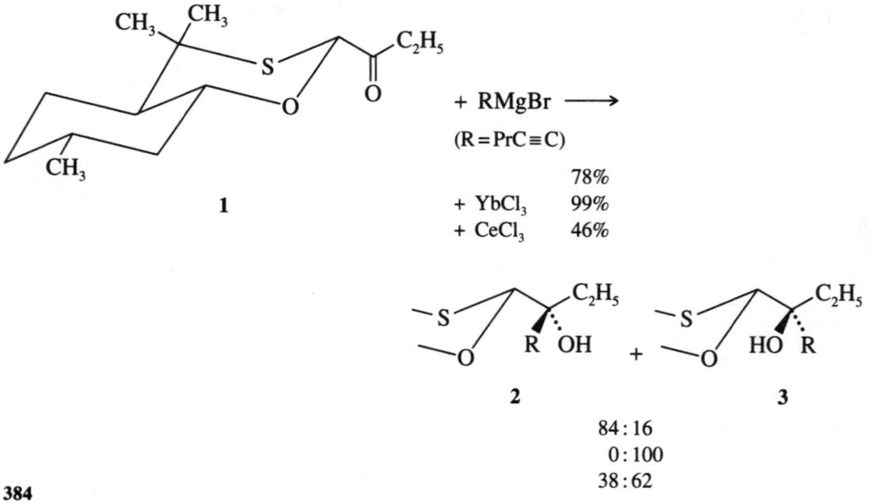

+ RMgBr $\longrightarrow$
(R = PrC≡C)

	78%
+ YbCl₃	99%
+ CeCl₃	46%

2   **3**

84 : 16
0 : 100
38 : 62

Grignard reagents because of Cram chelation to provide mainly adducts **2**, precursors to optically active tertiary alcohols. However, if a THF solution of the Grignard reagent is added to a suspension of $YbCl_3$ in THF and stored at 0° for 2 hours, the resultant species, $RMgBr \cdot YbCl_3$, reacts with **1** at $-78°$ to provide **3** in 95.4% de in quantitative yield.

Ytterbium(III) chloride can also reverse the diastereoselectivity of reactions of **1** with alkynyllithium reagents. Cerium(III) shows similar effects, but yields are also lowered.

[1] K. Utimoto, A. Nakamura, and S. Matsubara, *Am. Soc.*, **112**, 8189 (1990).

Z

Zinc.

Luche allylation (**13**, 298). One unusual feature of Luche allylation of aldehydes is the use of H_2O/THF as solvent. More recently Wilson and Guazzaroni[1] have effected this allylation with allyl halides and zinc dust in an aqueous NH_4Cl solution with a solid organic support in place of THF. The support can be reverse-phase C-18 silica gel (Custom Chem. Laboratories), biobeads SX8 (Bio-Red), or OV-101 on

$$n\text{-}C_6H_{13}CHO + CH_3CH{=}CHCH_2Cl \xrightarrow[\substack{H_2O, \text{ C-18 SiO}_2 \\ 88\%}]{NH_4Cl, \text{ Zn}} n\text{-}C_6H_{13}\underset{\underset{OH}{|}}{\overset{\overset{CH_3}{|}}{C}}HCHCH{=}CH_2$$

$$(1:1)$$

Chromosorb. One advantage of this new system is that a hydroxyl group does not require protection. An organozinc reagent is probably not involved since such species are unstable in water.

Reactions of activated zinc with RX.[2] Activated zinc, Zn*, obtained by reduction of $ZnCl_2$ with lithium naphthalenide in THF or DME, can add oxidatively to alkyl, aryl, and vinyl halides. This reaction tolerates a wide variety of substituents. The resulting species, RZnBr(I), on treatment with $CuCN \cdot 2LiBr$ is converted into a copper derivative RuCu(CN)ZnBr(I) (**15**, 225–226). These organocopper–zinc reagents couple directly with acid chlorides to furnish ketones in good yield (equation I) or undergo conjugate addition to enones (**15**, 229–230). Reaction of the copper-zinc reagents with allylic halides results mainly in S_N2' substitution (equation II).

$$(I) \quad Br(CH_2)_3COOC_2H_5 \xrightarrow[\substack{2) \text{ CuCN} \cdot 2LiCl}]{1) \text{ Zn*}} C_2H_5OOC(CH_2)_3Cu(CN)ZnBr$$

$$90\% \downarrow C_6H_5COCl$$

$$C_6H_5CO(CH_2)_3COOC_2H_5$$

(II) $Br(CH_2)_6CN \longrightarrow [NC(CH_2)_6Cu(CN)ZnBr]$

$$91\% \downarrow \overset{CH_3 \diagdown \diagup CH_2Cl}{}$$

$$\underset{NC(CH_2)_6}{\overset{CH_3\diagdown}{}} CHCH=CH_2 \quad + \quad CH_3 \diagdown\diagup\diagdown (CH_2)_6CN$$

$$S_N 2' \qquad\qquad 97:3$$

In the presence of a Pd(0) catalyst, organozinc halides can couple directly with aryl and vinyl halides to form novel arenes or biaryls.

(III)

$$\text{ZnBr} + p\text{-}BrC_6H_4CN \xrightarrow[82\%]{Pd(0)}$$

cis-*Reduction of propargylic alcohols*.[3] Rieke zinc (5, 753; 6, 679) in THF/ CH_3OH/H_2O (7:5:1) reduces simple propargylic alcohols to *cis*-allylic alcohols. It also reduces conjugate enynols and diynols to dienols. Activation by the alcohol group permits selective reduction of nonconjugated diynols.

$$\underset{CH_3}{\overset{HO\diagdown}{}} CHC\equiv C(CH_2)_2C\equiv C(CH_2)_3CH_3 \xrightarrow[95\%]{\underset{CH_3OH,\,H_2O,\,\Delta}{Zn^*}}$$

Reformatsky reaction. Fürstner[4] has reviewed literature (1887–1988) on this reaction, both inter- and intramolecular. A comparison of different forms of zinc for the reaction suggests that metal prepared by reduction of $ZnCl_2$ with silver–graphite (**13**, 348) is the most satisfactory.

[1] S. R. Wilson and M. E. Guazzaroni, *J. Org.*, **54**, 3087 (1989).
[2] L. Zhu, R. M. Wehmeyer, and R. D. Rieke, *ibid.*, **56**, 1445 (1991).
[3] W.-N. Chou, D. L. Clark, and J. B. White, *Tetrahedron Letters*, **32**, 299 (1991).
[4] A. Fürstner, *Synthesis*, 571 (1989).

Zinc–Copper couple.

Pinacol reductive coupling.[1] A couple prepared by sonication of zinc dust and $CuCl_2$ in acetone–water (4:1) promotes coupling of enones and acetone to α,β-unsaturated *vic*-diols. The reaction is slow in the absence of sonication.

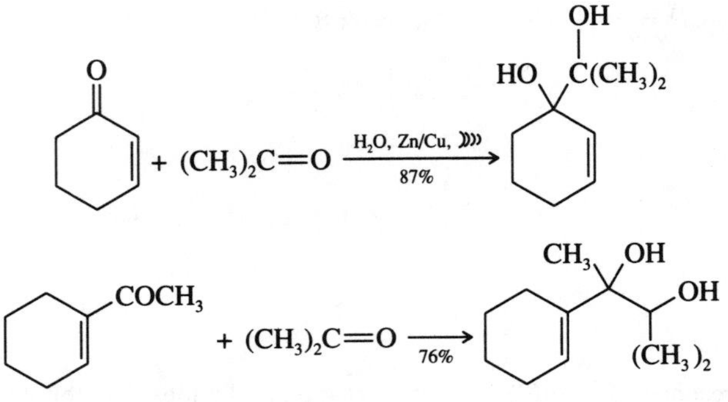

[1] P. Delair and J.-L. Luche, *J.C.S. Chem. Comm.*, 398 (1989).

Zinc–Copper/silver couple, 14, 350–351.

(E,E,Z)-Trienes.[1] Zinc activated by Cu(OAc)$_2$ and AgNO$_3$ effects stereoselective reduction of the dienyne 1 to the (E,E,Z)-triene 2 in CH$_3$OH/H$_2$O. Catalytic hydrogenation using Lindlar's catalyst gives 2, but only in 30% yield.

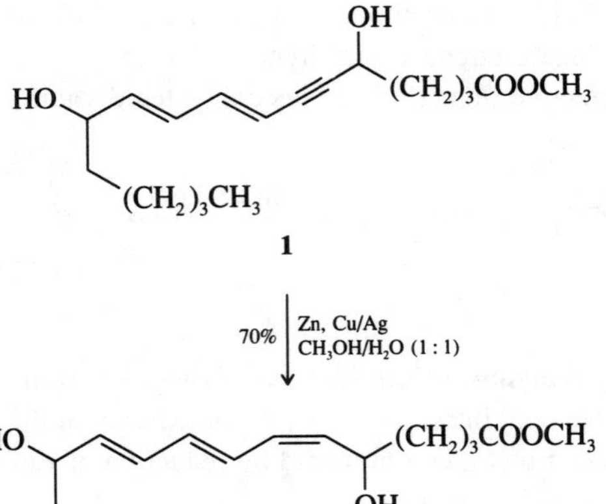

[1] M. Avignon-Tropis and J. R. Pougny, *Tetrahedron Letters*, 30, 4951 (1989).

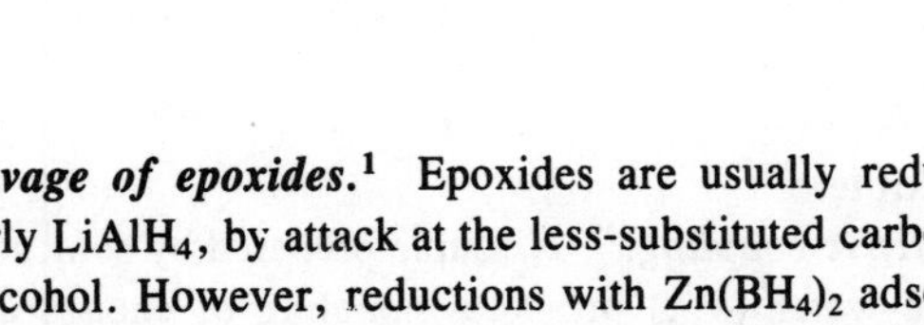

Zinc borohydride.

Reductive cleavage of epoxides.[1] Epoxides are usually reduced by hydride reagents, particularly LiAlH$_4$, by attack at the less-substituted carbon to provide the more-substituted alcohol. However, reductions with Zn(BH$_4$)$_2$ adsorbed on SiO$_2$ in THF at 25° can show the opposite regioselectivity. Thus the epoxide of a 1-alkene is

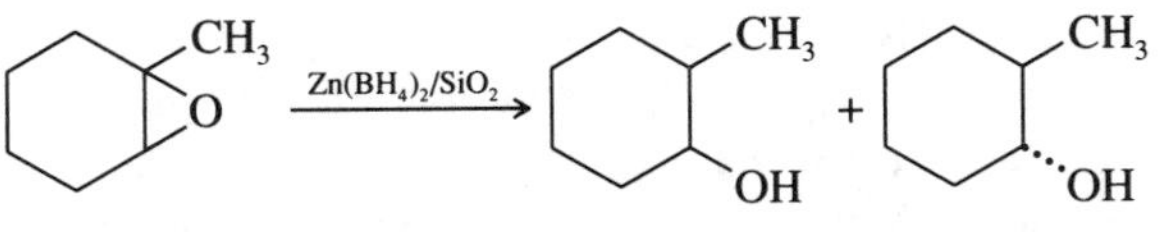

cis, 85%   trans, 10%

reduced to the corresponding primary alcohol in about 90% yield. Styrene oxide is reduced mainly to 2-phenylethanol (~85% yield). The reduction of cyclic oxides is regio- and stereoselective.

Selective reductions.[2] Zinc borohydride in DME can reduce saturated ketones and α,β-enals at −15° without effect on α,β-enones or saturated aldehydes.

Reduction of azomethines.[3] $Zn(BH_4)_2$ in ether reduces Schiff bases to secondary amines or the amine·BH_3 complex in high yield. This procedure can be applied to alkylation–reduction of nitriles to yield 1-phenylalkylamines.

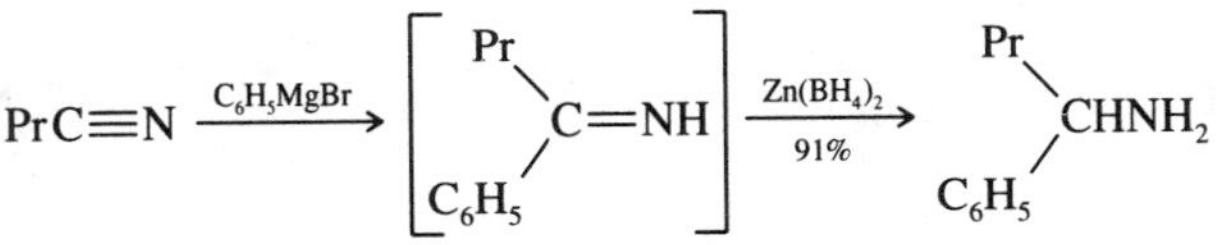

[1] B. C. Ranu and A. R. Das, *J.C.S. Chem. Comm.*, 1334 (1990).
[2] D. C. Sarkar, A. R. Das, and B. C. Ranu, *J. Org.*, **55**, 5799 (1990).
[3] H. Kotsuki, N. Yoshimura, I. Kadota, Y. Ushio, and M. Ochi, *Synthesis*, 401 (1990).

Zinc bromide.
Diels–Alder catalyst.[1] A novel, general route to tricyclic diterpenes involves a Diels–Alder reaction between a hindered diene such as 1 and a 1,4-benzoquinone.

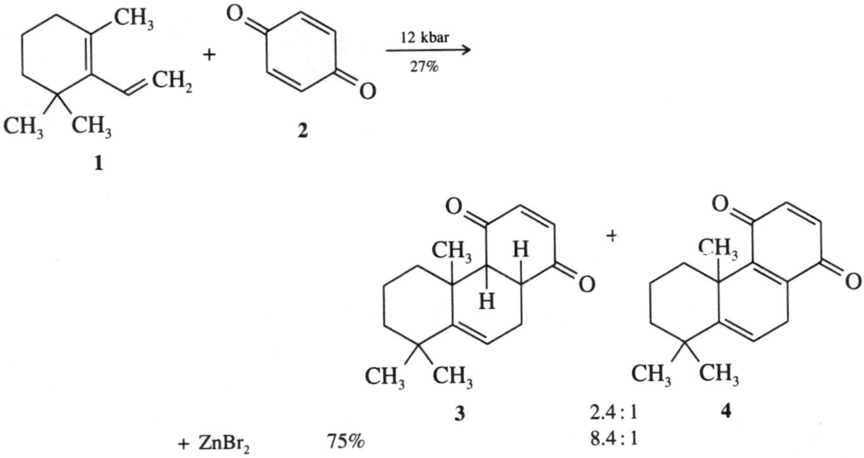

The steric difficulty inherent in a hindered diene is partly solved by use of high pressure. Lewis acids can also accelerate the rate of high-pressure reactions, but more importantly, can improve stereo- and regioselectivity. Thus the reaction of **1** with the quinone **2** under high pressure results in both **3** and **4**. The latter, undesired product is a result of pressure-promoted oxidation of **3**. Use of $ZnBr_2$ improves the yield of **3** and decreases formation of **4**. Yb(fod)$_3$ and Eu(fod)$_3$ are useful catalysts in reactions of the enol ether **5**.

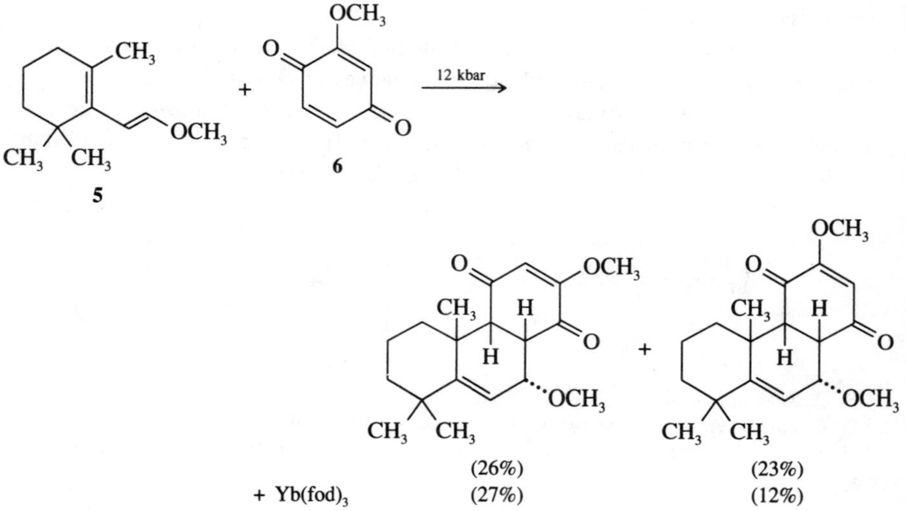

ZnBr$_2$ is an efficient catalyst for reaction of dienes with the dihydropyridinone **9** to give *cis*-hydroisoquinolines such as **10**.[2] The corresponding thermal reactions provide adducts in only 15–25% yield.

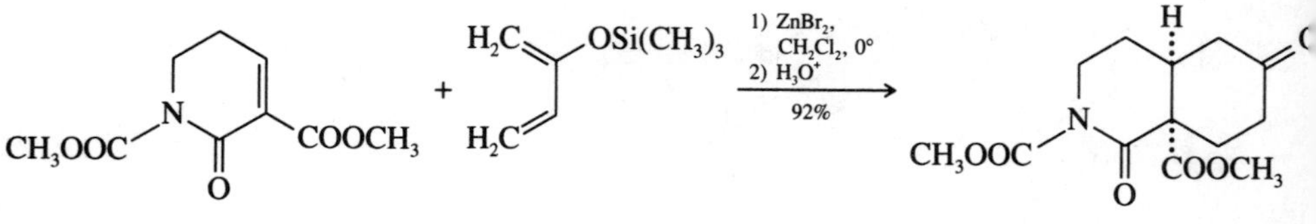

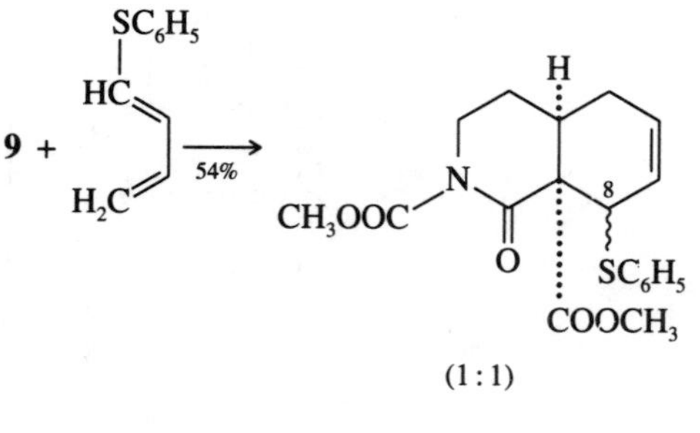

Reaction of acrylonitrile with ketene acetals.[3] Depending on the zinc salt and the solvent, ketene silyl acetals undergo [2 + 2]cycloaddition or a Michael-type addition with acrylonitrile. The former reaction occurs in CCl_4 with $ZnBr_2$, the latter in CH_2Cl_2 with ZnI_2, with no interconversion. 2-Chloroacrylonitrile can also be used in this way, but substituted acrylonitriles are inactive.

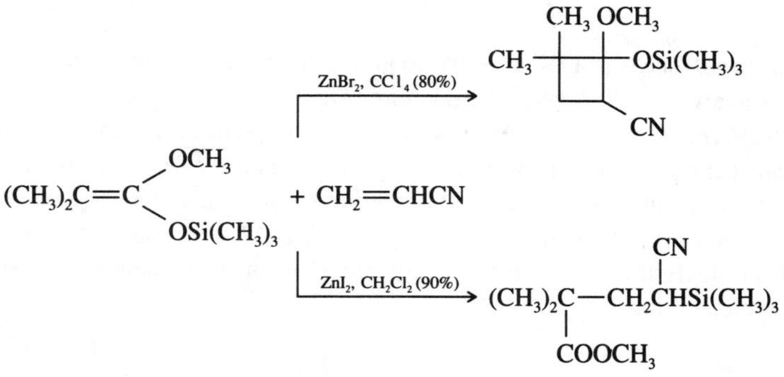

[1] T. A. Engler, U. Sampath, S. Naganathan, D. V. Velde, F. Takusagawa, and D. Yohannes, *J. Org.*, **54**, 5712 (1989).
[2] D. de Oliveira Imbroisi and N. S. Simpkins, *Tetrahedron Letters*, **30**, 4309 (1989).
[3] A. Quendo and G. Rousseau, *Syn. Comm.*, **19**, 1551 (1989).

Zinc chloride.

(S)-β-Amino acids.[1] The Mannich reaction of the Schiff base **1**, prepared from 2,3,4,6-tetra-O-pivaloyl-β-D-galactosylamine, reacts with the silyl ketene acetal **2** in

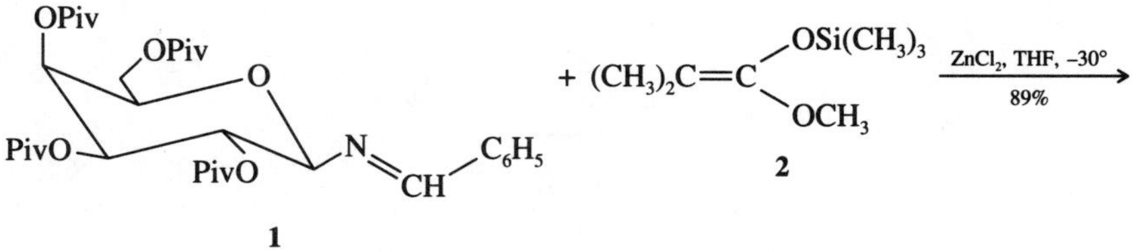

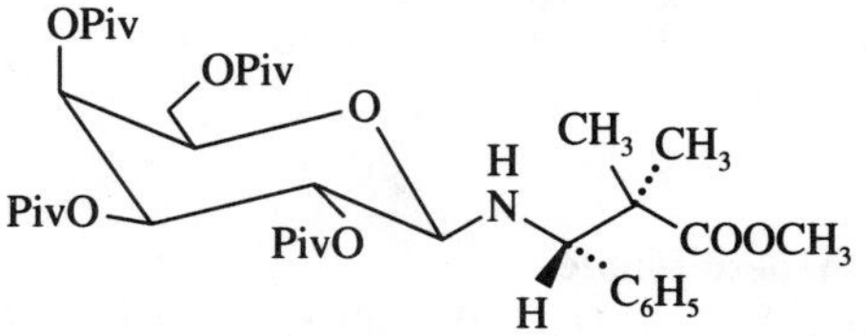

3 (150 : 1)

HCl, CH_3OH

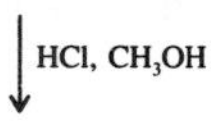

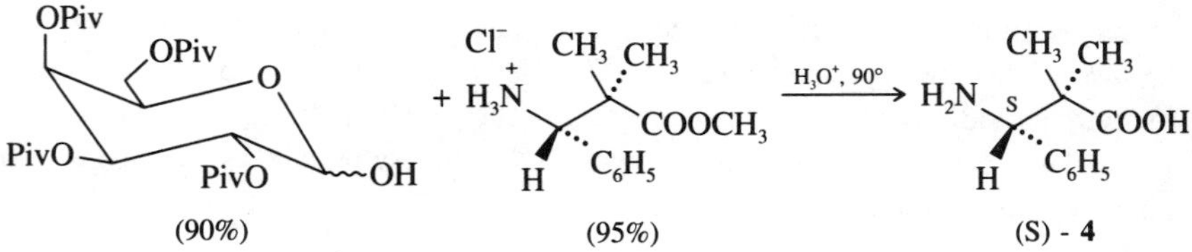

(90%)   (95%)   (S) - 4

the presence of $ZnCl_2$ in THF at $-30°$ to give essentially a single product (3), which can be converted to (S)-β-phenyl-β-alanine (4).

[3 + 2]Cycloaddition of styrene with p-quinone methides.[2] In the presence of this Lewis acid, p-quinone methides and styrenes undergo a formal [3 + 2]cycloaddition to form dihydro-1H-indenes. The reaction shows some stereoselectivity. Thus the geometry of the (E)-styrene is largely retained (17 : 1) and only two of the four possible products are formed. Presumably, any electron-rich alkene could participate in this cycloaddition.

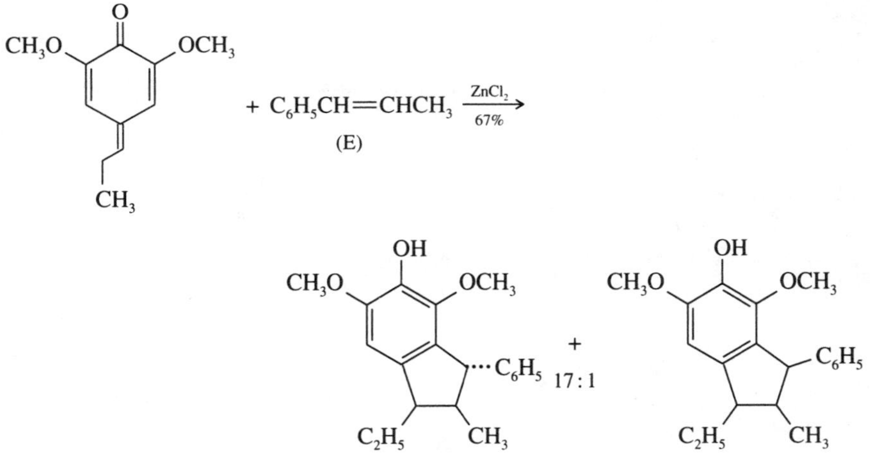

[1] H. Kunz and D. Schanzenbach, *Angew. Chem. Int. Ed.*, **28**, 1068 (1989).
[2] R. S. Angle and D. O. Arnaiz, *J. Org.*, **55**, 3708 (1990).

Zinc trifluoromethanesulfonate.

Amidoethylation.[1] $Zn(OTf)_2$ is the only Lewis acid of a large number examined that promotes reaction of (2R)- or (2S)-2-azirdinecarboxylates with indoles to obtain optically pure tryptophans, albeit in only moderate yield.

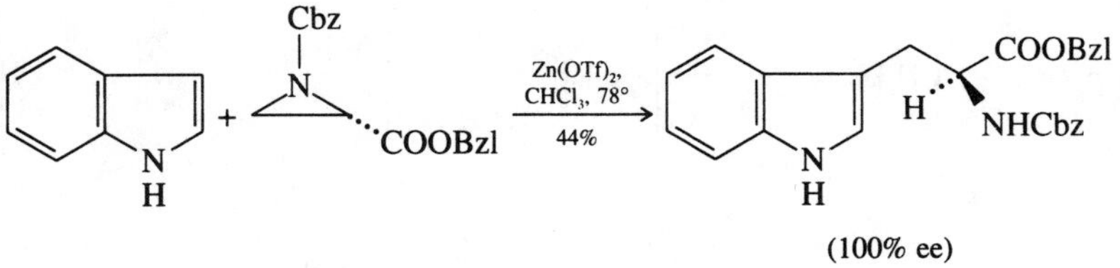

¹ K. Sato and A. P. Kozikowski, *Tetrahedron Letters*, **30**, 4073 (1989).

Zirconocene–4-Dimethylaminopyridine complexes, $Cp_2Zr[(CH_3)_2NPy]_2$, **1a**, or $Cp_2Zr[(CH_3)_2NPy]THF$, **1b**. The complexes are prepared by reaction of Cp_2ZrBu_2 with 1 or 2 equiv. of DMAP in THF at 25°. The burgundy complex **1a** is more stable than the cherry red **1b** and can be stored for several hours.

Reductive coupling of alkynes.[1] The reaction can be effected in good yield by reaction of 1-trimethylthioalkynes with **1a** or **1b** followed by addition of another alkyne to form coupled products.

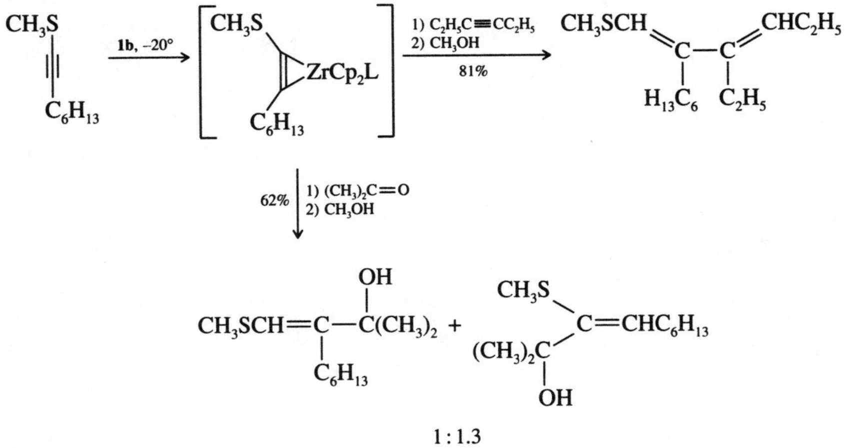

¹ B. C. Van Wagenen and T. Livinghouse, *Tetrahedron Letters*, **30**, 3495 (1989).

Zirconocene dichloride–Dibromomethane–Zinc.

Methylenation.[1] These three reagents form a complex, possibly **2**, which methylenylates aldehydes, ketones, or enones but not esters or lactones. CH_2Br_2 can be replaced by CH_2I_2 but not by CH_2Cl_2.

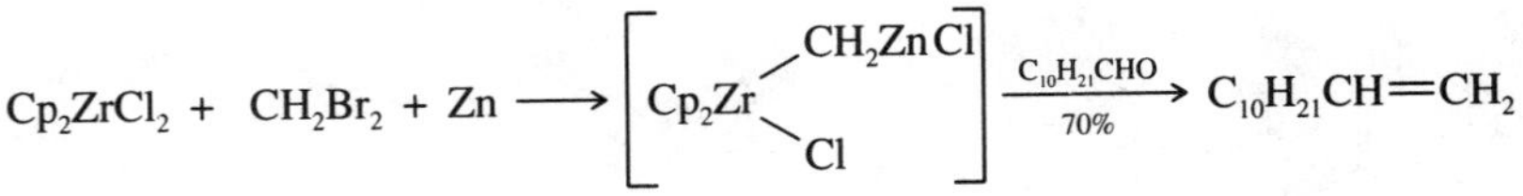

$$Cp_2ZrCl_2 \ + \ CH_2Br_2 \ + \ Zn \ \longrightarrow \ \left[Cp_2Zr \diagup{\begin{array}{c} CH_2ZnCl \\ \diagdown Cl \end{array}} \right] \xrightarrow[70\%]{C_{10}H_{21}CHO} C_{10}H_{21}CH{=}CH_2$$

2

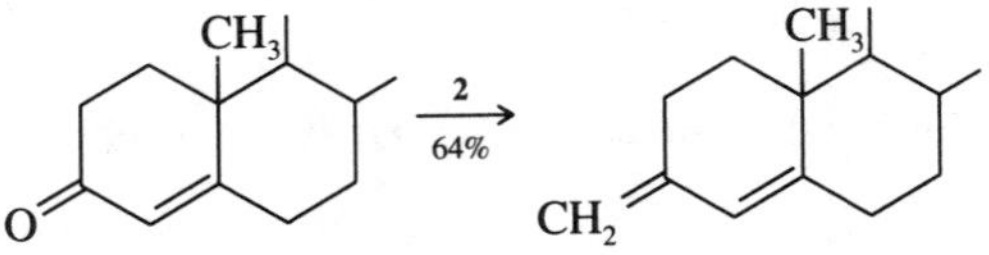

[1] J. M. Tour, P. V. Bedworth, and R. Wu, *Tetrahedron Letters*, **30**, 3927 (1989).

AUTHOR INDEX

Oohara, T., 80
Ooi, T., 211
Ooka, Y., 26
Oose, M., 285
Oppolzer, W., 58, 321, 322
Osakada, K., 131
Oshima, K., 350
Otera, J., 111, 112, 216
Ouimet, N., 289
Oumar-Mahamat, H., 200
Overman, L. E., 25, 219, 265
Owen, T. C., 99
Ozaki, S., 125
Ozawa, F., 35

Padmanabhan, S., 183
Padwa, A., 107, 289
Page, P. C. B., 161
Pagnoni, U. M., 202
Paguaga, E., 348
Pahuja, S., 179
Pakrashi, S. C., 366
Palkowitz, A. D., 135
Pandiarajan, P. K., 71
Panek, J. S., 8, 25, 45, 250
Paquette, L. A., 10, 71, 216
Parekh, S. I., 209
Park, J., 371
Park, J. H., 199
Park, S. Z., 357
Park, Y. J., 57
Passacantilli, P., 269
Patricia, J. J., 57
Pattenden, G., 94
Pearson, A. J., 352
Pearson, D. A., 356
Pearson, M. M., 292
Pearson, W. H., 279
Pedersen, S. F., 229, 370, 371
Pellon, P., 208
Peña-Matheud, C. A., 325
Pendalwar, S. L., 264, 354
Periasamy, M., 182
Pericàs, M. A., 113
Petasis, N. A., 150
Peterson, G. A., 88
Petre, J. E., 71
Petty, C. M., 99
Pfaltz, A., 39, 40, 55
Pflieger, P., 287
Pham, K. M., 305

Philippo, C. M. G., 71
Piantini, U., 137, 246
Piekstra, O. G., 96
Pieters, R. J., 292
Pietroni, B., 267
Pikul, S., 153, 154
Pilarski, B., 177
Pinetti, A., 202
Pini, D., 154
Piras, P. P., 194
Pizzo, F., 82
Plant, A., 199
Plumb, J. B., 272
Plumet, J., 199
Podesta, J. C., 373
Pokorski, U., 337
Poon, Y.-F., 279
Porta, O., 330
Porter, J. R., 67, 138, 139, 140, 364
Porter, N. A., 147, 148
Posner, G. H., 356
Pougny, J. R., 388
Poupaert, J. H., 13
Poupart, M.-A., 289
Powell, A. R., 43
Powers, D. B., 135
Pradilla, R. Fernandez de la, 235
Prakash, G. K. S., 50, 310
Prakash, O., 179
Prandi, J., 237
Prange, T., 20
Prasad, C. V. C., 58, 300
Prasad, J. S., 330
Price, M. E., 365
Pridgen, L. N., 257
Prodger, J. C., 161
Pulley, S. R., 110
Punzalan, E. R., 56, 57
Putman, D., 5

Qian, L., 160
Quendo, A., 391
Querci, C., 199
Quici, S., 199, 324
Quinn, R., 151

Racherla, U. S., 6
Raczko, J., 163
RajanBabu, T. V., 74
Ramasseul, R., 151
Rambaud, M., 164

SUBJECT INDEX